Contents

Preface

Fundamentals of Electricity and Electronics was written with the average student in mind. It is intended for use in first year courses within electronics and/or electricity training programs, and for those persons who desire to learn basic concepts on their own.

Every attempt has been made to keep the reading level as basic as possible, the mathematics as simple as possible, and to develop all subjects fully. The text was written with both students and teachers in mind. Frequent self-checks are included throughout the text. Each chapter is closed with a selection of objective questions which can be used by the teacher as test items or can be assigned to the students as homework.

All explanations proceed from the simple to complex and from the known to unknown. "Ideal components" are used as the subjects for all discussions. No attempt has been made to include all the intricacies of electronic phenomena that are needed by the design technician or engineer. This book was designed for, and is addressed to, the practical technician.

The sequence of chapters used here is the sequence that the authors view as being the most logical approach to development of the material. However, we realize that other people may have other ideas. Therefore, each chapter is developed so that it can be inserted into another sequence as desired.

For those of you who like to teach safety first, Chapter 27 is a short review of electronic safety and first aid. House wiring is discussed in Chapter 26 for those of you who include this subject in your programs. Appendix A contains material that can be used for a review of basic mathematics in programs that use this approach. Appendix B provides information that can be used for a short introductory course on soldering.

Acknowledgments

My most earnest thanks and appreciation are extended to my wife, Anne. She has proofread every chapter of this manuscript enough times to have memorized it. In addition, she has persevered and supported me throughout the year that I have devoted to this project. Many of the things we would normally have done have had to be postponed and must still be done.

Also my deep appreciation to my supporting authors and their families. They have contributed greatly to the completed project. Their families, too, have experienced considerable inconvenience.

My thanks to Maria Theodore, acquisitions editor, Rachel Hockett, project editor, and Paul Becker for their assistance with this project, and to all those people at Holt, Rinehart and Winston and Cobb/Dunlop Publisher Services who have been so helpful.

To people who have reviewed the manuscript at all stages of development, I extend my appreciation. Their feedback has been valuable.

Last, but not least, I would like to thank all of the industries, large and small, who have supported this book with materials, time, and photographs.

John E. Lackey

Introduction

Why Electronics?

I would suspect that most people reading this book already realize the importance of electronics to modern life. It would be a rare person who does not recognize the impact that electronics has had on our lives during the past 20 years. No matter where we look our eyes fall on some electronic marvel. Television has changed from a regionalized system to worldwide coverage through communications satellites. Medical treatment has incorporated many electronic devices. Computers are no longer available only to governmental agencies, but are now available to the home hobbyist. Hand-held calculators have revolutionized the study of mathematics. We could go on and on, but the advances are too numerous to imagine, much less mention.

Your use of this book implies an interest, on your part, in the world of electronics. Potential for growth in the field is unlimited. Your interest could easily be translated into a highly successful career. To reach that goal it will be necessary to study and learn many lessons pertaining to electronics.

These studies will aid you in learning the laws of electronics, the components used to build electronic systems, and how that equipment operates. Your studies will probably lead you to employment as an electronic technician, a title which is used to classify a wide variety of jobs. Its range spreads from the operation and repair of multiple systems in a small business, to specialization in a large industry, the testing and design of new and innovative systems, and the maintenance of home systems.

Regardless of your employment, it is a safe bet that your job will involve you in the practical application of your knowledge. The next few pages will be devoted to providing you with some insight into what this can mean.

As an electronic technician, your most valuable tools will be test equipment designed specifically for analysis of electronic equipment. Figure I.1 illustrates different instruments which are used to measure circuit values.

As can be seen from the pictures, instruments come in various shapes and sizes. The instruments pictured are all used to measure voltage, current, and resistance within electronic systems. Part of your duties will require you to identify the proper test instrument and to use it correctly in analysis of electronic equipment. Of the instruments shown, two are digital multimeters and one is an analog multimeter. The digital type provides its indication in numerical readout form, and the analog type uses a pointer and scale position for its indication.

Figure I.2 is the picture of an oscilloscope. This instrument is used for everything from checking electronic circuits to displaying the heartbeat of a hospital patient visually. The unit shown here is designed to portray actual pictures of electronic signals as they occur within a circuit. The display is shown on a small picture tube similar to the one used in a television.

Most technician jobs require a high level of proficiency in the use of these and other test instruments. Your studies will involve you in learning about the different instruments and how to use them.

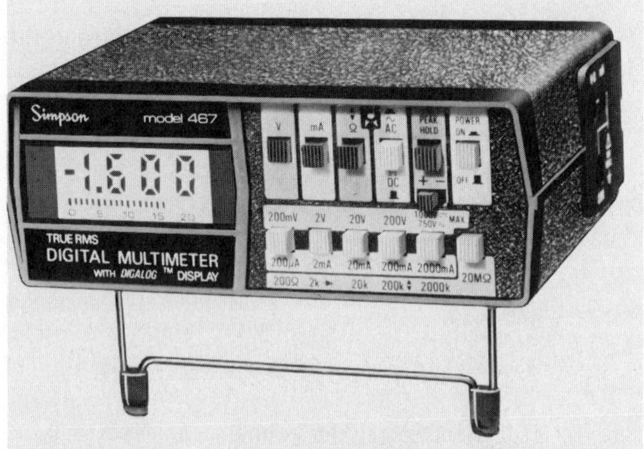

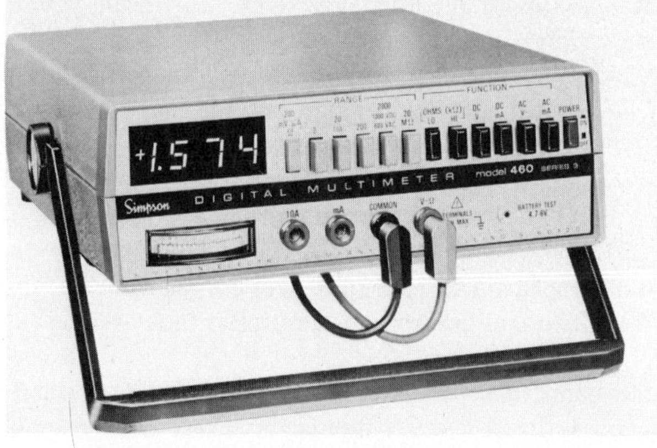

FIGURE I.1
Different Versions of Instruments Used to Measure Electrical Circuit Values (Compliments of Simpson Electric Company)

Where Are We Now?

Part of your education and job performance will require you to complete many mathematical calculations. Figure I.3 presents two kinds of calculators suitable for performing these calculations. This text was developed in such a way as to assist each student in becoming proficient in the use of calculators.

Many of the electronic systems in use today would have been considered science fiction only a few years ago. The computer pictured in Figure I.4 is one that is available for personal use within the home or office. As recently as the 1970s, a computer was priced out of reach of most school systems.

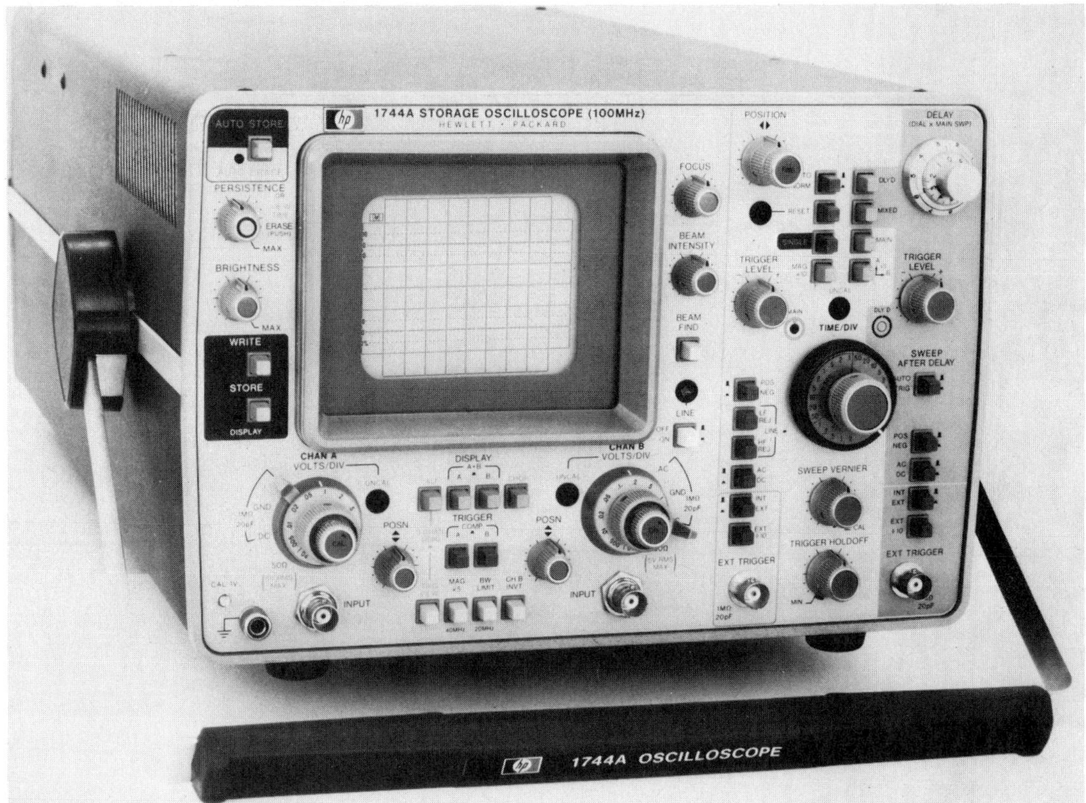

FIGURE I.2

Modern Oscilloscope Suitable for Use in Analyzing Electronic Circuits (Compliments of Hewlett-Packard Company)

Now they are found in many homes. This book was typed on a computer system like the one pictured in Figure I.5. Availability of systems like this make it possible to program many different applications. For typing, the computer is programmed to operate as a word processor. Once the text is stored, the computer will transfer the data to the printer which types the material in typewriter quality.

A more sophisticated computer is shown in Figure I.6. This system is capable of performing highly complex scientific computations that were almost impossible only a few years ago.

The area of biomedical electronics is exploding. Figure I.7 shows a patient area in the intensive care unit of a large modern hospital. Electronic equipment is available to monitor many body functions, to provide body stimulation, and to assist in all areas of treatment.

New advances occur each day. One of the most visible programs involves the NASA Space Shuttle. Figure I.8 shows a picture taken within the cockpit of Spaceship Columbia. This picture features the three onboard computers. Note the three big display screens. Each of these represents a different computer. A computer with the same capabilities would have required one or more rooms for its operation only a few years ago.

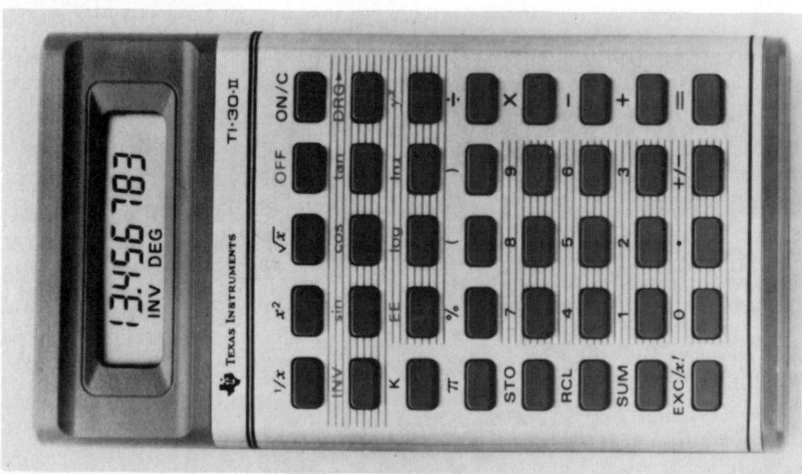

FIGURE I.3
Hand-Held Electronic Calculators Suitable for Use in Electronic Circuit Mathematical Analysis (Compliments of Texas Instruments, Inc.)

FIGURE I.4
Microcomputer of the Type that Is Now Available for Use in the Home and/or Office (Courtesy of Heath Company)

FIGURE I.5
Microcomputer Capable of Performing Numerous Functions. The Printer Produces Letter Quality Print that is Suitable for all Types of Correspondence (Courtesy of Heath Company)

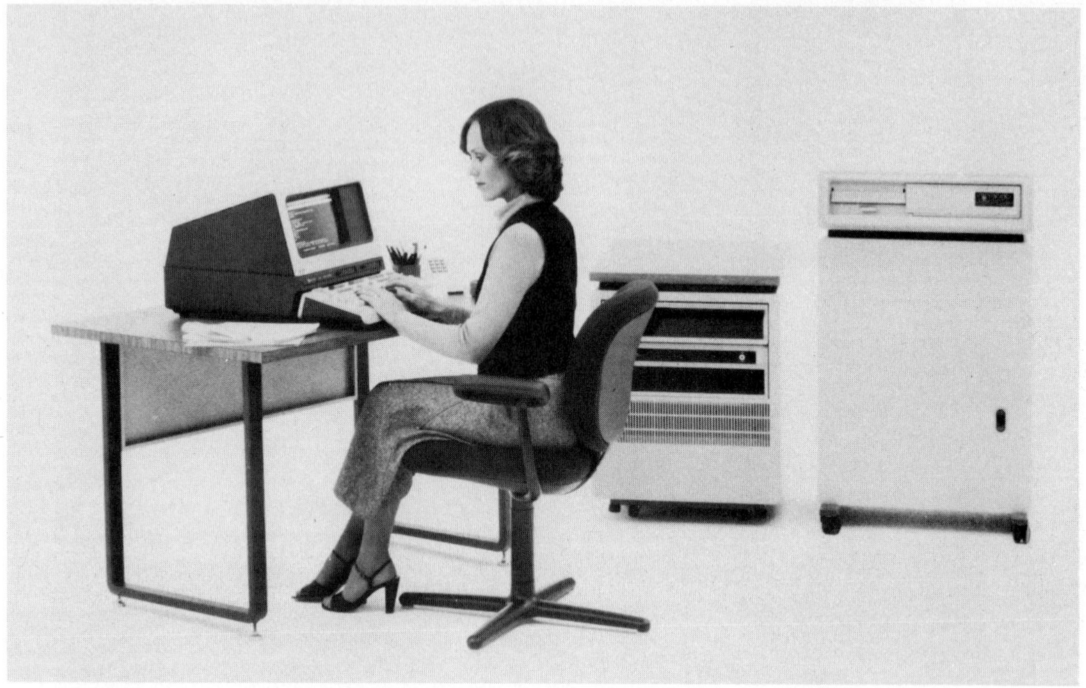

FIGURE I.6
More Sophisticated Computer System (Courtesy of Hewlett-Packard Company)

During the shuttle mission of November 1981, first tests were performed using the "Canada Arm" pictured in Figure I.9. In NASA news releases this arm is called the *Remote Manipulator System* (*RMS*). Note that the arm has an elbow joint, a wrist joint, and a mechanical hand that can be used to pickup and move objects. The entire system is controlled remotely by an astronaut in the shuttle's cabin. Television cameras mounted in the cargo bay and on the arm assist the astronaut with positioning the arm correctly. This arm will be used to offload objects that are to be inserted into space orbit and to retrieve satellites that have completed their mission or that need repair.

Industrial application of electronics is very widespread. Figure I.10 shows a modern computer-controlled milling machine, and its control panel.

Machines like this are now used throughout industry. Electronic technicians are employed to maintain the computer system used with machines of this type. The level of training needed is quite technical.

Manufacturers are replacing many workers with computer controlled robots like the one shown in Figure I.11. Assembly lines are becoming more automated every day. As industry strives to improve production more complex systems will be installed. Machines of this type can be used to perform repetitious and monotonous jobs or to lift and position heavy objects or materials. Newer models can be programmed to perform several different jobs. Once the programs are in place, the operator can use the machine for a different job by performing a predetermined switching operation.

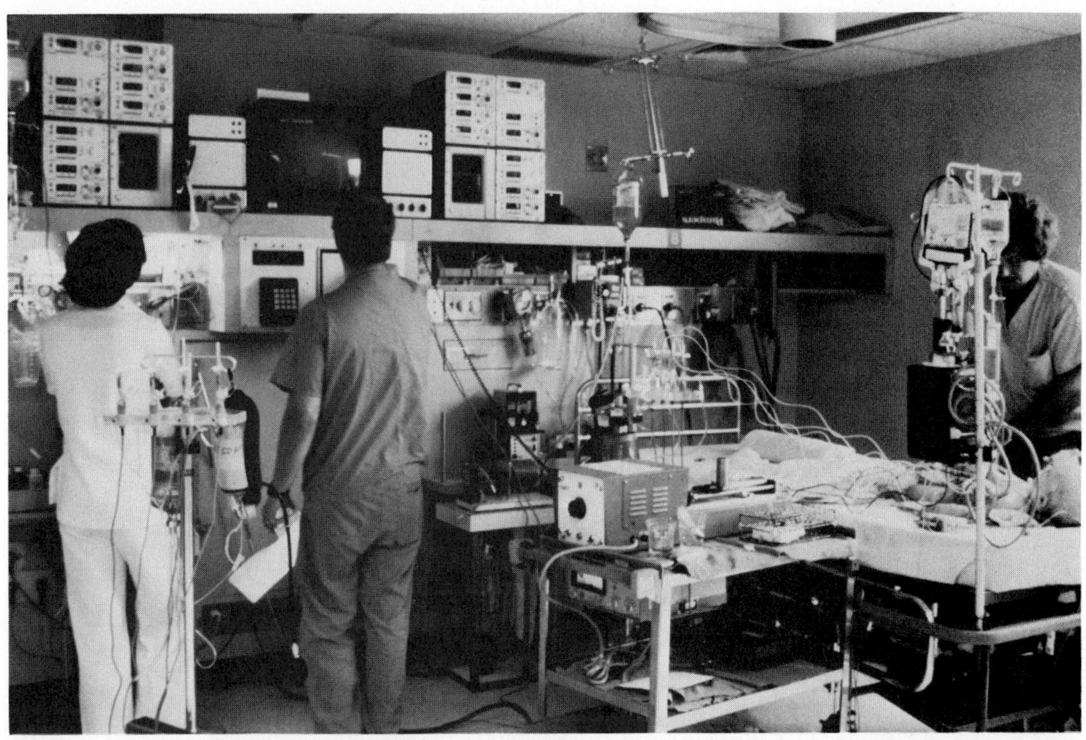

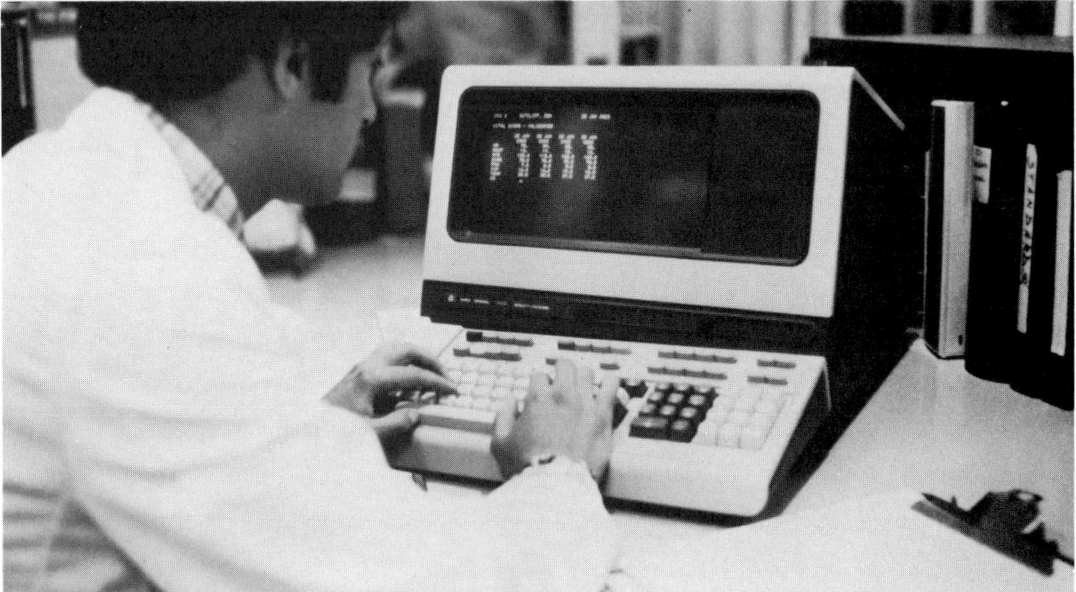

FIGURE I.7

Patient Data Management System Collects and Organizes the Large Amount of Data Generated in the Care of Critically Ill Patients. Independent Terminals Enable Hospital Staff to Request or Manually Enter Information Needed in Making Patient Care Decisions (Courtesy of Hewlett-Packard Company)

FIGURE I.8
Interior Shot of Computer Panel in the Cockpit of Space Shuttle Columbia (Courtesy of NASA)

FIGURE I.9
View of the Remote Manipulator System Through a Porthole from Within Space Shuttle Columbia with the Earth Also Visible (Courtesy of NASA)

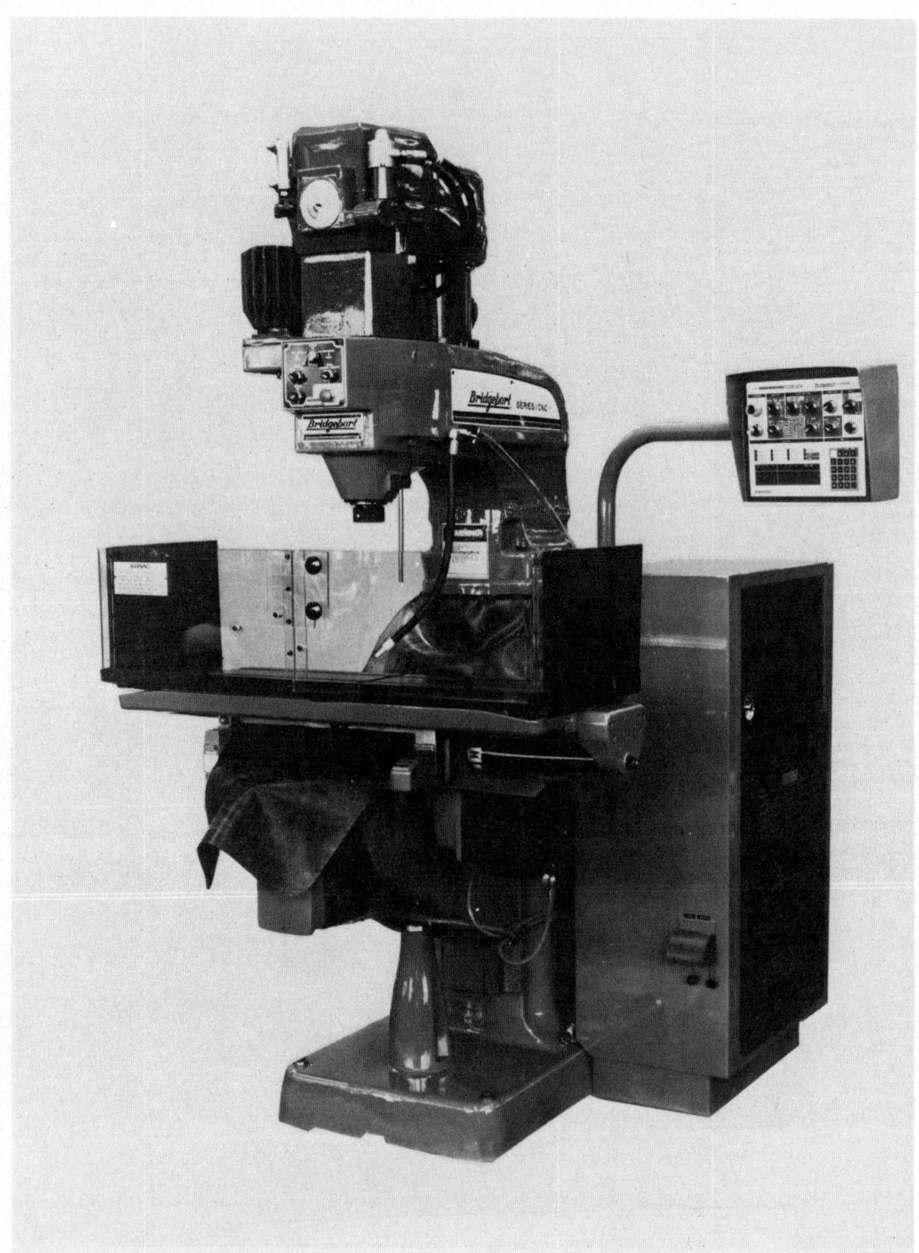

FIGURE I.10
Computer Numerically Controlled Milling Machine (Courtesy of Bridgeport Machines)

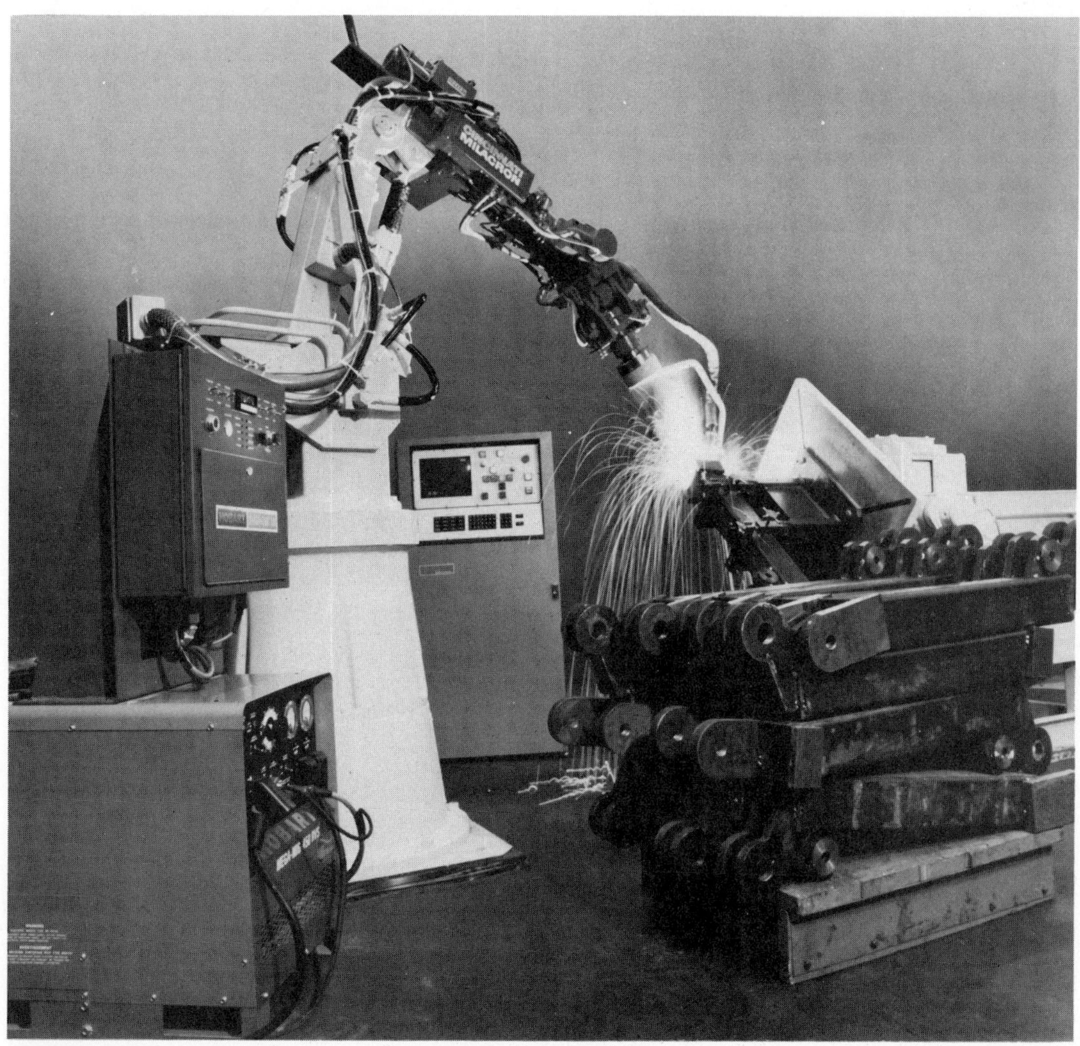

FIGURE I.11
Computer-Controlled Industrial Robot of the Type Being Used Extensively Within Industry (Courtesy Cincinnati Milacron, Inc.

As industry adds new systems, the training level of the work force steadily increases. The technical training level expected of the electronic technician will increase accordingly. Part of your job will be to stay abreast of these and other developments.

Where Are We Going from Here?

The systems of tomorrow will be just as revolutionary to those of you just entering the field as today's systems are to those who have been around for a while. Your challenge will be to learn the systems and to apply yourself to the furthering of this knowledge.

Future technicians will be expected to maintain home computers, microwave ovens, and so on, much as the television repairmen of today. Some technicians will be employed whose only responsibility is to insure the continual operation of complex computer-controlled manufacturing systems. Other technicians will travel with and perform alongside astronauts and scientists in space. In fact the odds are that if you can imagine it, someone will be doing it.

Congratulations on your choice, and good luck along the way.

1

Introduction to Electronics Theory

Introduction

The study of electronics is based on electron theory. All the effects of electricity can be explained by the existence of a tiny particle, called an *electron.* The effects of electricity can be predicted accurately using electron theory. Based on this theory, predictions have been made that have allowed, for example, entry into space and advances in medicine.

This book is designed to help students learn electron theory—what it is, what it does, how it does it. The information learned here can be used throughout the study of electronic devices and circuits to develop a practical approach to repair of electronic systems similar to the systems included in the introduction to this book.

Objectives

Each student is required to:

1. Write the definition of matter.
2. List the three states of matter.
3. Name the smallest particle to which an element and a compound can be reduced while retaining their original properties.
4. Name the two major particles found in an atom's nucleus.
5. Describe the relationship that exists between the atom's nucleus and its orbiting electrons.
6. Define *valence* and *valence electrons* as they pertain to an atom's chemical activity.
7. Explain the laws of attraction and repulsion.
8. Classify electrons, protons, and neutrons according to their electrical charges.
9. Explain the ionization process and how negative and positive ions are created.
10. Define a conductor in terms of its atomic structure.
11. Identify three common materials used to make conductors.
12. Define an insulator in terms of its atomic structure.

13. List three common materials used as insulators.
14. Define a semiconductor in terms of its atomic structure.
15. List the two elements commonly used as semiconductors.

Structure of Matter

Electron theory assumes that all electronic devices work because of the movement of electrons from place to place. For you to understand this theory, a knowledge of the electron is helpful. Since all electrons were originally part of an atom, the study of atomic structure is a good starting point. An atom is:

> ***The smallest particle to which an element can be divided and still retain the properties of the element.***

An element is the smallest particle to which matter can be reduced and still retain its atomic form. All atoms of a single element are identical. The definition of an element is:

> ***A substance which cannot be reduced to any simpler state by chemical means.***

By definition, matter is:

> ***Anything that occupies space and has weight or mass.***

From this definition, we can reach the conclusion that everything is matter. Matter exists in three states: *solid, liquid,* and *gas*. The state which matter takes depends on its atomic structure and other characteristics. Water is a compound formed from two gases, hydrogen and oxygen. Air consists of several gases, including oxygen, hydrogen, and nitrogen.

Matter can be further classified as *elements, molecules, compounds,* and *mixtures*. An element is the most basic of these classes, because it contains atoms that are exactly alike.

Molecules are chemical compounds formed by the chemical reaction of two or more elements. Matter which results from these reactions is much more complex than is a single element. The smallest part of a compound is a *molecule*. One molecule of water is formed by the chemical combination of two atoms of hydrogen and one atom of oxygen. Water resulting from this chemical reaction is much different from either element used to form it. Matter that results from a chemical reaction is called a *compound*. Matter can also exist in the form of a *mixture*. A mixture results from the physical mixing of elements. This might be compared to mixing gravel, sand, water, and cement, which results in concrete. Even though parts have been mixed together, each part has not lost its original properties.

Atomic theory would take considerable time to develop extensively. Here, we are interested in the compound, so let us examine that. A molecule of water is an unbelievably small part of a drop of water. That drop contains billions of molecules. The molecule is the smallest particle to which the drop of water can be divided and still retain the properties of water. It is possible to reduce the molecule to its basic atoms, where it again takes on the properties of hydrogen and oxygen. Atoms are the smallest particles to which an element can be divided. Further reduction would result in a nuclear reaction.

Hydrogen is the simplest of atoms; Figure 1-1 illustrates its structure.

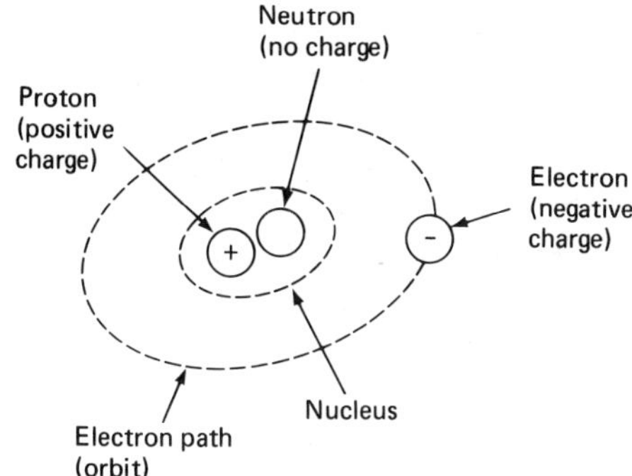

FIGURE 1-1

Examination of the figure reveals three subparticles. Near the center of the atom is an area called the *nucleus*. Two different particles are contained in the nucleus. These are a proton, represented by a positive (+) sign, and a neutron, represented by a neutral (o) sign. In orbit around the nucleus is an electron, represented by a negative (−) sign. These three particles are important in the study of electronics.

Electrical charges exist in one of three states: positive (+), negative (−), or neutral (o). The positive (+) proton, the neutral (o) neutron, and the negative (−) electron carry signs representing their electrical charges.

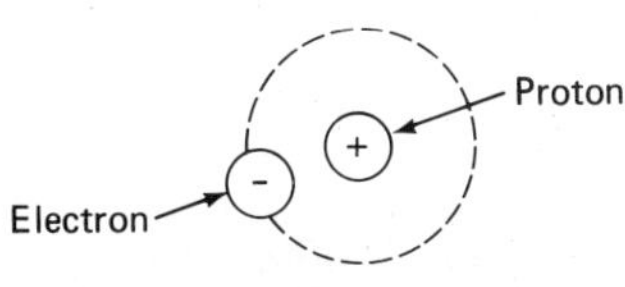

FIGURE 1-2

Figure 1-2 is a representation of the proton-electron relationship of a hydrogen atom. Notice that there is one of each. The electron's track, as it rotates around the nucleus, is called an *orbit*, a *shell*, or an *energy level*. Protons and neutrons are bound into the nucleus. The number of protons equals the number of electrons. The electrical charges of protons and electrons are opposite. Their electrical charges cancel each other, making the atom a neutrally charged particle, because the neutron has no electrical charge and will not affect the atom's charge. When compared to protons and electrons, however, the neutron is quite heavy, and its weight contributes to the weight of the atom.

Figure 1-3 shows a carbon atom. The atom contains electrons, protons, and neutrons, as does the hydrogen atom. But the two atoms can be distinguished from each other by the number of protons and electrons they contain. A carbon atom contains six protons, six electrons, and six neutrons. With larger atoms, it is not practical to display all the protons and neutrons in a diagram. Note that the quantities of protons and neutrons are represented by the letters *N* and *P*, followed by numbers giving the quantity of each.

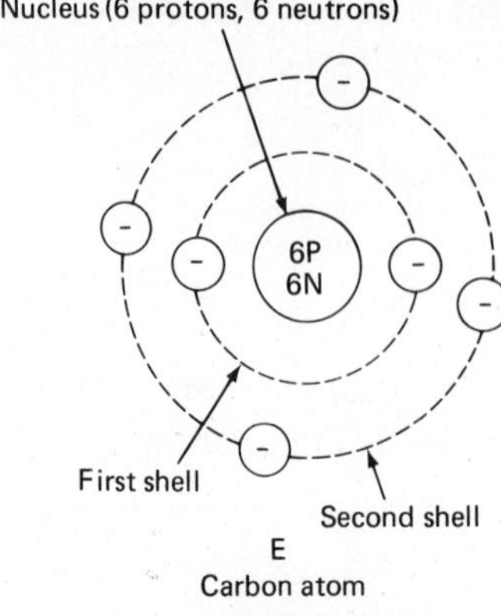

E
Carbon atom

FIGURE 1-3

As found in nature, all atoms have an equal number of protons and electrons. This results in a neutral charge. It is possible, however, for an atom to lose or add an electron to its orbit. When that happens, the atom is no longer neutrally charged. It is also possible to alter the number of electrons physically in the outermost orbit of many atoms. Ionization is the process used to do this. Ionization is possible because of the ability of electrons to absorb and release energy. When an atom is exposed to an energy source, its electrons absorb some of the energy. Eventually, electrons reach an energy level at which they break away from their orbit and become *free* electrons. The atom that gave up the electron now has more protons than electrons. Since protons have positive charges, the atom now has a positive (+) charge and is called a *positive* ion. A positive ion is an atom which has lost one or more electrons. If an atom accepts an extra electron into its orbit, it has too many electrons to remain neutrally charged. Because an electron has a negative electrical charge, the excess electron causes the atom to take on a negative charge; it is now called a *negative* ion. The ionization process can result in two types of ions.

Negative ions are those that have an excess of electrons, and positive ions are those that have a deficiency of electrons.

Definitions included in this section are:

matter.
Anything that occupies space and has weight or mass

states of matter.
Solids, liquids, and gases

elements.
Substances that cannot be further divided by chemical means

atom.
The smallest part of an element which retains all the properties of the element

nucleus.
The positively charged central portion of an atom

proton.
A positively charged particle found in the nucleus

neutron.
A neutrally charged particle found in the nucleus

electron.
A negatively charged particle which orbits the nucleus of an atom

molecule.
The smallest physical unit to which a *compound* can be reduced yet retain all properties of the compound

compound.
Matter formed by chemical activity of elements

ion.
An atom with an imbalance of electrons and protons

positive ion.
An atom with a deficiency of electrons

negative ion.
An atom with an excess of electrons

Self-Check

Complete each of the following items by inserting the words and/or numbers needed to make a true and complete statement.

1. ______________, ______________, and ______________ are the three main subparticles found in all atoms.

2. A ______________ ion has a deficiency of electrons.

3. ______________, ______________, and ______________ are the three states of matter.

4. Altering the number of electrons contained in orbit around an atom is called the ______________ process.

Conductors, Insulators, and Semiconductors

In the study of electronics it is important to understand the relationship between matter and electricity. Different types of matter allow electrons to move through them with varying degrees of difficulty. Every electrical device is made from materials available on today's market. The effect these materials have on electron movement must be understood before a piece of electronic equipment can be designed, built, or repaired.

For our purposes, all matter can be divided into three categories: *conductor, insulator,* or *semiconductor*. This classification is related to the ability of matter to support the movement of electrons. Conductors are made of material whose electrons exist at high energy levels and which are easily moved to free-electron status. Conductors have little opposition to the movement of electrons. They are made of such materials as gold, silver, copper, and aluminum. Insulators are made from materials that provide high opposition to electron movement. At normal operating ranges, insulators restrict electron movement to negligible levels. Some materials used to make insulators are rubber, glass, ceramic, and mica. Semiconductors are neither good conductors nor good insulators. Silicon and germanium are the two elements most often used to make semiconductor devices. Semiconductors are discussed fully later in your studies.

Electrons revolve around an atom's nucleus in orbits and at different energy levels. The electrons that orbit in the outermost shell contain the most energy. (In conductors, electrons exist at high energy levels, and many are free to drift from atom to atom without the aid of external force.) Electrons in the outermost orbit are called *valence electrons*; their orbit is the *valence orbit*. Valence electrons are responsible for all the chemical activity of an element. Conductors are made of materials having three or fewer valence electrons, and insulators of materials having seven or eight valence electrons. Semiconductors have four valence electrons. When we consider that an element having a valence of 1 is most active and the one with a valence of 8 is least active, we can see the significance this has for their classification as a conductor, insulator, or semiconductor.

Conductivity is the ability of a material to support electron movement. To have high conductivity, a material must have low opposition to electron movement. Applying external energy to a substance with high conductivity causes valence electrons to break the bonds that attach them to their atoms. Once released, these electrons become free electrons.

When not under the influence of an external force, free electrons drift from point to point within a material. When they are controlled by an external force, however, free electrons can be made to move in predictable directions and quantities. The effect of an external force is felt equally throughout a conductive material, causing free electrons to stop drifting and move along the conductor in an orderly manner. The orderly movement of electrons is known as *electron current flow*.

Electrons are conducted by several common elements. Silver is the best conductor, but its cost as well as some other disadvantages limit its use. Gold is a good conductor; it resists corrosion exceptionally well. But the cost of gold limits its use to special applications. Copper is almost as good a conductor as silver, and it is much less expensive. The use of copper and copper-alloy conductors is widespread. Alloy metals can be added to copper to increase its strength without drastically affecting its ability to carry electrons. Such copper alloys can be used where strength and heavy electron flow are required but where small physical size is also a requirement. For lightweight conductors such as those used in power transmission lines, it is common to use aluminum. The light weight and strength of aluminum make it ideal for suspension over considerable distances.

Self-Check

Answer each of the following items by indicating whether the statement is true or false.

5. ____ It is easy for electrons to pass through an insulator.

6. ____ Electrons in the outermost orbit of an atom are called valence electrons.

7. ___ An element whose atoms contain one valence electron would make a good conductor.

8. ___ An atom with four valence electrons is classified as a semiconductor.

9. ___ Copper is an example of a good semiconductor.

Charged Bodies and Coulomb's Law

As has been explained, an atom is a neutrally charged particle; however, it is possible for it to gain or lose an electron in its valence orbit. This causes it to no longer be neutrally charged. Figure 1-4 illustrates such an event. The atom losing the electron becomes a positive ion and the one gaining the electron becomes a negative ion. If a substance accumulates a large number of either type of ion, it is said to be *charged with that polarity.* Because this charge is stable (at rest), it is called a *static charge.* When a charge is moving, it is called a *dynamic charge.* Charged bodies may be either positive or negative. Each polarity is surrounded by a *field of force.* A French physicist named Charles Augustin de Coulomb found that when two charged bodies are brought close together they always react in the same way. From his experiments Coulomb developed a set of rules explaining the behavior of charged bodies:

Like charges repel each other.
Unlike charges attract each other.

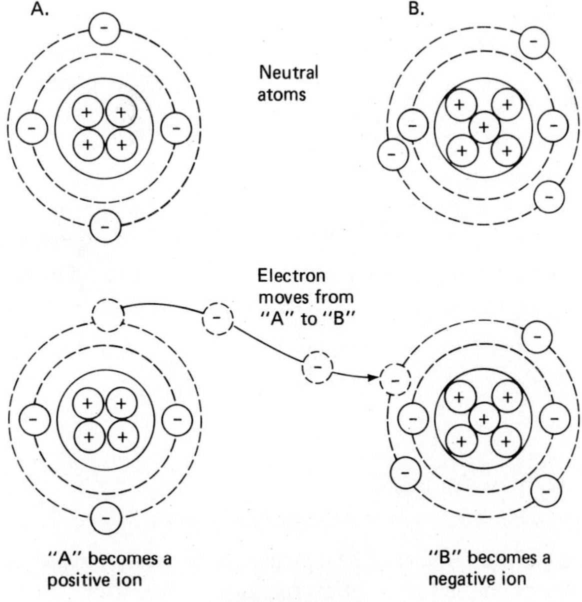

FIGURE 1-4

Figure 1-5 illustrates the laws of attraction and repulsion.

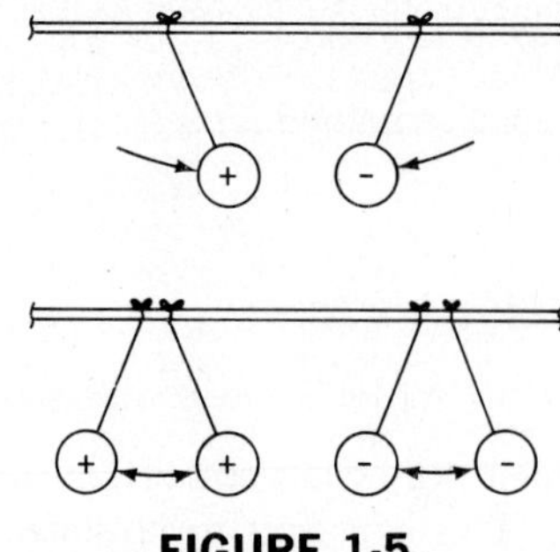

FIGURE 1-5

Coulomb's law explains the reaction of charged bodies to other charged bodies. This law states:

> ***The force existing between two charged bodies is directly proportional to the product of the two charges and inversely proportional to the square of the distance between the bodies.***

Coulomb further defined the reaction of charged bodies by developing a unit of measure that can be used to state the amount of charge located on a charged body. One coulomb of charge (Q) is equal to

$$Q = 6.28 \times 10^{18} \text{ electrons}$$

A more practical unit of measurement for the charge of a body is the *electrostatic unit* (ESU). The ESU has a direct relationship to 1 dyne of force. The definition of 1 dyne is:

> ***A force equal to an acceleration of 1 meter per second per second with a mass of 1 gram.***

An electrostatic unit is equal to:

$$\text{ESU} = 2.19 \times 10^{9} \text{ electrons}$$

Using the ESU, Coulomb devised a formula that can be used to calculate the force existing between two charged bodies:

$$F = \frac{Q_1 \times Q_2}{d^2}$$

where

F = force in dynes
Q_1 = charge on body one in ESUs
Q_2 = charge on body two in ESUs
d^2 = distance between bodies, in centimeters squared

Figure 1-6 represents two charged bodies. The charges for Q_1 and Q_2 are stated in ESUs. Positive (+) signs represent polarity of charge, not signs of mathematical operation. To determine the amount and type of force that exists between the two charged bodies, Coulomb's law is used as follows:

$$F = \frac{10 \times 20}{5^2}$$

$$F = \frac{200}{25}$$

$F = 8$ dynes of repulsion

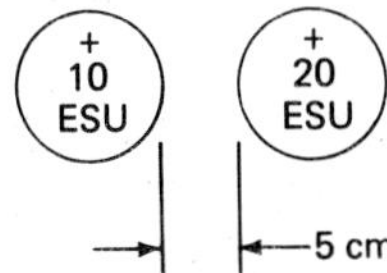

FIGURE 1-6

Whether the force is attraction or repulsion is determined by the polarities of charge on the bodies. Since both charges are positive, we apply the rule "like charges repel" to determine that this force is repulsion.

Charging an Object

Bodies can be charged by conduction, induction, or friction. Friction is a common cause of static charges. Combing your hair with a hard-rubber comb places a charge on the comb. Rubbing a glass rod with a silk cloth creates a positive charge on the rod; rubbing a rubber rod with a piece of cat's fur creates a negative charge on the rod. Walking across a nylon carpet can cause your body to accumulate a static charge.

Conduction occurs when two charged bodies having unequal or opposite charges are allowed to touch or are connected together by a conductor. Remember the time you reached for the doorknob and got a shock? A static charge was conducted between your body and the doorknob. When you see a flash of lightning in a thunderstorm, you are witnessing the results of the conduction (arcing) of electrons between static charges. Figure 1–7 illustrates the principal of conduction.

Touching a positively charged rod to a neutral bar causes some of the electrons on the bar to be attracted to the rod. When rod and bar are separated, both have positive charges. Note that only a portion of the positive charges on the rod have been neutralized by an electron. Loss of electrons by the bar has caused it to gain a positive charge.

Induction is used to charge a body indirectly. When a positively charged rod is placed near the bar (Figure 1-8), electrons on the bar are attracted to the end nearest the positive rod. This leaves the area at the other end of the bar with a positive charge because of the deficiency in electrons created by their being attracted to the position near the positively charged rod. The bar continues to have this charge-distribution pattern as long as the rod remains nearby. If you touch the bar with your finger, electrons will be attracted from your body. When you remove the rod and your finger, the bar will have a negative charge as a result of the rod's induction, even though the rod and bar never touched each other.

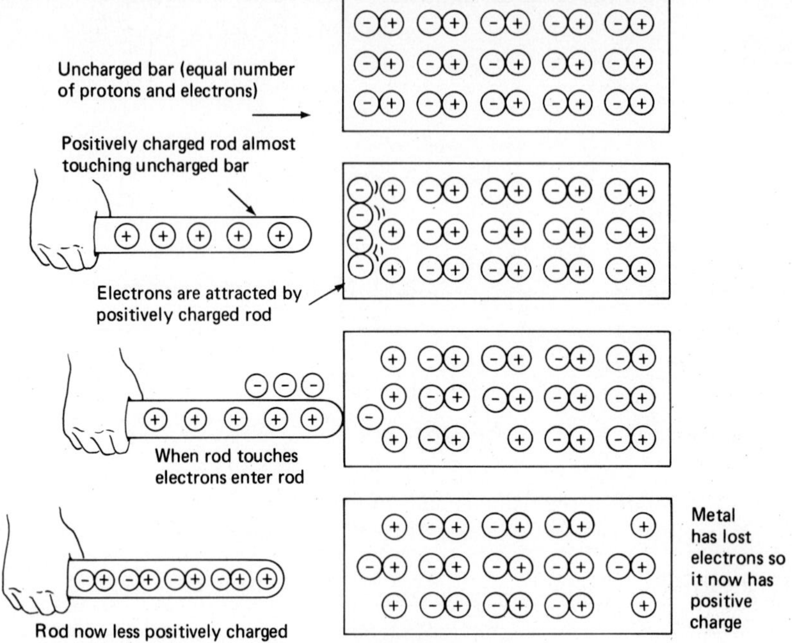

FIGURE 1-7

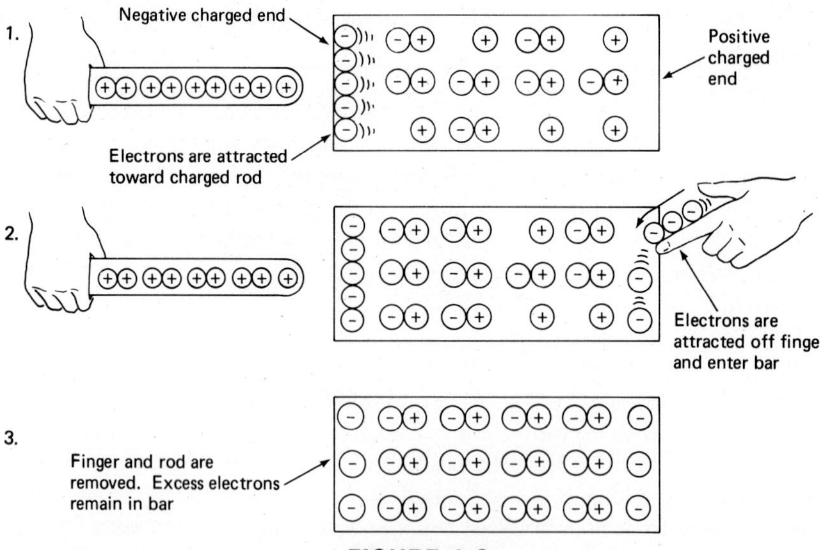

FIGURE 1-8

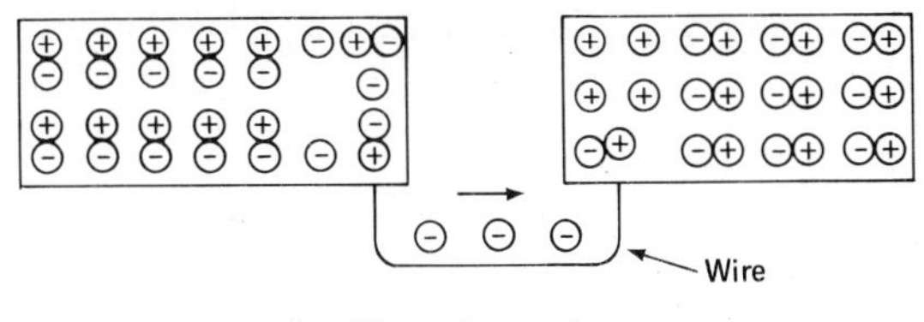

Through an arc

By contact

FIGURE 1-9

In the examples illustrated, one produces a positive charge and the other a negative charge. Figure 1-9 illustrates the way in which charges that exist between two bodies can be neutralized by conduction; by arcing; by physical contact. All are forms of conduction, with the type of conductor making the difference. In the case of the metallic conductor, small charges can cause conduction. In arcing, the charges must be extremely large to cause electrons to move through the air. When bodies physically touch, no other conductive medium is necessary.

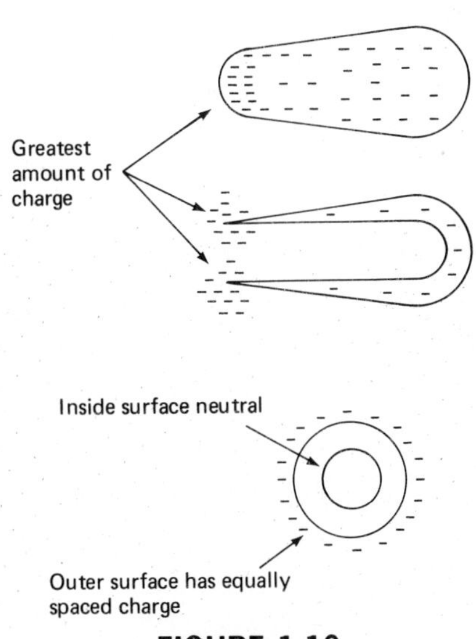

FIGURE 1-10

Distribution of Charges on Bodies

Once a body becomes charged, the charge distributes itself according to certain patterns and conditions (see Figure 1-10).

a. Objects having irregular shape will not have evenly spaced charge distribution. The charge concentration will be greatest at the points of greatest curvature.
b. A charge placed on a hollow sphere (ball). Notice that all of the charge is distributed on the outside of the sphere and is equally spaced throughout the surface.

Self-Check

Answer each item below by indicating whether the statement is true or false.

10. ___ An atom that has an excess of electrons in its orbits has a negative electrical charge.

11. ___ Like charges attract and unlike charges repel.

12. ___ The force that exists between two charged bodies is inversely proportional to the charge on each body.

13. ___ Induction can be used to charge a body indirectly.

14. ___ The maximum charge on a charged body exists at the points having the greatest curvature.

Summary

matter.
Anything that occupies space and has weight or mass

states of matter.
Solids, liquids, gases

elements.
Substances that cannot be divided further by chemical means

atom.
The smallest part of an element that retains all the properties of the element

nucleus.
The positively charged central portion of an atom

proton.
The positively charged particle located in the nucleus

neutron.
The neutrally charged particle located in the nucleus

electron.
The negatively charged particle in orbit around the nucleus

molecule.
The smallest physical unit to which a compound can be reduced and retain the properties of the compound

compound.
Matter formed by the chemical action of elements

ion.
An atom with an imbalance of electrons and protons

conductor.
A material capable of supporting the flow of electrons

insulator.
A material which, because of its structure, is able to limit electron flow to small amounts

semiconductor.
A material that is neither a good conductor nor a good insulator

valence.
A classification of the ability of different elements to combine chemically with other elements

positive ion.
An atom with a deficiency of electrons

negative ion.
An atom with an excess of electrons

electrical charge.
Positive or negative charge located on or in a body

law of electrostatic charges.
Unlike charges attract; like charges repel

coulomb.
A charge equal to 6.28×10^{18} electrons

electrostatic unit (ESU).
A charge equal to 2.19×10^{9} electrons

methods of charging.
Friction, conduction, induction

distribution of charge.
Concentrates at points of greatest curvature on irregularly shaped objects; evenly spaced on round objects

static charge.
An electrical charge that is at rest

dynamic charge.
An electrical charge that is moving

Review Questions and/or Problems

1. The process used to alter the number of electrons in the outer orbit of an atom is called the ______________ process.
   (a) ionization
   (b) current flow
   (c) neutralization
   (d) none of the above

2. When an atom has more electrons than protons, it is called a ______________.
   (a) positive ion
   (b) positive atom
   (c) negative ion
   (d) negative atom

3. An atom is defined as:
   (a) the smallest part of matter.
   (b) the smallest part of a compound.
   (c) the smallest part of an element.
   (d) the smallest part of a mixture.

4. A good conductor:
   (a) opposes the movement of electrons.
   (b) must have very few atoms.
   (c) has many free electrons.
   (d) has an atomic structure consisting of many electrons.

5. A coulomb is:
   (a) an element containing equal numbers of electrons and protons.
   (b) a unit of electrical charge.
   (c) equal to 6.28×10^{18} electrons.
   (d) both (b) and (c) are correct.

6. The paths electrons take around a nucleus are called ______________.
   (a) orbits
   (b) energy levels
   (c) shells
   (d) all the above

7. A good insulator:
   (a) has many free electrons.
   (b) conducts free electrons easily.
   (c) has few free electrons.
   (d) none of the above.

8. A semiconductor material is neither a good ______________ nor a good ______________.

9. List three ways in which bodies can become charged.
   (a) ______________
   (b) ______________
   (c) ______________

10. Valence refers to the ______________ activity of an atom, and valence electrons are found in the ______________ orbit.

11. Explain in your own words: the ionization process, how a positive ion is produced, and how a negative ion is produced.

12. Draw a pictorial representation of an atom and identify the following:
   (a) element family
   (b) nucleus
   (c) electrons
   (d) neutrons
   (e) protons
   (f) valence electrons
   (g) inner orbits
   (h) electrons that exist at the highest energy level within the atom

2

Calculators and Electronics

Introduction

In recent years the hand-held electronic calculator has become a valuable aid to students and technicians in the electronics field. Many computations can be performed faster and more easily with a calculator. Several manufacturers market calculators. Any calculator is suitable for use, assuming it has the capabilities required. The discussion in this chapter is not intended as one of the merits of a particular brand. When you choose a calculator for use in electronics, however, it should have certain functions. The explanation here approaches calculators from the standpoint of the algebraic operating system (AOS). Figure 2-1 shows a typical calculator of the type described in this chapter.

Objectives

Each student is required to:

1. Express numbers in scientifically notated form.
2. Perform mathematical solutions using scientifically notated quantities.
3. Convert numerical quantities to scientifically notated quantities, and vice versa.
4. Identify the electronic prefixes *giga*, *mega*, *kilo*, *milli*, *micro*, *nano*, and *pico*.
5. Express each prefix as a numerical quantity and as a scientifically notated quantity.
6. Convert either type of quantity to the corresponding prefix.
7. Use the calculator to:
   (a) solve simple mathematical equations.
   (b) convert numerical quantities to scientific notations.
   (c) convert scientific notations to numerical quantities.
   (d) solve complex problems, using numerical and scientifically notated quantities.

Figure 2-1
A Hand-Held Calculator that Operates on the Algebraic Operating System (Courtesy of Texas Instruments, Inc.)

Required Functions

At a minimum, any calculator purchased should provide the following *functions* and/or *operations*.

Key	*Function*
+	Addition
−	Subtraction
×	Multiplication
÷	Division
STO	Store display in memory
RCL	Recall memory to display
π	Pi (enters constant 3.141592654)
$\sqrt{X}$	Extracts square root of displayed quantity
X^2	Raises displayed quantity to its square
sin	Converts an angle displayed in degrees to its sine function
cos	Converts an angle displayed in degrees to its cosine function
tan	Converts an angle displayed in degrees to its tangent function

Key	*Function*
INV or ARC	For use with sine, cosine, and tangent functions. With a trigonometric function entered in the display, the depressing of this key, followed by either the sin, cos, or tan key, causes the function to be converted to the angle it represents. The quantity indicated will be in degrees. *Note:* Some calculators label this key *ARC*, others *INV*.
CA	Clear (erase) all data from the display and all memories. Some calculators have a single CLEAR key. On these, it is common practice to press this key two times to *clear all.*
CE	Clear (erase) last entry *only*. (To clear the last entry on some calculators, you must depress the CA key one time *only*. To clear all data on these models, you must depress the CLEAR key two times.)
CLR	Clears all data except that stored in memory.
EE	Enter Exponent—used when entering powers of 10 directly in the calculator. Also used in converting displayed data to scientific notation (powers of 10).
+/−	(1) Change the sign of operation of data in display, or (2) change the sign of operation for the exponent when using powers of 10. When used properly, it affects *only* the exponent.
1/X	Perform the reciprocal. Divides the display data into 1.
e^x	Enters the constant 2.718281828. This constant is used in the calculation of some AC circuit values.
SUM	Adds displayed data to memory. If the data in the display are to be subtracted from memory, press +/−, then the SUM key.

Most calculators have more functions than those listed. Many additional functions are helpful in electronics, but the ones listed are sufficient for your studies. Whether your career is in electronics or business management, there is a calculator that is *best* for that career. You will want to make sure the calculator you buy is suitable for the job anticipated.

When using the calculator for addition, subtraction, multiplication, and division, it is only necessary for you to enter the data, then press the prescribed signs of operation and the equal sign. The results are displayed in numerical quantities as long as the result does not exceed the display's capability (usually eight places). When that capability is exceeded, the result is displayed in scientific notation. To help you understand displays, a review of scientific notation is included.

Scientific Notation of Numbers (Powers of 10)

It is common for different skills to involve "tools of the trade"; to the electronics technician, scientific notation and prefixes are such tools. (Prefixes are discussed fully later in this chapter.) Scientific notations are used to identify numerical quantities, and prefixes are names given to specific quantities. Both help identify numbers and solve problems that contain numbers which are awkward to manipulate.

Scientific notation is based on the fact that the mathematical system contains 10 digits (0–9). It is referred to as a *base 10 numbering system.* Scientific notated quantities are often referred to as *powers of 10;* but various base numbers are used with their own numbering systems. Different computers may use any of the following numbering systems: binary (base 2), octal (base 8), or hexidecimal (base 16). All can be stated in scientific notation; at this point, though, we limit our discussion to powers of 10.

When the hand-held calculator was designed, it contained powers of 10 as a capability. Quantities displayed on the calculator are often larger than the whole number that could be displayed. The calculator automatically converts the quantities to a power of 10 and presents them in scientific notation. When entering a quantity in the calculator, it is often necessary to enter scientifically notated quantities. These are reasons for becoming familiar with the scientific notation system and learning to use it. Without a calculator, powers of 10 are even more valuable, since they allow us to manipulate large numbers more easily. First we explore the base 10 numbering system manually; later the discussion is centered on the calculator.

In any study of scientific notations, the terms used must first be defined. They are:

base number.
A number multiplied times itself (raised to a power); in powers of 10, this number is 10

exponent.
The power to which the base number is to be raised; the exponent indicates the number of times the base number must be multiplied times itself

numerical coefficient.
The number that precedes the base number and exponent

A typical, scientifically notated (power of 10) statement is as follows:

Sample statement:

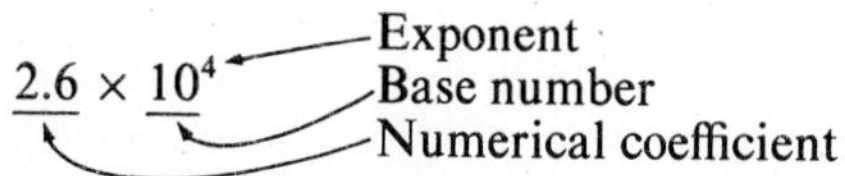

The fact that the base number is 10 and is raised to some power leads to the statement being called a power of 10. For further explanation, consider Table 2-1. In the table each exponent from 0 through 9 is listed and the effect of raising the base number to the exponent shown is presented; the numerical equivalent is the result. Notice that the number of zeros resulting from each example equals the power of the exponent. For example, $1 \times 10^0 = 1$ followed by no zeros; and $1 \times 10^4 = 1$ followed by four zeros. All the exponents are positive. Positive exponents are used to identify numbers that are 1 or larger.

It is often necessary to deal with numbers less than 1. To indicate that a power of 10 refers to a number smaller than 1, the exponent is negative. Numerical quantities of this type are preceded by a decimal point and are referred to as *decimal numbers* or *quantities.* Examples are illustrated in Table 2-2.

TABLE 2-1

Power of 10 Raised to the Power Indicated
$1 \times 10^0 = 1$
$1 \times 10^1 = 10$
$1 \times 10^2 = 10 \times 10 = 100$
$1 \times 10^3 = 10 \times 10 \times 10 = 1{,}000$
$1 \times 10^4 = 10 \times 10 \times 10 \times 10 = 10{,}000$
$1 \times 10^5 = 10 \times 10 \times 10 \times 10 \times 10 = 100{,}000$
$1 \times 10^6 = 10 \times 10 \times 10 \times 10 \times 10 \times 10 = 1{,}000{,}000$
$1 \times 10^7 = 10 \times 10 \times 10 \times 10 \times 10 \times 10 \times 10 = 10{,}000{,}000$
$1 \times 10^8 = 100{,}000{,}000$
$1 \times 10^9 = 1{,}000{,}000{,}000$

TABLE 2-2

Power of 10 Raised to the Power Indicated
$1 \times 10^{-1} = 0.1 = \frac{1}{10}$
$1 \times 10^{-2} = 0.01 = \frac{1}{100}$
$1 \times 10^{-3} = 0.001 = \frac{1}{1{,}000}$
$1 \times 10^{-4} = 0.0001 = \frac{1}{10{,}000}$
$1 \times 10^{-5} = 0.00001 = \frac{1}{100{,}000}$
$1 \times 10^{-6} = 0.000001 = \frac{1}{1{,}000{,}000}$
$1 \times 10^{-9} = 0.000000001 = \frac{1}{1{,}000{,}000{,}000}$
$1 \times 10^{-12} = 0.000000000001 = \frac{1}{1{,}000{,}000{,}000{,}000}$

Of the quantities listed, 10^{-3}, 10^{-6}, and 10^{-12} are used extensively in electronics. The use of these and other scientific notations (powers of 10) in your study of electronics can:

1. Simplify calculations that would otherwise involve quantities difficult to manipulate.
2. Provide a more convenient way of expressing very large and very small quantities.

Any number can be converted to a power of 10, and any power of 10 statement has a numerical coefficient, a base number, and an exponent. In all cases the numerical coefficient is stated as a number between 1 and 10. The statement tells you to multiply the numerical coefficient by the quantity that results when the base number is raised to the power indicated. When converting numbers it is necessary to move the decimal enough spaces to yield a number between 1 and 10 as the remainder; that number is the *numerical coefficient.*

For numbers larger than 10, the decimal must be moved left. The writing of this type number does not include the decimal point; it is understood that any number with no decimal point indicated is a whole number. An example of a whole number with a decimal is:

EXAMPLE

4000.← The decimal is located here.

Converting Numbers Larger than 10

To convert large numbers to scientifically notated quantities, you must:

Move the decimal point to the left until a number between 1 and 10 remains. Write this number as the numerical coefficient times the base number 10. The exponent to be used with the base number equals the number of spaces the decimal was moved.

To become proficient, practice converting quantities from numerical statements to powers of 10, and back again. Some examples are:

EXAMPLE 1:

Given the number 4000., convert it to scientific notation.
4.000 *Note:* The decimal is moved three spaces.

4 is the numerical coefficient.
10 is the base number.
3 is the exponent. (The decimal was moved three places to the left.)

This results in the statement 4×10^3

EXAMPLE 2:

Given Notation
$4{,}230{,}000 = 4.23 \times 10^6$

EXAMPLE 3:

Given Notation
$3190 = 3.19 \times 10^3$

EXAMPLE 4:

Given Notation
$3{,}600{,}000 = 3.6 \times 10^6$

Converting Decimal Numbers

In electronics we often work with numbers considerably smaller than 1. To convert these numbers to powers of 10, the process differs slightly. The rule for such conversions is:

Move the decimal point to the right until a number between 1 and 10 results. Write that number as the numerical coefficient multiplied times the base number 10. The exponent to be used equals the number of spaces the decimal was moved. To indicate that the quantity is smaller than 1, the exponent is identified by a minus (−) sign.

The following examples explain the process.

EXAMPLE 1:

Given 0.0 0 0 4 3, convert it to scientific notation.

0.0 0 0 4.3—decimal has moved four spaces.
4.3 is the numerical coefficient.
10 is the base number.
−4 is the exponent.

The statement that results is 4.3×10^{-4}
Other examples are:

EXAMPLE 2:

Given Notation
$0.000067 = 6.7 \times 10^{-5}$

EXAMPLE 3:

Given Notation
$0.00003 = 3.0 \times 10^{-5}$

EXAMPLE 4:

Given Notation
$0.1234 = 1.234 \times 10^{-1}$

An understanding of scientific notation can greatly benefit your work in math. Scientific notation can simplify multiplication, division, root extraction, and raising-to-a-power operations. To use the system for multiplication,

Combine the numerical coefficients by multiplication and combine the exponents by algebraic addition.

Multiplication Using Scientific Notations

Use of multiplication of powers of 10 can simplify the processes of handling long number statements. The accuracy required in electronics is tolerant to the extent that use of a three-digit number preceded or followed by a string of zeros is accurate enough; therefore, the numerical coefficient can be limited to three digits. Examples of multiplication are:

EXAMPLE 1:

Given the two statements below, perform the multiplication.

$(4 \times 10^6)(2 \times 10^3) = ?$

Multiply numerical coefficients	$4 \times 2 = 8$
Add exponents	$6 + 3 = 9$
Base number	$= 10$
Statement of results	8×10^9

Where large numbers are used, they must first be converted to powers of 10. Once converted, the procedures used for multiplication are the same as those in Example 1. See Example 2.

EXAMPLE 2:

Given the two numbers indicated, perform the multiplication.

(23400000) (0.000439)	$= ?$
Convert 23400000	$= 2.34 \times 10^7$
Convert 0.000439	$= 4.39 \times 10^{-4}$
Multiply numerical coefficients	$= 10.2726$
Combine exponents	$= 7 + (-4) = 3$
Base number	$= 10$
Statement of results	10.2726×10^3

Notice that the numerical coefficient (10.2726) is a number larger than 10. Remember: To be an accurate power-of-10 statement, the numerical coefficient must be less than 10. This number can be converted by moving the decimal point one space left and adding +1 to the exponent. The new, and now correct, statement is:

1.027×10^4

Note: The statement is rounded to three decimal places.

Other examples are:

EXAMPLE 3:

Given Notation

$(2.6 \times 10^4)(2 \times 10^{-6}) = 5.2 \times 10^{-2}$

EXAMPLE 4:

Given Notation

$(3.1 \times 10^{-2})(1.4 \times 10^{-6}) = 4.34 \times 10^{-8}$

Division with Scientific Notation

In performing division problems, the steps differ somewhat from those used in multiplication. The rule for division is:

Combine numerical coefficients by division and combine exponents by algebraic subtraction.

Note: To perform algebraic subtraction, change the sign of the exponent in the denominator and add the two exponents algebraically.

EXAMPLE 1:

given	$\dfrac{6 \times 10^3}{3 \times 10^2}$
divide numerical coefficients	$6 \div 3 = 2$
combine exponents	$3 - 2 = 1$
statement of results	2×10^1

EXAMPLE 2:

given	$\dfrac{8 \times 10^{-4}}{2 \times 10^7}$
statement of results	4×10^{-11}

EXAMPLE 3:

given	$\dfrac{9 \times 10^{-5}}{3 \times 10^{-5}}$
statement of results	3×10^0 or 3

Raising a Quantity to a Power

Electronic analysis involves numerous formulas which require that a quantity be squared or raised to another power. To use scientific notation for these operations, use the rule:

Square the numerical coefficient and multiply the exponent by 2.

The steps set forth by this rule are demonstrated in the following examples:

EXAMPLE 1:

given	$(2 \times 10^3)^2$	$= ?$
square the numerical equivalent	$(2)^2$	$= 4$
multiply exponent by 2	2×3	$= 6$
statement of results	4×10^6	

EXAMPLE 2:

given $(3 \times 10^0)^2$
statement of results 9×10^0

EXAMPLE 3:

given $(1.5 \times 10^{-3})^2$
statement of results 2.25×10^{-6}

EXAMPLE 4:

given $(3.2 \times 10^7)^2$
statement of results 10.24×10^{14}
converted to new statement 1.024×10^{15}

Extraction of the Square Root

Other formulas in electronic analysis requires extraction of the square root of quantities. The rule for using scientific notation when extracting the square root is:

Extract the square root of the numerical coefficient and divide the exponent by 2.

Note: In extracting the square root using this system, the exponent must be an even number. The statement is changed to ensure that the exponent is an even number by moving the decimal one space either left or right.

EXAMPLE 1:

given $\sqrt{9 \times 10^6}$
square root of numerical coefficient $\sqrt{9} = 3$
divide exponent by 2 $6 \div 2 = 3$
statement of results 3×10^3

EXAMPLE 2:

given $\sqrt{1.6 \times 10^5}$
change exponent $\sqrt{1.6 \times 10^{5(-1)}} = \sqrt{16 \times 10^4}$
numerical coefficient $\sqrt{16} = 4$
divide exponent by 2 $4 \div 2 = 2$
statement of results 4×10^2

EXAMPLE 3:

given $\sqrt{1.69 \times 10^6}$
statement of results 1.3×10^3

EXAMPLE 4:

given $\sqrt{1.69 \times 10^{-6}}$
statement of results 1.3×10^{-3}

Self-Check

Complete each item by inserting the words and/or numbers needed to make a true and complete statement.

1. The *base* number for our mathematical system is ______________.

2. When large whole numbers are converted to scientific notations, the exponent results from the number of spaces the ______________ is moved.

3. When two powers of 10 are multiplied, the ______________ are combined by addition.

4. A correctly stated numerical coefficient is equal to or larger than ______________ but less than ______________.

5. Numbers smaller than 1 are represented by a ______________ exponent.

Using the Calculator to Convert to Powers of 10

Calculators can convert numerical quantities to scientific notations (powers of 10) and scientifically notated quantities to numerical quantities. To convert a number to a power of 10, using the calculator, you must:

1. Enter the quantity.
2. Press the EE key.
3. Read the power of 10 quantity. Numbers larger than 1 are stated as a numerical coefficient and are followed by a blank space and a two-digit exponent. For quantities less than 1, the numerical coefficient is followed by a "–" and a two-digit exponent.

Table 2-3 contains some conversions. Use your calculator to duplicate each one.

TABLE 2-3

Entry	*Press*	*Press*	*Display*	*Power of 10*
6000	EE	=	6 03	6×10^3
0.0001	EE	=	1 – 04	1×10^{-4}
53000	EE	=	5.3 04	5.3×10^4
0.000024	EE	=	2.4 – 05	2.4×10^{-5}

Converting Powers of 10 to Decimal Numbers

When using a calculator to convert powers of 10 to numerical data, the process employed is:

1. Enter the power of 10 in the display.
2. Press the following keys in the order shown.
   (a) STO—store the display.
   (b) CLR—clear display but not memory.
   (c) RCL—recall data from memory.
3. Read—numerical data appears on the display.

Practice conversions are presented in Table 2-4; use your calculator to practice them. Repeat each conversion and verify its correctness. *Note:* These displays are the ones that resulted from Table 2-3.

TABLE 2-4

Display	*Press*	*Press*	*Press*	*Display*
6 03	STO	CLR	RCL	6000
1 – 04	STO	CLR	RCL	0.0001
5.3 04	STO	CLR	RCL	53000
2.4 – 05	STO	CLR	RCL	0.000024

In many calculations it will be necessary to enter information in the form of powers of 10. In these cases, use the rules presented in the section "Scientific Notation of Numbers (Powers of 10)." Table 2-5 presents practice problems which should be repeated in order to learn this operation.

TABLE 2-5

Desired Entry	*Enter*	*Press*	*Press*	*Press*	*Display*
6×10^3	6	EE	3	—	6 03
10×10^5	10	EE	5	—	10 05
10×10^{-3}	10	EE	+/–	3	10 – 03
15×10^{-6}	15	EE	+/–	6	15 – 06
1×10^{-12}	1	EE	+/–	12	1 – 12
5×10^6	5	EE	6	—	5 06

In the electronics career field, it is common for large and small quantities to be listed in some notation other than scientific. This notation is often called *prefixes*. A prefix is a name assigned to a specific power of 10. Since powers of 10 represent numerical quantities, prefixes can be converted to either a *numerical quantity* or a *power of 10*. The next section is presented to familiarize you with prefixes and their use.

Prefixes Used in Electronics

Learning electronics requires the performance of calculations that contain many different quantities and values varying from very large to very small. To assist in verbal identification of some quantities, a name, or *prefix,* has been assigned to specific quantities. A prefix can be used to identify numbers and powers of 10. Table 2-6 provides a cross-reference between numerical quantities, powers of 10, and prefixes. The identifier (symbol) used to identify each prefix is also listed.

TABLE 2-6

Number	*Fraction*	*Power*	*Prefix*	*Symbol*
1,000,000,000,000	$\frac{1,000,000,000,000}{1}$	10^{12}	tera	T
1,000,000,000	$\frac{1,000,000,000}{1}$	10^{9}	giga	G[a]
1,000,000	$\frac{1,000,000}{1}$	10^{6}	mega	M[a]
1,000	$\frac{1,000}{1}$	10^{3}	kilo	k[a]
1	$\frac{1}{1}$	10^{0}	unity	unity
.001	$\frac{1}{1,000}$	10^{-3}	milli	m[a]
.000001	$\frac{1}{1,000,000}$	10^{-6}	micro	μ[a]
.000000001	$\frac{1}{1,000,000,000}$	10^{-9}	nano	n
.000000000001	$\frac{1}{1,000,000,000,000}$	10^{-12}	pico	p[a]

[a] *Denotes prefixes used most often in electronics.*

Numbers less than 1000 are identified in the units they refer to, for example, 100 volts, 10 amperes. For numbers between 1000 and 1,000,000, the prefix *kilo* is used; this prefix is usually identified by the letter k. Examples of the application of *kilo* are:

1000 = 1 kilo = 1 k
1200 = 1.2 kilo = 1.2 k
10,100 = 10.1 kilo = 10.1 k
670,000 = 670 kilo = 670 k
354,000 = 354 kilo = 354 k

Note: The prefix *kilo* is used to represent the power-of-10 statement, 1×10^3.

Numbers 1,000,000 or larger are identified with the prefix *mega* and the identifier M. Examples are:

1,100,000 = 1.1 mega = 1.1 M
1,230,000 = 1.23 mega = 1.23 M
11,000,000 = 11 mega = 11 M

Note: The prefix *mega* represents the power-of-10 statement, 1×10^6.

Examination of Table 2-6 reveals that numbers smaller than 1 can also be stated as prefixes. For numbers between 0.999 and .0001, the prefix *milli*, symbol m, is used. For numbers between .0001 and .000000001 the prefix is *micro*, and its symbol is μ. For numbers smaller than .000000001, the prefix *pico*, symbol p, is used. Examples are presented in Table 2-7.

TABLE 2-7

Number	*Prefix*	*Power-of-10 Statement*
.1	100 m (milli)	100×10^{-3}
.01	10 m	10×10^{-3}
.001	1 m	1×10^{-3}
.0001	100 μ (micro)	100×10^{-6}
.00001	10 μ	10×10^{-6}
.000001	1 μ	1×10^{-6}
.000000000001	1 p (pico)	1×10^{-12}
.00000000001	10 p	10×10^{-12}
.0000000001	100 p	100×10^{-12}

A conversion table is presented as Table 2-8. This table arranges the prefixes and their corresponding power-of-10 values on a scale from mega to pico. Note that 1 appears as *unity* and with an exponent of 0. Positive exponents are arranged to the left of unity, negative exponents to the right.

TABLE 2-8
Conversion Table

giga			*mega*			*kilo*			*unity*			*milli*			*micro*			*nano*			*pico*
	8	7		5	4		2	1		−1	−2		−4	−5		−7	−8		−10	−11	
10^9			10^6			10^3			10^0			10^{-3}			10^{-6}			10^{-9}			10^{-12}

Using powers of 10, the statement 1×10^0 is the same as the statement 1×1. This is done in order to include the 0 exponent position in the system. Let's use this *conversion table* to identify prefixes, their powers of 10, and identification statements and to perform some conversions. Following are examples of these conversions and the procedures used to convert quantities using this method:

EXAMPLE 1:

Given 6000, convert it to the correct prefix.

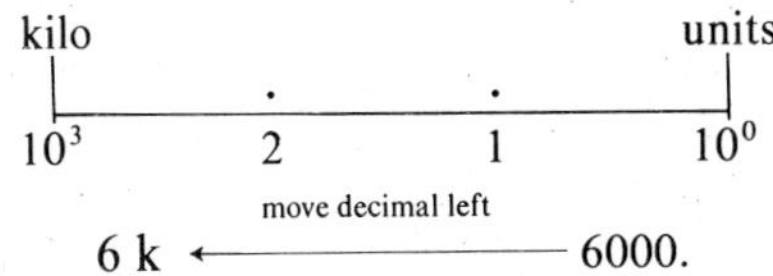

The number 6000. is followed by the decimal point as shown. This number represents 6000 units. To convert it to a prefix, move it along the scale until a number between 1 and 10 is left, as in scientific notation. Moving to the left and dropping a zero each time you pass a number results in a prefix statement of 6 k. Therefore, 6000. converts to 6 k.

EXAMPLE 2:

Convert 1 ampere to milliamperes.

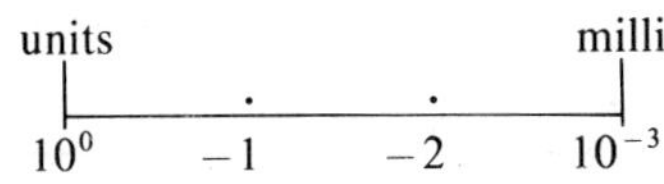

move decimal right

1 ampere ⟶ 1000 milliamperes

EXAMPLE 3:

Convert 2000 μA to amperes.

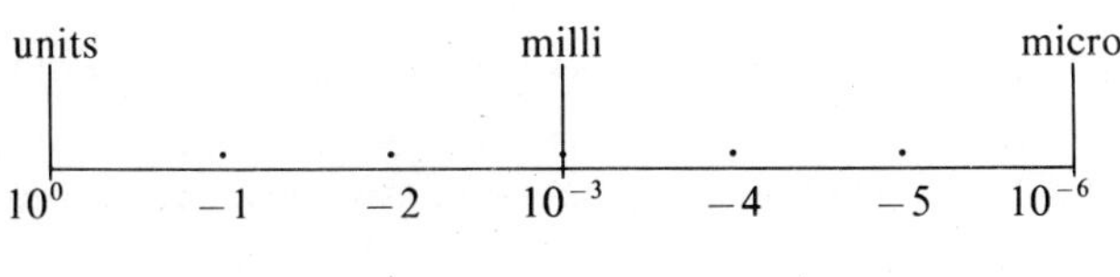

0.002 A ⟵ 2000. μA

In each example presented, you are required to:

1. Identify a quantity.
2. Locate its position on the line.
3. Locate the position of the desired prefix.
4. Count the number of spaces between the two locations.
5. Write the statement in the new prefix by
   (a) adding that number of zeros after the number for movement to the right
   Example: 40 k converts to 40.0 0 0.

   Example: 0.4 mA converts to 0.4 0 0. μA

(b) Movement of a quantity to the right in order to insert a number of zeros after the decimal to convert the quantity to a different prefix
Example: 400 Ω converts to .000400. MΩ

Example: 4.000. μA converts to 4 mA

This list can be used to convert any quantity to a prefix or a prefix to another prefix.

Some students understand the following conversion table in Table 2-9 better than they do Table 2-8:

TABLE 2-9

Conversions						
Original value	Desired value					
	mega	kilo	units	milli	micro	pico
mega		3→	6→	9→	12→	18→
kilo	←3		3→	6→	9→	12→
units	←6	←3		3→	6→	9→
milli	←9	←6	←3		3→	6→
micro	←12	←9	←6	←3		3→
pico	←18	←12	←9	←6	←3	

In Table 2-9, the main prefixes are listed across the top and on the far left. Inserted in the appropriate blocks are arrows indicating the direction of movement of the decimal and the number of zeros that must be added for conversion. Upon close examination of the table, you should be able to understand the application of the table to these conversions.

Self-Check

Complete each item by inserting the words and/or numbers needed to make a true and complete statement.

6. The name used to identify a specific power of 10 is called a ____________.

7. Make a list of the five most important prefixes and the numerical value of each.
   (a) __________, __________
   (b) __________, __________
   (c) __________, __________
   (d) __________, __________
   (e) __________, __________

8. The number 123,000 converts to the prefix value ______________.

9. The symbol used to designate *micro* is ______________.

10. The prefix 15 mega converts to the number ______________.

11. Complete the blank spaces in the chart:

Item Number	*Decimal Quantity*	*Scientific Notation*	*Assigned Prefix*
(a)	1500		
(b)			5 μ
(c)		4×10^6	
(d)			3.6 m
(e)	.0095		
(f)		1×10^{-12}	

Prefixes and the Calculator

Table 2-10 presents prefixes, their numerical value and power-of-10 value for each, and an entry sequence for entering each power of 10 in the calculator. Practice performing these operations on your calculator.

TABLE 2-10

				Calculator Entries				
Prefix name	Noted as	Numerical value	Power-of-10 value	Enter	Press	Press	Press	Display
1 mega	1 M	1000000	1×10^6	1	EE	6	—	1 06
1 kilo	1 k	1000	1×10^3	1	EE	3	—	1 03
1 unit	1	1	1×10^0	1	EE	0	—	1 00
1 milli	1 m	.001	1×10^{-3}	1	EE	+/−	3	1 − 03
1 micro	1 μ	.000001	1×10^{-6}	1	EE	+/−	6	1 − 06
1 pico	1 p	.000000000001	1×10^{-12}	1	EE	+/−	12	1 − 12

Note that when you enter quantities carrying a negative exponent, it is necessary to press the "change sign" (+/−) key. On some calculators, it doesn't matter whether this key is pressed before or after the exponent is entered; however, it must be pressed after the "EE" key has been pressed. A good practice is to enter the (+/−) immediately after pressing EE; if you press the key (+/−) at any time other than during exponent entry, the sign of operation for the entire statement will be changed.

Once entries have been made and the calculator has been instructed to perform the operation, it does so quickly. When entering information, it is not necessary to convert all data to valid powers-of-10 statements; the calculator will accept and process intermixed entries. Four examples are:

EXAMPLE 1:

Given the formula $R_t = R_1 + R_2 + R_3$

where $R_1 = 500\ \Omega$, $R_2 = 10\ \text{k}\ \Omega$, and $R_3 = 1\ \text{M}\ \Omega$, solve for $R_t = ?$. To use the calculator with this solution, do the steps listed in Table 2-11.

TABLE 2-11

Desired Entry	*Enter*	*Press*	*Press*	*Display*
500 ohms	500	—	—	500
+	—	—	—	500
10 kilo ohms	10	EE	3	10 03
+	—	—	—	1.05 04
1 meg ohms	1	EE	6	1 06
=	—	—	—	1.0105 06

Answer $R = 1.01$ meg ohms

EXAMPLE 2:

Given: $E = IR$, $I = 10$ m, and $R = 5$ M
Solve for $E = ?$ Steps for solution appear in Table 2-12.

TABLE 2-12

Desired Entry	*Enter*	*Press*	*Press*	*Press*	*Display*
I = 10 m	10	EE	+/−	3	10 − 03
multiply	X	—	—	—	1 − 02
$R = 5$ M	5	EE	6	—	5 06
equals	=	—	—	—	5 04

Answer $E = 50{,}000$, or 50 kilounits

EXAMPLE 3:

Given: Formula is $X_C = 1/2\pi fC$, $f = 10k$ and $C = 5\ \mu$. Solve for X_C. The steps appear in Table 2-13.

TABLE 2-13

Desired Entry	*Enter*	*Press*	*Press*	*Press*	*Display*
2	2	—	—	—	2
multiply	X	—	—	—	2
pi	π	—	—	—	3.14159
multiply	X	—	—	—	6.28318
$f = 10$ k	10	EE	3	—	10 03
multiply	X	—	—	—	6.28318 04
$C = 5\mu$	5	EE	+/−	6	5 − 06
equals	=	—	—	—	3.14159 − 01
reciprocal	$1/X$	—	—	—	3.183 00

Answer: $X_C = 3.183$ ohms

EXAMPLE 4:

If $R_1 = 60$ k, $R_2 = 30$ k, and $R_3 = 20$ k, use the following formula and solve for R_t. The steps are given in Table 2-14.

Formula: $\frac{1}{R_t} = \frac{1}{R_1} + \frac{1}{R_2} + \frac{1}{R_3}$

TABLE 2-14

Action	*Entry*	*Display*
Enter	60, EE, 3	60 03
Press	1/X	1.67 − 05
Press	+	1.67 − 05
Enter	30, EE, 3	30 03
Press	1/X	3.33 − 05
Press	+	5 − 05
Enter	20, EE, 3	20 03
Press	1/X	5 − 05
Press	=	1 − 04
Press	1/X	1 04

Answer $R_t = 10\ \text{k}\Omega$

Practice these procedures until you can do them without hesitation. The better you can use your calculator, the better you will do in your electronics classes.

Self-Check

Answer each item by indicating whether the statement is true or false.

12. ____ You must convert all data to powers of 10 before entering them on your calculator.

13. ___ Calculators are designed so they display large quantities in powers of 10.

14. ___ The change sign key can be used to change the displayed data from − to + or from + to −.

15. ___ Some keys on a scientific calculator are used to enter fixed quantities in your calculations.

16. ___ To enter a power of 10 on the calculator, press the EE key before entering the numerical coefficient.

Summary

Calculators come in numerous types. The discussion of calculators is centered around those using the Algebraic Operating System (AOS), a system similar to the conventional mathematical processes you have already learned. No attempt was made to include calculators of the other type. Scientific notation is defined and related to powers of 10. The rules for converting numerical quantities to powers of 10 were introduced. Practice was provided in the conversion of quantities between scientific notations and numerical quantities. Power-of-10 statements were used to solve several different problems.

Electronics prefixes were introduced and explained. The need for this system of identification was presented along with the procedures used to convert numerical or power-of-10 statements to prefixes.

The hand-held calculator can be a valuable asset in your study of electronics. Its use allows you to proceed at a faster pace. Use of a calculator to solve simple math problems, convert quantities to powers of 10, convert powers of 10 to numerical quantities, and solve more complex problems is discussed and examples presented. The remainder of this book is written around the use of AOS-type calculators and their use in electronics.

Review Questions and/or Problems

1. Convert the following quantities to scientific notation (powers of 10).
   (a) 2000
   (b) 5.678
   (c) 390,000
   (d) 0.00039
   (e) 18,000
   (f) 0.00046
   (g) 0.1971
   (h) 0.321,000,000,000
   (i) 0.000,000,000,069
   (j) 0.5
   (k) 1,400,000

2. Multiply the following. Express answers in powers of 10 and in prefixes.

Problem	Power of 10	Prefix
(a) 713,000 × 1,810,000	________	________
(b) $(2.3 \times 10^{6})(1.3 \times 10^{4})$	________	________
(c) $(6.5 \times 10^{5})(6.3 \times 10^{3})$	________	________
(d) $(3.2 \times 10^{-4})(1.8 \times 10^{-6})$	________	________
(e) $(7.6 \times 10^{-3})(8.2 \times 10^{-3})$	________	________
(f) $(1.2 \times 10^{-8})(9.3 \times 10^{5})$	________	________

3. Divide the following, expressing answers in powers of 10 and in prefixes.

(a) $\frac{299{,}200}{88}$ ________ ________

(b) $\frac{1.8 \times 10^{8}}{3 \times 10^{-5}}$ ________ ________

(c) $\frac{42 \times 10^{6}}{6 \times 10^{3}}$ ________ ________

(d) $\frac{0.0018}{0.0009}$ ________ ________

(e) $\frac{6 \times 10^{12}}{2 \times 10^{4}}$ ________ ________

(f) $\frac{8 \times 10^{-9}}{20 \times 10^{3}}$ ________ ________

(g) $\frac{8 \times 10^{6}}{0.003}$ ________ ________

(h) $\frac{27{,}000}{300}$ ________ ________

4. For each prefix shown, provide the information requested in the three columns.

Prefix	Power of 10	Quantity (in units)	Calculator Entry
Units			
Micro			
Mega			
Milli			
Kilo			
Pico			

3

Electronic Devices and Symbols

Introduction

In Chapter 1 you were introduced to conductors, insulators, and electron flow. A short review is now in order.

conductors—materials with many free electrons
insulators—materials with few free electrons
electron flow—orderly movement of electrons in a conductor

In this chapter we expand on these definitions and introduce several other areas of electronics.

Objectives

Each student is required to:

1. List the requirements for current flow.
2. Explain the principle of electron flow versus conventional current.
3. Describe current flow within a conductor in terms of electron movement.
4. Describe the direction electrons flow in an external circuit.
5. Identify the practical unit of current and its symbol.
6. Define the term *ampere* and state its relation to the coulomb.
7. Define electromotive force (EMF) and identify its symbol.
8. Define the term *volt* and write its symbol.
9. List five common sources of electromotive force.
10. Define the term *cell* and identify its schematic symbol.
11. Explain the difference between dry cells and wet cells, on one hand, and primary cells and secondary cells, on the other.
12. Explain the result of connecting cells in series, in parallel, and in parallel-series.

13. Draw simple schematics of cells connected in series, in parallel, and in series-parallel.
14. Define *resistance* and state its use in electronics.
15. Draw the schematic symbols for fixed, adjustable, and variable resistors.
16. Describe how fixed-carbon, fixed-wirewound, and variable resistors are made.
17. Use the Electronic Industries Association (EIA) resistor color code to determine the value and tolerance of unknown fixed resistors.
18. Identify standard schematic symbols and draw simple schematics using appropriate symbols.

Electron Current

In Chapter 1 we discussed the coulomb as a unit of static charge. The coulomb is used to indicate a specific amount of charge. The difference between a *charge* and *electron current* is one of movement and time. A charge in motion is *current.*

An understanding of electron current must involve understanding the free electron within a conductor and how it is affected by another electron or other external force. You have already learned that:

Like charges repel, and unlike charges attract.

All electrons have negative charges and thus repel each other. An electron, however, is attracted by a positive charge.

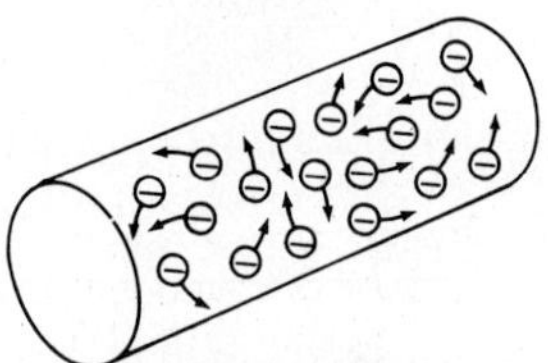

FIGURE 3-1

Consider the cutaway view of the piece of copper conductor in Figure 3-1. Notice that free electrons are illustrated and that they are equally spaced because of the repelling force that exists between them. Still, electrons will be drifting from place to place within the conductor. The drift continues as long as the energy level of the electrons is high enough for them to remain free of their valence bands.

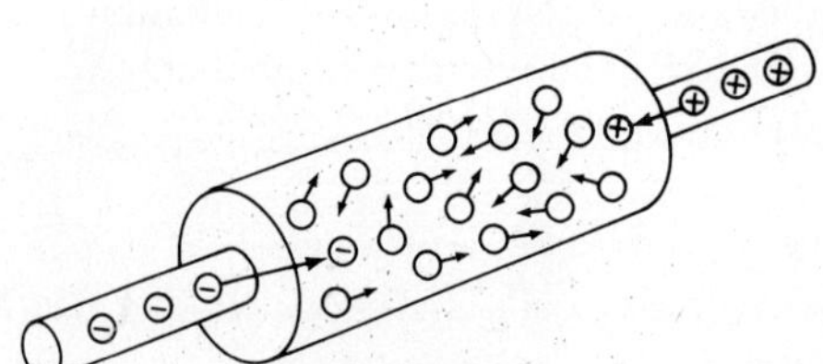

FIGURE 3-2

When an external force is connected to the conductor such that its positive charge is connected to the right end and its negative charge is connected to the left end, movement of the electrons becomes predictable (Figure 3-2).

The negative charge repels the conductor's free electrons, and the positive charge attracts them. Any positive charges that are in the conductor are repelled to the left, where they are attracted to the negative pole.

Long before the electron theory had been developed, electricity had been recognized and studied. Previously, the logical assumption was that work is performed by positive charges moving down to a lower, or negative level. Today, the movement of positive charges in a circuit is known as *conventional current flow*. After discovery of the electron, experimentation revealed that the particle moving is the negative charge, or electron. From this discovery, the electron theory was developed. This theory states that electrons leave the negative pole of an electrical source, move through a closed loop, and return to the source at its positive pole.

Analysis techniques are identical regardless of the type of current used. For our purposes, however, current and current flow are associated with electron movement.

When the external source described above is connected, electron movement becomes orderly and movement is from left to right. Connecting an external source, as in Figure 3-3, causes the movement to be in the direction of the arrow.

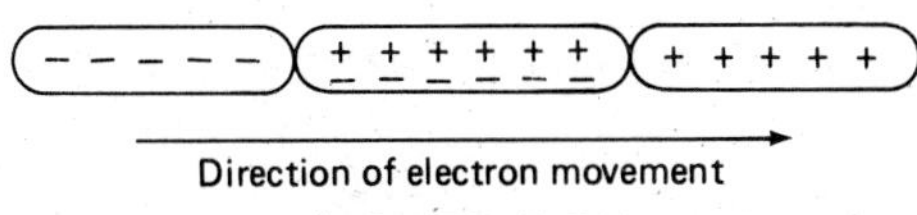

FIGURE 3-3

Free electrons located in the negative source move into the neutral conductor and combine with a positive charge. At the same time, an electron is attracted from the neutral conductor and combines with a positive charge in the positive source. Electrons continue to move in this direction as long as the two external sources can support the process. Notice that the negative source supplies the electrons and that the positive source accepts them.

Consider a situation in which this occurs—an automobile battery. Figure 3-4 illustrates use of the battery as an external source.

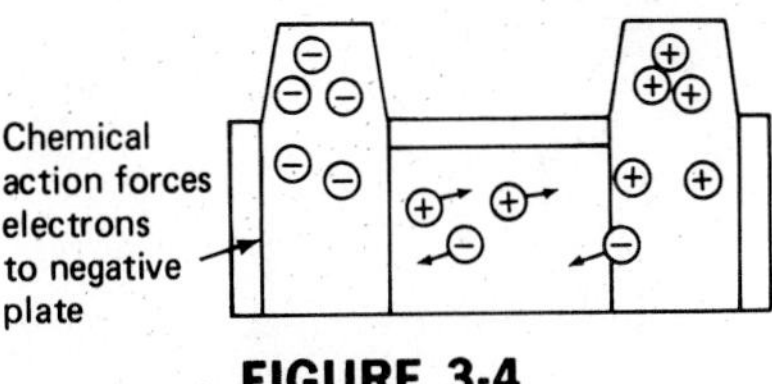

FIGURE 3-4

Chemical action within the battery causes negative ions (excess electrons) to concentrate at the negative (−) pole of the battery and positive ions (deficient electrons) to concentrate at the positive (+) pole. Depending on the pole's polarity, the charge located at these poles will attract or repel electrons.

Requirements for Current Flow

When a conductor links positive and negative poles, we can predict the result (Figure 3-5).

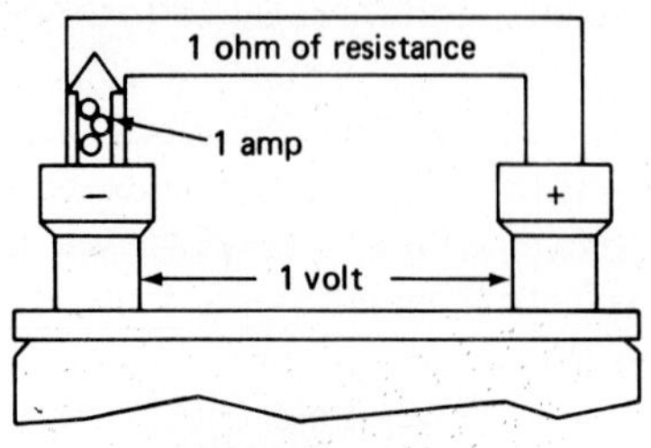

FIGURE 3-5

The negative pole repels electrons and attracts positive ions, whereas the positive pole attracts electrons and repels positive ions. Electrons enter the conductor at the left and leave it at the right. The result is a flow of electrons from the negative pole of battery through the external circuit (conductor) and back to the positive pole of the battery.

Note: Any connection of conductors and/or components that form a complete path for the flow of electrons from negative to positive poles of a power source is a *circuit*. Figure 3-5 illustrates a circuit. This circuit is of no practical use, since it isn't designed to perform useful work.

Electrons move approximately at the speed of light. In the circuit shown in Figure 3-5, current begins to flow instantaneously; it continues to flow as long as the source continues to supply electrons. As one electron enters the conductor, another leaves. The result is a steady flow of electrons through the conductor.

Within a battery, chemical action causes electrons entering the positive pole to move back to the negative pole. Outside the battery, electrons are forced to move from negative to positive; internally, electrons move from positive to negative.

The terms *negative* and *positive*, along with the plus (+) and minus (−) symbols, are for reference only; they are not mathematical signs of operation. This is best illustrated in Figure 3-6.

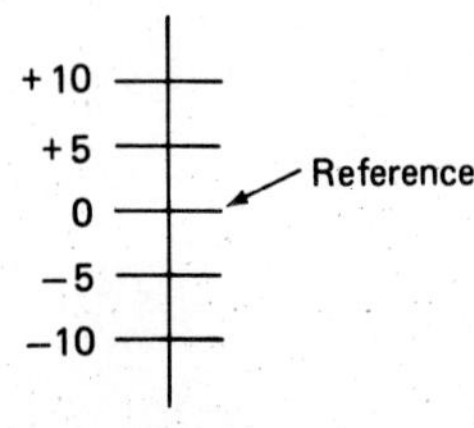

FIGURE 3-6

If neutral is the reference (0) on the graph, then any point above zero is positive, whereas points below zero are negative. But, +10 is positive when compared to +5, and −5 is positive when compared to −10.

In Chapter 1 we saw that a positive charge results from a deficiency of electrons and that a negative charge results from an excess of electrons. Relating this to the battery, we see that the positive pole represents a deficiency of electrons, and the negative pole an excess. The imbalance of electrons is a potential source of energy. When free electrons are recombined with atoms, they release the energy they absorbed as they become free electrons. The energy released can be used to do work. The greater the number of excess electrons, the more potential energy is available.

Referring to Figure 3-5, we see that it is easy to understand why connecting a conductor between the two poles of the battery allows electrons to flow. Potential sources of energy with larger amounts of electrical charge cause larger amounts of electrons to flow. Two conditions are necessary for current to flow:

1. There must be a difference in potential (an energy source).
2. Conductors must be connected to form a continuous path between the positive pole and the negative pole of the energy source.

Once these conditions are met, the closed loop becomes a *circuit*. If the conductor is not continuous between the negative pole and the positive pole, the circuit is said to be *open,* and electrons will not flow. The potential energy of the source is still available, but it cannot be used until the open circuit is closed.

Current is a measure of the number of electrons that pass a given point in one second. In Chapter 1 the coulomb of charge was explained as being equal to 6.28×10^{18} electrons. The measure of current is stated in amperes.

1 ampere of current is flowing when 1 coulomb of electrons passes a given point during 1 second.

Note: The flow of current involves both time and amount of charge (coulombs).

For example, 2 amperes are said to be flowing if 2 coulombs of charge pass the point in 1 second. In schematics, the symbol used to indicate current is *I* (intensity of flow). Amounts of current are referred to as amperes (A) or by the prefixes milliamperes (mA) or microamperes (μA).

Self-Check

Complete each item by inserting the words and/or numbers needed to make a true and complete statement.

1. When current flows from a positive pole to a negative pole, it is called ____________ current flow.

2. Electrons flow from a ____________ charge to a ____________ charge.

3. If two coulombs of charge pass a given point in 1 second, current flow equals ____________ amperes.

4. ____________ is the symbol used to identify current.

5. The unit of measure for current flow is the ____________.

EMF, Voltage, and Difference in Potential

We have discussed the availability of an energy source but not the origin of the source or what it is. Batteries are used to exert an electromotive force (EMF) on the circuit, which causes electrons to move in a predictable manner. Batteries are only one source of EMF; several other types are discussed later in this book. All sources of EMF have some things in common. All are capable of providing an EMF (designated by the letter *E* in a schematic); all have positive and negative poles, and all can support current flow. Any electrical charge, whether negative or positive, is a potential source of energy. The difference between the two poles of a battery is called a *difference in potential.* Difference in potential is measured in voltage (symbol *V*), and one volt is defined thus:

> ***The difference in potential needed to cause 1 ampere of current flow through 1 ohm of resistance.***

Often the terms *volt, EMF,* and *difference in potential* are used interchangeably. The symbols *E* and *V* are also used interchangeably, since both refer to a difference in potential.

Production of Electromotive Force

Of all the energy consumed in the world, electrical energy has the widest application. For this reason, electrical energy has been, and will continue to be, an important factor in world development.

EMF is produced by one of five different methods; each method is discussed:

1. *mechanical.* Mechanical energy is converted into electrical energy. The alternator in an automobile is an example. Some of the mechanical energy of the car's engine is used to produce electrical energy, which keeps the battery charged. Some mechanical systems are steam, nuclear, water, gasoline, and windmill. Mechanical energy is used to generate a large portion of the electrical energy used today.
2. *chemical.* A controlled chemical reaction is used to produce electrical energy. This method is used in batteries (discussed later in this chapter).
3. *thermoelectric.* Heat is used to produce an EMF. When two dissimilar metals are heated, they produce a difference in potential. A thermocouple is one example. Heat applied to the common (joined) area of the metals causes them to produce opposite charges. A potential difference exists between the two separated terminals. An electric thermometer such as that used by a nurse works according to this principle.
4. *photoelectric.* Light striking a light-sensitive surface (called a photo cathode) causes electrons to be emitted, which form an energy source. Solar cells, TV cameras, and satellite electrical systems use this principle.
5. *piezoelectric.* Some ionic crystals produce an EMF when stress is applied to their surface. Squeezing a quartz crystal causes voltage to appear between the two opposite sides of the crystal; reversing the pressure causes the polarity of the EMF to reverse. This is called the *piezoelectric effect.* The quartz crystal in a digital watch operates on the piezoelectric effect.

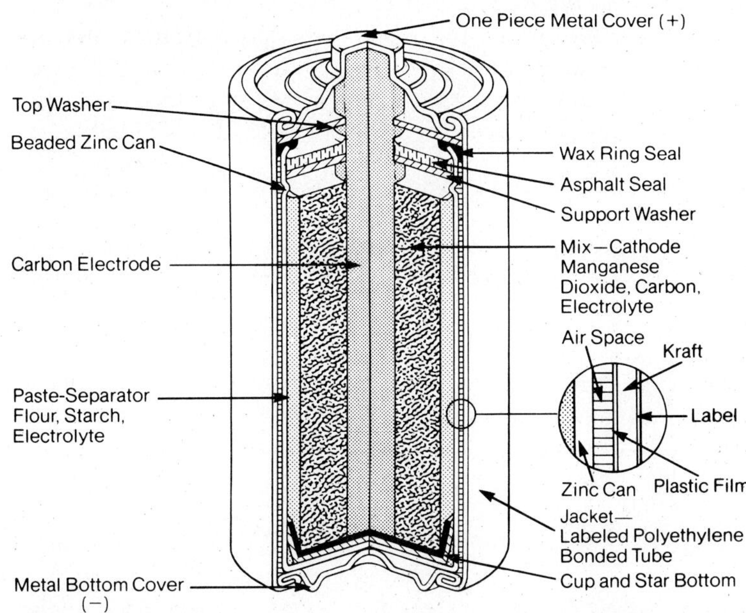

FIGURE 3-7
"D" Cell—1.5 V DC (Courtesy of Eveready, Union Carbide)

Self-Check

Complete each item by inserting the words and/or numbers needed to make a true and complete statement.

6. Other names for *voltage* are ____________ and ____________.

7. The practical unit of measure for current is the ____________.

8. The practical unit of measure for electromotive force is the ____________.

9. ____________, ____________, ____________, and ____________ are methods used to produce an EMF.

10. An automobile battery is an example of a ____________ means of producing voltage.

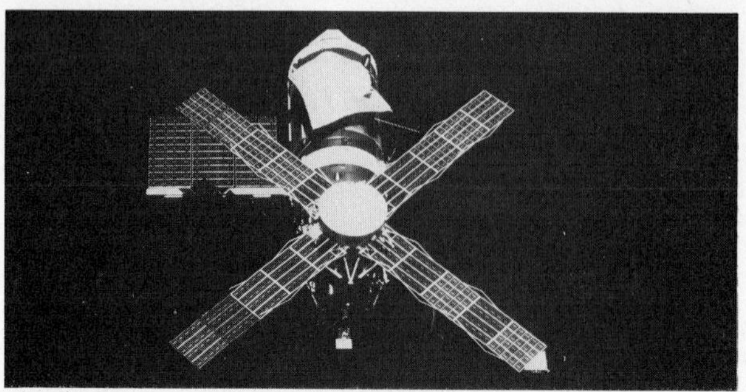

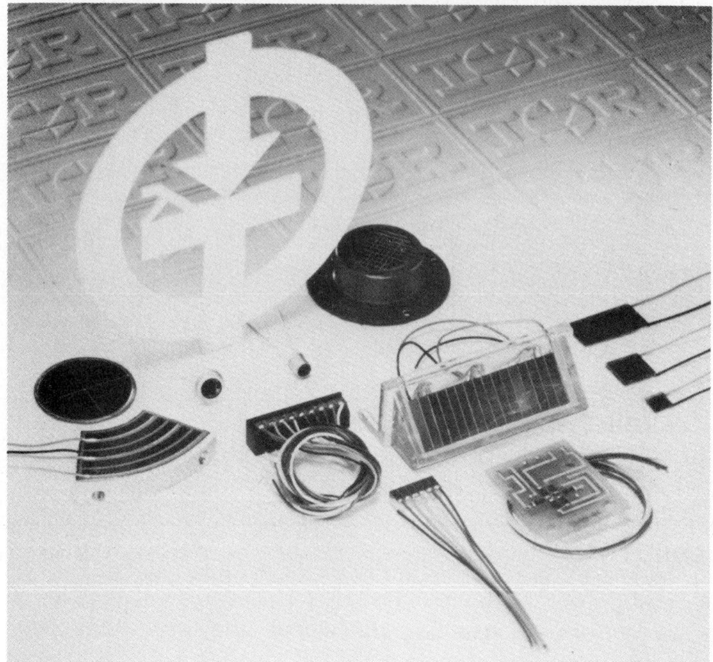

FIGURE 3-8
(a) NASA Space Lab with Solar Cells Displayed (Courtesy of NASA) (b) Commercial Photo Cells (Courtesy of International Rectifier)

Conductors and Resistance

Electromotive force and current flow would be useless if we had not learned how to use them. To be useful, voltage and current must produce heat, turn motors, transmit sound waves, and do other work. Devices used to perform work are connected to the power source (EMF) by conductors. A voltage source, conductors, and a work-producing device are the three components needed to perform electrical work. A *resistive device* is used to do the work. When all three components are present, they are said to be connected in a *practical circuit.* A circuit is present when there is a complete loop for electron-current flow. To better understand the practical circuit, it is necessary to understand resistance.

Resistors are devices which oppose the flow of electrons.

Resistors are represented in a schematic drawing by the letter *R*. The international unit of measure for resistance is the ohm, which is symbolized by the Greek letter omega (Ω). Many resistors have large values requiring thousands or even millions of ohms; they are identified by the prefixes kilo ohm (kΩ) or mega ohm (MΩ). The ohm is defined as follows:

1 ohm of resistance is required to limit current flow within a circuit to 1 ampere when 1 volt of EMF is applied.

Conductors

Different materials have different amounts of resistance. Therefore, the material used to manufacture a component affects its resistance. The ease with which materials release their free electrons determines their ability to carry current. Conductors must have very low resistance, whereas insulators must have very high resistance. Two conductors produced from the same material have the same resistance, assuming they are of the same size and shape. Changing the size of a conductor causes it to have a different resistance, which, in turn, causes it to allow a different amount of current flow.

Table 3-1 illustrates the resistance of various materials compared to silver. Silver is a good conductor, but a nichrome wire is not. Nichrome is often used as the heating element in hair dryers and toasters. The table compares sections of each material of identical size and shape. The numbers listed are ratios of comparison with the standard (silver).

TABLE 3-1

Substance	*Relative Resistance (ohms per section)*
Silver (standard)	1.0
Copper	1.08
Gold	1.5
Aluminum	1.8
Tungsten	3.5
Brass	4.8
Steel	9.3
Nichrome	64.8
Carbon	5400

Note: Table values are approximations.

Length Versus Resistance

Let us assume that a roll of copper wire has been manufactured so that a 1-foot *section* has 1 ohm of resistance. Ten sections (10 ft) equal 10 ohms, and 100 sections (100 ft) equal 100 ohms. The length-to-resistance relationship is expressed as follows:

The resistance of a conductor is directly proportional to its length.

Diameter or Cross-Sectional Area Versus Resistance

The resistance of a conductor is also affected by its cross-sectional area (the diameter squared). Two wires made of identical material, but having different diameters, are illustrated in Figure 3-9.

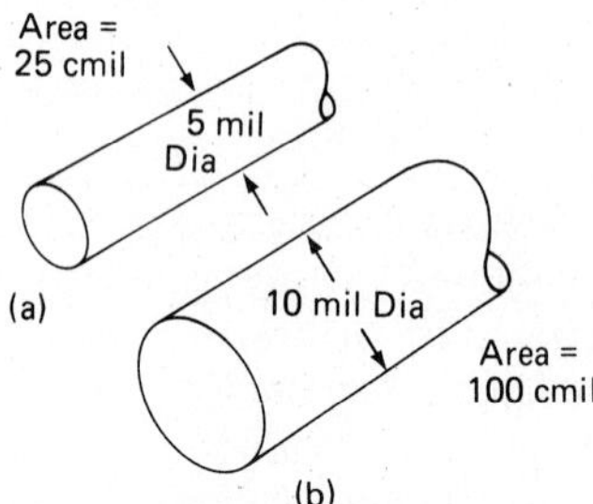

FIGURE 3-9

The face of each wire is shown. This face is the cross-sectional area. Area is equal to the diameter squared. One mil is equal to 0.001 in. Therefore, if *a* has a diameter of 5 mils, it has an area of 25 circular mills (cmils). Conductor *b* has a diameter of 10 mils; this gives it an area of 100 cmils. Therefore, conductor *b* has a diameter double that of conductor *a* and a cross-sectional area four times that of conductor *a*. As the cross-sectional area of a conductor increases, the conductor's resistance decreases:

The resistance of a conductor is inversely proportional to its cross-sectional area.

When this rule is applied, we see that conductor *b* has one-fourth the resistance of conductor *a*. This comparison is similar to the one of 2-lane and 4-lane highways: the more lanes (area), the more traffic that can flow.

Temperature Versus Resistance

Another factor that affects a conductor's resistance is temperature. The resistance of some materials increases as the temperature increases. When the temperature is cooler, the conductor can carry more current. The variance of resistance according to temperature is known as the temperature *coefficient*. If an increase in temperature causes a conductor's resistance to increase, the material is said to have a "positive temperature coefficient"; if an increase in temperature causes a decrease in conductor resistance, the material has a "negative temperature coefficient." Conductors made of metal have positive coefficients, but semiconductor materials (carbon, silicon, germanium) have negative coefficients.

Resistivity

Resistance to the flow of electrical current due to the type of conductive material used is called *resistivity*. The resistivity of a material, in turn, is inversely proportional to the number of free electrons available in 1 cubic centimeter of the material. Resistivity is represented by the Greek letter (ρ).

Let's compare copper and carbon as to resistivity. It takes more energy to move electrons from the carbon atom than it does in copper, even though carbon has four valence electrons and copper has only one. Carbon could conceivably provide four electrons to copper's one; but the copper atom releases its electron with a fraction of the energy required to move a carbon electron. Hence, copper makes a better conductor and has lower resistivity.

The resistivity of a material is measured under strict, standard dimensions of weight, density, and temperature. The *specific resistance* of several materials is shown in Table 3-2. Specific resistance is stated with respect to *circular mil per foot* and in terms of metric cubic centimeter.

TABLE 3-2

Material	*Circular mil/foot at 20 degrees C*	*Cubic Centimeter at 20 degrees C ($\times 10^{-6}$)*
Silver	9.56	1.629
Copper	10.4	1.724
Aluminum	17.0	2.828
Tungsten	34.0	5.510
Brass	42.0	7.500
Nickel	60.0	10.00
Iron	61.00	9.800
Manganin	264.00	48.00
Constantan	294.00	49.00
Nichrome	675.00	108.0
Carbon	22,000	3700

Printed Circuits

Most modern circuitry is designed using a process called *printed circuitry*. In this process, copper-clad phenolic or epoxy glass board is used to make the conductors. The desired conductor pattern is printed on the solid-clad face with a protective film. The entire board is then treated with a substance that removes all copper other than that protected by the printed coating. Holes are drilled in the board at points where components will be located, and the components are mounted on the side of the board opposite the conductive foil, with their leads inserted through the board. The leads are then connected to the copper foil by a process called *soldering* (explained more fully in Appendix B).

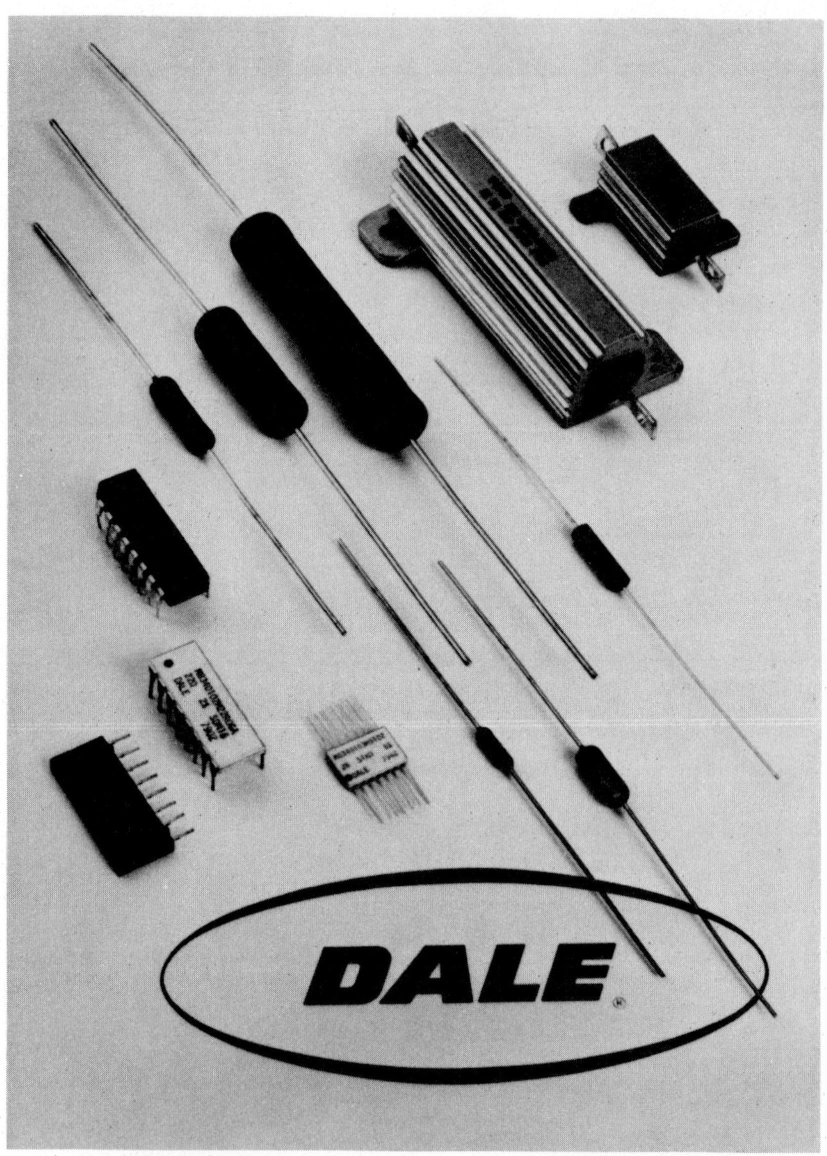

FIGURE 3-10
Resistor Assortment (Courtesy of Dale Electronics)

Fixed Resistors

Conductors are part of every circuit; ideally, they should have zero ohms of resistance. Conductors may be large wires or they may be small strips of metal foil. The type of conductor chosen is a compromise between the ideal and what can best be used in a specific application. Components to which the conductors are connected is another matter. The resistance of a component is often very high. Many components (called *resistors*) are manufactured to provide specific resistance values.

Resistors are manufactured from materials having known resistance values and are designed for operation at a standard temperature of 77 degrees Fahrenheit. A material is selected according to its diameter, length, and temperature coefficient—all of which provide predictable opposition to current flow. The material is then formed into various resistances, shapes, and sizes. Materials commonly used in the manufacture of resistors are carbon, wire, and metal film. Most resistors have a fixed value, but some are variable. Wirewound resistors are common in circuits having high current or that require precise values.

Wirewound Resistors

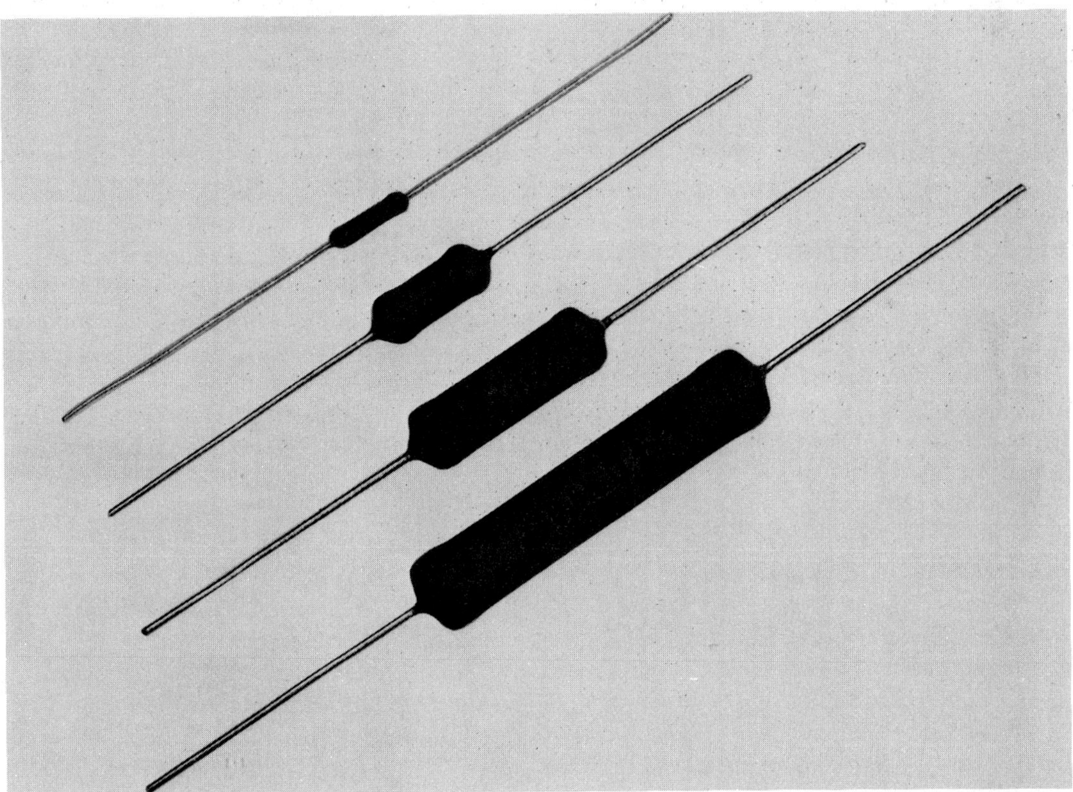

FIGURE 3-11
Precision Wirewound Resistors (Courtesy of Ohmite, a North American Phillips Company)

The type of resistor known as a wirewound resistor is made by wrapping a resistive wire around an insulator frame a sufficient number of turns to use the amount of wire required to provide a desired ohmic value. Nichrome wire is one type of wire used. Wire ends are connected to metal "tabs" at each end, which aid connection of the device in a circuit. Figure 3-12 shows various types of wirewound resistors. Once the wire is in place, the entire unit is insulated to prevent shorting. The larger size of the resistor and the fact that it can be hollow, plus the higher heat capability of the resistive wire, allows the resistor to operate in circuits with high currents and heat.

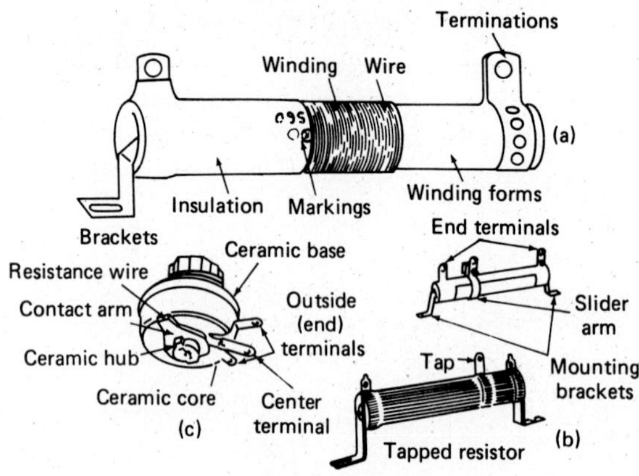

FIGURE 3-12

Wirewound, adjustable, variable, and tapped resistors are usually stamped to show resistance, wattage, and tolerance value. These important data can be read directly from the resistor body.

Some wirewound resistors are actually several resistors in one. Figure 3-12b shows a *tapped resistor*. At intervals along the wire used to wind the resistor, taps are made which are connected to metal tabs. These tabs can then be wired in a circuit, providing two or more resistances from one wirewound device.

An adjustable wirewound resistor is shown in Figure 3-12c. This resistor has a movable tap, or slider, which can be positioned along the body of the resistor until the desired resistance is obtained. The resistor can then be locked in place. This type of resistor is usually adjusted during assembly, and requires no further adjustment.

Precision Resistors

Precision resistors are shown in Figure 3-13. Precision resistors are used when a resistor's value must be extremely accurate. Precision resistors often have an accuracy of 1% or greater.

Fixed-film resistors are made of either metal film or Cermet film. Metal-film resistors consist of a conductive metal film coating over a glass base; Cermet-film resistors are made by firing a ceramic metal (Cermet) film coating on a substrate base. Both can be manufactured with fairly precise amounts of resistance. Recent manufacturing changes have reduced the cost of precision film resistors. Due to the competitive pricing of these resistors, many applications that once used carbon resistors now use fixed-film resistors.

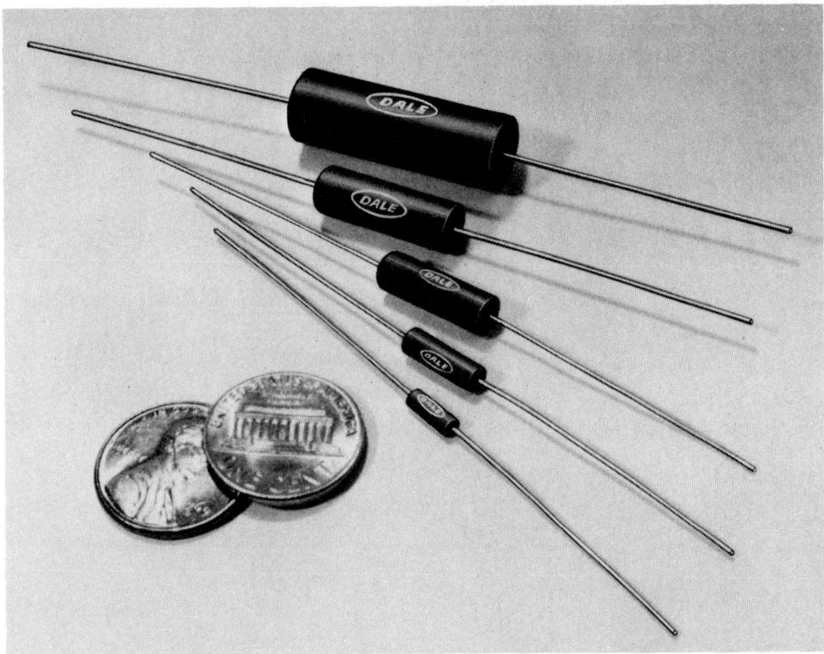

FIGURE 3-13
Assorted Fixed-Film Resistors (Courtesy of Dale Electronics)

Cermet-film resistors have another advantage: they can be manufactured in square or rectangular shapes. This feature makes them especially suitable for use with printed circuits. It is also possible to include several resistors in one flat package, with enough wire leads to connect each resistor in the printed circuit—an application similar to that in which tapped, wirewound resistors are used.

In recent years the cost of manufacturing fixed-film resistors has changed enough to make them competitive with fixed-carbon resistors. The EIA color code is used to mark fixed-film resistors. In the manufacture of film resistors, a laser is used for more accurate trimming. This process is so accurate that tolerances of $\pm 1\%$ and $\pm 2\%$ are available. With the advent of such tolerances, the color code chart has been revised. A *brown* fourth band indicates that a resistor has a tolerance of 1%, and a *red* fourth band indicates a tolerance of 2%. In the following discussion of the color code, resistors having this tolerance are discussed.

Variable Resistors

Some circuits provide the operator with the means of controlling circuit operation. To accomplish this, a variable resistor is used. The volume control on a stereo receiver is one example. Different people like to set the volume at different levels; the variable resistor allows them to do so. Two types of variable resistors are shown in Figure 3-15.

FIGURE 3-14
Assorted Potentiometers (Courtesy of Allen Bradley)

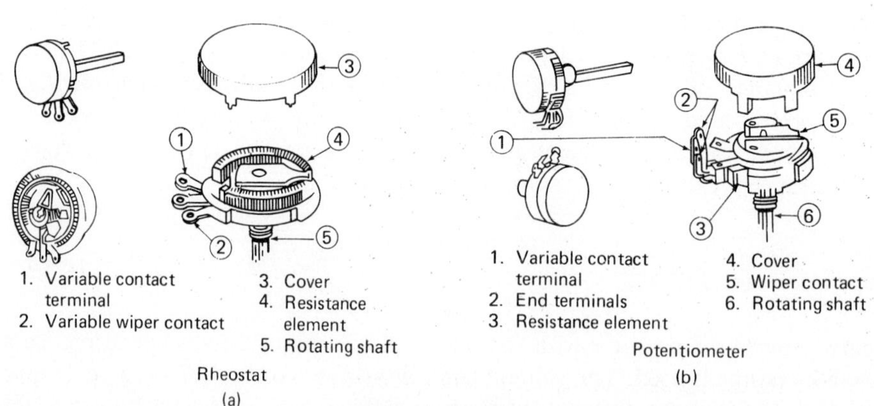

FIGURE 3-15

For controlling circuits with high currents, the wirewound resistor shown in Figure 3-15 is used. Because the wire used to wind variable resistors is quite heavy, it can withstand a heavy current flow. The resistive wire is wrapped around a circular insulator. A movable contact, or wiper, is connected to a rotating shaft; the wiper can be positioned along the resistor to select the resistance desired. This device is called a *rheostat*. It has two tabs (electrical connections) and is used to vary the current flow within a circuit. Operator controlled, it is used, for example, in the dimmer control of a car's dashboard lights.

Variable resistors designed to operate in low-current circuits use a carbon-disc resistance element. The movable wiper can be positioned along its disc to obtain a particular resistance. Used most often as a potentiometer, this type of resistor is used to select a voltage for application to a subsequent circuit. The potentiometer can be identified by its three electrical connections. The one shown in Figure 3-15 can be used as a volume control in a stereo.

It is possible to convert the potentiometer to a rheostat for use in low-current circuits. To do that, the center tab (wiper) is connected to either end tab. This procedure converts the potentiometer from a three-connection device to one with two electrical connections. Schematic symbols for fixed, variable, and tapped resistors are shown in Figure 3-16.

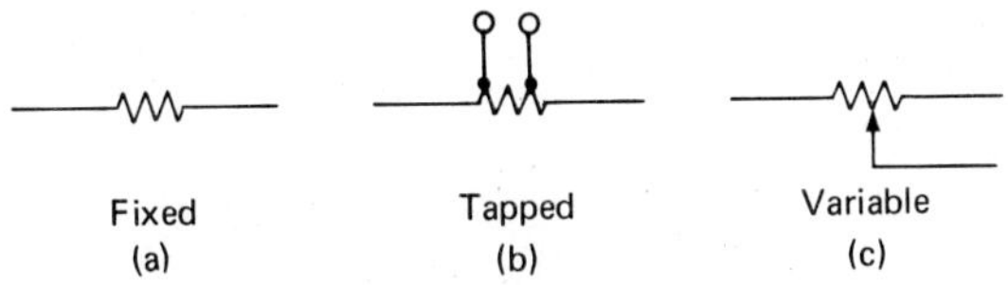

Fixed (a) | Tapped (b) | Variable (c)

FIGURE 3-16

Fixed-Carbon Resistors

Fixed-carbon resistors have numerous applications that call for fixed values of resistance. These resistors, made of carbon graphite, can be manufactured to fairly precise values. Wire leads are formed into each end of the resistor so the resistor can be connected to a circuit. A color code is painted on the body of the fixed-carbon resistor. (Figure 3-17 depicts fixed-carbon resistors.)

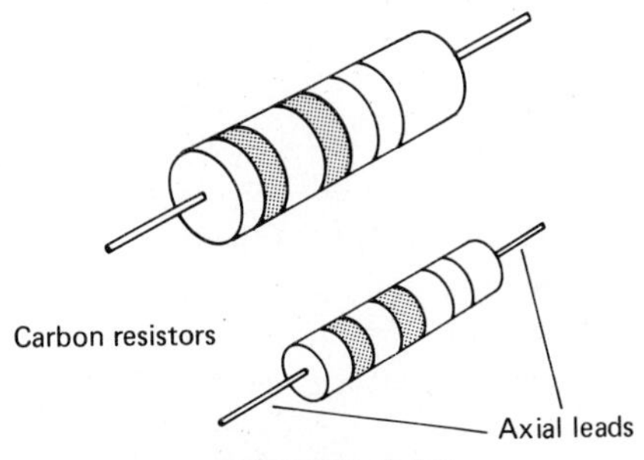

FIGURE 3-17

The Electronic Industries Association (EIA) has adopted a standardized coding for fixed-carbon resistors. The standard is identical to the international standard recognized by the International Electrotechnical Commission (IEC). The code standardizes marking both fixed-carbon resistors and fixed-film resistors. The use of color bands allows resistor value, tolerance, and other data to be visible on the resistor body, regardless of how small the body may be. Table 3-3 contains the data used in the color coding system.

TABLE 3-3
Resistor Color Code

Color of band	Significant Figures of Ohmic Value			Tolerance	Failure rate per 1000 h (%)
	1st Number	*2nd Number*	*Multiplier*		
Black	0	0	1	—	L 5
Brown	1	1	10	1%	M 1
Red	2	2	100	2%	P 0.1
Orange	3	3	1000	—	R 0.01
Yellow	4	4	10000	—	S 0.001
Green	5	5	100000	—	T 0.0001
Blue	6	6	1000000	—	—
Violet	7	7	10000000	—	—
Gray	8	8	—	—	—
White	9	9	—	—	—
Gold	—	—	0.1	5%	—
Silver	—	—	0.01	10%	—
No fourth band		—	—	20%	—

Resistor Tolerance

The EIA color code allows for five tolerances, 1%, 2%, 5%, 10%, and 20%. Learning the color code is quite easy. Resistors manufactured using this code are circled by three, four, or five color bands. The bands start near one end of the resistor and read toward the other end.

Failure Rate

On some newer resistors, a fifth band is present. For a given batch of resistors, the fifth band indicates the number of resistors that can be expected to fail during 1000 hours of operation. For example, a *black* fifth band indicates a probability of 50 failures in each 1000 resistors used. A *green* fifth band indicates 1 failure per 1000 hours of operation for each 10,000 resistors used. The failure rate is not a big problem in day-to-day operation, but, for projects requiring great endurance, such as space satellite or remote operations, the failure rate is important. The fifth band is not discussed here.

Interpretation of Color Codes

Remember that the color code we are discussing is used on *fixed-carbon* and *fixed-film* resistors. Examine the resistor in Figure 3-18.

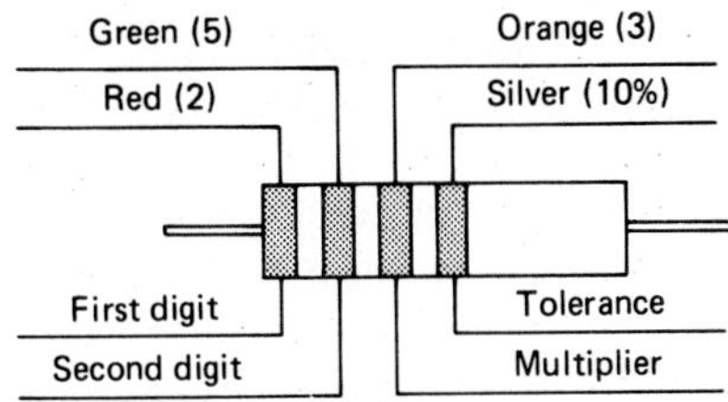

FIGURE 3-18

The band nearest the end is coded *red.* In the table we find that red represents the number 2 when it is located in bands 1 or 2. The *green* second band represents the number 5, and the *orange* third band represents a multiplier of 1000. The fourth band is *silver,* meaning that the tolerance can be plus or minus (±)10% above or below the coded value and the resistor can still be used.

When the numerical values have been determined, the resistor's ohmic value is determined:

First Digit	*Second Digit*	*Multiplier*	*Tolerance*
2	5	1000	10%

The resistor value can be stated as 25,000 Ω ±10% or, using the correct prefix, as 25 kΩ ± 10%. By using the tolerance (10%), we can calculate the maximum and minimum resistance values:

	Mathematical solution	*Calculator Solution*		
Tolerance =	*25,000 × 10% = 2500*	*Action*	*Entry*	*Display*
		ENTER	25000	25,000
		PRESS	×	25,000
		ENTER	10	10
		PRESS	%	0.1
		PRESS	=	2500
Maximum =	25,000 + 2500 = 27,500	ENTER	25000	25,000
Maximum =	27,500 Ω	PRESS	+	25,000
		ENTER	10	10
		PRESS	%	2500
		PRESS	=	27,500
Minimum =	25,000 − 2500 = 22,500	ENTER	25000	25,000
Minimum =	22,500 Ω	PRESS	−	25,000
		ENTER	10	10
		PRESS	%	2500
		PRESS	=	22,500

Examine these solutions and you will find that the last two calculator solutions do not require solving the first solution. They automatically calculate the 10%, then add or subtract as needed to determine maximum and minimum resistor values. These data tell you that, when replacing this resistor you can use any fixed resistor that has at least 22,500 Ω and not more than 27,500 Ω, and the replacement will work as well as the original.

Many people, including the authors, find it helpful to memorize a jingle, which can be recalled when reading a resistor's color code. A jingle commonly used in electronics with the resistor color code is:

Bad Beer Ruins Our Young Guts But Vodka Goes Well. Get Some—Get Some Now.

Notice that when you take the first letter of each word, you have the first letters of the colors listed in Table 3-3. These letters occur in the same order as they do in the table and can be related directly:

B ad	***black***	***0***
B eer	***brown***	***1***
R uins	***red***	***2***
O ur	***orange***	***3***
Y oung	***yellow***	***4***
G uts	***green***	***5***
B ut	***blue***	***6***
V odka	***violet***	***7***
G oes	***gray***	***8***
W ell	***white***	***9***

Note: The next two codes are used as third-band multipliers:

G et	***gold***	***0.1***
S ome	***silver***	***0.01***

Note: The remaining codes are fourth-band tolerance codes.

G et	***gold***	***5%***
S ome	***silver***	***10%***
N ow	***no fourth band***	***20%***
—	***brown***	***1% (as shown in Table 3-3)***
—	***red***	***2% (as shown in Table 3-3)***

The procedures used in reading resistor color codes are covered in the next few figures. Remembering the colors listed above, and their values, you should be able to learn the color code quickly. Calculator solutions are duplicates of those above, and are not presented for each resistor.

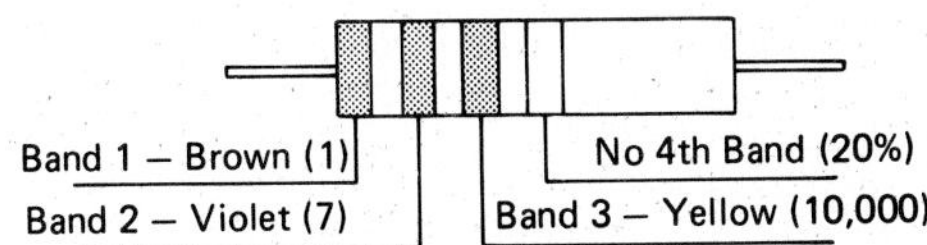

FIGURE 3-19

The determination of resistor values using the color bands from Figure 3-19 results in the following:

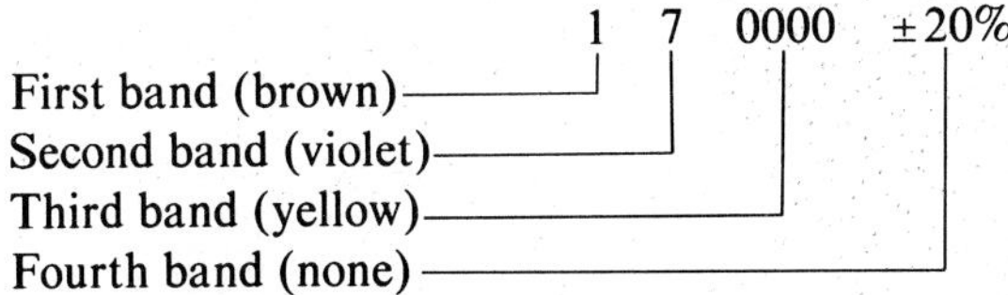

This resistor is rated at 170 kΩ ± 20%. Solving for the tolerance range, we get:

$$\begin{aligned} \text{Tolerance} &= 170{,}000 \times 20\% = 34{,}000\ \Omega \\ \text{Maximum} &= 170{,}000 + 34{,}000 = 204{,}000\ \Omega \\ \text{Minimum} &= 170{,}000 - 34{,}000 = 136{,}000\ \Omega \end{aligned}$$

A replacement resistor could vary from 136 to 204 kΩ and still be within tolerance.

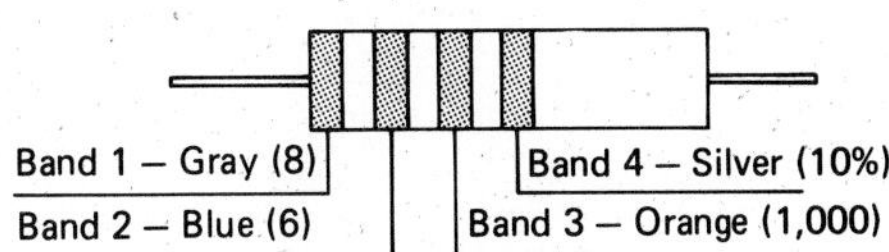

FIGURE 3-20

Refer to Figure 3-20. Color bands convert to a color-code value of 86,000 Ω ±10%. Computing the tolerance, we get:

$$\begin{aligned} &\text{Tolerance} && 86{,}000 \times 10\% = 8600 \\ &\text{Maximum} && 86{,}000 + 8600 = 94{,}600\ \Omega \\ &\text{Minimum} && 86{,}000 - 8600 = 77{,}400\ \Omega \end{aligned}$$

A replacement resistor could vary from 136 to 204 kΩ and still be within tolerance.

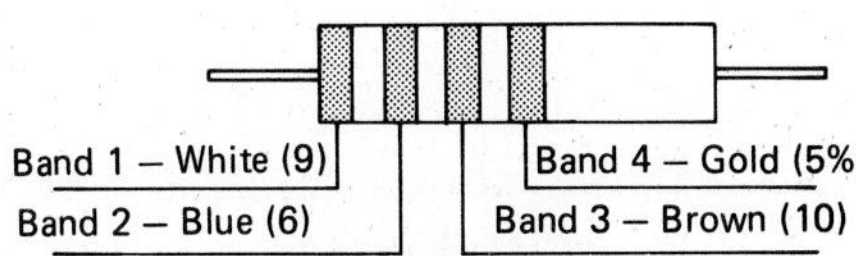

FIGURE 3-21

Refer to Figure 3-21. Color bands convert to a color-code value of 960 Ω ± 5%. Computing the tolerance, we get:

Tolerance	$960 \times 5\% = 48\ \Omega$
Maximum	$960 + 48 = 1008\ \Omega$
Minimum	$960 - 48 = 912\ \Omega$

A replacement resistor could vary from 912 to 1008 Ω and still be within tolerance.

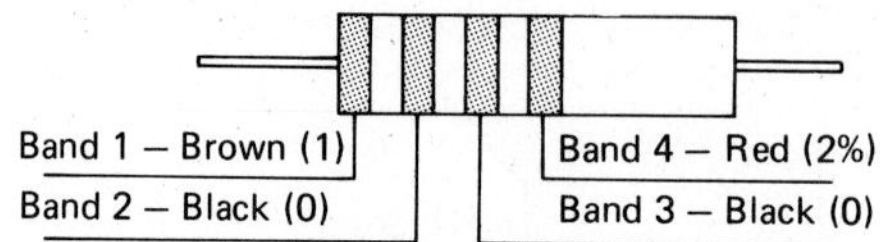

FIGURE 3-22

Refer to Figure 3-22. Color bands convert to a color-code value of 10 Ω ± 2%. Computing the tolerance, we get:

Tolerance	$10 \times 2\% = 0.2\ \Omega$
Maximum	$10 + 0.2 = 10.2\ \Omega$
Minimum	$10 - 0.2 = 9.8\ \Omega$

A replacement resistor could vary from 9.8 to 10.2 Ω and still be within tolerance.

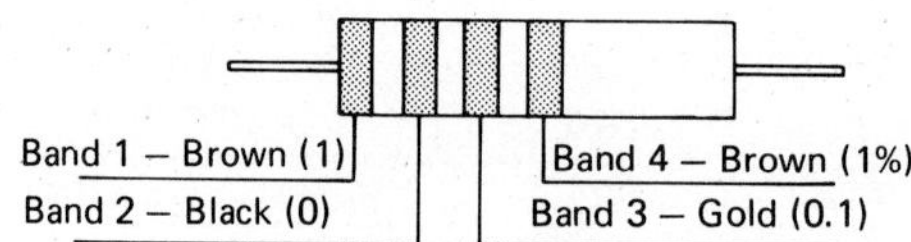

FIGURE 3-23

Refer to Figure 3-23. Color bands convert to a color-code value of 1 Ω ± 1%. Computing the tolerance, we get:

Tolerance	$1 \times 1\% = 0.01\ \Omega$
Maximum	$1 + 0.01 = 1.01\ \Omega$
Minimum	$1 - 0.01 = 0.99\ \Omega$

The replacement resistor could vary from 0.99 to 1.01 Ω and still be within tolerance.

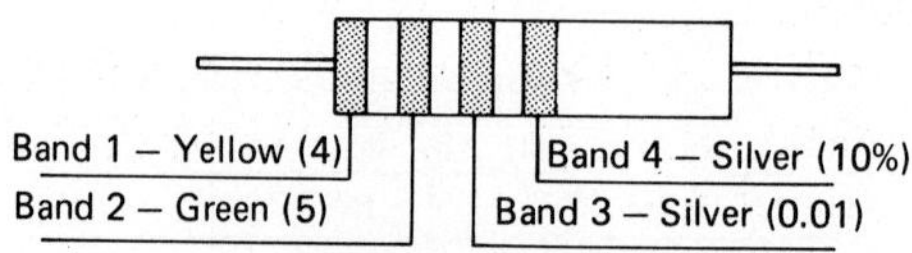

FIGURE 3-24

Refer to Figure 3-24. Color bands convert to a color-code value of 0.45 Ω ±10%. Computing the tolerance, we get:

Tolerance	0.45 × 10%	= 0.045 Ω
Maximum	0.45 + 0.045	= 0.495 Ω
Minimum	0.45 − 0.045	= 0.405 Ω

The replacement resistor could vary from 0.405 Ω to 0.495 Ω and still be within tolerance.

Self-Check

Answer each item by indicating whether the statement is true or false.

11. ____ When the fourth band on a color-coded resistor is red, you must add three zeros to the right of the second digit.

12. ____ When you use the resistor color code, the number six is represented by the color blue.

13. ____ The minimum value acceptable in replacing a 27 kΩ ± 5% tolerance resistor is 25,650 Ω.

14. ____ The volume control for your stereo or radio is a good example of a precision-fixed resistor.

15. ____ A resistor is coded blue–red–green–silver; thus its ohmic value is 6.2 MΩ, with ±10% tolerance.

16. ____ A tapped resistor can actually be several resistors in one.

Conductance

Conductance is defined as

The ability of a conductor to carry electrons.

Note that this definition is the opposite of the definition of resistance. Here, we refer to the ease with which current can flow, whereas *resistance* refers to opposition to current flow. The more resistance a material has, the less its conductance.

By now you should begin to realize that resistance and conductance are opposites of each other. In fact, the formula for conductance is the reciprocal of resistance. The letter G is used to designate conductance. The conductance formula is:

$$G = \frac{1}{R}$$

The internationally recognized unit for conductance is the Siemens (S), but the term *mho* is often used. The symbol for mho is the inverted ohms symbol, ℧. This extends the thought that resistance and conductance are opposites, since mho is "ohm" spelled backward, and the omega is inverted to symbolize mho.

A mathematical example of this conversion is: A circuit has 10 Ω of resistance.

$$G = \frac{1}{R}$$

$$G = \frac{1}{10}$$

$$G = 0.1 \text{ mho}$$

The conductance of the circuit is 0.1 mho.

Table 3-4 presents a comparison of relative resistance and relative conductance of several conductive materials. Annealed-copper conductors are used as a standard. Note that copper has arbitrarily been assigned a relative resistance and a relative conductance of 1 and that all the other materials are compared to this, as to whether they have higher or lower resistance or conductance.

TABLE 3-4

Material	*Relative Resistance*	*Relative Conductance*
Silver	0.92	1.08
Copper	1.00	1.00
Gold	1.38	0.725
Aluminum	1.59	0.629
Tungsten	3.20	0.312
Zinc	3.62	0.275
Brass	4.40	0.227
Platinum	5.80	0.172
Iron	6.67	0.149
Nickel	7.73	0.129
Tin	8.20	0.121
Steel	8.62	0.116
Lead	12.76	0.081
Mercury	54.60	0.018
Nichrome	60.00	0.0166
Carbon	2030.00	0.0004

Electronic circuitry can be explained from the standpoint of either resistance or conductance. In series circuits, resistance is generally easier to explain, since series voltages are proportional to resistances. Conductance can be convenient in parallel circuits and is discussed again at that point.

Schematic Symbols

Symbols are used to represent electronic components in drawings of circuits. Each component is assigned a symbol which is used to represent that component in the drawing. These symbols are called *schematic symbols* and the drawings *schematic drawings.*

Drawing a representation of each component would be impractical; the schematic drawing for even a small circuit would be large and bulky. Figure 3-25 illustrates the relative sizes of pictorial representation and schematic symbols. Once learned, these symbols are easily recognized and interpreted.

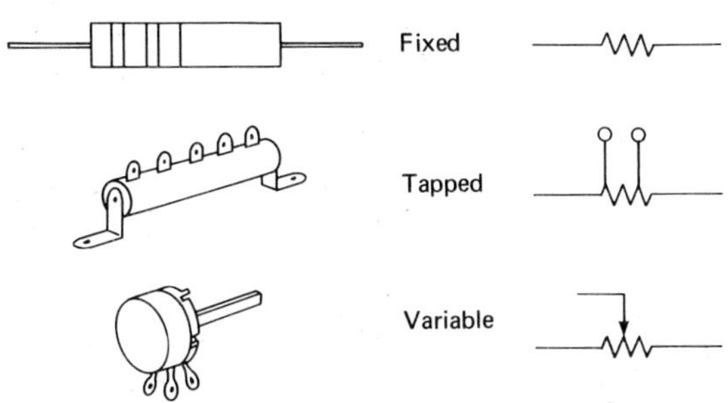

FIGURE 3-25

The recognition of schematic symbols and the ability to interpret them is important to the electronics technician. Schematic symbols that are widely used in electronic circuitry appear in Figure 3-26; because other components are covered later, their symbols are introduced.

Using Figure 3-26 as a reference, locate the diagrams and learn the symbols corresponding to these definitions:

fixed resistor.
(diagram a). This symbol is introduced in this chapter. A fixed resistor is manufactured to provide a fixed opposition to the flow of electrons. Fixed resistors are available in fixed-carbon, fixed-film, or wirewound form.

tapped resistor.
(diagram b). A wirewound resistor tapped at specific resistance values, which can be used as two or more resistors when mounted in a circuit.

potentiometer.
(diagram c). A variable resistor with a wiper contact that can be moved along the resistive element to select resistance values as needed. Possessing three electrical connections, it is used to select voltage for application to another circuit.

rheostat.
(diagram d). A variable resistor having two electrical connections, used to vary resistance and current within a circuit. Resistance variation is accomplished by rotating a wiper along a circular resistance element.

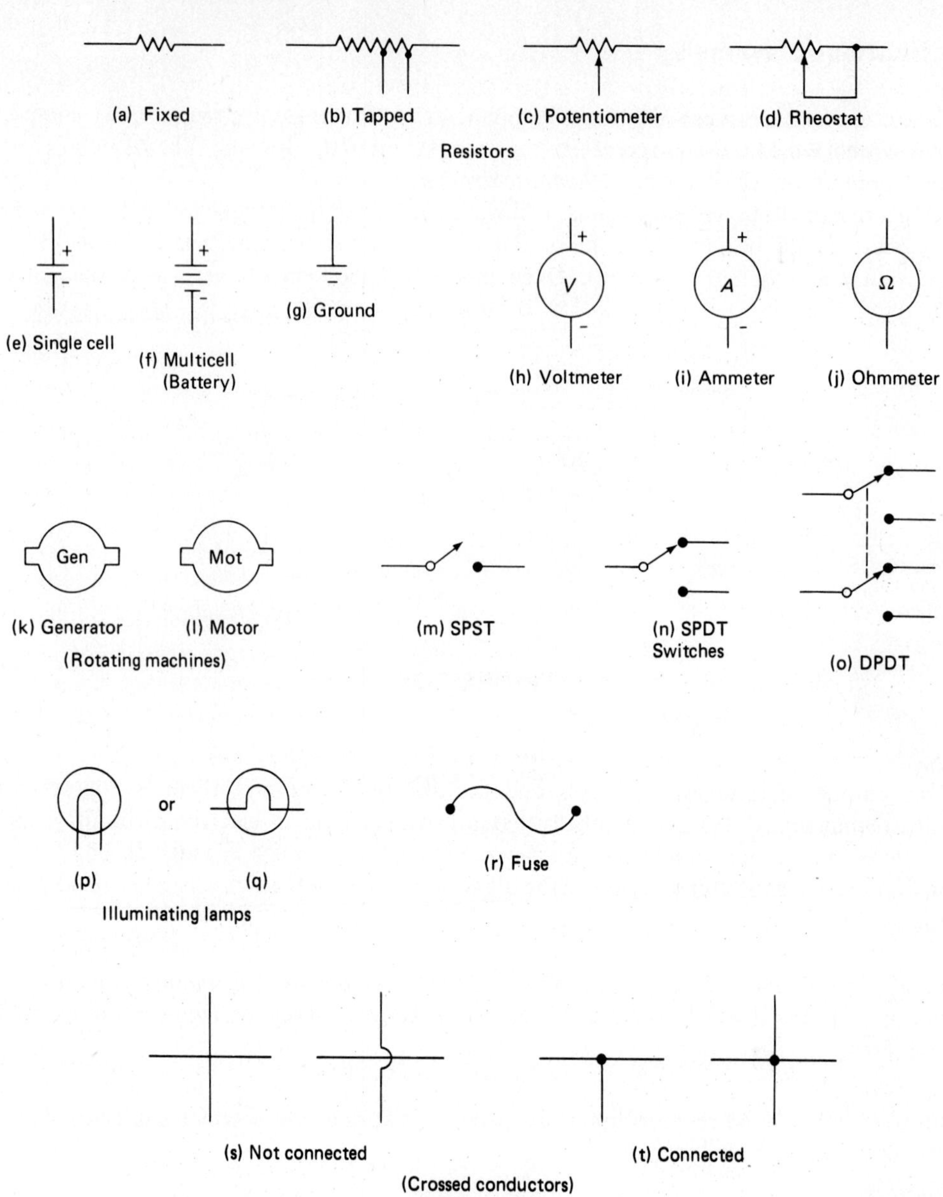

FIGURE 3-26

cell.
(diagram e). Can represent either dry or wet cells. The cell is a device that provides an EMF to a circuit because of the chemical reaction that occurs within it. Cells are manufactured from several different substances and can supply 1.2 to 2 volts EMF, depending on the type. The plus sign (+) identifies the positive pole and the minus sign (−) identifies the negative pole.

battery.
(diagram f). Two or more cells of similar construction, connected together. Connecting cells in different configurations can provide higher voltages and/or make larger amounts of current available than could be obtained from a single cell.

ground.
(diagram g). A point in a circuit designated as the reference point. Voltage measurements are made with respect to ground.

voltmeter.
(diagram h). An instrument used to measure voltage. Positive (+) and negative (−) signs indicate the polarity required for connection in a circuit.

ammeter.
(diagram i). An instrument used to measure current. Positive (+) and negative (−) signs indicate the polarity required for connection in a circuit.

ohmmeter.
(diagram j). An instrument used to measure resistance (ohms) or to check a circuit for continuity.

generator.
(diagram k). A device that converts mechanical energy into electrical energy.

motor.
(diagram l). A device used to convert electrical energy into mechanical energy.

single-pole-single-throw (SPST) switch.
(diagram m). A device used to control the operation of an electrical circuit.

single-pole-double-throw (SPDT) switch
(diagram n). A device used to control the operation of two separate circuits.

double-pole-double-throw (DPDT) switch.
(diagram o). A device used to control the operation of up to four different circuits.

lamp.
(diagrams p and q). A device that emits light when sufficient current flows through it.

fuse.
(diagram r). A protective device used to protect circuit components from damage in case of a current overload.

crossed-not-connected-conductors.
(diagram s). Wires crossing each other but where there is no electrical connection.

crossed and connected conductors.
(diagram t). Crossed wires that have been connected physically and electrically.

Schematic symbols are used to identify actual components and the way they are connected in a circuit. As has been stated, a practical circuit contains the following components:

1. a battery or other source of voltage
2. a resistive device to do the work
3. conductors connected so as to form a complete loop

Practical circuitry is discussed fully, starting in Chapter 4. For now, an illustration of the use of schematic symbols is presented (Figure 3-27). Several symbols are used in the circuit shown. Each of the symbols used in Figure 3-27 is explained below.

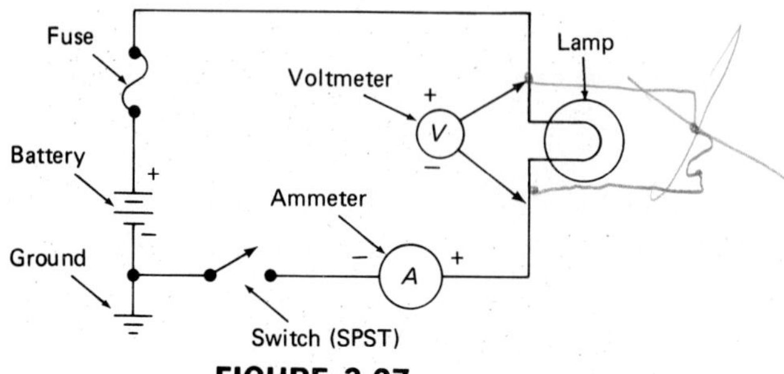

FIGURE 3-27

lamp.
Placed in the circuit because it provides resistance to current flow. When sufficient current flows through the lamp, heat will cause the filament to glow.

fuse.
Used to protect the circuit from a current overload. Fuses are selected to meet specific current and voltage needs of the device being protected. In the lamp circuit above, the fuse's rating is less than the amount of current that would damage the lamp. If the circuit current exceeds the rating of the fuse, a metal filament in the fuse becomes so hot that it melts and divides into two parts, thus stopping all current flow.

FIGURE 3-28
An Assortment of Switches (Courtesy EATON Corporation/Cutler Hammer Products)

switch.
The switch controls circuit operation. When closed, a switch allows current to flow. When it is opened, current stops. Switches are rated according to their voltage and current capabilities.

voltmeter.
A test instrument connected to measure the voltage drop on the lamp. In this case, it is also monitoring the voltage of the battery. *Note:* The positive leads of the meter and the battery are connected together and the negative leads are connected together.

ammeter.
A test instrument connected to monitor current flow in the circuit. The negative lead is connected toward the negative pole of the battery and the positive lead toward the positive pole.

battery.
Supplies the EMF for operation of the circuit.

ground.
The common voltage reference point.

FIGURE 3-29
Assorted Batteries (Courtesy of Eveready, Union Carbide)

Self-Check

Answer each item by indicating whether the statement is true or false.

17. ___ When a switch is connected in a circuit, it will provide protection from a current overload.

18. ___ A rheostat is a device for controlling voltage.

19. ___ Potentiometers have only two electrical connections.

20. ___ A fuse allows the user to control current flow in a circuit.

21. ___ Fixed resistors can be used to limit current flow in a circuit.

Cells and Batteries

We have discussed using a battery to supply the EMF for a circuit. The simplest form of battery is one containing a single cell. Batteries and cells are made of several different materials and will supply voltages ranging from 1.25 to 2.2 V. Table 3-5 lists important information regarding the most common cells.

TABLE 3-5
Cells by Type and Voltage

Name	*Voltage*	*Wet or Dry*	*Class*	*Notes*
Carbon-zinc	1.5	Dry	Pri	Lowest priced, flashlight batteries, short shelf life.
Manganese (alkaline)	1.5	Dry	Both	Manganese-dioxide and zinc in hydroxide; can supply currents above 300 mA.
Silver-oxide	1.5	Dry	Pri	Silver-oxide and zinc in hydroxide; hearing aids, and digital watches.
Mercury	1.35	Dry	Both	Mercury-oxide and zinc in hydroxide; constant voltage; long shelf life; hearing aids and watches.
Nickel-Cadmium (NiCAD)	1.25	Dry	Sec	Constant voltage; long shelf life; used in rechargeable calculators, cameras, power tools, and flashlights.
Edison	1.4	Wet	Sec	Nickel and iron in hydroxide; industrial uses such as trolley cars.
Lead-acid	2.2	Wet	Sec.	Low internal resistance; very high current ratings 6- and 12-V batteries for autos.

Two types of cell are available, dry and wet. The dry cell is sealed. It depends on chemicals sealed inside to produce electrons for operating circuits. Wet cells contain a fluid that promotes the release of electrons for use by the circuit. Table 3-5 identifies cells as being either dry or wet.

In many cases, dry cells are manufactured for one-time use, then are thrown away. Throwaway cells, which cannot be recharged, are classified as primary cells. Some dry cells and all wet cells can be recharged many times; they are classified as secondary cells. To further clarify this point, primary cells cannot be recharged, but secondary cells can be recharged repeatedly.

Examine Table 3-5 closely. The table gives cell classifications, type, voltage, and other information. A single cell provides the EMF shown in the voltage column for each cell. In most cases, this voltage is insufficient to power the circuit to be operated. By connecting cells, we can obtain higher voltage and thus make available more current. To increase the EMF (voltage), cells are connected in series (end-to-end) as shown in Figure 3-30. The rule is:

When cells are connected in series, the EMF available will equal the sum of the voltages of the individual cells.

When connecting cells in series, it is necessary to connect the negative pole of one cell to the positive pole of the next; then that cell's negative pole to the positive pole of the next cell; and so on.

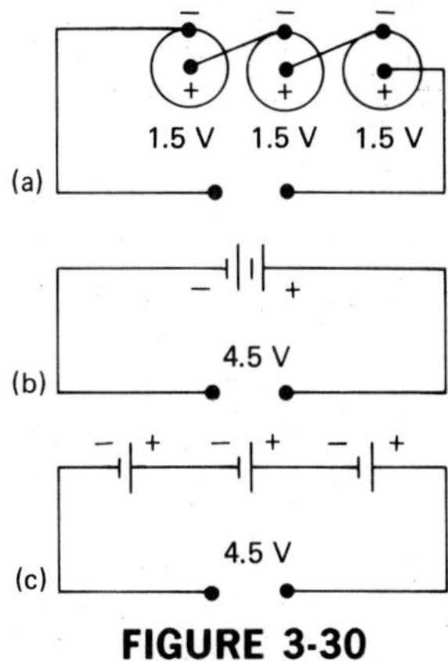

FIGURE 3-30

Cells connected as shown in Figure 3-30 form a battery with an EMF of 4.5 V. Diagram a is a pictorial representation, diagram c is a schematic showing the three symbols connected together, and diagram b is the symbol for a battery with its total voltage noted. By connecting as many cells as needed, we can obtain many voltage values. Note that, for cells and batteries, the long side of the symbol represents the positive pole and the short side represents the negative pole.

Cells are capable of supplying relatively small currents. Connecting them in series does not increase the current available, but connecting them in parallel (side by side) does make more current available. For an example of parallel connections, see Figure 3-31.

Diagram a is the pictorial, b the cell symbols, and c the battery symbol. When we connect batteries in parallel, all negative poles are connected together, as are all positive poles. The rule is:

When cells are connected in parallel, the available current is increased.

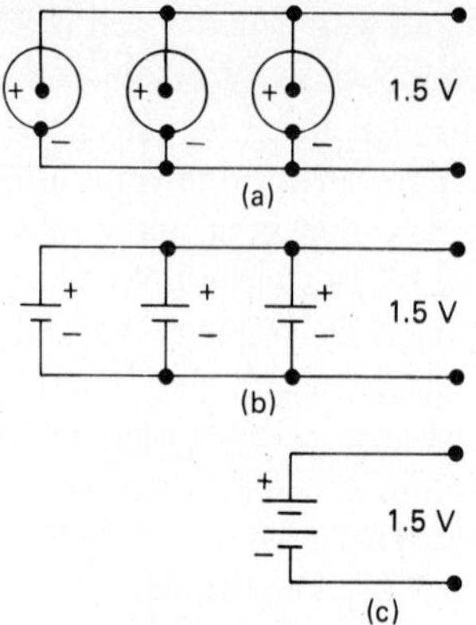

FIGURE 3-31

Note that when the cells are connected in parallel, the terminal voltage remains 1.5 V and does not change, but that the battery can supply approximately triple the current of one cell alone.

Many circuits need voltages higher than that supplied by a single cell, and much more current than one cell can deliver. To provide this power, it is necessary to connect cells in series until the total voltage equals the voltage needed. Construct several strings of series-connected cells, then connect the strings in parallel. These connections are shown in Figure 3-32.

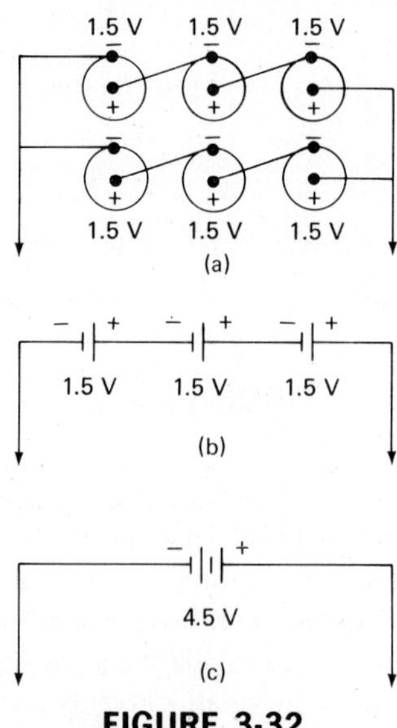

FIGURE 3-32

The rule for a series-parallel connection of cells is:

When cells are connected in parallel series, the terminal voltage will equal the sum of the individual cell voltages of one string. As each string is added, the battery's available current will be increased.

Self-Check

Answer each item by indicating whether the statement is true or false.

22. ____ When batteries are connected in series, available current increases.

23. ____ Connecting batteries in parallel provides higher output voltage.

24. ____ Cells and/or batteries that cannot be recharged are called secondary cells.

25. ____ When batteries are connected in parallel, available current increases.

26. ____ When batteries are connected in series, the output voltage equals the sum of the battery voltages.

27. ____ Batteries produce an EMF by chemical means.

28. ____ Cells and batteries that can be recharged are called secondary cells or secondary batteries.

29. ____ Wet cells are limited in the places where they can be used because of the fluid they contain.

30. ____ Battery voltage is more economical than generated voltage.

31. ____ Series-connected cells provide higher output voltage and greater availability of current.

Summary

In this chapter we discussed the requirements for current flow and how current is identified in a schematic diagram, as well as the direction of current flow in a circuit. The ampere was established as the unit of measure for current, with 1 ampere equal to 1 coulomb of electrons passing a given point in 1 second.

Voltage and current were defined and discussed. Various types of resistors were introduced and their construction was described; their schematic symbols were identified and their use in circuits discussed. The EIA color code was presented, and sample solutions were used to discuss color-band markings and tolerance.

Schematic symbols most often used were presented, identified, and discussed. Cells and batteries were introduced. Different types of cells were discussed, along with their capacity for recharge. The type of connections required for increasing output voltage and available current were discussed, and sample circuits were presented.

Review Questions and/or Problems

1. What are two requirements for current flow?
   (a) ____________
   (b) ____________

2. Describe current flow within the conductor, in terms of electron movement.

3. In the external circuit, current flows from the ____________ pole to the ____________ pole.

4. The symbol for current is ____________.

5. The symbol for voltage is ____________.

6. One ____________ of current is flowing when 1 ____________ of electrons pass a given point in 1 ____________.

7. The practical unit of resistance is ____________.

8. Resistance of a conductor is directly proportional to its ____________.

9. The symbol for resistance is ____________.

10. The resistance of a conductor is inversely proportional to its ____________.

11. Opposition to current flow is called ____________.

12. Which statement is true?
    (a) Batteries produce EMF by thermal means.
    (b) Generators produce EMF by mechanical means.
    (c) A thermocouple operates on chemical reaction.
    (d) The piezoelectric effect refers to an EMF produced by nuclear means.

13. All three bands of a resistor are yellow; its resistance is ____________.

14. List the color codes for the following resistors. If no tolerance is stated, assume 20%.

Band	*1*	*2*	*3*	*4*
(a) 2500 Ω				
(b) 9.0 Ω 1%				
(c) 470k 10%				
(d) 1 meg 5%				
(e) 130 k 5%				
(f) 100 Ω 2%				

15. Determine the value and tolerance of resistors coded as shown below. State the values in prefixes.

Band 1	*Band 2*	*Band 3*	*Band 4*	*Value*	*Tolerance*
(a) brown	blue	yellow	gold		
(b) brown	green	green	silver		
(c) orange	black	yellow	none		
(d) blue	violet	gray	gold		
(e) brown	green	black	brown		
(f) gray	red	gold	red		
(g) blue	green	silver	brown		

16. The three requirements for a practical circuit are:
    (a) conductor, ohmmeter, battery
    (b) switch, battery, conductor
    (c) battery, resistive device, conductors
    (d) all the above

17. What is the schematic symbol for a voltmeter?

18. Which of the symbols in Figure 3-33 represents an ammeter?

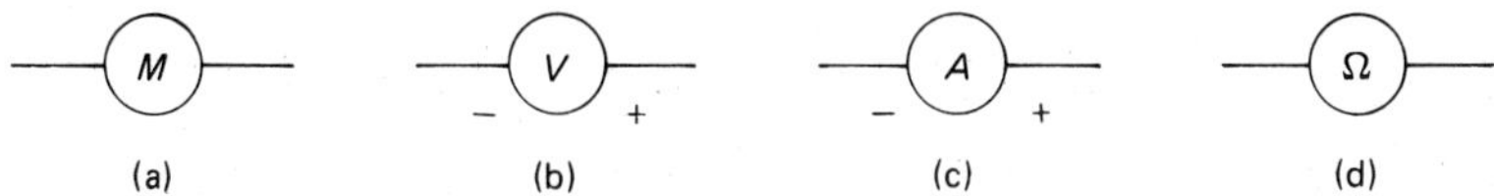

FIGURE 3-33

19. Which of the symbols in Figure 3-33 represents an ohmmeter?

20. Identify the schematic symbols shown in Figure 3-34.

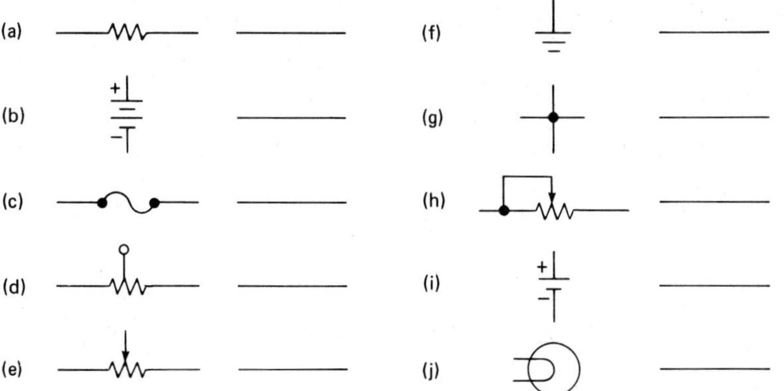

FIGURE 3-34

21. Cells are connected in series to ______________ ______________.

22. Cells are connected in parallel to ______________ ______________ ______________.

23. When cells are connected parallel-series output voltage will ______________ and current available will ______________.

24. If all the cells in Figure 3-35 are 1.5 V, what is the terminal voltage (E) for each circuit?

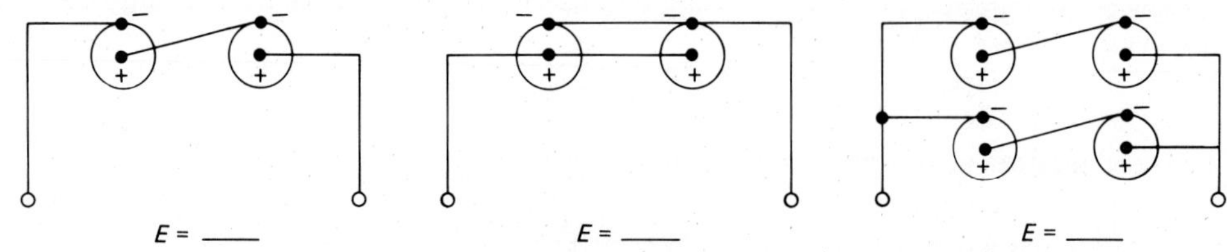

FIGURE 3-35

25. In the external circuit, electron current flows from the ______________ pole of a battery to the ______________ pole of the same battery.

4

Multimeter

Introduction and Applications

Of all the types of test equipment available to a technician, the VOM, or *multimeter,* is probably the most useful. The design of this instrument makes it suitable for measuring four electrical values, and some new types have a fifth function. A conventional multimeter (VOM) has:

1. A DC voltmeter
2. A DC ammeter
3. An AC voltmeter
4. An ohmmeter

In recent years, the electronic multimeter and the digital multimeter have gained a larger segment of the market. These multimeters usually have an AC ammeter function in addition. Some also have a low-power ohms function. With the exception of the low-power ohms function, these functions are discussed in this chapter.

Objectives

Each student is required:

1. When provided with a multimeter, test leads, and various fixed resistors, to:
    (a) Correctly connect the test leads to the ohmmeter.
    (b) Set the meter to the ohmmeter function.
    (c) Select the proper range for measuring each resistor.
    (d) Zero the meter before each use.
    (e) Identify the ohms scale, read the value indicated on the scale, and perform any interpretation necessary to arrive at the correct value.
    (f) Make a list of the precautions that must be observed when preparing the ohmmeter for use and during use.
2. Given a multimeter, test leads, and a variable DC power supply, to:
    (a) Correctly connect the leads to the multimeter for measuring +DC or −DC voltage.
    (b) Set the meter to the DC voltmeter function and select the correct range for measuring a known voltage.

   (c) Identify the scale to be used when measuring +DC or −DC voltage.
   (d) Use safe procedures when connecting the meter to the power source and when adjusting the source to the prescribed voltage.
   (e) Demonstrate the ability to measure unknown voltages.
   (f) List the precautions associated with use of a DC voltmeter.
3. Given a multimeter, test leads, and DC circuitry, to:
   (a) Set up the multimeter for use as a DC ammeter.
   (b) Correctly connect test leads to the meter.
   (c) Connect the DC ammeter into an operating circuit correctly and safely.
   (d) Identify the correct scale and measure various DC currents.
   (e) List the precautions pertaining to use of a DC ammeter.
4. Given a multimeter, test leads, and a variable source of AC voltage, to:
   (a) Set up the multimeter for use as an AC voltmeter.
   (b) Correctly connect the AC voltmeter to the power supply.
   (c) Identify, read, and interpret the AC voltmeter scale.
   (d) List the precautions pertaining to use of an AC voltmeter.
5. Explain, in writing, the procedures appropriate to use of an AC ammeter.

The Ohmmeter

An ohmmeter contains its own power source and does not require current from the test circuit as part of its use. A combination of batteries and circuits is part of the design of the ohmmeter, which provides the correct range of current for operating the meter in each ohms range.

For ohmmeter readings to be accurate, the meter should be calibrated (zeroed) each time a check is made. The scale used on this type of ohmmeter is called an *inverse scale.* An inverse scale is one in which the zero is located at the right end of the scale; all other scales have a left-side zero. In adjusting the meter for *ohms zero,* you must:

1. Short the leads together. This allows current from the internal batteries to start flowing. The pointer will deflect to the right (near ohms zero). A *zero adjust* is located on the meter. To ensure the correct amount of current, adjust the pointer to the zero (0) position.
2. When you open the current path (that is, separate the test leads), the pointer moves back to the left, or infinity, side of the scale.

The meter is now calibrated for use.

It is not necessary to observe polarity when connecting the test leads to an ohmmeter. Because the current source is internal, the direction of current flow is established by the ohmmeter. It is a good practice, however, to form the habit of always connecting the *black* lead to the black (−, or COM) jack and the *red* lead to the red (+, or POS) jack.

Figure 4-1 shows a multimeter with the functions volts, amps, and ohms; plus (+) and minus (−) jacks; a zero ohms adjust; a switch labeled "+DC, −DC, AC"; and a switch used to select different ranges for measurement. The topmost scale on the meter's face is used to make all measurements of ohms. Notice that the zero is at the right end of the scale and that infinity is at the left end. Also note that the scale is highly nonlinear. The most accurate readings are taken from the right one-third of the scale.

FIGURE 4-1
Typical Volt-Ohm-Milliammeter Used in Repair of Electronic Circuitry (Courtesy of Simpson Electric Company)

Resistance Checks

The ohmmeter is used to measure opposition, or resistance, to the current flow. To make accurate measurements of resistance, you must perform steps in the following sequence:

1. Set the +DC, −DC, AC switch to the +DC position. Refer to Figure 4-2.
2. Connect the test leads to the plus and minus jacks.
3. Select the desired range.
4. Touch (short) the leads together and adjust the pointer to coincide with the zero at the right-hand end of the ohms scale.
5. Disconnect (open) the test leads and check that the pointer returns to infinity. The meter is now ready for use.
6. Make sure the voltage is turned off on the unit you wish to check. If the component to be checked is connected in a complex circuit, you must isolate the component from all other components so the ohmmeter current will flow only through that component; otherwise, erroneous readings will result.

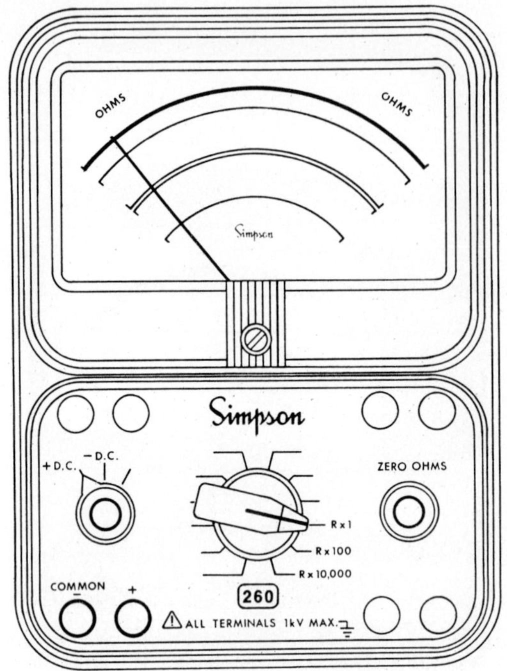

FIGURE 4-2

Multimeter Set to Jack and Switch Positions to Measure Resistance with Full Scale at a Scale Selection of RX1 (Courtesy of Simpson Electric Company)

7. Connect a lead to each end of the component you wish to test. Figure 4-3 illustrates the ohmmeter's connection to an isolated resistor.
8. The pointer should deflect (move) right. Read and interpret the point on the scale indicated by the pointer. If you selected the RX100 range in step 3, multiply the reading by 100. If some other range was selected, multiply by that number.
9. If it is necessary to change to a different range, go back to step 3 and repeat each step between 3 and 9.

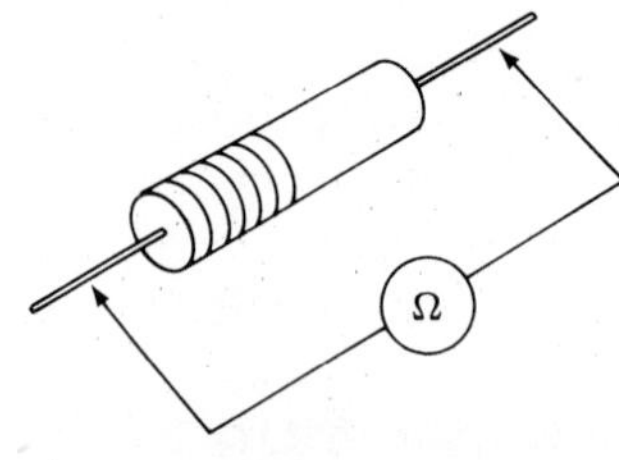

FIGURE 4-3

Continuity Checks

An ohmmeter is used to check continuity within a circuit. All of the steps listed above also apply to continuity checks. In contrast to checking an individual component, a continuity check determines the circuit's ability to conduct current. We are not concerned here with the amount of resistance, only the completed path for current flow.

When the ohmmeter is connected across the circuit or part of the circuit we wish to check, three indications are possible:

1. *The infinity reading.* Should be considered reliable only when the measurement is made on the highest *ohms* range available. A reading of infinity indicates that the path for current is broken (open).
2. *The zero reading.* Reliable only when it results from measurements made using the smallest possible *ohms* range. A zero indication means that there is zero resistance (a *short*) between the test leads. This is possible only in the case of a solid piece of wire. Other components lying in the path are an indication of a defective circuit.
3. *Some resistance value.* The most common indication. It confirms that there is continuity but that the path has some resistance.

Examine the ohms scale in Figure 4-4. It is the type of scale used on most portable multimeters. Note that the scale is an *inverse* scale and is highly nonlinear. The best readings are made on the right one-third of the scale.

Many meter manufacturers use the color green to distinguish the ohms scale from other scales.

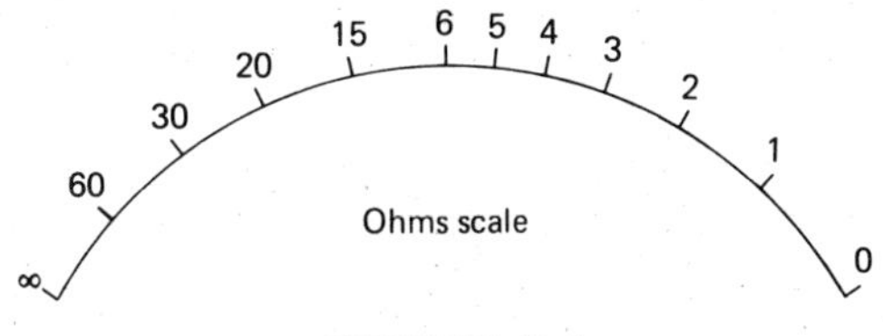

FIGURE 4-4

Self-Check

Answer each item by indicating whether the statement is true or false.

1. ___ It is possible to read zero ohms when performing a continuity check and to have a correct indication.

2. ___ An ohmmeter must be zeroed each time its range is changed.

3. ___ The ohmmeter uses a linear scale.

4. ___ An ohmmeter can supply its own current.

5. ___ When connecting the leads of an ohmmeter, the user observes polarity.

Voltmeters

The voltmeter in this multimeter (Figure 4-1) is actually three voltmeters in one. Located in the upper left-hand corner of the instrument is a switch labeled +DC, −DC, AC. By positioning this switch properly, it is possible to measure any of these voltages. The internal circuitry of the meter is not important to our discussion; what *is* important is the correct use of the voltmeter section of a multimeter.

Before making voltage checks in a circuit, be aware of the type of voltage, the point to be checked, and the normal voltage at that point. To measure particular types of voltage, the +DC, −DC, AC switch should be set as follows (see Figure 4-5):

1. To measure positive DC voltage, set the switch to +DC.
2. To measure negative DC voltage, set the switch to −DC. *Note:* Reversing switch positions causes the pointer to deflect left and go off the scale.
3. To measure AC voltage, set the switch to AC. *Note:* Measurement of DC with the switch in the AC position causes highly erroneous indications, as does attempting to measure AC with the switch in either DC position.

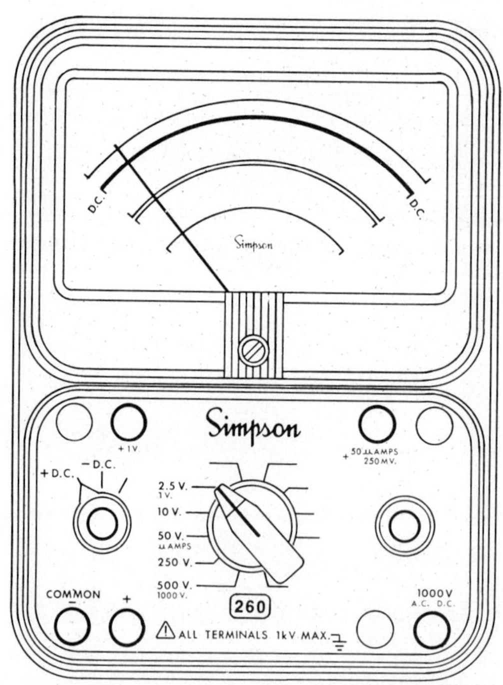

FIGURE 4-5

Multimeter Set to Measure +DC Volts with Full Scale Value Equal to 2.5 V (Courtesy of Simpson Electric Company)

The range for each voltage is selected by positioning the *range* switch to a DC position: 2.5 V, 10 V, 50 V, 250 V, 500 V. Readings are taken from indications along the DC scale. Note that the scale is divided linearly; thus the scale is a *linear scale.* The divisions along the scale are marked in scalar quantities. The quantities marked at the right end of the scale are the full-scale deflection (FSD) quantities. Ranges selected by the switch correlate with these FSD values. Notice that the first two digits of each scale selected correspond to one of the FSD values. This is the series of numbers used to make readings when the meter is set to that range. For example, the 50 V and 500 V ranges correspond to each other. Readings on the 50 V range are taken directly from the series of numbers leading up to the number 50. For 500 V, the same group of numbers is used; but this time, to obtain the correct value, the reading is multiplied by 10. The 2.5 V and 250 V ranges share a similar relationship, except that when you use the 250-V scale, all readings must be multiplied by 100.

Voltmeters are connected in parallel with the component or circuit to be measured. For the instrument to measure voltages accurately, it must fit the following description.

Voltmeters have very high internal resistance and are connected in parallel with the voltage to be measured.

Measuring DC Voltage

To measure either positive or negative DC voltage, follow these procedures. (Refer to Figure 4-5.)

1. Connect the *black* test lead to the *black* jack and the *red* test lead to the *red* jack.
2. To measure positive DC, set the +DC, −DC, AC switch to +DC; for negative DC, set it to −DC.
3. Set the range switch to 500 V.
4. Connect the *black* test lead to the negative side of the voltage and the *red* lead to the positive side. Figure 4-6 contains a diagram of this connection. Notice that the meter is connected at the side of the component under test. We say that the meter is connected in parallel with the component.

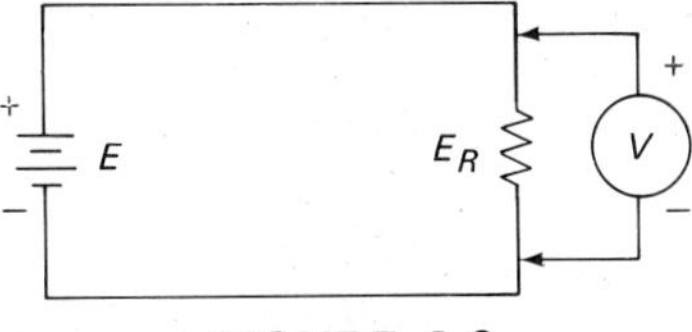

FIGURE 4-6

5. Read the value indicated on the scale and apply the multiplier. Since the 500 V range was used, we must now use the series of numbers that lead to the 50 V FSD value on the DC scale in Figure 4-7. On many meters, the DC scale is black. Once the indication has been determined, it must be multiplied by 10 (500 divided by 50), to obtain the correct value. *Note:* The largest range is selected for measurement of any unknown voltage. If the deflection is small, decrease the range until an accurate reading can be made.

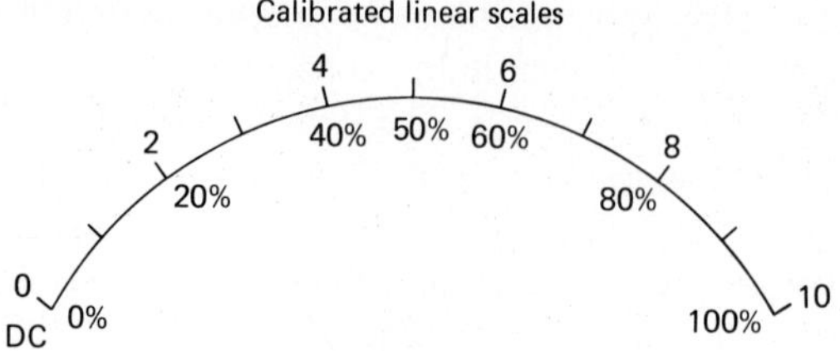

FIGURE 4-7

6. Should the pointer deflect to the left when connected to the circuit, it can be made to deflect right by:
   (a) Reversing the position of the +DC, −DC, AC switch. As presently set, the switch would be moved to −DC. If the switch is in −DC, and the pointer deflects left, move the switch to +DC.
   (b) Reversing the position of the test leads—putting the *red* lead where the black is and the *black* where the red lead is connected.

Measuring AC Voltage

To measure AC voltage, the +DC, −DC, AC switch is set to the AC position (Figure 4-8). All other procedures are the same as for DC measurement, except that observance of polarity is not necessary

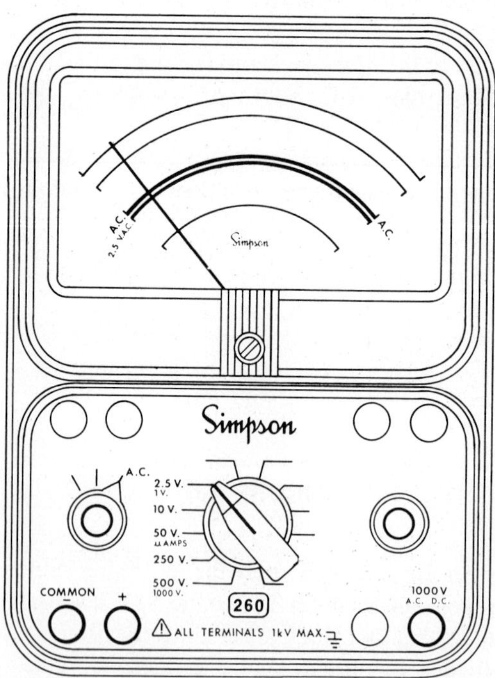

FIGURE 4-8
Multimeter Set to Measure AC Volts with Full Scale Value Equal to 2.5 V (Courtesy of Simpson Electric Company)

and the pointer cannot deflect to the left. An important point is:

> ***Indications received for AC voltage measurements are read on the AC voltage scale. On many meters this scale is red, to distinguish it from other scales.***

Caution: When using the voltmeter, you must work in a manner that best avoids electrical shock.

Self-Check

Answer each item by indicating whether the statement is true or false.

6. ___ When measuring AC voltage, polarity can be changed by reversing a switch or by reversing the test leads.

7. ___ The voltmeter is connected in parallel with the component to be tested.

8. ___ AC and DC voltages are measured by use of the DC scale.

9. ___ Most voltage readings require that the scale interpretation be multiplied by some quantity.

10. ___ Attempting to measure +DC voltage with the switch set to −DC causes the pointer to deflect right.

11. ___ A pointer deflecting left can be corrected by changing the +DC, −DC, AC switch to its opposite DC position.

Ammeters

Ammeters are devices used to measure the electron flow (current flow) in a circuit. Most multimeters can measure *DC amperage only*. Using a DC ammeter to measure alternating current, or an AC ammeter to measure DC current, can damage the instrument.

> ***Ammeters have extremely small internal resistances and must be connected in series with the circuit to be checked.***

The method used to insert an ammeter in a circuit is illustrated in Figure 4-9. Note that this connection requires considerable work on your part. The steps are:

1. Disconnect the circuit at point *A*.
2. Insert the ammeter in the circuit at point *A*, with its positive lead toward the battery's positive pole and the negative lead toward the negative pole. *Note:* For AC ammeters, observance of polarity is not necessary.
3. Make the reading. Readings are made on the same scale used for the DC voltmeter readings, a scale that is often *black*.
4. Remove the ammeter from the circuit and repair the break in the circuit.

Remember: Use of either an AC ammeter or DC ammeter is identical, except for the type of current and the facts that an AC ammeter requires no observance of polarity and that AC current values are indicated on the AC volts scale.

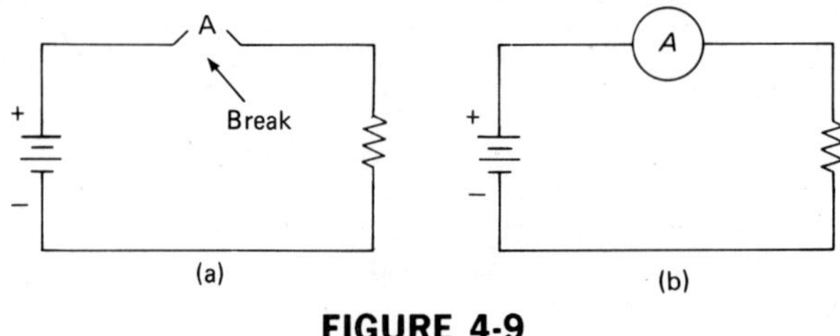

FIGURE 4-9

Using the DC Ammeter

To measure current with an ammeter, you must observe certain precautions. Failure to do so can result in severe damage to ammeters (their low internal resistance makes them delicate). As is true of the DC voltmeter, either +DC amperes or −DC amperes can be measured. You can change the operation of the meters from one type to another by performing one of the following steps:

1. Reverse the DC position of the +DC, −DC, AC switch.
2. Reverse the polarity of the leads as they are connected to the circuit.

After you have determined the direction of current flow in the circuit, the steps to be followed in using the DC ammeter are:

1. Turn *off* the voltage source.
2. Disconnect the circuit (Figure 4-9a).
3. Observe Figure 4-10.
4. Set the ammeter to +DC and the highest DC range available. Notice that the ammeter has some ranges calibrated in mA (milliamperes; 1 mA equals one-thousandth ampere).
5. Insert the ammeter as shown in Figure 4-9b.
6. Locate the DC scale. It is the same one used for DC voltage. (For AC, use the AC, or *red,* scale; the same methods as used with AC-voltage measurement.)
7. Turn *on* the circuit voltage.
8. Check the scale indication of the pointer.
9. Use the same multiplier system you used in making DC-voltage measurements for determining the true value of current flow.
10. Turn *off* the circuit voltage.
11. Disconnect the ammeter from the circuit.
12. Repair the break in the circuit.
13. Turn the circuit voltage *on.*

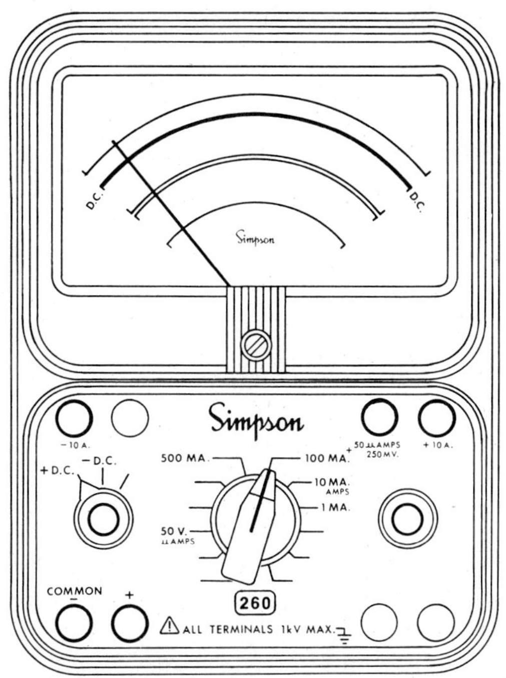

FIGURE 4-10
Multimeter Set to Jack and Switch Positions to Measure Direct Current with Full Scale Value Equal to 100 mA (Courtesy of Simpson Electric Company)

Precautions

Always remember that the ammeter has a very low resistance and can be damaged by small currents. The ammeter must have low resistance because it is connected in series in the circuit. High resistance would interfere with normal operation of the circuit. Let us reemphasize the following:

Ammeters have low resistance and must be connected in series with the circuit under test.

When measuring unknown currents, always begin with the highest possible range and decrease the range setting until an accurate reading can be made.

Never insert an ammeter in a "live" circuit unless there is no way to avoid it. Turn the voltage off.

Self-Check

Answer each item by indicating whether the statement is true or false.

12. ___ A DC ammeter indicates the amount of current flowing on a linear scale.

13. ___ An ammeter is connected in series with the circuit under test.

14. ___ Voltage must be *off* while a current test is being made.

15. ___ An ammeter has high internal resistance.

16. ___ AC ammeter indications are taken from the DC voltage scale.

Summary

The multimeter is a valuable instrument. It is durable and portable and is reasonably accurate. These factors, plus the multimeter's cost, make it popular among electronics technicians. With the newer electronic and digital types this is even more so. The multimeter allows several circuit values to be measured with one indicating device. In many cases, it is called a VOM (for volts, ohms, milliamperes; multimeters usually have at least those functions), but some models have additional functions.

All the scales on a multimeter are linear, with the exception of the ohms scale. This scale is an inverse scale and is highly nonlinear. Most accurate measurements of ohms are taken on the right one-third of the scale.

The measurement desired is selected by a function switch, and the range is selected by a range switch. On many meters these may be selected by using one switch, but both function and range will be shown on the switch markings.

Review Questions and/or Problems

1. When using a multimeter to measure resistance, the function switch must be set to:
   (a) DC volts.
   (b) AC volts.
   (c) ohms.
   (d) DC current.

2. When using the ohmmeter, you can expect the pointer to deflect further when you are measuring:
   (a) a large resistor.
   (b) a small resistor.
   (c) a shorted component.
   (d) an open component.

3. The ______________ section of a multimeter is used to measure current.
   (a) ohmmeter
   (b) ammeter
   (c) voltmeter
   (d) wattmeter

4. When measuring unknown voltage, the ______________ range is used.
   (a) largest
   (b) smallest
   (c) top
   (d) middle

5. When measuring voltage drop on a component, the voltmeter must be connected in ______________ with the component under test.
   (a) series
   (b) parallel
   (c) the circuit
   (d) none of the above

6. To measure an unknown current, first set the range switch to the ______________ range.
   (a) desired
   (b) smallest
   (c) middle
   (d) largest

7. An ammeter has ______________ internal resistance.
   (a) high
   (b) low
   (c) medium

8. A voltmeter has ______________ internal resistance.
   (a) high
   (b) low
   (c) medium

9. When an ohmmeter is used to check a shorted component (zero resistance), its pointer:
   (a) deflects full scale.
   (b) deflects to midscale.
   (c) does not move.

10. When using an ammeter, the meter must be connected in series with the circuit under test.
    (a) true
    (b) false

11. When an ohmmeter is used to check an open component (infinite resistance), its pointer:
    (a) deflects full scale.
    (b) deflects to midscale.
    (c) does not move.

12. It is not necessary to observe polarity when connecting the test leads for AC volt and ammeters.
    (a) true
    (b) false

13. When you use the ohmmeter to measure the resistance of a *good* resistor, the pointer deflects to some position along the ohms scale.
    (a) true
    (b) false

14. Multipliers are used with most ranges on the multimeter.
    (a) true
    (b) false

15. Voltmeter polarity can be reversed by using the +DC, −DC, AC switch on the multimeter shown in Figure 4-11.
    (a) true
    (b) false

FIGURE 4-11

Typical Volt-Ohm-Milliammeter of the Type Used to Measure Resistance, Voltage, and DC Current (Courtesy of Simpson Electric Company)

5

Ohm's and Kirchhoff's Laws

Introduction

The information presented up to now is in the form of an introduction. To understand electronics, it is necessary to become familiar with complex applications; only then can technicians be confident of their ability to repair electronic equipment.

This chapter introduces you to the basic practical circuit. Schematic symbols are used which signify that a specific device is connected to a battery. A circuit is present any time a battery and conductors are connected so that current can flow. The requirements for a circuit are:

1. A power source (battery)
2. Continuously connected conductors that form a complete loop

The addition of a resistive device, to perform practical work, converts the circuit into a "practical" circuit.

Few devices are perfect, and electronics is no exception. The materials and procedures used in the manufacture of electronic components are all subject to imperfections. Electronics, however, can be learned by approaching circuits as though they consisted of ideal (perfect) components.

For our purposes, all solutions to circuits are based on the use of ideal components. All cases are explained where tolerances are part of the knowledge needed to understand the application of a component or a circuit. The main intent of this text is the development of technicians capable of gaining employment in situations that require the ability to analyze and repair electronic circuitry.

Objectives

Each student is required to:

1. List three requirements for a practical circuit.
2. Write three forms of Ohm's law.

3. Use the three forms of Ohm's law to analyze theoretically simple, series-resistive circuits.
4. Write Kirchhoff's voltage and current laws in both verbal and formula form.
5. Apply both Kirchhoff laws to an analysis of simple, series-resistive circuits.
6. State the rule for calculating total resistance in a series-resistive circuit.
7. Analyze series-resistive circuits to determine total resistance.

The Practical Circuit

A practical circuit is a circuit designed and built to meet a specific need. It can, and does, meet that need. See Figure 5-1 for a schematic drawing.

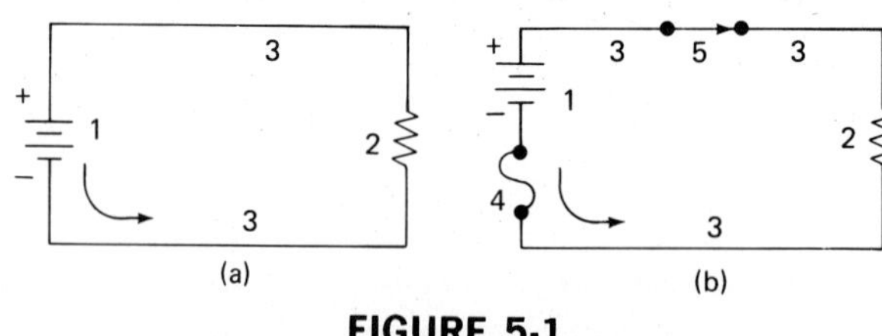

FIGURE 5-1

Regardless of the complexity of a circuit, thorough analysis results in an equivalent circuit that contains a power source and one resistor. Any complex circuit can be reduced to an equivalent resistance and a power source. The equivalent circuit is identical to that in Figure 5-1a. Voltage, current, and resistance indicated on the equivalent circuit represent equivalent values obtained after complete analysis of a circuit.

Figure 5-1a presents a practical circuit which contains the three components necessary for operation:

1. A voltage source
2. A resistive device
3. Conductors that connect components, forming a closed path, or loop, for current flow

It is a series circuit because it contains only one path for current flow. The components are connected end to end; any current that flows must flow through each component *in series.*

Many practical circuits contain additional components. Two components that often appear are the fuse and the switch (Figure 5-1b). The five components used to connect the circuit are:

1. A voltage source (battery)
2. A load device (resistor), which does the work
3. Conductors (wires), which connect components
4. A safety device (fuse)
5. A control device (switch)

Self-Check

Answer each item by indicating whether the statement is true or false.

1. ____ An equivalent circuit uses a single resistor to represent a circuit's total resistance.

2. ____ The three requirements for a practical circuit are voltage, current, and conductors.

3. ____ A switch is used to limit the amount of current flowing in a circuit.

4. ____ A fuse is designed to oppose current flow in a circuit.

5. ____ Current flows from the negative pole of a battery to the positive pole through an external circuit.

Ohm's Law

In analyzing a resistive circuit, the technician is concerned with three values: voltage, current, and resistance. These values are interrelated. When any two are known, the third can be calculated accurately.

Accurate calculations are possible because of the experiments of the German physicist Georg Simon Ohm (1787–1854). Ohm stated the law that now bears his name:

Current in a circuit is directly proportional to the applied voltage, and inversely proportional to the resistance.

Stated as formula, Ohm's law is:

$$I = \frac{E}{R}$$

I = current

E = voltage

R = resistance

Using this formula, we can say that, if any two values are known, the third can be calculated. Figure 5-2 illustrates the application of the law.

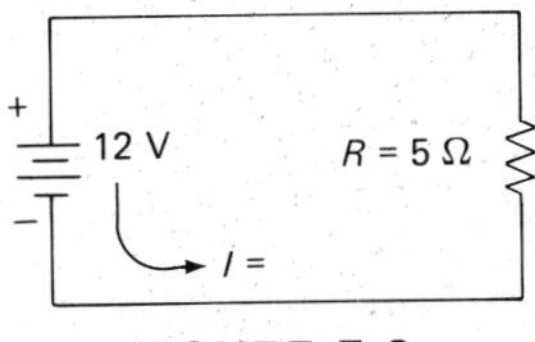

FIGURE 5-2

Let us assume that the resistive device in the circuit is an automobile's starter motor and that it has a resistance of 5 Ω. With a battery voltage of 12 V, we can use Ohm's law to determine the total current (I_t) flowing in the circuit:

$$I_t = \frac{E_t}{R_t}$$

$$I_t = \frac{12\ \text{V}}{5\ \Omega}$$

$$I_t = 2.4\ \text{amperes}$$

Having completed the solution, we know all three values—voltage, current, and resistance. For practice, though, we use these values to examine two other applications of Ohm's law.

Transposition of Ohm's law results in the formulas,

$$R_t = \frac{E_t}{I_t}$$

$$E_t = I_t \times R_t$$

Substituting 12 V for E_t and 2.4 A for I_t, we get:

$$R_t = \frac{12\ \text{V}}{2.4\ \text{A}}$$

$$R_t = 5\ \Omega$$

With the values for current ($I_t = 2.4$ A) and resistance ($R_t = 5\ \Omega$) known, we can solve for voltage (E_t):

$$E_t = I_t \times R_t$$

$$E_t = 2.4\ \text{A} \times 5\ \Omega$$

$$E_t = 12\ \text{V}$$

Three forms of Ohm's law have been presented and explained. They were used to solve for three units (voltage, current, and resistance) in the circuit. The formulas are:

Resistance: $R = \frac{E}{I}$, answer expressed in ohms

Voltage: $E = I \times R$, answer expressed in volts

Current: $I = \frac{E}{R}$, answer expressed in amperes

Self-Check

Complete each item by inserting the words and/or numbers needed to make a true and complete statement.

6. Ohm's law states: Current is ______________ proportional to applied voltage and is ______________ proportional to resistance.

7. In verbal form, Ohm's law states that voltage is equal to ______________ multiplied by ______________.

8. Using Ohm's law, we can say that if the voltage applied to a circuit increases, then circuit current will ______________.

9. If current equals 10 A and voltage equals 50 V, circuit resistance equals ______________ Ω.

10. Circuit current is 5 mA and resistance is 5 kΩ; applied voltage is ______________ V.

11. Voltage applied to a circuit is 50 V; if circuit resistance is doubled, circuit current will ______________.

12. Voltage applied to a circuit is 100 V; if resistance of the circuit is 50 Ω, total current will equal ______________.

13. In one form, Ohm's law states that: ______________ equals applied voltage divided by total current.

Total Resistance in a Series Circuit

Total resistance, symbol R_t, of a series circuit is equal to the sum of all the individual resistances contained in the circuit (closed loop). For a circuit with two series resistors, the formula is:

$$R_t = R_1 + R_2$$

Refer to Figures 5-3 and 5-4. Note that there are two resistors in the circuit, one with a value of 5 Ω and the other with a value of 10 Ω.

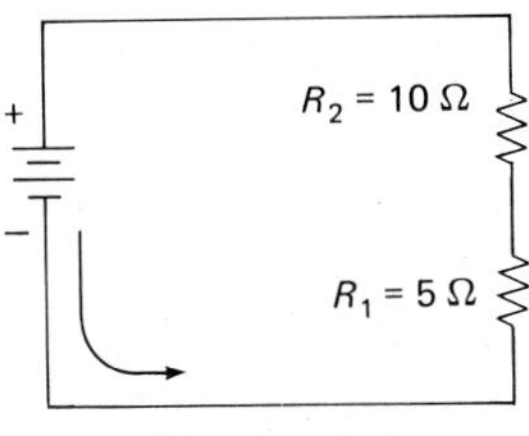

FIGURE 5-3

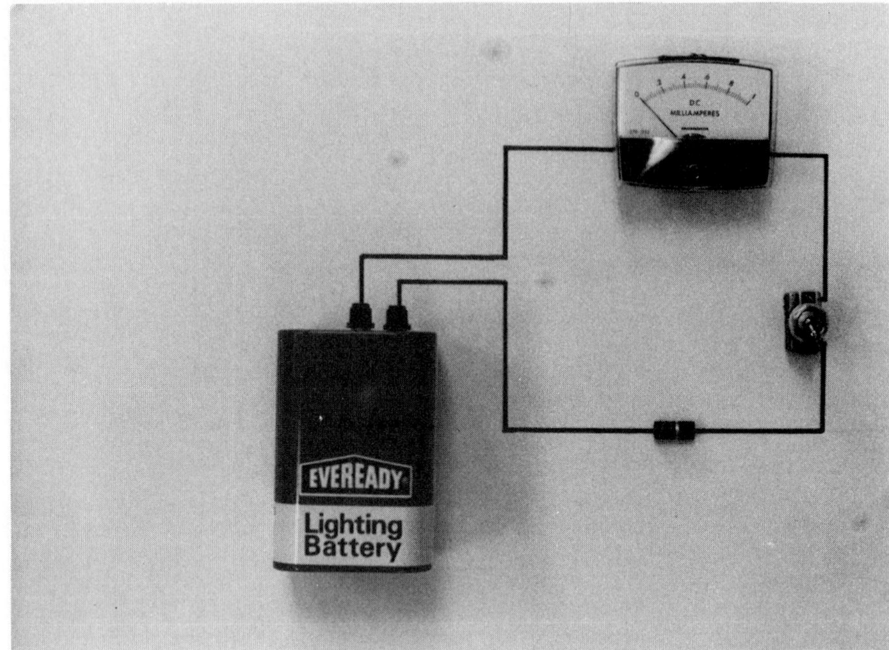

FIGURE 5-4
Practical Circuit (Photo by Riddell Allen)

To solve for total resistance in this circuit, we apply the formula as shown in the table:

	Calculator Solution		
Mathematical Solution	*Action*	*Entry*	*Display*
$R_t = R_1 + R_2$	ENTER	5	5
$R_t = 5\ \Omega + 10\ \Omega$	PRESS	+	5
$R_t = 15\ \Omega$	ENTER	10	10
	PRESS	=	15

Regardless of the number of resistors in a series circuit, the formula is the same. All resistor values are added to obtain the total resistance of a series circuit. Stated as a formula, it is:

$$R_t = R_1 + R_2 + R_3 + R_4, \ldots, R_n$$

The schematic in Figure 5-5 shows a circuit with three series-connected resistors. To solve for R_t, we use the following:

	Calculator Solution		
Mathematical Solution	*Action*	*Entry*	*Display*
$R_t = R_1 + R_2 + R_3$	ENTER	10	10
$R_t = 10 + 20 + 30$	PRESS	+	10
$R_t = 60\ \Omega$	ENTER	20	20
	PRESS	+	30
	ENTER	30	30
	PRESS	=	60

$R_3 = 30\ \Omega$
$R_2 = 20\ \Omega$
I $R_1 = 10\ \Omega$

FIGURE 5-5

An examination of Figure 5-6 reveals that two resistors are connected in series. To solve for R_t in this circuit, we proceed as follows:

	Calculator Solution		
Mathematical Solution	*Action*	*Entry*	*Display*
$R_t = R_1 + R_2$	ENTER	13,EE,3	13 03
$R_t = 13{,}000 + 12{,}000$	PRESS	+	1.3 04
$R_t = 25{,}000\ \Omega$	ENTER	12,EE,3	12 03
	PRESS	=	2.5 04

Converting result to a prefix results in $R_t = 25\ \text{k}\Omega$.

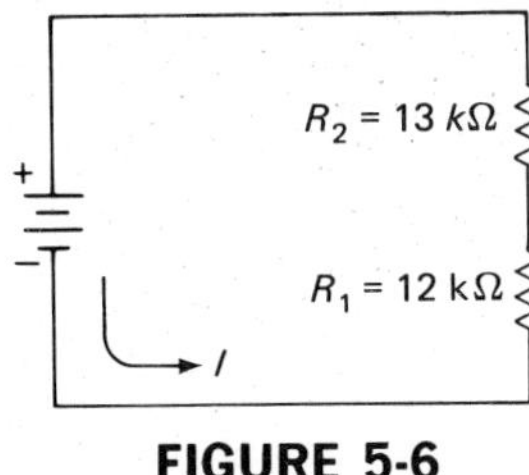

FIGURE 5-6

In Figure 5-7, when solving for total resistance, it is necessary to proceed differently:

Mathematical Solution	Calculator Solution		
	Action	Entry	Display
$R_t = \dfrac{E_t}{I_t}$	ENTER	100	100
$R_t = \dfrac{100\text{ V}}{5\text{ mA}}$	PRESS	÷	100
$R_t = 20\text{ k}\Omega$	ENTER	5,EE,+/−,3	5 − 03
	PRESS	=	2 04

Converting 2 04 to prefix yields 20 kΩ.

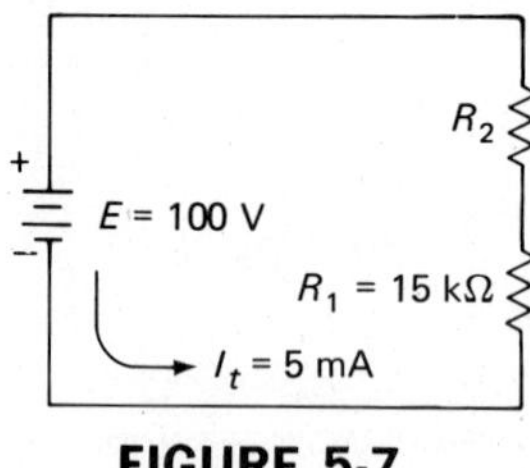

FIGURE 5-7

Figure 5-8 requires that the ohmic value of R_1 be calculated. To do this, we proceed as follows. First we solve for R_t:

Mathematical Solution	Calculator Solution		
	Action	Entry	Display
$R_t = \dfrac{E_t}{I_t}$	ENTER	60	60
$R_t = \dfrac{60\text{ V}}{5\text{ mA}}$	PRESS	÷	60
$R_t = 12\text{ k}\Omega$	ENTER	5,EE,+/−,3	5 − 03
	PRESS	=	1.2 04

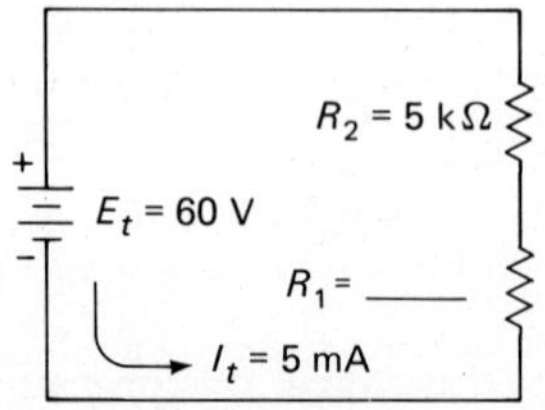

FIGURE 5-8

When converted to a prefix 1.2 04 equals 12 kΩ. R_1 can then be calculated thus:

	Calculator Solution		
Mathematical Solution	*Action*	*Entry*	*Display*
$R_1 = R_t - R_2$	ENTER	12,EE,3	12 03
$R_1 = 12\text{ k} - 5\text{ k}$	PRESS	−	1.2 04
$R_1 = 7\text{ k}\Omega$	ENTER	5,EE,3	5 03
	PRESS	=	7 03

When converted to a prefix, 7 03 equals 7 kΩ.

The resistance of a battery, switch, or fuse is so small that we do not consider it as part of the total resistance in circuits of this type. Some circuits covered in later studies, however, are so delicate that their resistances must be taken into consideration in the analysis.

Self-Check

Answer each item by indicating whether the statement is true or false.

14. ____ The rule for total resistance in a series-resistive circuit is that total resistance equals the sum of all resistors.

15. ____ Total resistance in a series circuit is always larger than the value of the largest resistor in the circuit.

16. ____ To solve for an unknown resistor in a circuit, we must add all other resistors to the total resistance.

17. ____ The formula for solving total resistance in a circuit is $R_t = R_1 + R_2 + R_3$.

18. ____ The calculator display 1 05 converts to 10 kΩ.

Kirchhoff's Current Law

Figure 5-9 illustrates a circuit that contains one resistor and one ammeter. (Note that all components are connected in series.) Figure 5-10 also illustrates this circuit.

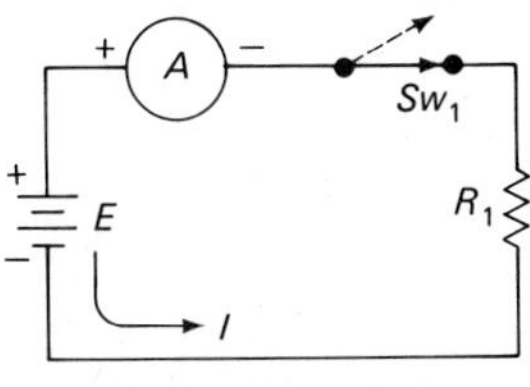

FIGURE 5-9

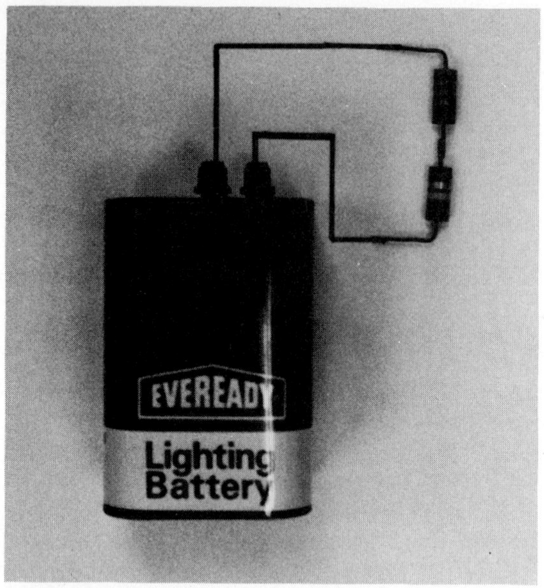

FIGURE 5-10
Two Fixed Resistor Circuit (Photo by Riddell Allen)

When the switch is closed in this circuit, the ammeter measures the current flowing through that part of the circuit. Since all current that leaves the battery must flow through the closed loop and return to the battery, current flowing in the meter equals the total current. This rule was stated and proved by the German physicist Gustav Robert Kirchhoff (1824–87). He stated his law thus:

The algebraic sum of all currents at any point in an electrical network is equal to zero.

This statement can be expressed in formula form, as follows:

$$I_t - I_{R1} = 0$$

where

I_t = total current
I_{R1} = current through R_1

This law is known as Kirchhoff's current law or Kirchhoff's first law. It is important in the analysis of all electronic circuits.

When we examine Kirchhoff's current law, we find that in a series loop,

current is the same at all points in a series circuit.

In Figure 5-9, this can be stated in formula form as follows:

$$I_t = I_1 = I_{R1}$$

The total resistance of the circuit has no effect on the application of Kirchhoff's current law. Figure 5-11 shows three resistors and three ammeters.

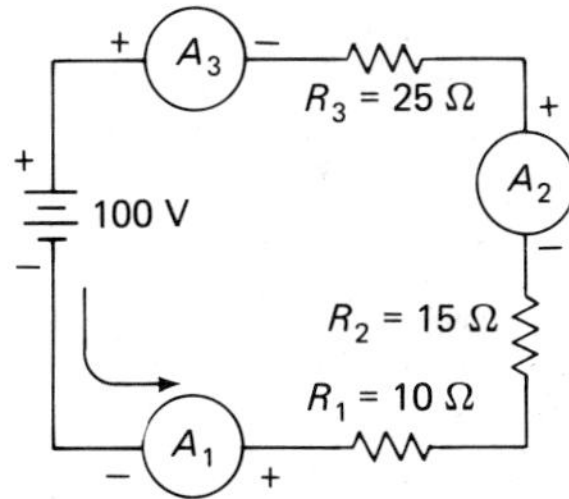

FIGURE 5-11

With additional ammeters connected in the circuit, the formula becomes:

$$I_t = I_1 = I_2 = I_3 = I_4 = \ldots I_n$$

Circuits containing more complex connections than those shown in series circuits also satisfy the provisions of the law, though application of the law may vary slightly (the variations are discussed as required for understanding more complex circuitry).

Refer to Figure 5-12. In the discussion of Kirchhoff's current law, we did not concern ourselves with the amount of current flowing. Our conclusion was: Whatever current is flowing will be measured by the ammeter at each location. Based on this conclusion, we can state that "current is the same at all points within a series circuit."

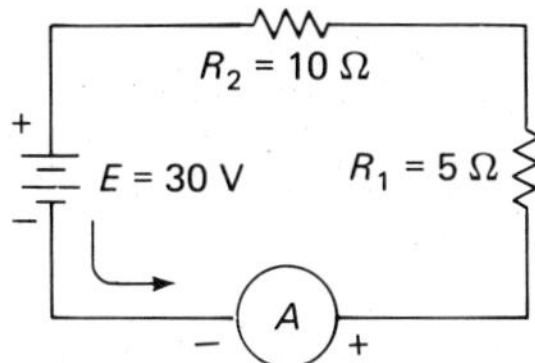

FIGURE 5-12

The circuit contains two resistors, R_1 and R_2; therefore, it presents a different problem from earlier Ohm's-law solutions. In those solutions, we used the form of Ohm's law, $I = E/R$. Part of the information given was total resistance, and the formula worked perfectly.

When a circuit contains two or more resistors, we solve for total resistance; then we use Ohm's law to solve for total current. To solve for total resistance, R_t, we apply this rule:

Total resistance of a series circuit equals the sum of the individual resistance values.

Stated in formula form:

$$R_t = R_1 + R_2$$

With the values R_1 and R_2 inserted:

$$R_t = 5\ \Omega + 10\ \Omega$$

$$R_t = 15\ \Omega$$

If the circuit had contained 5, 10, 25, or 100 resistors, their values would be added together to find the total circuit resistance. Again, we treat fuse, battery, and switch as though they were zero ohms.

With the total resistance (R_t) known, we can calculate the total current (I_t), using Ohm's law:

$$I_t = \frac{E_t}{R_t}$$

$$I_t = \frac{30\ \text{V}}{15\ \Omega}$$

$$I_t = 2\ \text{A}$$

Two amperes of current are flowing in the circuit. Relating this to Kirchhoff's current law, we see that current flow through R_1 and R_2 is also 2 A.

What would happen if the resistors were replaced by resistors having different values? Note that Figure 5-13 has two resistors, each with a value of 15 Ω. How would this change affect the total current, I_t?

Mathematical Solution	Calculator Solution		
	Action	Entry	Display
$R_t = R_1 + R_2$	ENTER	15	15
$R_t = 15\ \Omega + 15\ \Omega$	PRESS	+	15
$R_t = 30\ \Omega$	ENTER	15	15
	PRESS	=	30

Now that we know R_t equals 30 Ω, we can solve for total current, I_t.

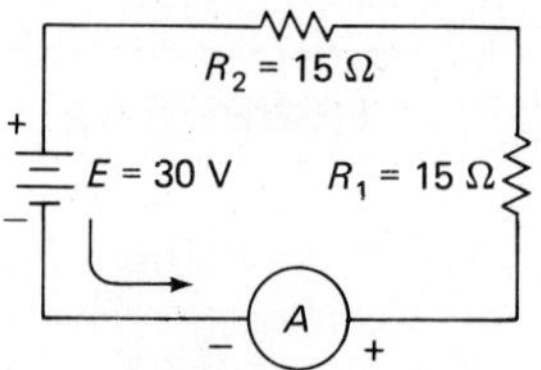

FIGURE 5-13

Mathematical Solution	Calculator Solution		
	Action	Entry	Display
$I_t = \frac{E_t}{R_t}$	ENTER	30	30
$I_t = \frac{30\ \text{V}}{30\ \Omega}$	PRESS	÷	30
$I_t = 1\ \text{A}$	ENTER	30	30
	PRESS	=	1

Refer to Ohm's law: current and resistance are inversely proportional. Therefore, with this increase in total resistance, total current must decrease. Had the total resistance decreased, total current would have increased.

What would happen if resistance were returned to 15 Ω but total voltage was doubled, to 60 V? See Figure 5-14. We have solved $R_t = 15\ \Omega$; thus:

	Calculator Solution		
Mathematical Solution	*Action*	*Entry*	*Display*
$I_t = \dfrac{E_t}{R_t}$	ENTER	60	60
$I_t = \dfrac{60\ \text{V}}{15\ \Omega}$	PRESS	÷	60
$I_t = 4\ \text{A}$	ENTER	15	15
	PRESS	=	4

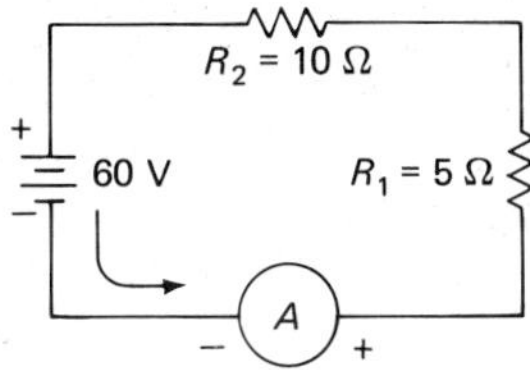

FIGURE 5-14

When resistance is held constant and voltage increases, circuit current increases. As total voltage is decreased, circuit current decreases. When voltage applied to a circuit increases, it causes the total current to increase. There is no effect on resistance, since it is a fixed value.

Consider Figure 5-15:

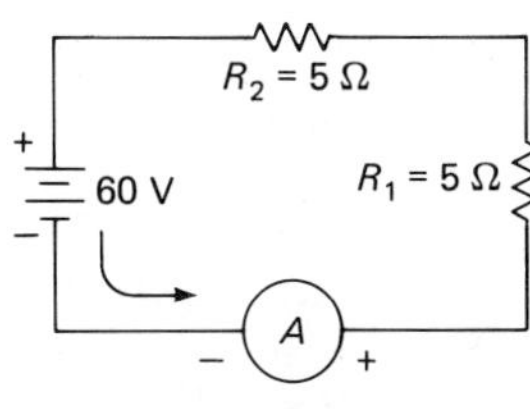

FIGURE 5-15

In this circuit, R_t is 10 Ω and the total voltage is 60 V. In what way is total current affected when the voltage remains constant and resistance is decreased?

Mathematical Solution	Calculator Solution		
	Action	Entry	Display
$I_t = \dfrac{E_t}{R_t}$	ENTER	60	60
$I_t = \dfrac{60\text{ V}}{10\ \Omega}$	PRESS	÷	60
$I_t = 6\text{ A}$	ENTER	10	10
	PRESS	=	6

Four rules for analysis have been developed in this discussion:

1. If resistance remains constant and voltage increases, current must also increase.
2. If resistance remains constant and voltage decreases, current must also decrease.
3. If voltage remains constant and resistance increases, current must decrease.
4. If voltage remains constant and resistance decreases, current must increase.

When doing a complete analysis of the circuit shown in Figure 5-15, we must determine the part of E_t that is dropped, or used, by each resistor during the process of causing current to be limited. (You have already learned that when one resistor is in the circuit, its voltage drop equals E_t.) With two or more resistors in the circuit, the total voltage is divided (or dropped) according to the ratio of the resistor values. To determine voltage drops in a circuit of this type, we use Ohm's law, $E = I \times R$. To solve for E_{R1} we must:

Mathematical Solution	Calculator Solution		
	Action	Entry	Display
$E_{R1} = I_t \times R_1$	ENTER	6	6
$E_{R1} = 6\text{ A} \times 5\ \Omega$	PRESS	×	6
$E_{R1} = 30\text{ V}$	ENTER	5	5
	PRESS	=	30

and for E_{R2}:

Mathematical Solution	Calculator Solution		
	Action	Entry	Display
$E_{R2} = I_t \times R_2$	ENTER	6	6
$E_{R2} = 6\text{ A} \times 5\ \Omega$	PRESS	×	6
$E_{R2} = 30\text{ V}$	ENTER	5	5
	PRESS	=	30

Notice that each resistor drops exactly half the applied voltage. This is logical, since the resistors have the same value and can be expected to provide equal opposition to the 60 V force.

Self-Check

Answer each item by indicating whether the statement is true or false.

19. ____ Stated in simple form, Kirchhoff's current law means that current is equal at all points in a closed-series loop.

20. ____ For a two-resistor circuit, Kirchhoff's current law can be stated by the formula, $I_t = I_{R1} + I_{R2}$.

21. ____ With the total resistance of a circuit known, the total current can be calculated, using Ohm's law: $I = E/R$.

22. ____ As resistors are added to a closed loop, total current increases.

23. ____ If voltage remains constant and resistance decreases, total current will increase.

24. ____ Any time resistance increases, current must decrease.

Kirchhoff's Voltage Law

When two resistors are connected in a series circuit, both resistors oppose the flow of electrons. The opposition of each resistor is proportional to its ohmic value. The amount of pressure, or voltage, dropped by a resistor is part of the total voltage applied. Kirchhoff's voltage law is used to explain this:

The algebraic sum of all the voltages in a closed loop equals zero.

Stating this law in the form of a formula, we have:

$$E_t - E_{R1} - E_{R2} = 0$$

But how do we find E_{R1} and E_{R2}?

In Figure 5-16, Ohm's law can be used to solve for individual resistor voltage drops in the same way that the law is used to calculate total voltage. Note that this circuit is identical to the one in Figure 5-12. An analysis of Figure 5-12 reveals that $I_t = 2$ A. Using this value for current, we apply Ohm's law to the solution of Figure 5-16. Solve for E_{R1}.

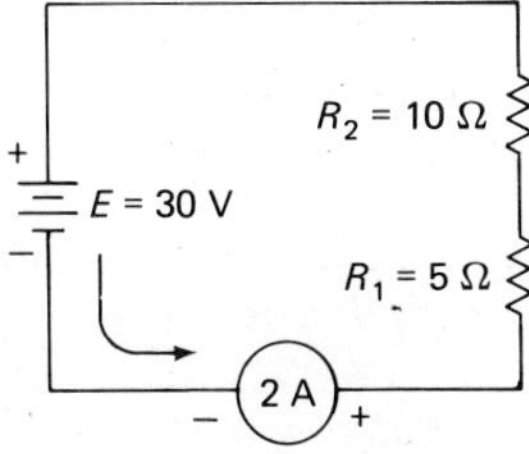

FIGURE 5-16

	Calculator Solution		
Mathematical Solution	*Action*	*Entry*	*Display*
$E_{R1} = I_t \times R_1$	ENTER	2	2
$E_{R1} = 2\ \text{A} \times 5\ \Omega$	PRESS	×	2
$E_{R1} = 10\ \text{V}$	ENTER	5	5
	PRESS	=	10

and solve for E_{R2}:

	Calculator Solution		
Mathematical Solution	*Action*	*Entry*	*Display*
$E_{R2} = I_t \times R_2$	ENTER	2	2
$E_{R2} = 2\ \text{A} \times 10$	PRESS	×	2
$E_{R2} = 20\ \text{V}$	PRESS	10	10
	PRESS	=	20

To prove Kirchhoff's voltage law, use:

	Calculator Solution		
Mathematical Solution	*Action*	*Entry*	*Display*
$E_t - E_{R1} - E_{R2} = 0$	ENTER	30	30
$30\ \text{V} - 10\ \text{V} - 20\ \text{V} = 0$	PRESS	−	30
$0 = 0$	ENTER	10	10
	PRESS	−	20
	ENTER	20	20
	PRESS	=	0

Quite often, Kirchhoff's voltage law is stated thus:

$$E_t = E_{R1} + E_{R2}$$

For many people working in electronics, this form seems easier to remember. It is derived by transposing Kirchhoff's voltage law to arrive at the above formula. Stated verbally, this version is:

The sum of all the voltage drops in any closed loop equals the applied voltage.

We can use Ohm's and Kirchhoff's laws to solve for unknown quantities in any series circuit. Consider voltage, current, and resistance; if any two are known, the third can be found using Ohm's law.

More rules have been developed that aid in the analysis of series circuits:

1. The sum of all the voltage drops in any closed loop equals the applied voltage.
2. Individual resistor voltage drops are computed, using the formula $E_R = I_R \times R$.
3. When a series circuit contains only one resistor, that resistor's voltage drop equals the applied voltage.
4. When a series circuit contains two or more resistors, the larger resistor drops the most voltage.

To gain practice using Ohm's and Kirchhoff's laws, let us solve the circuit in Figure 5-17, for I_t, R_t, E_{R1}, E_{R2}, and E_{R3}.

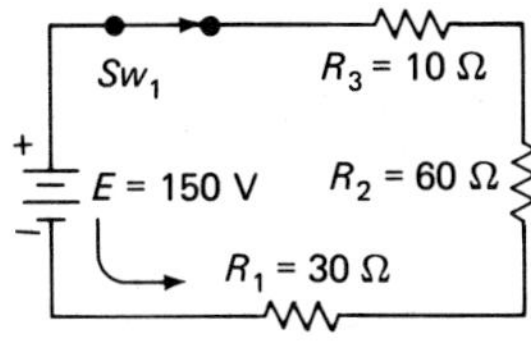

FIGURE 5-17

Total voltage and resistor sizes are given in the figure. We use this data to solve the circuit for the items listed above. When the switch is closed, 150 V is applied, which causes electron current to flow. To use Ohm's law in calculating total current, I_t, we need total resistance, R_t. R_t is solved by adding the resistor values:

	Calculator Solution		
Mathematical Solution	*Action*	*Entry*	*Display*
$R_t = R_1 + R_2 + R_3$	ENTER	30	30
$R_t = 30 + 60 + 10$	PRESS	+	30
$R_t = 100\ \Omega$	ENTER	60	60
	PRESS	+	90
	ENTER	10	10
	PRESS	=	100

With R_t known, solve for total current:

	Calculator Solution		
Mathematical Solution	*Action*	*Entry*	*Display*
$I_t = \frac{E_t}{R_t}$	ENTER	150	150
$I_t = \frac{150\text{ V}}{100\ \Omega}$	PRESS	÷	150
$I_t = 1.5\text{ A}$	ENTER	100	100
	PRESS	=	1.5

With total current, I_t, known, individual voltage drops can be calculated:

Mathematical Solution	Calculator Solution		
	Action	Entry	Display
$E_{R1} = I_t \times R_1$	ENTER	1.5	1.5
$E_{R1} = 1.5\ \text{A} \times 30\ \Omega$	PRESS	×	1.5
$E_{R1} = 45\ \text{V}$	ENTER	30	30
	PRESS	=	45

and

Mathematical Solution	Calculator Solution		
	Action	Entry	Display
$E_{R2} = I_t \times R_2$	ENTER	1.5	1.5
$E_{R2} = 1.5\ \text{A} \times 60\ \Omega$	PRESS	×	1.5
$E_{R2} = 90\ \text{V}$	ENTER	60	60
	PRESS	=	90

and

Mathematical Solution	Calculator Solution		
	Action	Entry	Display
$E_{R3} = I_t \times R_3$	ENTER	1.5	1.5
$E_{R3} = 1.5\ \text{A} \times 10\ \Omega$	PRESS	×	1.5
$E_{R3} \times 15\ \text{V}$	ENTER	10	10
	PRESS	=	15

Now we can use the three voltage drops to check the problem, using Kirchhoff's voltage (or second) law:

Mathematical Solution	Calculator Solution		
	Action	Entry	Display
$E_t - E_{R1} - E_{R2} - E_{R3} = 0$	ENTER	150	150
$150\ \text{V} - 45\ \text{V} - 90\ \text{V}$	PRESS	−	150
$- 15\ \text{V} = 0$			
$0 = 0$	ENTER	45	45
	PRESS	−	105
	ENTER	90	90
	PRESS	−	15
	ENTER	15	15
	PRESS	=	0

Another method for checking our work is to use the alternate form of Kirchhoff's voltage law:

$$E_t = E_{R1} + E_{R2} + E_{R3}$$

$$150\ \text{V} = 45\ \text{V} + 90\ \text{V} + 15\ \text{V}$$

$$150\ \text{V} = 150\ \text{V}$$

Different problems in the analysis of series circuits require the use of formulas in their transposed form. For voltage and resistance solutions, these formulas are:

Voltage solutions:

Formula known: $E_t = E_{R1} + E_{R2} + E_{R3}$

Solve for E_{R3}: $E_{R3} = E_t - E_{R1} - E_{R2}$

Solve for E_{R2}: $E_{R2} = E_t - E_{R1} - E_{R3}$

Solve for E_{R1}: $E_{R1} = E_t - E_{R2} - E_{R3}$

Resistance solutions:

Formula known: $R_t = R_1 + R_2 + R_3$

Solve for R_1: $R_1 = R_t - R_2 - R_3$

Solve for R_2: $R_2 = R_t - R_1 - R_3$

Solve for R_3: $R_3 = R_t - R_1 - R_2$

The circuit in Figure 5-18 presents a different problem. Here, we have a circuit in which total voltage, total current, and the value of R_1 are known. We are asked to solve for the value of R_2 needed for this circuit to operate properly. To do this, we proceed as follows:

	Calculator Solution		
Mathematical Solution	*Action*	*Entry*	*Display*
$R_t = \dfrac{E_t}{I_t}$	ENTER	100	100
$R_t = \dfrac{100\ \text{V}}{10\ \text{mA}}$	PRESS	÷	100
$R_t = 10\ \text{k}\Omega$	ENTER PRESS	10,EE,+/−,3 =	10 − 03 1 04

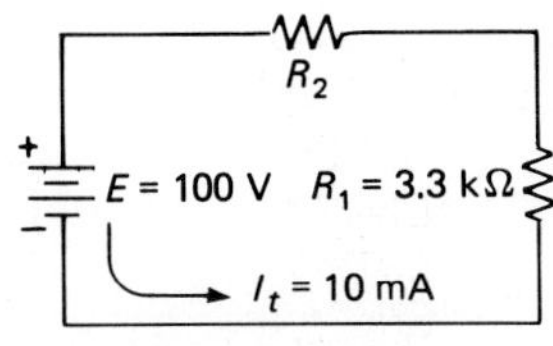

FIGURE 5-18

When converted to a prefix, 1 04 equals 10 kΩ.

Mathematical Solution	Calculator Solution		
	Action	Entry	Display
$R_2 = R_t - R_1$	ENTER	10,EE,3	10 03
$R_2 = 10{,}000 - 3300$	PRESS	−	1 04
$R_2 = 6700\ \Omega$	ENTER	3.3,EE,3	3.3 03
	PRESS	=	6.7 03

When converted to a prefix, each solution equals 6.7 kΩ.

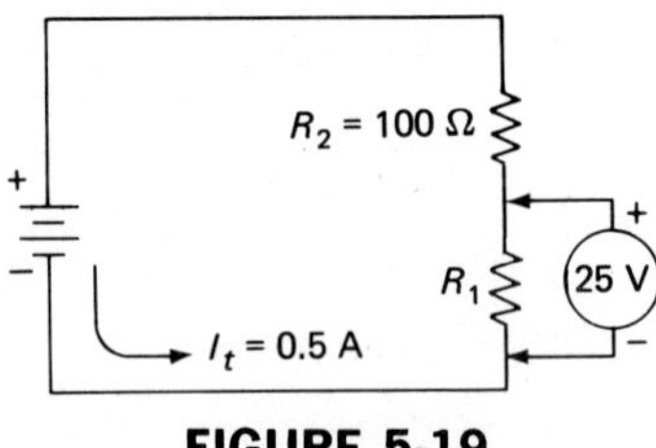

FIGURE 5-19

In Figure 5-19, note that R_2, E_{R1}, and I_t are given. Solve for E_t. The unknown you are told to solve for will always provide you with a choice of formulas to use. We have covered only two formulas for solving E_t. They are:

$$E_t = I_t \times R_t$$

$$E_t = E_{R1} + E_{R2}$$

Examination of the circuit in Figure 5-19 reveals that the data needed for use in either formula are not given. Notice, however, that we do have two known quantities about R_2.

You have learned that current is the same at all points in a series circuit. Therefore, if the total current is 0.5 A, the current through R_2 must be 0.5 A. This gives you two knowns for R_2, current and resistance. Use these data to solve for E_{R2} as follows:

Mathematical Solution	Calculator Solution		
	Action	Entry	Display
$E_{R2} = I_t \times R_2$	ENTER	.5	0.5
$E_{R2} = 0.5\ \text{A} \times 100\ \Omega$	PRESS	×	0.5
$E_{R2} = 50\ \text{V}$	ENTER	100	100
	PRESS	=	50

Once E_{R2} is known, E_t can be found:

Mathematical Solution	Calculator Solution		
	Action	Entry	Display
$E_t = E_{R1} + E_{R2}$	ENTER	25	25
$E_t = 25\text{ V} + 50\text{ V}$	PRESS	+	25
$E_t = 75\text{ V}$	ENTER	50	50
	PRESS	=	75

We have covered several more rules for analyzing series-resistive circuits, which we can now apply in the solution of Figure 5-20.

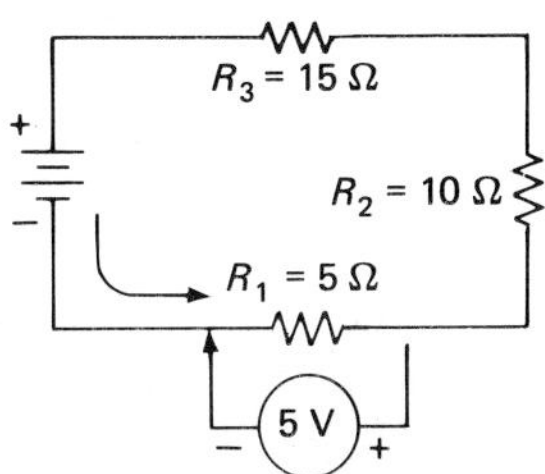

FIGURE 5-20

Examine the circuit in the figure thoroughly. Locate a component about which two known quantities are available. E_{R1} is represented by a voltmeter reading, and R_1 resistance is given. Given voltage and resistance, we can solve for current at that point. We know that current at R_1 is the same as at any other point in the circuit.

We have also covered calculator solutions. Try to solve this circuit by developing your own calculator programs.

$$I_{R1} = \frac{E_{R1}}{R_1}$$

$$I_{R1} = \frac{5\text{ V}}{5\ \Omega}$$

$$I_{R1} = 1\text{ A}$$

To solve for R_t:

$$R_t = R_1 + R_2 + R_3$$

$$R_t = 5\ \Omega + 10\ \Omega + 15\ \Omega$$

$$R_t = 30\ \Omega$$

To solve for E_{R2}:

$$E_{R2} = I_t \times R_2$$

$$E_{R2} = 1\text{ A} \times 10\ \Omega$$

$$E_{R2} = 10\text{ V}$$

To solve for E_{R3}:

$$E_{R3} = I_t \times R_3$$

$$E_{R3} = 1\ \text{A} \times 15\ \Omega$$

$$E_{R3} = 15\ \text{V}$$

To check your results:

$$E_t = E_{R1} + E_{R2} + E_{R3}$$

$$E_t = 5\ \text{V} + 10\ \text{V} + 15\ \text{V}$$

$$E_t = 30\ \text{V}$$

Kirchhoff's current law states that current is equal at all points in a series circuit. Figure 5-21 illustrates an application of this law. Given I_{R1} equals 15 mA, and Kirchhoff's law, we know that I_t is 15 mA.

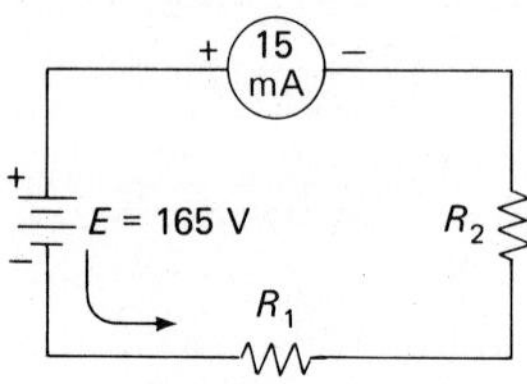

FIGURE 5-21

Therefore

$$I_t = I_{R1} = 15\ \text{mA}$$

With the current known, solve for R_t, using Ohm's law:

	Calculator Solution		
Mathematical Solution	*Action*	*Entry*	*Display*
$R_t = \dfrac{E_t}{I_t}$	ENTER	165	165
$R_t = \dfrac{165\ \text{V}}{15\ \text{mA}}$	PRESS	÷	165
$R_t = 11{,}000\ \Omega$	ENTER	15,EE,+/−,3	15 − 03
$R_t = 11\ \text{k}\Omega$	PRESS	=	1.1 04

Note: 1.1 04 and 11,000 convert to 11 kΩ when the prefix is used.

Try another problem, but this time let E_t be the unknown. Ohm's and Kirchhoff's laws are used in achieving the solution. Refer to Figure 5-22 for the schematic.

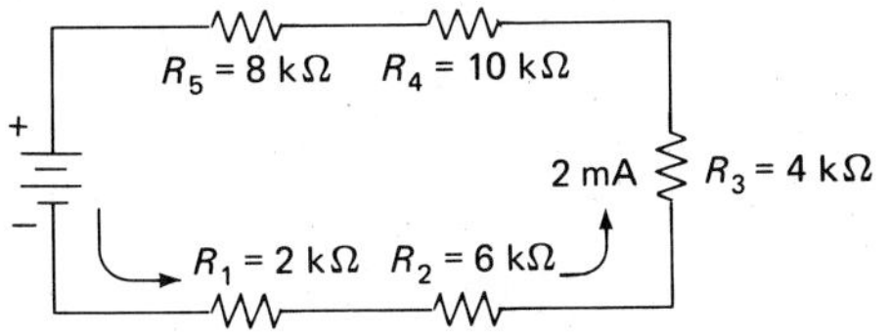

FIGURE 5-22

In this circuit we are given I_t equals 2 mA. The ohmic value of all resistors is given. To solve for E_t, we use:

$$E_t = I_t \times R_t$$

Before we can use this equation, however, we must solve for R_t.

	Calculator Solution		
Mathematical Solution	*Action*	*Entry*	*Display*
$R_t = R_1 + R_2 + R_3$	ENTER	2,EE,3	2 03
$+ R_4 + R_5$	PRESS	+	2 03
$R_t = 2\ k\Omega + 6\ k\Omega + 4\ k\Omega$	ENTER	6,EE,3	6 03
$+ 10\ k\Omega + 8\ k\Omega$	PRESS	+	8 03
$R_t = 30\ k\Omega$	ENTER	4,EE,3	4 03
	PRESS	+	1.2 04
	ENTER	10,EE,3	10 03
	PRESS	+	2.2 04
	ENTER	8,EE,3	8 03
	PRESS	=	3 04

When 3 04 is converted to a prefix R_t equals 30 kΩ.

Use Ohm's law to solve for E_t:

	Calculator Solution		
Mathematical Solution	*Action*	*Entry*	*Display*
$E_t = I_t \times R_t$	ENTER	2,EE,+/−,3	2 − 03
$E_t = 2\ mA \times 30\ k\Omega$	PRESS	×	2 − 03
$E_t = 60\ V$	ENTER	30,EE,3	30 03
	PRESS	=	6 01

When 6 01 is converted to a prefix, $E_t = 60$ V.

Kirchhoff's voltage law can also be used to solve for E_t. The formula is:

$$E_t = E_{R1} + E_{R2} + E_{R3} + E_{R4} + E_{R5}$$

To use this formula, we must know the voltage drop for each resistor. Because they weren't given, you must solve for them, using the following procedures.

	Calculator Solution		
Mathematical Solution	*Action*	*Entry*	*Display*
$E_{R1} = I_t \times R_1$	ENTER	2,EE,+/−,3	2 − 03
$E_{R1} = 2\text{ mA} \times 2\text{ k}\Omega$	PRESS	×	2 − 03
$E_{R1} = 4\text{ V}$	ENTER	2,EE,3	2 03
	PRESS	=	4 00

When 4 00 is converted to a prefix, $E_{R1} = 4$ V.

Using either method, the results are:

$$E_{R1} = 4\text{ V}$$

$$E_{R2} = 12\text{ V}$$

$$E_{R3} = 8\text{ V}$$

$$E_{R4} = 20\text{ V}$$

$$E_{R5} = 16\text{ V}$$

Now we can solve for E_t:

$$E_t = 4\text{ V} + 12\text{ V} + 8\text{ V} + 20\text{ V} + 16\text{ V}$$

$$E_t = 60\text{ V}$$

Self-Check

Complete each item by inserting the words and/or numbers needed to make a true and complete statement.

25. Kirchhoff's voltage law states that the algebraic sum of all voltages in a closed loop equals __________.

26. Voltage drops in a circuit are E_{R1}, 20 V, and E_{R2}, 30 V; the applied voltage is __________ volts.

27. Kirchhoff's voltage law states that the sum of the __________ drops equals the __applied__ voltage.

28. A circuit has an applied voltage of 90 V and 2 resistors. R_1 drops 35 V; R_2 drops __________.

29. Adding a third resistor to the problem stated in number 28 causes E_{R1} to __________.

30. Removing a resistor from a circuit with two resistors causes the voltage drop on the remaining resistor to __RISE__.

Summary

The following rules for series-resistive circuits were developed in this chapter.

1. Current in a circuit is directly proportional to total voltage, and is inversely proportional to resistance.
2. Total voltage in a circuit is equal to the product of current times resistance.
3. Current is equal at all points within a closed loop.
4. Total resistance in a closed loop is equal to the sum of all the individual resistances contained in the loop.
5. The sum of all the voltage drops in a closed loop equals the applied voltage.
6. The voltage drop across a resistor in a series circuit is directly proportional to the resistor's ohmic value.

Review Questions and/or Problems

1. Solve for E_t in the circuit shown in Figure 5-23.
   (a) 50 V
   (b) 100 V
   (c) 200 V
   (d) 400 V

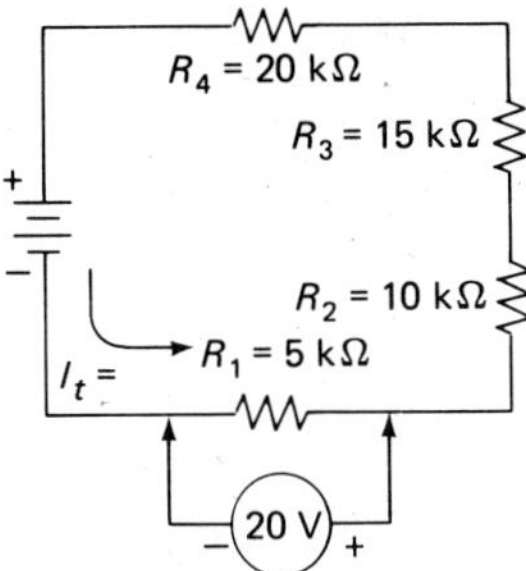

FIGURE 5-23

2. Solve for R_t in the circuit shown in Figure 5-24.
   (a) 100 Ω
   (b) 20 Ω
   (c) 500 Ω
   (d) 2500 Ω

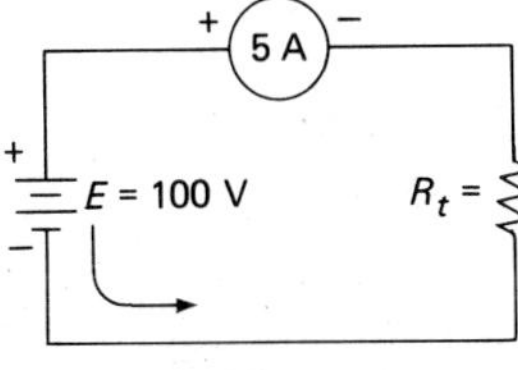

FIGURE 5-24

3. Solve for I_t in the circuit in Figure 5-25.
   (a) 10 A
   (b) 1000 A
   (c) 10 mA
   (d) 1000 mA

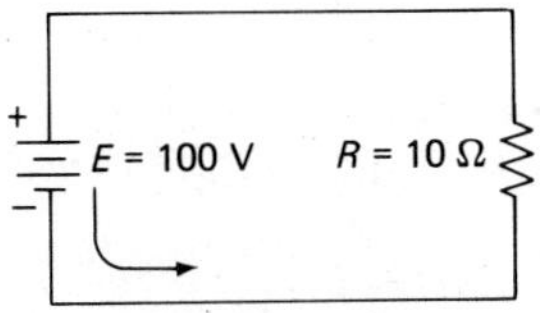

FIGURE 5-25

4. Which of the following is *not* a form of Ohm's law?
   (a) $I = E \times R$
   (b) $I = \frac{E}{R}$
   (c) $R = \frac{E}{I}$
   (d) $E = I \times R$

5. When voltage applied to a series circuit is increased, and the resistance remains constant, circuit current:
   (a) increases.
   (b) decreases.
   (c) remains the same.

6. If the resistance in a circuit is doubled, and the voltage is doubled, the circuit current:
   (a) increases.
   (b) decreases.
   (c) remains the same.

7. What is the value of I_t in the circuit shown in Figure 5-26?
   (a) 50 mA
   (b) 10 mA
   (c) 8 mA
   (d) 5 mA

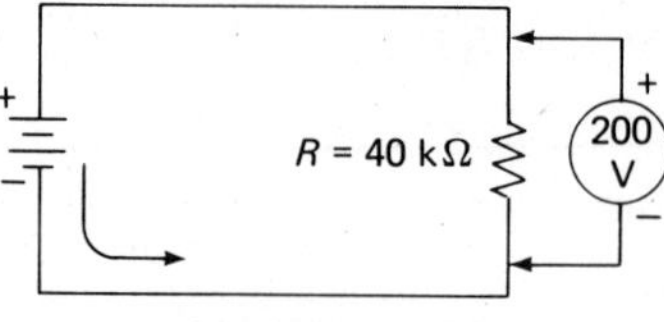

FIGURE 5-26

8. What is the total current flowing in the circuit shown in Figure 5-27?
   (a) 25 A
   (b) 25 mA
   (c) 3 A
   (d) 3 mA

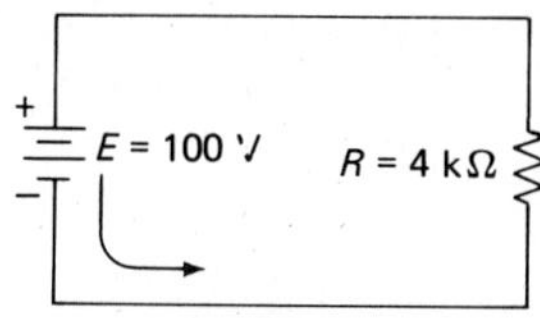

FIGURE 5-27

9. In a series circuit with three resistors, the ______________ resistor has the ______________ voltage drop.
   (a) largest, largest
   (b) largest, smallest
   (c) smallest, largest
   (d) none of the above

10. What is total resistance (R_t) in the circuit shown in Figure 5-28?
(a) 5 kΩ
(b) 10 kΩ
(c) 20 kΩ
(d) 40 kΩ

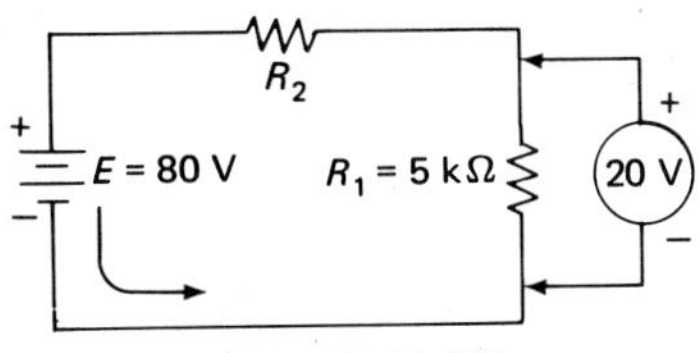

FIGURE 5-28

11. In the circuit shown in Figure 5-29, what is the value of E_t?
(a) 400 V
(b) 200 V
(c) 50 V
(d) 40 V

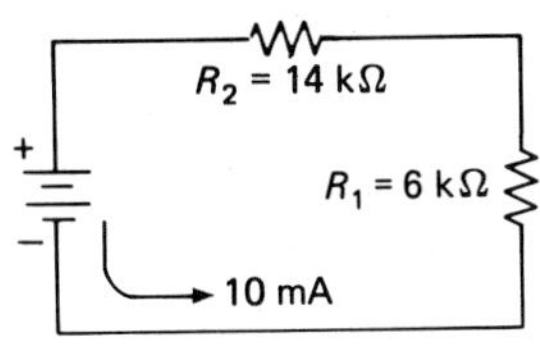

FIGURE 5-29

12. What is the value of R_t in the circuit shown in Figure 5-30?
(a) 10 kΩ
(b) 16 kΩ
(c) 5 kΩ
(d) 20 kΩ

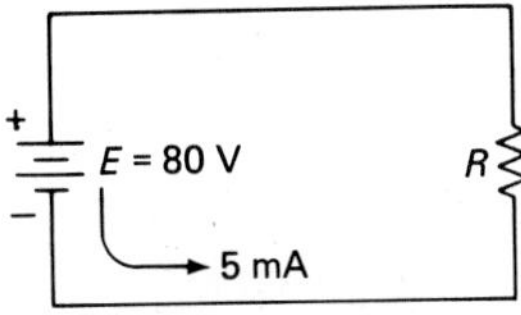

FIGURE 5-30

13. What is total resistance of the circuit in Figure 5-31?
(a) 13 kΩ
(b) 25 kΩ
(c) 192 kΩ
(d) 8 kΩ

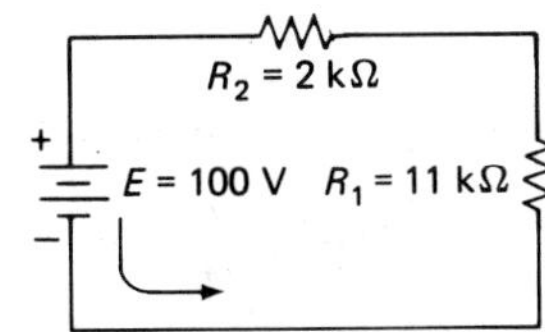

FIGURE 5-31

14. What is the current in the circuit shown in Figure 5-32?
(a) 0.91 A
(b) 91 mA
(c) 0.2 A
(d) 20 mA

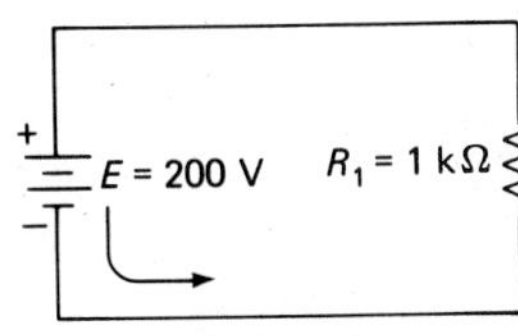

FIGURE 5-32

15. The total current in the circuit shown in Figure 5-33 equals which of the following?
(a) 5 mA
(b) 50 mA
(c) 25 mA
(d) 10 mA

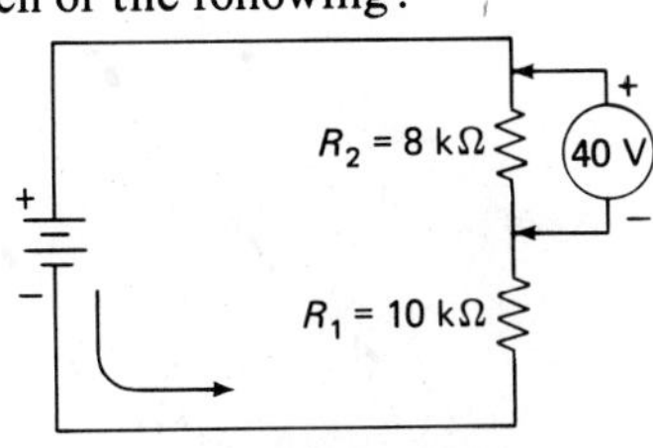

FIGURE 5-33

16. Solve for total voltage in the circuit shown in Figure 5-34.
    (a) 2 mV
    (b) 400 V
    (c) 300 V
    (d) 200 V

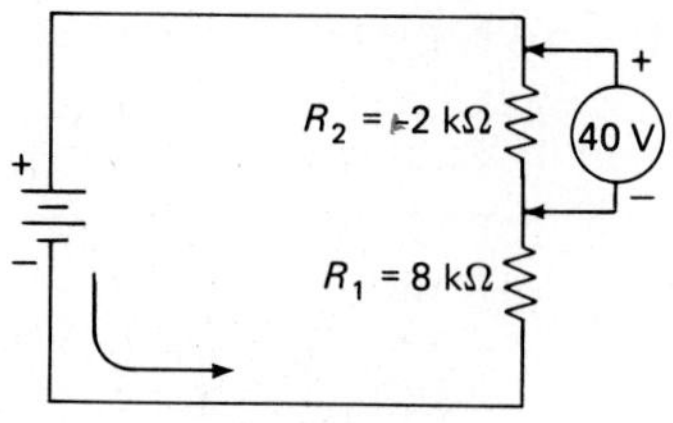

FIGURE 5-34

17. In the circuit shown in Figure 5-35, I_t = 25 mA, what does E_t equal?
    (a) 50 V
    (b) 12.5 V
    (c) 25 V
    (d) 100 V

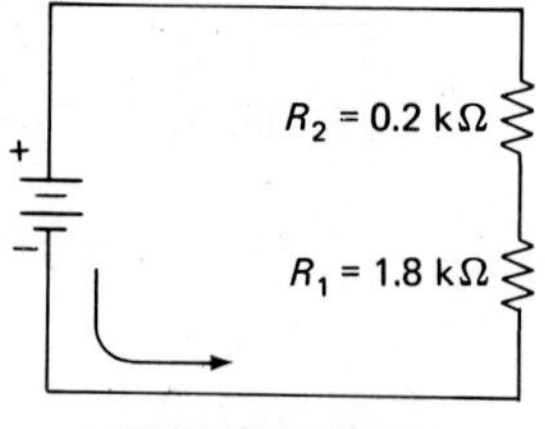

FIGURE 5-35

18. Refer to Figure 5-36 and solve for R_2.
    (a) 20 Ω
    (b) 20 kΩ
    (c) 12 Ω
    (d) 12 kΩ

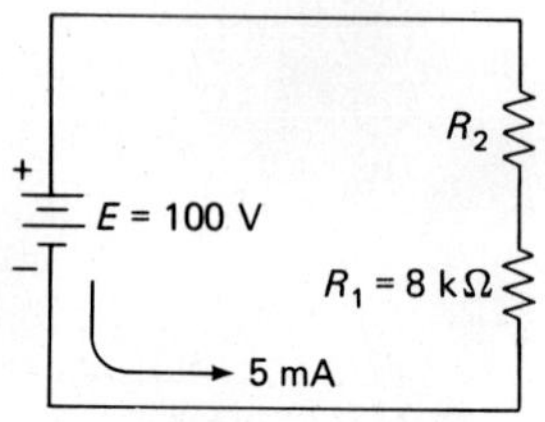

FIGURE 5-36

19. Refer to Figure 5-37 and solve for E_{R2}.
    (a) 75 V
    (b) 30 V
    (c) 45 V
    (d) 150 V

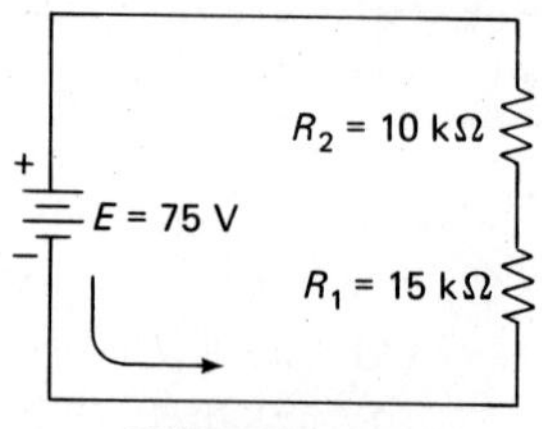

FIGURE 5-37

20. Refer to Figure 5-38 and solve the circuit for E_{R1}.
    (a) 24 V
    (b) 3 mA
    (c) 9 mA
    (d) 9 V

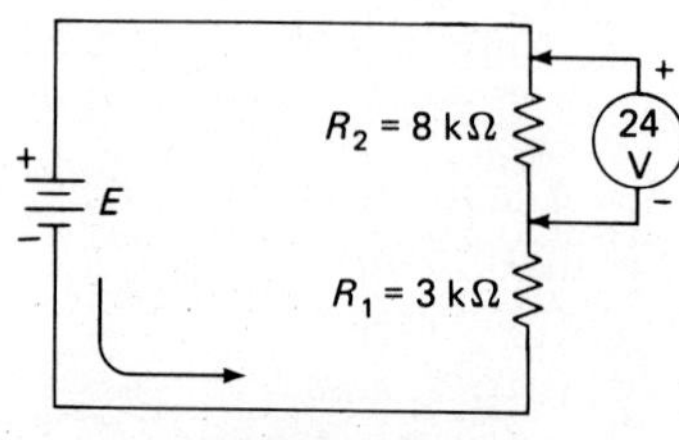

FIGURE 5-38

6

Series-Resistive Circuits

Introduction

In this chapter we continue the discussion of series-resistive circuits. Ohm's and Kirchhoff's laws continue to be of special interest. As the chapter develops, we take up the topic of power dissipation in a series-resistive circuit.

It is important to remember that power-dissipating devices are used to perform work in electrical circuits. These devices are resistive ones. Analysis of series-resistive circuits involves the distribution of voltage and current, along with their heat effect, because of the action of voltage and current on resistors in a closed loop.

Objectives

Each student is required to:

1. Analyze series-resistive circuits and determine total resistance, total current, total voltage, and individual voltage drops.
2. Define the terms *energy, power, work,* and *dyne* as they apply to electronic circuitry.
3. Write, in formula form, the basic rule for computing electrical power.
4. Use the power formula $P = I^2 \times R$ to derive eight alternate forms of the formula.
5. Apply these versions to analysis of series-resistive circuits for total power, and individual resistor power dissipations.
6. Analyze series-resistive circuits for resistance, voltage, current, and power values.
7. Demonstrate an ability to apply Ohm's law, Kirchhoff's laws, and power formulas to the analysis of circuits.
8. Troubleshoot series circuits for opens and shorts.
9. List the correct procedures used to connect voltmeters, ohmmeters, and ammeters in operating circuits.

10. Explain, in writing, the indications that can be expected when voltmeters, ohmeters, and ammeters are connected in open, or shorted, circuits.

DC and AC Comparisons

As has been discussed, two types of current are used in electronics, direct (DC) and alternating (AC). The analysis that results from the application of either type is identical. When total voltages (AC or DC) are equal, total current is the same, assuming total resistance is the same. Variations in applied voltage and total resistance affect total current equally, regardless of the type voltage applied. Whether the voltage applied is AC or DC, in the external circuit, electron current flow is from the negative pole of the power source, through the circuit, and back to the positive pole of the source.

All the rules developed in Chapter 5, regarding Ohm's and Kirchhoff's laws, are equally true in this chapter. Any differences that arise are discussed later in the book.

All analyses regarding resistors in DC circuits are equally true for resistors in AC circuits.

Self-Check

Complete each item by inserting the words and/or numbers needed to make a true and complete statement.

1. A circuit is operating with alternating current; it has electrons flowing from the __________ pole of the power source to the __________ pole.

2. In an AC-powered circuit, as voltage increases, total current __________.

3. In an AC-powered circuit, as stated by Ohm's law, current is directly proportional to __________.

4. Total resistance in an AC-powered circuit equals the __________ of all individual resistors.

5. Whether the circuit is powered by AC or DC, the circuits total __________ and total __________ are the same.

Electrical Power and Energy

Power is a term common throughout industry. In electronics, power is defined as the *rate of doing work*. *Work,* by whatever means, is defined as *the result of a force acting on a mass over a distance.*

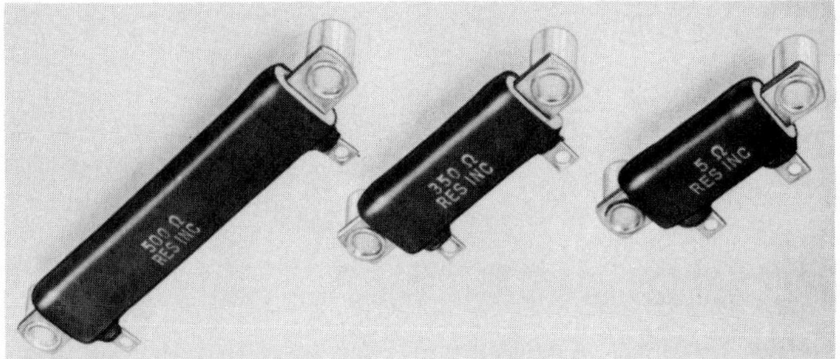

FIGURE 6-1
Low-Resistance Stackable Power Resistors (Photo from Resistors, Inc. Brochure)

In electrical circuitry, power involves force (voltage) acting on a mass (electrons) over a distance. The amount of time required to complete the work determines the amount of power expended. This relationship can be expressed as a formula:

$$P = \frac{W}{t} \qquad \begin{aligned} P &= \text{power, in watts} \\ W &= \text{work, in joules} \\ t &= \text{time, in seconds} \end{aligned}$$

Energy is the capacity to do work. In electrical circuits, electrical energy is converted into heat energy. The amount of heat energy used by an electrical circuit equals the amount of energy supplied by the electrical source.

Experimentation in electrical power was performed in 1843 by James Prescott Joule, an English physicist. Joule discovered a relationship between current, resistance, and voltage in the development of heat by an electrical circuit. Joule's law states:

> ***The amount of heat produced by a circuit element is directly proportional to resistance, the square of the current, and time.***

Expressed as a formula, Joule's law is:

$$\text{Work} = I^2 \times R \times t = 1 \text{ joule}$$

We have already established that:

$$P = \frac{\text{W}}{t}$$

Substituting Joule's law for work, we have:

$$P = \frac{(I^2 \times R \times t)}{t}$$

The two time (t) entries cancel, leaving the power formula:

$$P = I^2 \times R$$

Transposing this formula results in:

$$I^2 = \frac{P}{R} \quad \text{or} \quad I = \sqrt{\frac{P}{R}}$$

$$R = \frac{P}{I^2}$$

One form of Ohm's law states that $R = E/I$. When this form is substituted for R in the power formula, we have:

$$P = I^2 \times \left(\frac{E}{I}\right)$$

When the terms are combined, we have:

$$P = I \times E$$

This results in an alternate form of the power formula. Transposing this formula, we get the two formulas:

$$I = \frac{P}{E}$$

$$E = \frac{P}{I}$$

By substituting the Ohm's-law equivalent ($I = E/R$) for I in the formula $P = I \times E$, we have:

$$P = \left(\frac{E}{R}\right) \times E$$

Combining terms, we get:

$$P = \frac{E^2}{R}$$

Having developed this formula, we have three forms of the power formula.

By transposing the terms included in the third form of the power formula, we get:

$$E^2 = P \times R \quad \text{or} \quad E = \sqrt{P \times R}$$

$$R = \frac{E^2}{P}$$

Nine versions of the power formula have now been introduced which are useful in the analysis of circuits. To be a capable technician, you must first know the power formulas, then be able to transpose them. Power is an important part of electronics; the more effort you put into learning and using these formulas, the more proficient you become.

The practical unit used in measuring electrical power is the *watt,* noted in the circuit by the letter W.

For 1 watt of power to be dissipated, 1 volt of pressure is exerted on 1 ohm of resistance, which causes 1 ampere of current to flow.

The watt represents the rate at which work is done. In the electric meter used by commercial power companies, power is measured in kilowatts per hour (kWh), which means that 1000 W of power is used continuously for 1 hour.

The circuit in Figure 6-2 is used to illustrate the effect produced when 5 V causes 5 A to flow

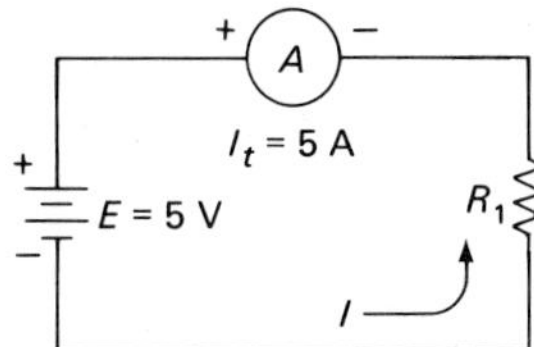

FIGURE 6-2

From the circuit we get the values $E_t = 5$ V and $I_t = 5$ A. With these values given, we are asked to solve for $P_t = ?$.

A review of the nine versions of the power formula reveals that $P = I \times E$ fits the need:

Mathematical Solution	*Action*	*Entry*	*Display*
$P_t = I_t \times E_t$	ENTER	5	5
$P_t = 5\text{ A} \times 5\text{ V}$	PRESS	×	5
$P_t = 25\text{ W}$	ENTER	5	5
	PRESS	=	25

Let us consider another form of the power formula. Refer to Figure 6-3.

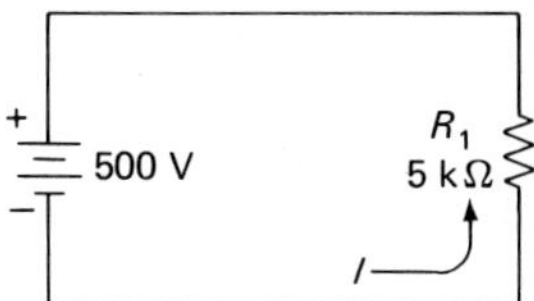

FIGURE 6-3

Here we are given $E_t = 500$ V and $R_t = 5$ kΩ and wish to solve $P_t = ?$. The procedure we follow is:

Mathematical Solution	*Action*	*Entry*	*Display*
$P_t = \dfrac{(E_t)^2}{R_t}$	ENTER	500	500
$P_t = \dfrac{(500)^2}{5\text{ k}\Omega}$	PRESS	X^2	250,000
$P_t = \dfrac{250{,}000}{5\text{ k}\Omega}$	PRESS	÷	250,000
$P_t = 50\text{ W}$	ENTER	5,EE,3	5 03
	PRESS	=	5 01

Converting 5 01 to a prefix yields 50 W.

Use Figure 6-4 to apply another version of the power formula:

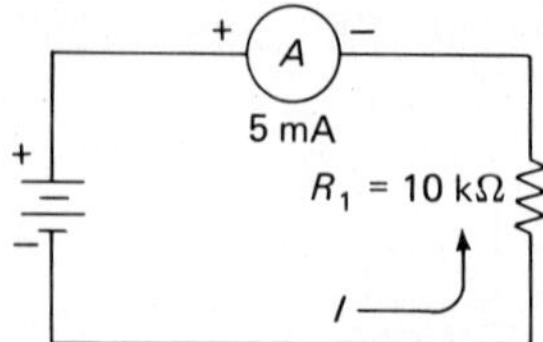

FIGURE 6-4

In the circuit in Figure 6-4, we are given $I_t = 5$ mA and $R_1 = 10$ kΩ. Solve $P_{R1} = ?$.

Mathematical Solution	*Action*	*Entry*	*Display*
$P_{R1} = I^2 \times R_1$	ENTER	5,EE,+/−,3	5 − 03
$P_{R1} = 5\text{ mA}^2 \times 10\text{k }\Omega$	PRESS	X^2	2.5 − 05
$P = 0.25$ W or 250 mW	PRESS	×	2.5 − 05
	ENTER	10,EE,3	10 03
	PRESS	=	2.5 − 01

When converted, 2.5 − 01 = 0.25 W, or 250 mW. This illustrates that the three versions of the power formula can be used to solve for electrical power; alternate forms of the power formula can be used to solve for other unknown quantities. For example, if power and one other quantity are known, a third quantity can be solved, using one of the formulas developed in this section.

In Figure 6-5, note that power and resistance are given.

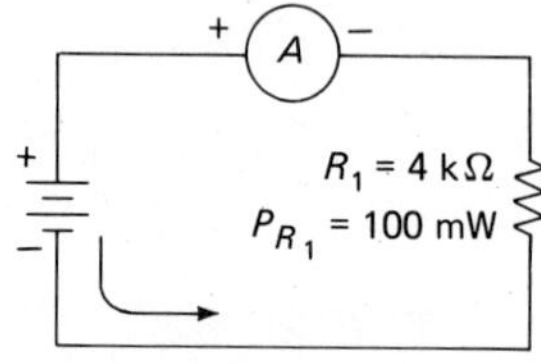

FIGURE 6-5

Two alternate versions of the power formula are available, which use P and R as their known quantities. Let's take a look at them.

Mathematical Solution	*Action*	*Entry*	*Display*
$E = \sqrt{P \times R}$	ENTER	100,EE,+/−,3	100 − 03
$E = \sqrt{100\text{ mW} \times 4\text{ k}\Omega}$	PRESS	×	1 − 01
$E = 20$ V	ENTER	4,EE,3	4 03
	PRESS	=	4 02
	PRESS	$\sqrt{X}$	2 01

When converted, 2 01 = 20 V.
To solve for I:

Mathematical Solution	Action	Entry	Display
$I = \sqrt{\frac{P}{R}}$	ENTER	100,EE,+/−,3	100 − 03
$I = \sqrt{\frac{100\text{ mW}}{4\text{ k}\Omega}}$	PRESS	÷	1 − 01
$I = 5$ mA or 0.005 A	ENTER	4,EE,3	4 03
	PRESS	=	2.5 − 05
	PRESS	√X	5 − 03

When converted, 5 03 = 5 mA.

By transposing Ohm's law, Kirchhoff's laws, and the power formulas, we have a selection of formulas that make solving series-resistive circuits quite easy. Remember: If two quantities of any one component are known, the third (unknown) quantity can be calculated.

To help you learn the formulas associated with Ohm's law and power, Figure 6-6 has been included. The twelve formulas already discussed, relating to these circuits, are represented in Figure 6-6. Note that the hub of the wheel consists of the designators power (P), voltage (E), current (I), and resistance (R). The quarter circle that fans out from each letter provides the three formulas available for solving that quantity when two quantities (watts, volts, amps, or ohms) are known.

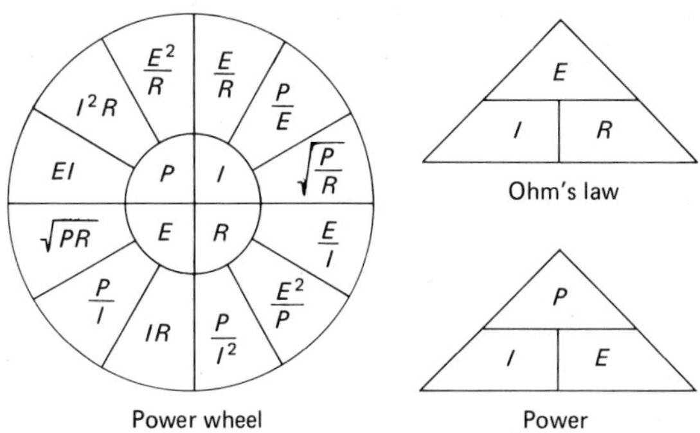

FIGURE 6-6

Use the wheel in Figure 6-6 in learning the formulas, as well as the way the quantities are interrelated. The bottom portion of the figure is devoted to Ohm's law and power formulas. Each formula is displayed so that when an unknown quantity is covered, the two quantities remaining are shown in the proper position to solve for the unknown. If E is covered in the Ohm's-law triangle, I and E are portrayed exactly as they would appear in the formula $E = I \times R$. With R covered, the formula is E/I; with I covered, the formula is E/R.

We have discussed the watt and the milliwatt. In industrial operations it is common to work in kilowatts (kW; 1000 watts) and even megawatts (MW; 1 million watts). Power is the rate of doing work or consuming energy; therefore, the length of time energy is used is a truer measure of the amount used than is the watt. It is common practice to purchase electricity in watt/hours. Because the watt is such a small unit of power, the amount of power purchased for a house is usually stated in kilowatt/hours. When 1000 W of power is consumed over a period of 1 hour, 1 kilowatt/hour (kWh) has been used.

An example is a 100-W light bulb that operates continuously for 10 hours. The light bulb consumes 1 kWh of power. If the average cost of electrical power is 5 cents per kWh, it would cost 0.5 cent per hour to operate the lamp.

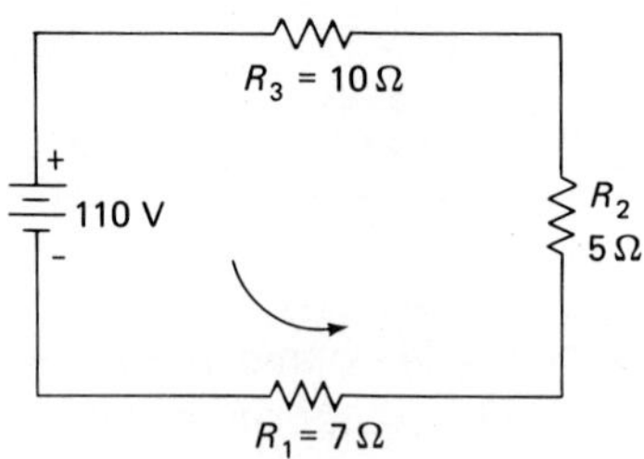

FIGURE 6-7

The circuit shown in Figure 6-7 contains three resistors, with their values and E_t given. Use these data to solve for R_t, I_t, E_{R1}, E_{R2}, E_{R3}, P_{R1}, P_{R2}, P_{R3}, and P_t. Math solutions are presented; calculator solutions are found in the appropriate section.

Solve for R_t:

$$R_t = R_1 + R_2 + R_3$$

$$R_t = 7\ \Omega + 5\ \Omega + 10\ \Omega$$

$$R_t = 22\ \Omega$$

Solve for I_t:

$$I_t = \frac{E_t}{R_t}$$

$$I_t = \frac{110\ \text{V}}{22\ \Omega}$$

$$I_t = 5\ \text{A}$$

Because $I_t = I_{R1} = I_{R2} = I_{R3}$, all currents equal 5 A.
Solve for E_{R1}:

$$E_{R1} = I_t \times R_1$$

$$E_{R1} = 5\ \text{A} \times 7\ \Omega$$

$$E_{R1} = 35\ \text{V}$$

Solve for E_{R2}:

$$E_{R2} = I_t \times R_2$$

$$E_{R2} = 5\ \text{A} \times 5\ \Omega$$

$$E_{R2} = 25\ \text{V}$$

Solve for E_{R3}:

$$E_{R3} = I_t \times R_3$$

$$E_{R3} = 5\ \text{A} \times 10\ \Omega$$

$$E_{R3} = 50\ \text{V}$$

Solve for P_{R1}:

$$P_{R1} = I_t \times E_{R1}$$

$$P_{R1} = 5\ \text{A} \times 35\ \text{V}$$

$$P_{R1} = 175\ \text{W}$$

Solve for P_{R2}:

$$P_{R2} = I_t \times E_{R2}$$

$$P_{R2} = 5\ \text{A} \times 25\ \text{V}$$

$$P_{R2} = 125\ \text{W}$$

Solve for P_{R3}:

$$P_{R3} = I_t \times E_{R3}$$

$$P_{R3} = 5\ \text{A} \times 50\ \text{V}$$

$$P_{R3} = 250\ \text{W}$$

Before solving for total power, P_t, let us discuss circuit characteristics. The rule for total power is:

Total power consumed by a circuit equals the sum of all its individual power dissipations.

This can also be stated thus:

$$P_t = P_{R1} + P_{R2} + P_{R3}$$

Still other formulas are:

$$P_t = I_t \times E_t$$

$$P_t = I_t^2 \times R_t$$

$$P_t = \frac{(E_t)^2}{R_t}$$

Solve for P_t:

$$P_t = P_{R1} + P_{R2} + P_{R3}$$

$$P_t = 175\ \text{W} + 125\ \text{W} + 250\ \text{W}$$

$$P_t = 550\ \text{W}$$

or

$$P_t = I_t \times E_t$$

$$P_t = 5\ \text{A} \times 110\ \text{V}$$

$$P_t = 550\ \text{W}$$

To determine the number of kilowatt/hours this circuit dissipates in 24 h, use the formula:

$$\text{kWh} = \frac{\text{watts } (W) \times \text{time } (t)}{1000}$$

$$\text{kWh} = \frac{550\ \text{W} \times 24\ \text{hours } (h)}{1000}$$

$$\text{kWh} = \frac{13{,}200}{1000}$$

$$\text{kWh} = 13.2$$

To convert your result to the cost of operating the circuit for one 24-h period, multiply the number of kilowatt/hours by the cost for 1 kWh. If the average cost of power is 5 cents per kWh, then:

$$\text{Cost} = \text{number kWh} \times \text{price per unit}$$

$$\text{Cost} = 13.2 \times \$0.05$$

$$\text{Cost} = \$0.66 \text{ or } 66 \text{ cents}$$

Electrical components are often rated according to the amount of power they can safely dissipate in the form of heat. This rating is referred to as the *power rating,* which is defined as the

Maximum power that the device can dissipate without damage to itself.

Resistors have power ratings. If a resistor is connected in a circuit which requires the resistor to dissipate more power than it can handle, the resistor will be destroyed.

In some cases, the term *watt* is used to state the amount of heat or light a device can produce. The resistor is assigned a power rating in watts. A resistor can be operated at any combination of voltage and current as long as its power rating is not exceeded. In most circuits the actual power dissipation expected of the resistor is much lower than its rating. It is common practice to select a resistor with a 100-percent safety factor for use in a circuit. For example, if the resistor dissipates a half watt, insert a resistor rated at 1 W.

Resistors of the same ohmic value are available in several different wattage ratings. The most common ratings are 1/8, 1/4, 1/2, 1, and 2 W. These are usually for fixed resistors. To increase a fixed resistor's ability to dissipate power, its physical size is increased. The larger size gives the resistor more surface area from which to radiate heat. Figure 6-8 shows the relative shapes and sizes of fixed resistors with different wattage ratings.

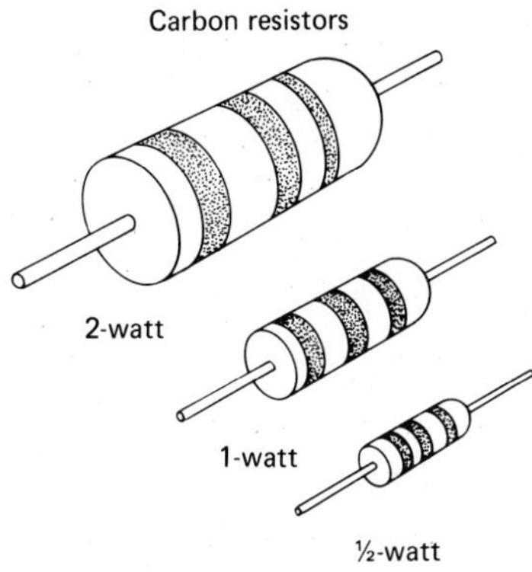

FIGURE 6-8

Resistors of more than 2 W are usually wirewound. Stock sizes are 5 to 200 W. Special types that operate above 200 W are available on special order, but the likelihood of encountering resistors of this size is remote.

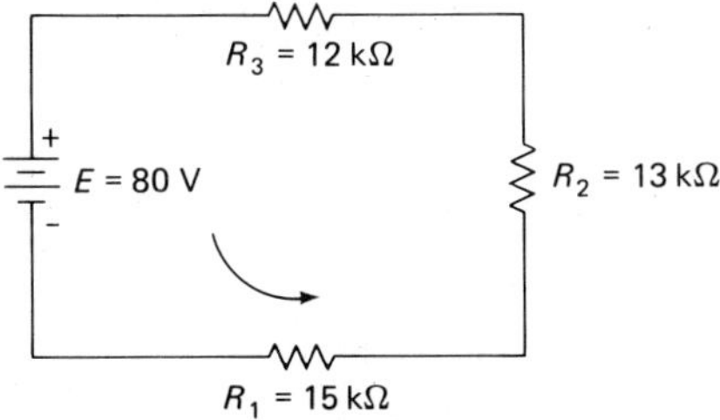

FIGURE 6-9

Examine Figure 6-9; we will use the circuit in that figure to solve for P_t, P_{R1}, P_{R2}, and P_{R3}.

Mathematical Solution	*Action*	*Entry*	*Display*
$R_t = R_1 + R_2 + R_3$	ENTER	15,EE,3	15 03
$R_t = 15k + 13k + 12k$	PRESS	+	1.5 04
$R_t = 40\ k\Omega$	ENTER	13,EE,3	13 03
	PRESS	+	2.8 04
	ENTER	12,EE,3	12 03
	PRESS	=	4 04

When converted to a prefix, 4 04 equals 40 kΩ.
Now to find I_t!

Mathematical Solution	*Action*	*Entry*	*Display*
$I_t = \frac{E_t}{R_t}$	ENTER	80	80
$I_t = \frac{80\text{ V}}{40\text{k}}$	PRESS	÷	80
$I_t = 2$ mA	ENTER	40,EE,3	40 03
	PRESS	=	2 − 03

Converted to a prefix, 2 − 03 equals 2 mA.
and P_{R1}!

Mathematical Solution	*Action*	*Entry*	*Display*
$P_{R1} = [I_t]^2 \times R_1$	ENTER	2,EE,+/−,3	2 − 03
$P_{R1} = [2\text{ mA}]^2 \times 15\text{k}\Omega$	PRESS	X^2	4 − 06
$P_{R1} = 60$ mW	PRESS	×	4 − 06
	ENTER	15,EE,3	15 03
	PRESS	=	6 − 02

Converted to a prefix, 6 − 02 equals 60 mW.
and P_{R2}!

Mathematical Solution	*Action*	*Entry*	*Display*
$P_{R2} = [I_t]^2 \times R_2$	ENTER	2,EE,+/−,3	2 − 03
$P_{R2} = [2\text{ mA}]^2 \times 13\text{k}\Omega$	PRESS	X^2	4 − 06
$P_{R2} = 52$ mW	PRESS	×	4 − 06
	ENTER	13,EE,3	13 03
	PRESS	=	5.2 − 02

Converted to a prefix, 5.2 − 02 equals 52 mW.
also P_{R3}!

Mathematical Solution	*Action*	*Entry*	*Display*
$P_{R3} = [I_t]^2 \times R_3$	ENTER	2,EE,+/−,3	2 − 03
$P_{R3} = [2\text{ mA}]^2 \times 12\text{k}\Omega$	PRESS	X^2	4 − 06
$P_{R3} = 48$ mW	PRESS	×	4 − 06
	ENTER	12,EE,3	12 03
	PRESS	=	4.8 − 02

Converted to a prefix, 4.8 − 02 equals 48 mW.
and P_t!

Mathematical Solution	*Action*	*Entry*	*Display*
$P_t = I_t \times E_t$	ENTER	2,EE,+/−,3	2 − 03
$P_t = 2\text{ mA} \times 80\text{ V}$	PRESS	×	2 − 03
$P_t = 160\text{ mW}$	ENTER	80	80
	PRESS	=	1.6 − 01

Converted to a prefix, 1.6 − 01 equals 160 mW.
This can be proved:

$$P_t = P_{R1} + P_{R2} + P_{R3}$$

$$P_t = 60\text{ mW} + 52\text{ mW} + 48\text{ mW}$$

$$P_t = 160\text{ mW}$$

Self-Check

Answer each item by indicating whether the statement is true or false.

6. ___ Power is the ability to do work.

7. ___ The unit of measure for electrical power is the watt.

8. ___ Power sold as 1000 *watts per hour* is measured in Wh.

9. ___ The power formula can be stated in three basic forms.

10. ___ The term *power rating* refers to the ability of a component to dissipate power.

Troubleshooting Series-Resistive Circuits

The term *troubleshooting* includes all actions taken to identify, locate, and repair an electrical circuit so the circuit can be put back in operation. Our initial venture in troubleshooting involves the study of series-resistive circuits. Most troubles that occur in electronic circuits result from one of two problems, opens and shorts.

> ***A word of caution should be inserted at this point. In rare instances, resistors change resistance value without shorting or opening. In these cases, finding the trouble can be frustrating. You should realize that this condition can, and does, exist. Opens and shorts are not the only problems, although they probably cause more than 95% of all failures.***

To ensure that we all start at the same point, let us define some terminology associated with troubleshooting.

open circuit.
Extremely high resistance. It reflects the loss of continuity in a circuit: the path for current flow is not complete, the wire is broken, a switch is open, a component has broken in two. The path for current flow is open, that is, broken. A component may have burned open. (An example is a burned-out light bulb.) An ohmmeter reads infinity when connected across an open.

short circuit.
Extremely low resistance. The ohmmeter reads zero ohms when connected across a short. A short circuit is a circuit in which a defect causes current to return to the source without having passed through all of the resistance. The short can cause the circuit to stop working because the heavy current flow that results can melt a conductor or component, causing it to break in two. A short results in a circuit breaker opening or a fuse blowing.

continuity.
The state of being continuous, or connected, a circuit that is not open. A complete path exists for electrons to flow from the negative pole of the battery, pass through the circuit, and return to the positive pole of the battery.

symptoms.
Whatever can be observed that tells you something is wrong in a circuit—a light that doesn't glow or the motor that doesn't turn.

Figure 6-10 contains examples of opens. Note that loose connections can occur anywhere in the circuit and still affect current flow. Resistors burn open when their power rating is exceeded and they cannot dissipate the heat generated. Fuses and lamps burn out because of excess current; broken conductors result from physical damage and vibration. Whatever the cause,

An open circuit has lost continuity and no current will flow through it.

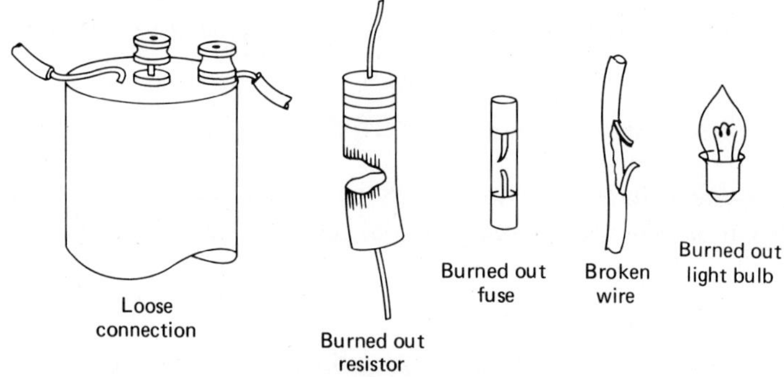

FIGURE 6-10

In actual experience most troubles cannot be located visually. This refers only to those already discussed. In cases where opens are not readily visible, a voltmeter, ammeter, or ohmmeter must be used. If the circuit were normal, the lamp in Figure 6-11 would glow when the switch is closed. If the circuit is open, as shown in Figure 6-11a, the symptom is easily visible; the lamp does not glow when the switch is closed. If you replace the bulb, and the light still does not glow, the bulb is not the

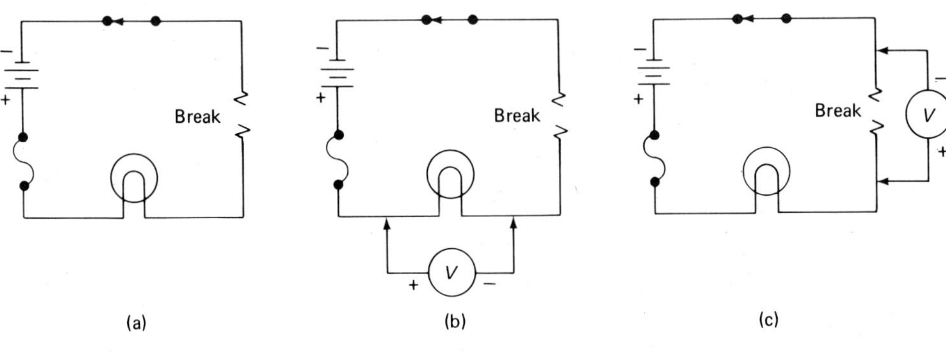

FIGURE 6-11

problem; a trouble is located in some part of the circuit other than the bulb. By connecting the voltmeter across the lamp (Figure 6-11b), we find that zero volts is applied to the bulb regardless of the switch position. Had the lamp been open, by connecting the voltmeter across the lamp, we would have found that the total voltage was dropped across the lamp when the switch was closed. With the lamp open, the voltmeter completes the circuit's continuity, and current can flow through the meter. Voltmeters have extremely high internal resistance. This resistance is connected in series with the circuit, which causes the meter to read approximately total voltage. When the voltmeter is connected across an open resistor, as shown in Figure 6-11c, and the switch is closed, the meter indication is approximately equal to the voltage applied. This occurs because the meter completes a path around the open, with the same results. This leads to a point that must be remembered:

> ***A voltmeter connected across an open component indicates an amount approximately equal to the voltage applied to that portion of the circuit.***

For an ammeter to be used in checking the circuit in Figure 6-12, the circuit must be broken (Figure 6-12a).

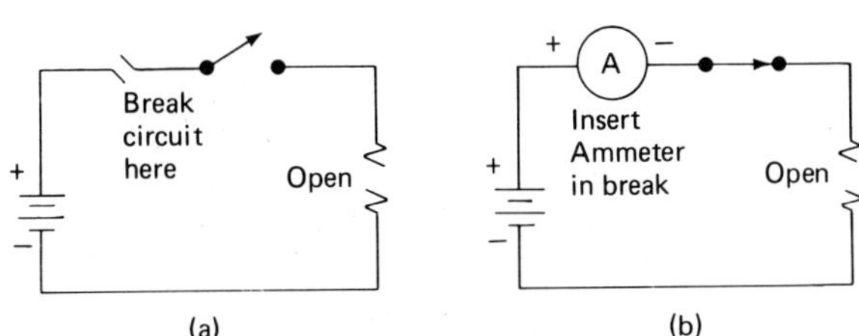

FIGURE 6-12

The point at which the circuit is broken is the point where the ammeter must be connected; the problem does not lie at that point. Figure 6-12b illustrates connection of the ammeter in the circuit. Remember—the ammeter is always connected in series with the circuit. Regardless of where the ammeter is connected in an open circuit, it will read zero when the switch is closed. Difficulty in connecting and disconnecting the ammeter, plus the fact that electrons are not flowing (the light does not glow), make the ammeter check one that is seldom used. The point to remember about using an ammeter is:

When connected into an open circuit, the ammeter will read zero.

An ohmmeter can be used to locate the open component, but you must remember to isolate the component to be checked from the power source. In Figure 6-13 it is possible to open the switch and isolate the resistor, but in other circuits you may have to physically disconnect the component from the rest of the circuit.

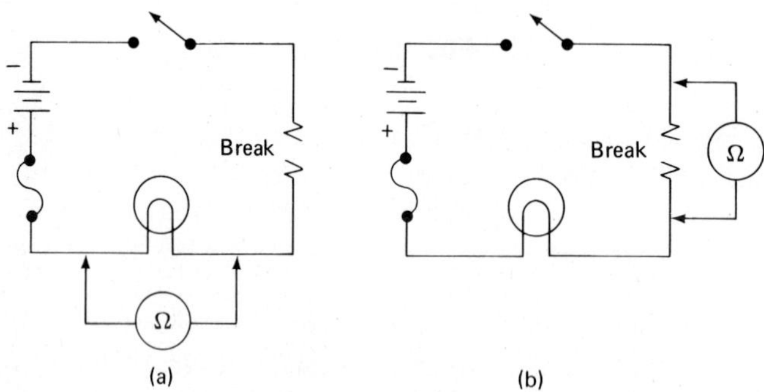

FIGURE 6-13

When the ohmmeter is connected across the lamp, as shown in Figure 6-13a, the meter reads the resistance of the lamp, because there is a complete path for the meter's current to flow through the lamp's resistance. When the ohmmeter is connected across the resistor (Figure 6-13b), no current can flow through the break; in this case, the ohmmeter will read infinity. Two more points have been developed which must be remembered:

An ohmmeter connected across a good component will read the ohmic value of the component being tested.

An ohmmeter connected across an open component will read infinite resistance.

From this discussion you should know the effects of opens in a circuit, the methods used to locate the opens, and the symptoms caused by the opens.

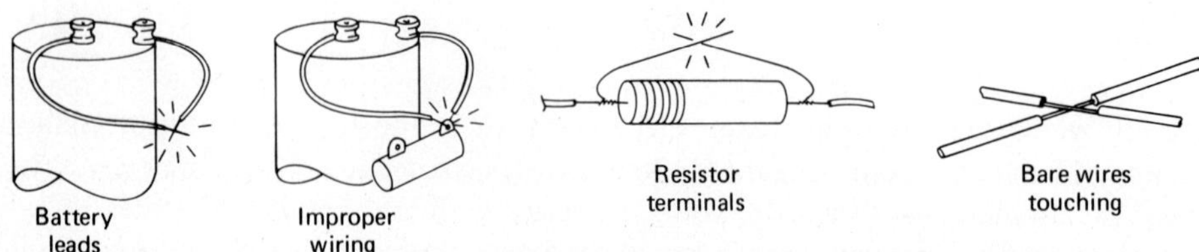

FIGURE 6-14

Let us now discuss another common problem with electronic equipment—the short. A short is the opposite of an open. In a shorted circuit, the flow of current is higher than normal and will remain that way until either the defect is fixed or the extra current causes a device to open. Some examples appear in Figure 6-14.

In all cases the trouble occurs because conductors make contact, which allows the current to bypass one or more resistive devices. When two wires short, or touch, their resistance is very low. Most of the current follows the path of least resistance and will flow through the short (Figure 6-15).

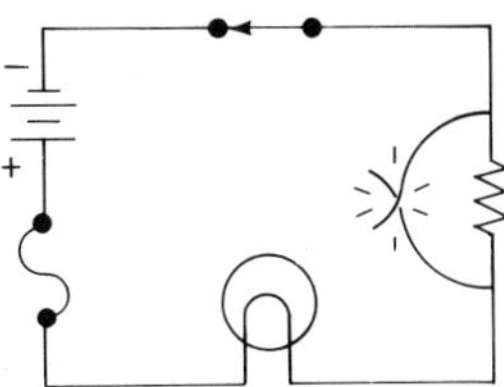

FIGURE 6-15

This circuit is designed to operate a lamp. A variable resistor is placed in series with the lamp and the battery. The resistor reduces the amount of current that can flow through the lamp. (An example is the dimmer switch for the dashboard lights of a car.) If the resistor shorts, its resistance immediately drops to zero ohms; but there is still a path for current to flow. This allows more current than normal to flow through the lamp. The symptom is this: The lamp glows much brighter than normal, and the dimmer has no control. In most cases, the bright light is only temporary; the higher current causes the lamp to burn open. This is what happens when a light goes out at home; there is a bright flash, the filament burns open, and the lamp becomes dark.

You have already been introduced to the fuse. The purpose of a fuse in a circuit is to prevent damage. Suppose the maximum current the lamp can conduct is 1.25 A; in that case, a 1 A fuse in the circuit would open before the excess current could damage the lamp.

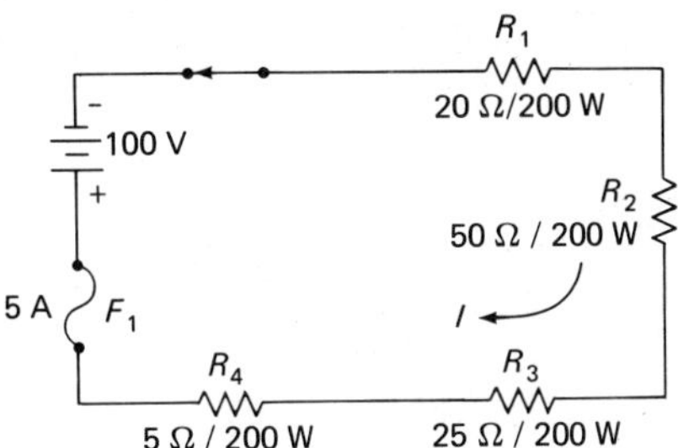

FIGURE 6-16

A circuit more representative of what can happen is shown in Figure 6-16. There are four resistors connected in series, each with a power rating of 200 W. In this circuit it is possible for a resistor to become shorted and not blow the fuse, or to exceed the power rating of the other resistors. We analyze the circuit to see what happens as various troubles are imposed on the circuit.

For us to recognize a symptom, we must know what conditions existed when everything was normal. If these resistors represented 200 W table lamps, the normal condition would be visible: Each light would shine. We have already established that when one light in a series circuit is open, all lights stop glowing. One lamp shorting does not prevent current from flowing, since there is a path (short) around it. To illustrate what can happen, we first establish the normal condition of the circuit.

Solve for R_t:

Mathematical Solution	*Action*	*Entry*	*Display*
$R_t = R_1 + R_2 + R_3 + R_4$	ENTER	20	20
$R_t = 20\ \Omega + 50\ \Omega + 25\ \Omega$	PRESS	+	20
$+ 5\ \Omega$			
$R_t = 100\ \Omega$	ENTER	50	50
	PRESS	+	70
	ENTER	25	25
	PRESS	+	95
	ENTER	5	5
	PRESS	=	100

Solve for I_t:

Mathematical Solution	*Action*	*Entry*	*Display*
$I_t = \dfrac{E_t}{R_t}$	ENTER	100	100
$I_t = \dfrac{100\ \text{V}}{100\Omega}$	PRESS	÷	100
$I_t = 1\ \text{A}$	ENTER	100	100
	PRESS	=	1

Summarizing the circuit's normal indications, we have:

$$
\begin{array}{rlrl}
E_t &= 100\ \text{V} & P_{R2} &= 50\ \text{W} \\
R_t &= 100\ \Omega & E_{R3} &= 25\ \text{V} \\
I_t &= 1\ \text{A} & P_{R3} &= 25\ \text{W} \\
E_{R1} &= 20\ \text{V} & E_{R4} &= 5\ \text{V} \\
P_{R1} &= 20\ \text{W} & P_{R4} &= 5\ \text{W} \\
E_{R2} &= 50\ \text{V} & P_t &= 100\ \text{W}
\end{array}
$$

When the circuit is operating normally, the power dissipated by each resistor is much less than its 200-W rating. What would happen if R_2 were to short? Figure 6-17a illustrates the conditions that exist when this happens.

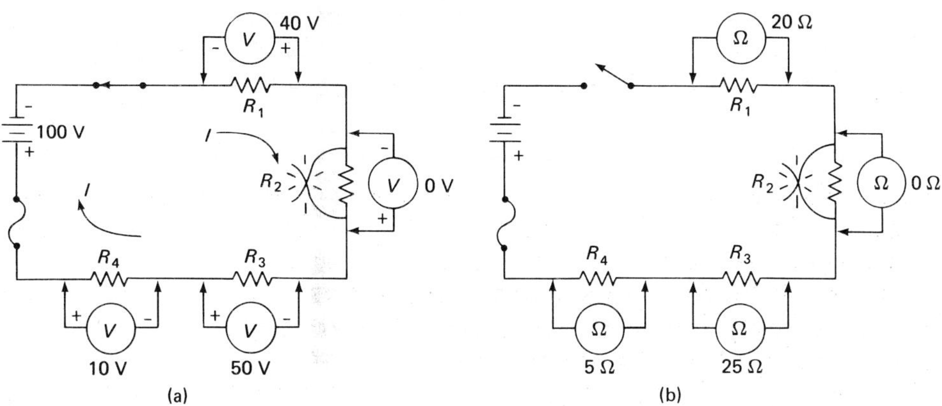

FIGURE 6-17

Mathematical and calculator solutions for problems of this type have been presented. By way of illustrating the conditions that exist within the circuit when R_2 is shorted, the results are stated below; use a calculator to confirm the correctness of the data. Because R_2 is shorted, it now has zero resistance. Therefore:

$R_t = 50\ \Omega$	$P_t = 200$ W
$I_t = 2$ A	$P_{R1} = 80$ W
$E_{R1} = 40$ V	$P_{R2} = 0$ W
$E_{R2} = 0$ V	$P_{R3} = 100$ W
$E_{R3} = 50$ V	$P_{R4} = 20$ W
$E_{R4} = 10$ V	

By comparing these values to the normal values already given, we can see that each resistor dissipates more power than before. Total resistance has decreased, total current has increased, the voltage drop on each resistor has increased, and total power dissipation has increased. Even though $I_t = 2$ A, current is still less than the fuse's rating. The three lights remaining glow brighter than normal, but the fuse has not blown. These symptoms tell us that current is too high, but visual inspection fails to locate the problem.

Now is the time to rely on test equipment for assistance in finding the trouble. Figure 6-17a represents this circuit, with voltmeters connected across each resistor; E_{R1}, E_{R3}, and E_{R4} are higher than normal, and E_{R2} is zero, which is less than normal. From these indications, we can state a rule for using the voltmeter to detect shorted components:

When a voltmeter is connected across a good component in a series circuit, which contains a short, the meter will indicate higher than normal voltage.

When the voltmeter is connected across a shorted component in a series circuit, the meter will indicate zero volts.

Figure 6-17b illustrates the use of ohmmeters in checking this circuit. Notice that the switch has been opened to isolate the ohmmeters from the battery voltage. You would expect readings across each resistor, as is noted in the drawing. Analysis of these indications tells us:

When an ohmmeter is connected across a good component in a series circuit, it indicates that component's ohmic value. When connected across a shorted component, the ohmmeter indicates zero ohms.

Again, because it is difficult to insert the ammeter in the circuit, use of the ammeter is limited. Should an ammeter be connected, however, it will indicate 2 A (twice the normal value) and can be of value in locating the defective component. Often, in cases where maintenance of a level current is necessary, an ammeter is designed into the circuit; this allows continuous monitoring of the current, as well as a quick reference when trouble is suspected.

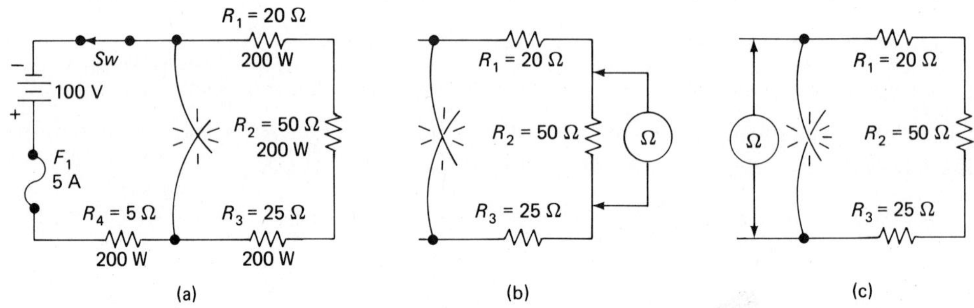

FIGURE 6-18

Suppose the short is positioned such that R_1, R_2, and R_3 are shorted (Figure 6-18a). That leaves R_4 as the only resistor in the circuit. With R_4 the only resistor not shorted, circuit conditions are:

$$R_t = R_4 = 5\ \Omega \qquad E_{R4} = 100\text{ V}$$

$$I_t = 20\text{ A} \qquad P_{R4} = 2000\text{ W}$$

If the circuit contained no fuse, the resistor would overheat and be destroyed. The metal filament in the 5-A fuse becomes so hot it melts in two as soon as the current exceeds 5 A. The fuse's opening stops current flow and prevents damage to the rest of the circuit. Now we have two troubles: an open fuse and a short.

These procedures are now used to pinpoint the trouble. Replacing the fuse merely results in "blowing" it again; therefore, the ohmmeter must be used. When the ohmmeter is connected to read R_1, R_2, or R_3, it has a parallel path similar to that shown in Figure 6-18b, depending on which resistor it is connected across. This procedure results in an erroneous reading that is less than the reading of the resistor the meter is connected across.

When the ohmmeter is connected across R_4, the indication will be the ohmic value of R_4, but when the ohmmeter is connected from the left side of R_1 to the left side of R_3, as shown in Figure 6-18c, the meter indicates zero ohms. This indicates that the short occurs from one side of R_1 to the opposite side of R_3 and that all the ohmmeter's current is flowing through the short. When the short is removed, and the fuse is replaced, the circuit returns to normal operation.

Of the several rules that have been established in this discussion, two are especially important in successful troubleshooting:

When an open occurs in any circuit, total current decreases.

When a short occurs in any circuit, total current increases.

Other rules are:

A voltmeter connected across a short indicates zero volts.

A voltmeter connected across an open indicates an amount approximately equal to the total applied voltage.

An ohmmeter connected across an open component indicates infinite ohms.

An ohmmeter connected across a short indicates zero ohms.

An ohmmeter connected across a good component indicates that component's resistance value.

An ammeter connected in an open circuit indicates zero amperes.

An ammeter connected in a circuit containing a short indicates current higher than normal.

Self-Check

Complete each item by inserting the words and/or numbers needed to make a true and complete statement.

11. The term ______________ denotes the process of identifying, locating, and repairing a defective electrical circuit.

12. An open component indicates ______________ Ω on an ohmmeter.

13. A voltmeter must be connected in ______________ with the component being checked.

14. A shorted component drops ______________ V.

15. An ______________ must be connected in series with the component being checked.

16. The ______________ has its own power source.

17. If a circuit contains an open, its total current will equal ______________ A.

Summary

1. Rules for series resistive circuits:
   (a) Total voltage in a series circuit is equal to the product of current and resistance.
   (b) Circuit current is equal at all points within a series circuit.
   (c) Total resistance in a series circuit is equal to the sum of all individual resistances.
   (d) The sum of all voltage drops in a closed loop equals the applied voltage for that portion of the circuit.
   (e) The voltage drop across a resistor in a series circuit is directly proportional to the resistor's ohmic value.
   (f) Total power dissipated in a circuit is equal to the sum of all the individual dissipations.
   (g) In a series circuit containing two or more resistors, the largest resistor will drop the largest voltage.
   (h) In a series circuit containing two or more resistors, the largest resistor will dissipate the most power.
2. Rules for troubleshooting series-resistive circuits:
   (a) A voltmeter connected across an open component indicates the applied voltage of that part of the circuit. Across a shorted component, the voltmeter will indicate 0 V.
   (b) An ammeter connected in an open circuit indicates zero current. In a circuit that contains a shorted component, the ammeter reads higher than normal.
   (c) Using an ohmmeter for troubleshooting yields the following indications: (1) infinity when connected across an open; (2) zero ohms when connected across a short; and (3) normal ohmic value when connected across a good component.
3. A blown fuse tells the technician that something is wrong in the circuit; which caused too much current to flow. When replacing a fuse, care should be taken to make sure the replacement has the same current and voltage ratings as the one being replaced. A fuse should never be replaced with one having a higher current rating or a lower voltage rating.

Review Questions and/or Problems

1. Solve for E_t in the circuit shown in Figure 6-19.
   (a) 50 V
   (b) 100 V
   (c) 200 V
   (d) 400 V

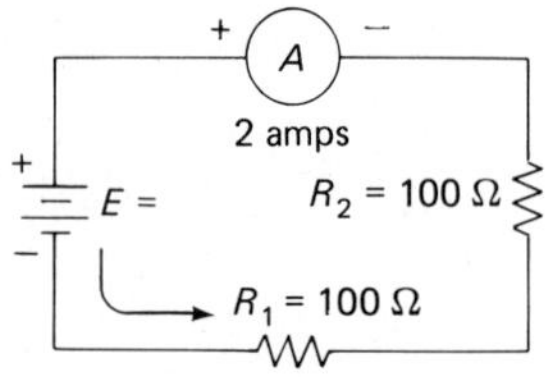

FIGURE 6-19

2. Solve for R_2 in the circuit shown in Figure 6-20.
   (a) 150 Ω
   (b) 15 Ω
   (c) 20 Ω
   (d) 1500 Ω

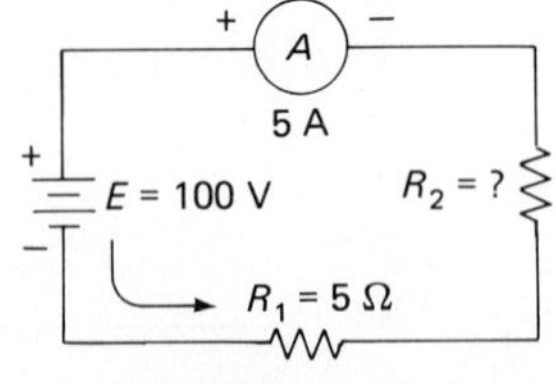

FIGURE 6-20

3. Solve for I_t in the circuit in Figure 6-21.
(a) 4 A
(b) 4000 A
(c) 4 mA
(d) 4000 mA

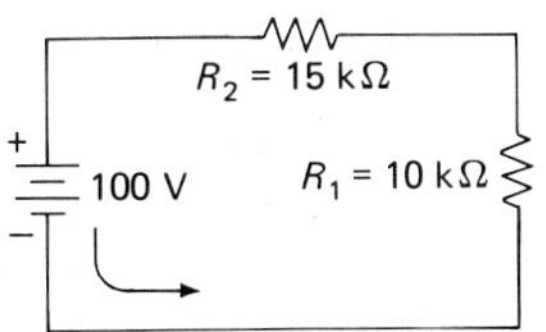

FIGURE 6-21

4. Which of the following is *not* a form of the power formula?
(a) $P = I \times E$
(b) $P = I^2 \times R$
(c) $P = \dfrac{E^2}{R}$
(d) $P = E^2 \times R$

5. When voltage applied to a series circuit is increased, and the resistance remains constant, circuit current:
(a) increases.
(b) decreases.
(c) remains the same.

6. If the resistance in a circuit is doubled, and the voltage is doubled, the circuit current:
(a) increases.
(b) decreases.
(c) remains the same.

7. What is the value of R_1 in the circuit shown in Figure 6-22?
(a) 10 kΩ
(b) 20 kΩ
(c) 8 kΩ
(d) 2 kΩ

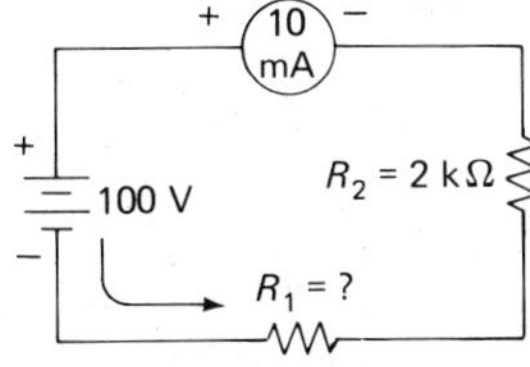

FIGURE 6-22

8. What is the total current flowing in the circuit shown in Figure 6-23?
(a) 0.3 A
(b) 30 mA
(c) 3 A
(d) 3 mA

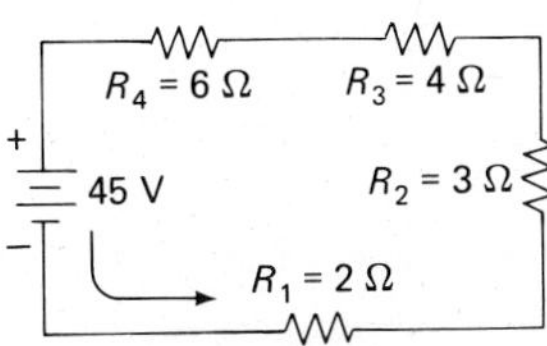

FIGURE 6-23

9. In a series circuit with three resistors, the ______________ resistor will dissipate the ______________ power.
   (a) largest, smallest
   (b) largest, most
   (c) smallest, largest
   (d) none of the above

10. What is the resistance of R_1 in the circuit shown in Figure 6-24?
   (a) 5 kΩ
   (b) 10 kΩ
   (c) 20 kΩ
   (d) 40 kΩ

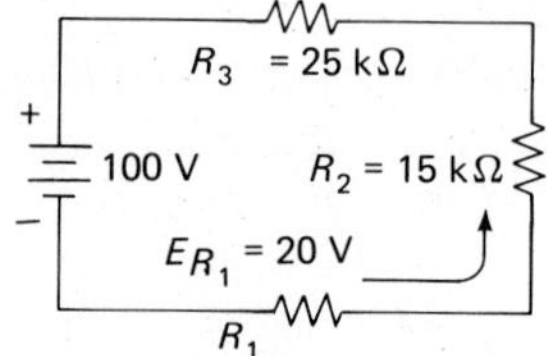

FIGURE 6-24

11. In the circuit in Figure 6-25, what is the value of R_1?
   (a) 400 Ω
   (b) 200 Ω
   (c) 50 Ω
   (d) 40 Ω

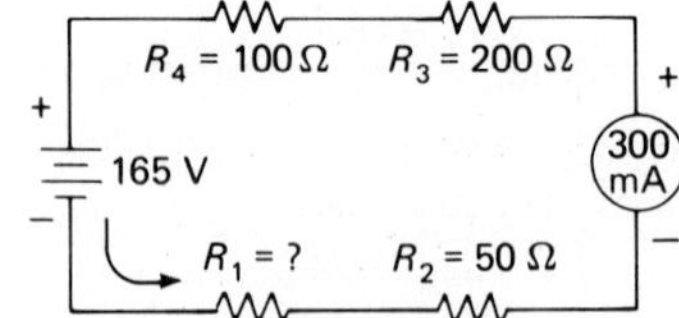

FIGURE 6-25

12. What is the value of R_5 in the circuit in Figure 6-26?
   (a) 10 kΩ
   (b) 16 kΩ
   (c) 5 kΩ
   (d) 20 kΩ

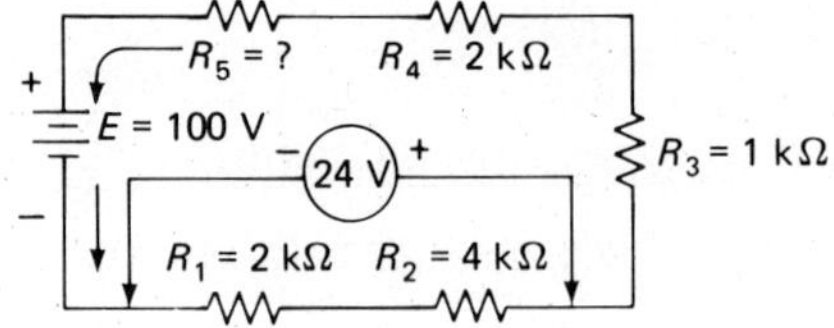

FIGURE 6-26

13. What is the total resistance of the circuit shown in Figure 6-27?
   (a) 13 kΩ
   (b) 25 kΩ
   (c) 192 kΩ
   (d) 8 kΩ

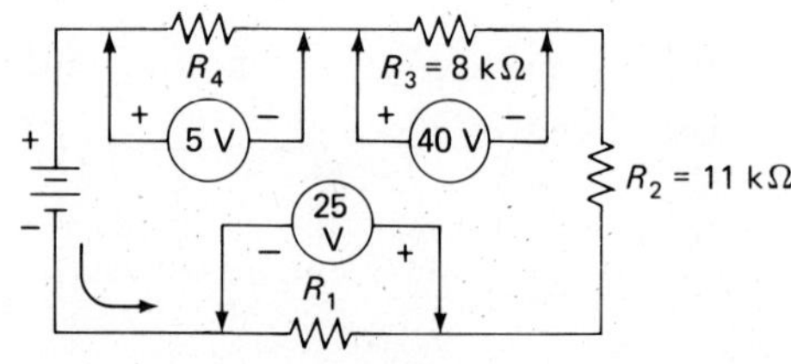

FIGURE 6-27

14. What is the current in the circuit in Figure 6-28?
(a) 0.91 A
(b) 91 mA
(c) 0.2 A
(d) 20 mA

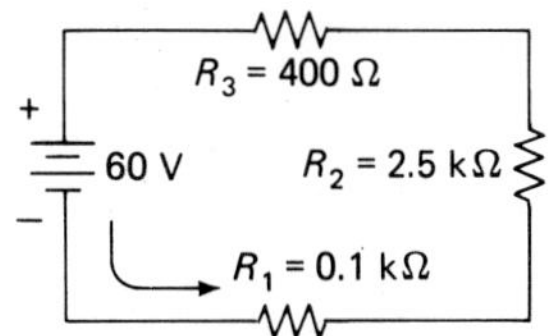

FIGURE 6-28

15. The total current in the circuit in Figure 6-29 equals
(a) 5 mA.
(b) 50 mA.
(c) 25 mA.
(d) 10 mA.

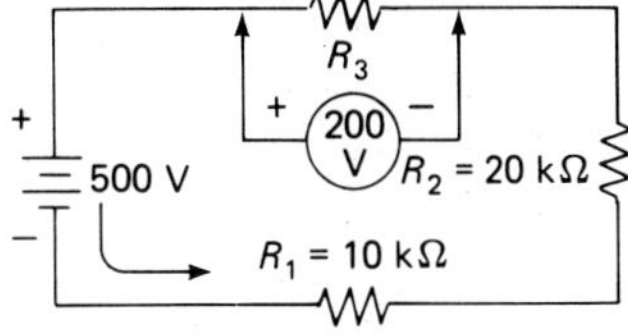

FIGURE 6-29

16. Solve for R_1 in the circuit in Figure 6-30.
(a) 1.3 kΩ
(b) 2.0 kΩ
(c) 1.0 kΩ
(d) 0.5 kΩ

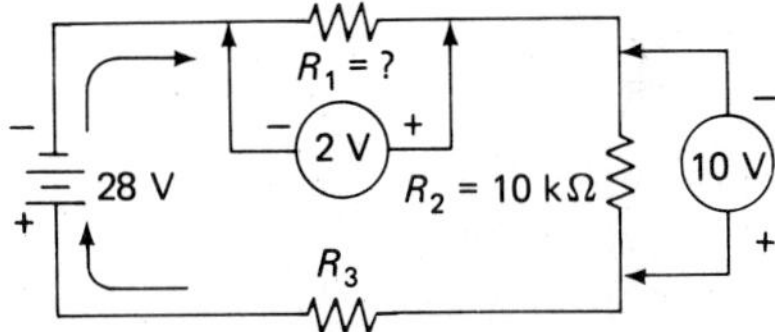

FIGURE 6-30

17. Solve for the total current in the circuit in Figure 6-31.
(a) 0.2 mA
(b) 15 mA
(c) 10 mA
(d) 0.5 mA

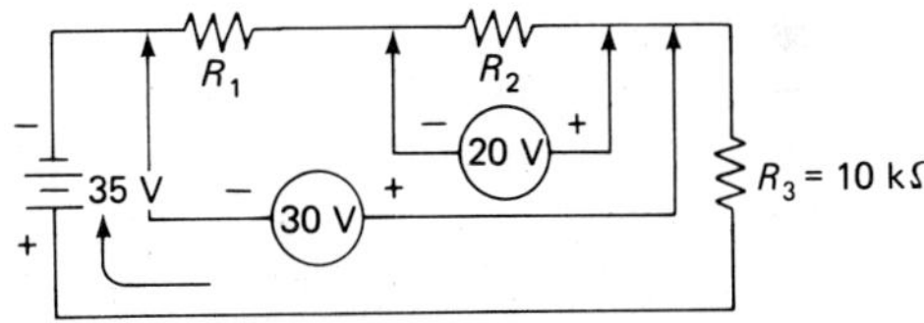

FIGURE 6-31

18. What is the applied voltage in the circuit in Figure 6-32?
(a) 138 V
(b) 72 V
(c) 150 V
(d) 210 V

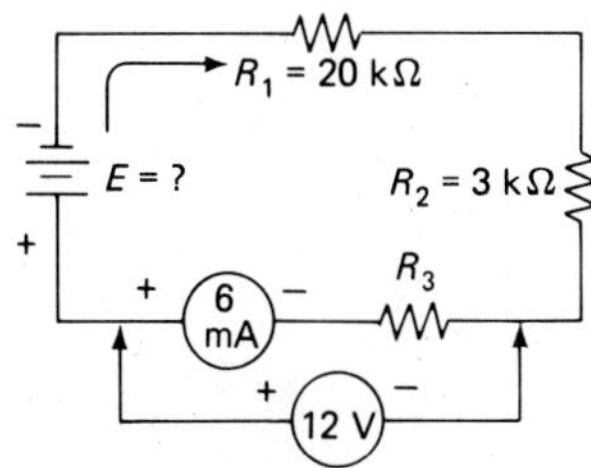

FIGURE 6-32

19. Solve for total voltage in the circuit shown in Figure 6-33.
    (a) 2 mV
    (b) 400 V
    (c) 400 mV
    (d) 200 V

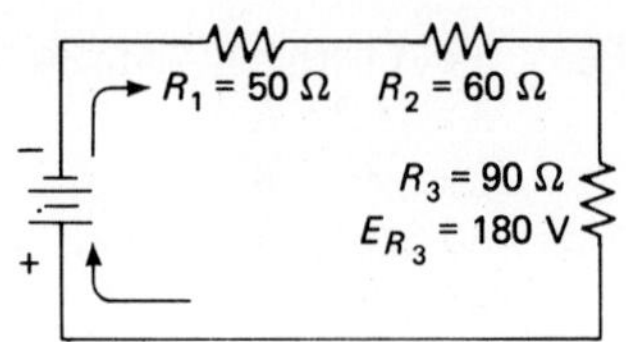

FIGURE 6-33

20. In the circuit shown in Figure 6-34, $I_t = 25$ mA, what does E_t equal?
    (a) 50 V
    (b) 12.5 V
    (c) 25 V
    (d) 100 V

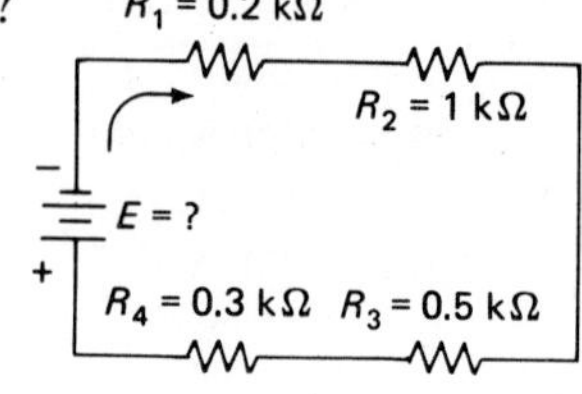

FIGURE 6-34

21. In the circuit shown in Figure 6-35, the lowest power rating that could be used for R_1 is
    (a) 20 W.
    (b) 1 W.
    (c) 100 mW.
    (d) 200 mW.

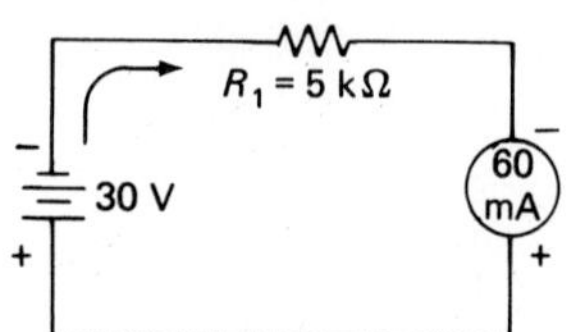

FIGURE 6-35

22. What is the power dissipated by R_L in the circuit in Figure 6-36?
    (a) 160 W
    (b) 60 W
    (c) 120 W
    (d) 200 W

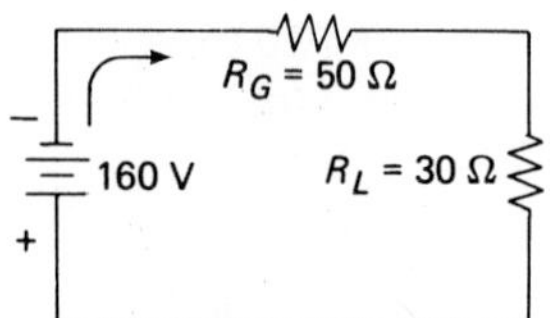

FIGURE 6-36

23. How much power is dissipated by R_1 in Figure 6-37?
    (a) 480 mW
    (b) 192 mW
    (c) 48 mW
    (d) 50 mW

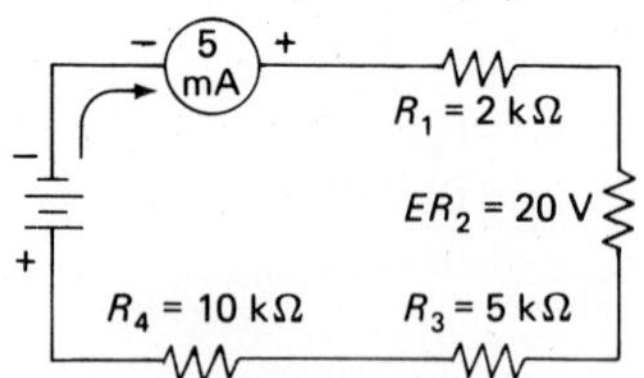

FIGURE 6-37

24. The power dissipated by R_1 in the circuit shown in Figure 6-38 equals
(a) 400 mW.
(b) 192 mW.
(c) 48 mW.
(d) 48 W.

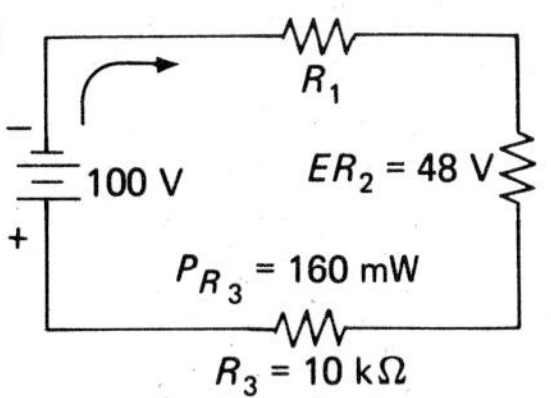

FIGURE 6-38

25. In the circuit shown in Figure 6-39, if R_4 dissipates half the power of R_3, what will the total power equal?
(a) 1008 mW
(b) 800 mW
(c) 528 mW
(d) 504 mW

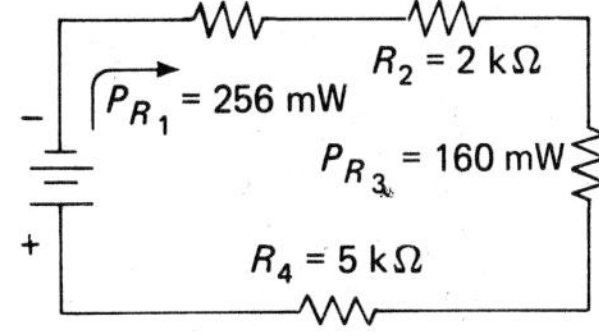

FIGURE 6-39

7

Parallel-Resistive Circuits

Introduction

It is impossible to use series circuits such as those discussed in Chapter 6 to wire complex systems like the radio, television set, computer, or spacecraft. It is equally impossible to provide a separate power source for each resistive device. There is a better way! By using parallel-connected resistive devices, we can connect many devices to a single voltage source. The main limitation on the number of devices that can be connected to a single power source is the ability of the source to provide the total current needed to operate all the devices. Figure 7-1 is an example of a parallel-resistive circuit.

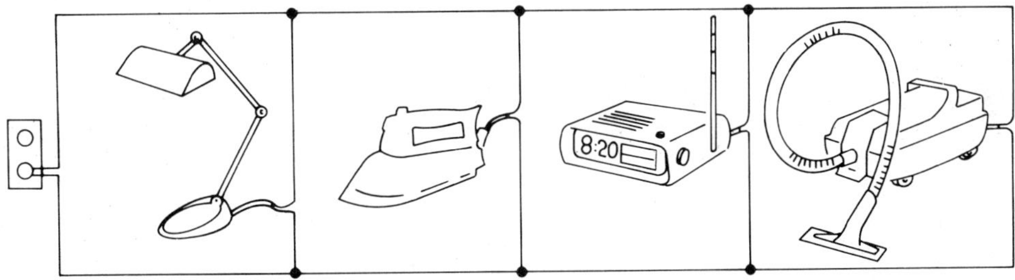

FIGURE 7-1

Remember that each device has its own on-off switch. With this circuit, any appliance can be used separately, all can be in use at the same time, or all can be off at the same time. The amount of current required from the power source varies from zero (when all are off) to maximum (when all are on). A single voltage source is connected directly to each appliance. Operation of the appliance becomes a matter of flipping a switch. What about the radio? When it is turned on, current flows from the power source to the radio, through the radio, and back to the power source. Current can flow through this path because the top of the radio circuit is connected directly to the power source by a conductor. The other side of the radio circuit is also connected to the power source by a conductor; together, they form a closed loop, with the radio as the resistive device. All other appliances are connected to the power source in the same way. Turning any switch on completes a loop through which current flows. The result is that an appliance can operate independently of the others.

All systems, regardless of their complexity, are connected in this manner. A system as small as a stereo has hundreds of individual circuits connected thus. Internally, a stereo is a mass of parallel-connected circuits. The external power supply sees the stereo as a single resistor whose resistance is equal to the system's equivalent resistance.

The three requirements for a practical circuit are discussed in Chapter 5. They are:

1. A voltage source capable of supplying current
2. A resistive device to perform the work
3. A continuous path for current flow

Figure 7-1 is different, in that four paths for current flow are present. This schematic illustrates a parallel circuit, which is defined as follows:

A parallel circuit contains two or more paths for current flow.

Objectives

Each student is required to:

1. Review the requirements for a practical circuit.
2. State the rules governing voltage and current distribution in a parallel-resistive circuit.
3. Write the formulas used to calculate total voltage, total current, and total resistance in a parallel circuit.
4. Analyze the theoretical operation of a parallel-resistive circuit to find branch currents, branch resistance, and branch-voltage drops.
5. Solve for total voltage when given branch current and branch resistance.
6. Calculate total voltage when given branch resistance and branch current.
7. Solve for branch resistance when given branch current and total voltage.
8. Analyze a parallel-resistive circuit for:
   (a) Total power when individual resistor power dissipations are given.
   (b) Power dissipated by each resistor when given circuit voltage, current, and/or resistance.
   (c) Total power when given total current, total voltage, and/or total resistance.
   (d) The relationship between total power and branch power dissipations.
   (e) Effect on total voltage, total current, and total power when parallel branches are added or deleted.
9. State the effect on total current and total power when total voltage is increased or decreased.
10. Apply theoretical troubleshooting techniques to parallel-resistive circuits as follows:
    (a) Use a voltmeter to locate opens and shorts.
    (b) Use an ammeter to locate opens and shorts.
    (c) Use an ohmmeter to identify good components, open components, and shorted components.

Simple Parallel-Resistive Circuits

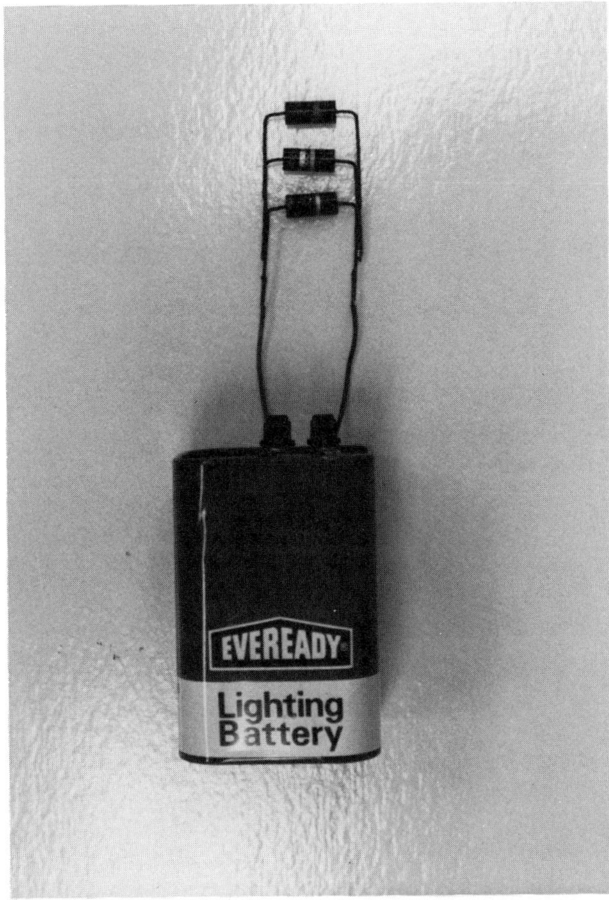

FIGURE 7-2
Parallel-Resistive Circuit (Photo by Riddell Allen)

Figure 7-3 represents the circuit shown in Figure 7-1, except that, in Figure 7-3, schematic symbols are used to represent the appliances, the battery, and the switch. The bottom of each resistor is connected to the battery through points A–B–C–D–E. With the switch open, voltage is applied to the left side of the switch but not to the top of the resistors. When the switch is closed, the top of each resistor is connected to the power source through points F–G–H–I–J, and the battery's total voltage is applied across each resistor. From this analysis, we can state the rule:

When resistors are connected in parallel across a voltage source, each branch will have total voltage applied.

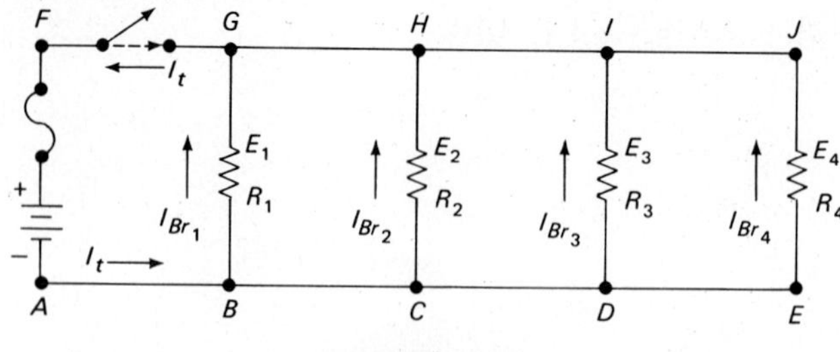

FIGURE 7-3

The voltage relationship shown in Figure 7-3 can be stated as a formula:

$$E_t = E_{R1} = E_{R2} = E_{R3} = E_{R4}$$

In simple terms, this formula states that the voltage drop on each resistor (branch) is equal to the battery voltage. Figure 7-4 illustrates this circuit, with the values for current, resistance, and voltage noted.

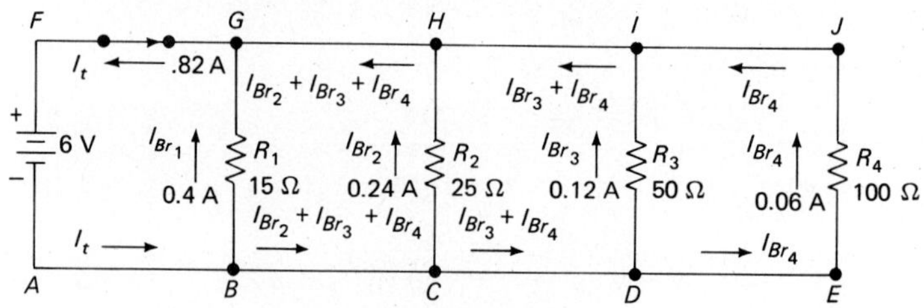

FIGURE 7-4

Note that the total current indicated is larger than that of any resistor (branch) current. Total current equals the sum of all the branch currents. At points B, C, D, and E, a portion of the total current flows into that branch, goes through the resistor, and rejoins the other branch currents at the top of the branch. Total current flows only between points A–B and G–F.

This is not to say that Kirchhoff's *current* law is no longer correct. This law states:

The algebraic sum of all currents flowing into a point must equal the algebraic sum of the currents leaving the point.

For parallel circuits, this law can be stated by the following formula:

$$I_t = I_1 + I_2 + I_3 + I_4$$

$$I_t = 0.4 \text{ A} + 0.24 \text{ A} + 0.12 \text{ A} + 0.06 \text{ A}$$

$$I_t = 0.82 \text{ A or } 820 \text{ mA}$$

where

I_t = total current

$I_1 = R_1$ (branch 1) current

$I_{Br2} = R_2$ (branch 2) current

$I_{Br3} = R_3$ (branch 3) current

$I_{Br4} = R_4$ (branch 4) current

To solve for branch currents, use Ohm's law exactly as it is applied in series circuits. Examples are:

Mathematical Solution	*Action*	*Entry*	*Display*
$I_{R1} = \frac{E_{R1}}{R_1}$	ENTER	6	6
$I_{R1} = \frac{6\text{ V}}{15\ \Omega}$	PRESS	÷	6
$I_{R1} = 0.4$ A or 400 mA	ENTER	15	15
	PRESS	=	0.4

Applying this formula to the other branches, we find their currents to be:

$I_{R2} = 0.24$ A or 240 mA

$I_{R3} = 0.12$ A or 120 mA

$I_{R4} = 0.06$ A or 60 mA

These solutions confirm the values noted in Figure 7-4. To solve for total current, we add the four branch currents.

To solve for I_t:

$I_t = I_{R1} + I_{R2} + I_{R3} + I_{R4}$

$I_t = 400\text{ mA} + 240\text{ mA} + 120\text{ mA} + 60\text{ mA}$

$I_t = 0.82$ A or 820 mA

Self-Check

Answer each item by indicating whether the statement is true or false.

1. ____ The voltage across each branch of a parallel circuit equals the applied voltage.

2. ____ Total current in a parallel circuit flows through each branch.

3. ____ There is no limit to the number of parallel branches a parallel circuit can contain.

4. ____ Branch currents must be added together to equal total current.

5. ____ Kirchhoff's voltage law does not apply to parallel-resistive circuits.

Total Resistance in a Parallel-Resistive Circuit

We have seen that when total voltage and total current are known, their values can be used to solve for total resistance, using Ohm's law.

Mathematical Solution	*Action*	*Entry*	*Display*
$R_t = \frac{E_t}{I_t}$	ENTER	6	6
$R_t = \frac{6\ \text{V}}{820\ \text{mA}}$	PRESS	÷	6
$R_t = 7.317\ \Omega$	ENTER	820,EE,+/−,3	820 − 03
	PRESS	=	7.317 00

Here is another important point in parallel-resistive circuit analysis:

In a parallel circuit, total resistance is less than the resistance contained in the smallest branch.

This statement appears to contradict statements already made; when you compare the circuits, however, you can see why this statement is true. The characteristics of a series circuit are:

1. Current is the same at all points within the circuit.
2. Total voltage is divided among all resistors.
3. Total resistance is the sum of all individual resistors.
4. Total resistance is larger than any individual resistor.

Examination of parallel circuits reveals:

1. Voltage is the same across all branches.
2. Total current is divided among all branches.
3. Total current is the sum of all branch currents.
4. As stated in Ohm's law, current and resistance are inversely proportional. Therefore, because total current is larger than the largest branch current, total resistance must be less than the resistance of the smallest branch.

Comparing the two sets of statements we see that the conditions within the circuits have been reversed. In mathematical terminology, they are *reciprocals* of each other. This indicates that the formula for finding total resistance in the circuit is the reciprocal of that used for series resistance. In fact, the formula best suited for use with a calculator is one called the *reciprocal formula.* In Figure 7-3, it is:

$$\frac{1}{R_t} = \frac{1}{R_1} + \frac{1}{R_2} + \frac{1}{R_3} + \frac{1}{R_4}$$

If we insert the resistor values from Figure 7-4 in the formula and solve for total resistance, we have:

Mathematical solution:

$$\frac{1}{R_t} = \frac{1}{15\ \Omega} + \frac{1}{25\ \Omega} + \frac{1}{50\ \Omega} + \frac{1}{100\ \Omega}$$

Find the lowest common denominator, LCD = 300:

$$300\frac{1}{R_t} = 300\frac{1}{15} + 300\frac{1}{25} + 300\frac{1}{50} + 300\frac{1}{100}$$

Simplify the terms by dividing each denominator into 300 and multiplying the result times the numerator:

$$\frac{300}{R_t} = 20 + 12 + 6 + 3$$

$$\frac{300}{R_t} = 41$$

$$41\ R_t = 300$$

$$R_t = 7.317\ \Omega$$

and

Calculator solution:

Action	Entry	Display
ENTER	15	15
PRESS	1/X	0.0666666667
PRESS	+	0.0666666667
ENTER	25	25
PRESS	1/X	0.04
PRESS	+	0.1066666667
ENTER	50	50
PRESS	1/X	0.02
PRESS	+	0.1266666667
ENTER	100	100
PRESS	1/X	0.01
PRESS	=	0.1366666667
PRESS	1/X	7.317073171

Again, $R_t = 7.317\ \Omega$.

Both the mathematical solution and the calculator solution confirm the Ohm's-law solution. Using all three approaches, we find that the total resistance equals 7.317 Ω. The reciprocal formula can be used to solve any parallel-resistive circuit for total resistance. Other forms of the total resistance formula are available, which apply to the special applications discussed later in this chapter. The reciprocal formula already discussed, however, can be used to solve all parallel-resistance problems. It should be committed to memory for use with a calculator.

The first special-application formula is one used with circuits containing two branches. A variation of the reciprocal formula that results from several steps of transposition, is called the *product-over-sum formula* and can be used to solve the circuit shown in Figure 7-5.

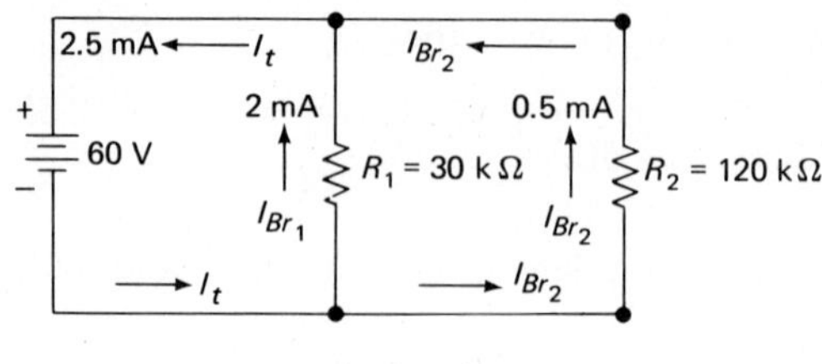

FIGURE 7-5

Formula and problem solution follow:

Formula:

$$R_t = \frac{R_1 \times R_2}{R_1 + R_2}$$

Mathematical solution:

$$R_t = \frac{(30\text{k} \times 120\text{k})}{(30\text{k} + 120\text{k})}$$

$$R_t = \frac{3600\ \text{M}}{150\text{k}}$$

$$R_t = 24\ \text{k}\Omega$$

Action	*Entry*	*Display*
ENTER	30,EE,3	30 03
PRESS	×	3 04
ENTER	120,EE,3	120 03
PRESS	÷	3.6 09
PRESS	(	3.6 09
ENTER	30,EE,3	30 03
PRESS	+	3 04
ENTER	120,EE,3	120 03
PRESS	)	1.5 05
PRESS	=	2.4 04

Convert 2.4 04 to prefix = 24 kΩ.

From facts already presented, we know that R_1 and R_2 each will have voltage drops equal to the applied voltage—or 60 V. These voltage drops can, with Ohm's law, be used to solve for current flow in each branch, as follows:

$$I_{R1} = \frac{E_t}{R_1} \qquad I_{R2} = \frac{E_t}{R_2}$$

$$I_{R1} = \frac{60}{30\text{k}} \qquad I_{R2} = \frac{60}{120\text{k}}$$

$$I_{R1} = 2 \text{ mA} \qquad I_{R2} = 0.5 \text{ mA}$$

With the branch currents known, we can use Kirchhoff's current law to solve for total current in the circuit:

$$I_t = I_{R1} + I_{R2}$$

$$I_t = 2 \text{ mA} + 0.5 \text{ mA}$$

$$I_t = 2.5 \text{ mA}$$

Using E_t and I_t, we double-check R_t, as follows:

$$R_t = \frac{E_t}{I_t}$$

$$R_t = \frac{60 \text{ V}}{2.5 \text{ mA}}$$

$$R_t = 24 \text{ k}\Omega$$

Calculator solutions for these problems are identical to those already presented.

In some cases, parallel circuits have several branches where branch resistances are equal (Figure 7-6).

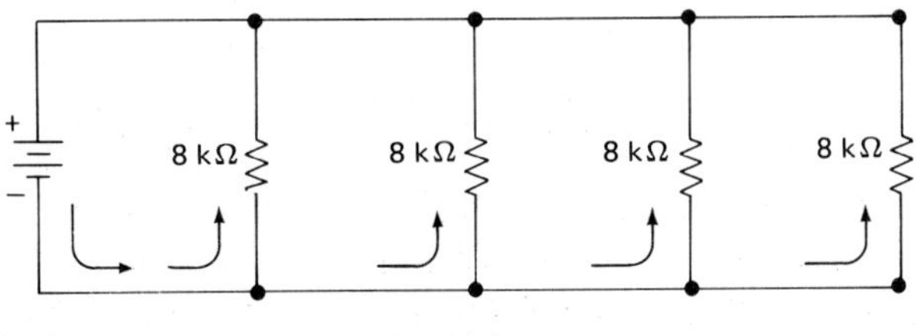

FIGURE 7-6

Another version of the reciprocal formula is available for use in calculating the total resistance of this type of parallel-resistive circuit. In many cases, it is possible to use the formula and perform total-resistance computations mentally. You must understand, however, that the formula is used *only* when branch resistances are equal. The R/N formula is:

$$R_t = \frac{R} {N} \qquad R = \text{resistance of one branch}$$

$$R_t = \frac{8\text{k}}{4} \qquad N = \text{number of equal branches}$$

$$R_t = 2 \text{ k}\Omega$$

Remember, the R/N formula can be used only when branch resistances are equal. It is possible, however, to intermix all three formulas for speed in solving total-resistance problems (see Figure 7-7).

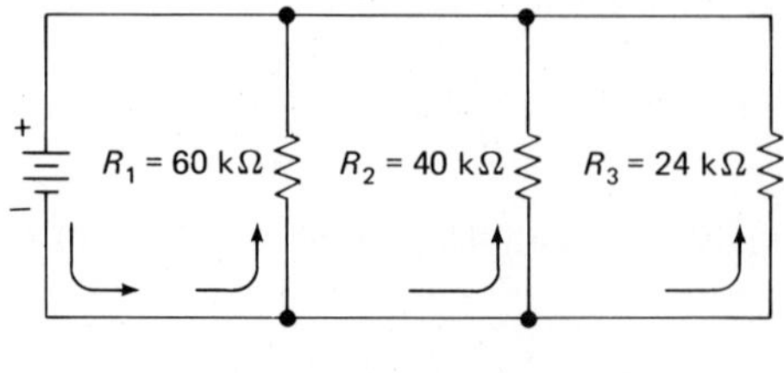

FIGURE 7-7

First, solve for the *equivalent* resistance of R_1 and R_2, using the product-over-sum formula:

$$R_{eq} = \frac{R_1 \times R_2}{R_1 + R_2}$$

$$R_{eq} = 24\ \text{k}\Omega$$

The resistances included in branches 1 and 2 are equivalent to 24 kΩ. Note that R_3 is a 24-kΩ resistor and is equal to R_1 and R_2 equivalent, R_{eq}. Total resistance can be found by using the R/N formula as follows:

$$R_t = \frac{R}{N}$$

$$R_t = \frac{24\text{k}}{2}$$

$$R_t = 12\ \text{k}\Omega^*$$

Refer to Figure 7-7 and consider the effect on R_t if a branch is either added or removed. Adding a fourth branch containing a 20-kΩ resistor causes the circuit to have four branches: 60 kΩ; 40 kΩ; 24 kΩ; and 20 kΩ. Using a calculator, solve for R_t:

Action	*Entry*	*Display*
ENTER	60,EE,3	60 03
PRESS	1/X	1.66667 − 05
PRESS	+	1.66667 − 05
ENTER	40,EE,3	40 03
PRESS	1/X	2.5 − 05
PRESS	+	4.166667 − 05
ENTER	24,EE,3	24 03
PRESS	1/X	4.166667 − 05
PRESS	+	8.333333 − 05
ENTER	20,EE,3	20 03
PRESS	1/X	5 − 05
PRESS	=	1.333333 − 04
PRESS	1/X	7.5 03 or 7.5 kΩ

*Realize, though, that the problem could be solved using the reciprocal formula and a calculator.

When we compare this result to that obtained with the three-branch circuit, we find that:

Adding a branch to a parallel circuit causes total resistance to decrease.

If total resistance decreases, total current must increase.

Now let us examine this circuit when it has only two branches, containing R_1 and R_2. The calculator solution for the circuit is:

Action	*Entry*	*Display*
ENTER	60,EE,3	60 03
PRESS	1/X	1.66667 – 05
PRESS	+	1.66667 – 05
ENTER	40,EE,3	40 03
PRESS	1/X	2.5 – 05
PRESS	=	4.16667 – 05
PRESS	1/X	2.4 04 or 24 kΩ

Comparing this result to the three-branch circuit in Figure 7-7, we find that:

When a branch is removed from a parallel circuit, total resistance increases and total current decreases.

Self-Check

Complete each item by inserting the words and/or numbers needed to make a true and complete statement.

6. Solve for total resistance in Figure 7-8.

   R_t = ______________

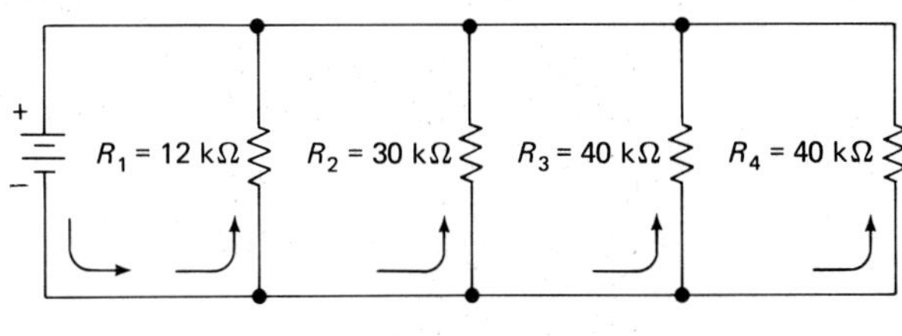

FIGURE 7-8

7. Use the R/N formula to solve for R_t in Figure 7-9.

   R_t = ______________

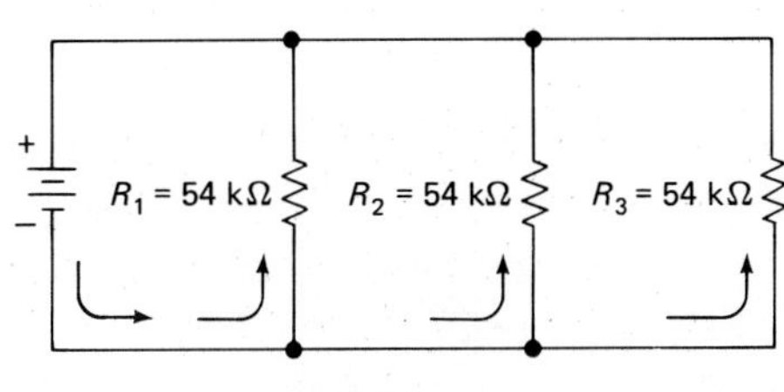

FIGURE 7-9

8. Use the product-over-sum formula to solve for R_t in Figure 7-10.

   R_t = ______________

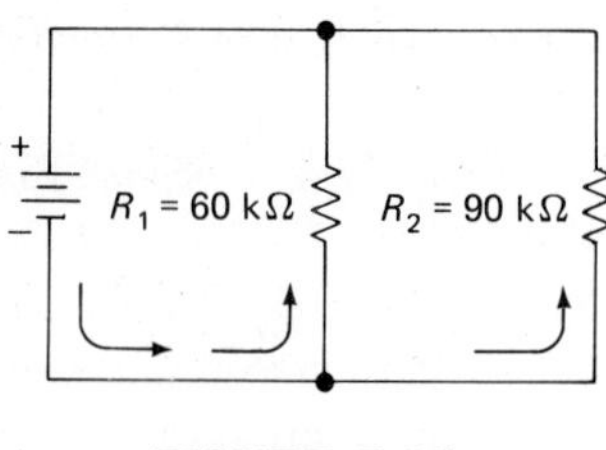

FIGURE 7-10

9. Adding a branch to a parallel-resistive circuit causes the total resistance to ______________.

10. A parallel circuit has ______________ path for current flow.

Current Distribution in Parallel Circuits

Conductance in Parallel Circuits

Remember, conductance is equal to $1/R$, which is the reciprocal of resistance. To state the conductance formula for a parallel circuit, we use:

$$G_t = G_1 + G_2 + G_3 + \ldots + G_n$$

Resistance is stated in ohms and conductance either in mhos or Siemens units.

Some electronics technicians prefer to analyze parallel circuits using conductance, since the math doesn't involve reciprocal operations. Each value of G, however, is already the reciprocal of resistance.

For each branch of a parallel circuit we can state current using either method, for instance;

Branch current is directly proportional to branch conductance.

or

Branch current is inversely proportional to branch resistance.

Stating the current in terms of conductance is comparable to stating voltage in terms of resistance in series circuits, where:

Voltage drop on a resistor is directly proportional to the size of the resistor.

Except for providing familiarity with the conductance parameter, we continue to examine circuits from the standpoint of resistance. Conductance is valuable in a discussion of the division of current between two or more branches of a parallel circuit.

already present. The result is:

When branches are added to a parallel circuit, total current increases.

As branches are removed from a parallel circuit, the opposite happens, with the result that:

When branches are removed from a parallel circuit, current decreases.

In summary, we can say that the greater the number of branches connected to a power source, the larger the current demand on the source and the less the total resistance. When total resistance decreases and total current increases, Ohm's law is confirmed. This law states:

Current and resistance are inversely proportional.

Division of Current Among Branches

We must often calculate the amount of current flowing in a branch, or we may need to predict what branch carries the most current. It is possible to do this even when we do not know the voltage applied to that part of the circuit. For a two-branch circuit such as the circuit shown in Figure 7-12, where we know resistance values and total current, we can use the formulas,

$$I_{Br1} = \frac{R_{Br2}}{(R_{Br1} + R_{Br2})} \times I_t$$

or

$$I_{Br2} = \frac{R_{Br1}}{(R_{Br1} + R_{Br2})} \times I_t$$

Applying these formulas to Figure 7-11a, we find that:

$$I_{Br1} = 20 \text{ mA}$$

and

$$I_{Br2} = 10 \text{ mA}$$

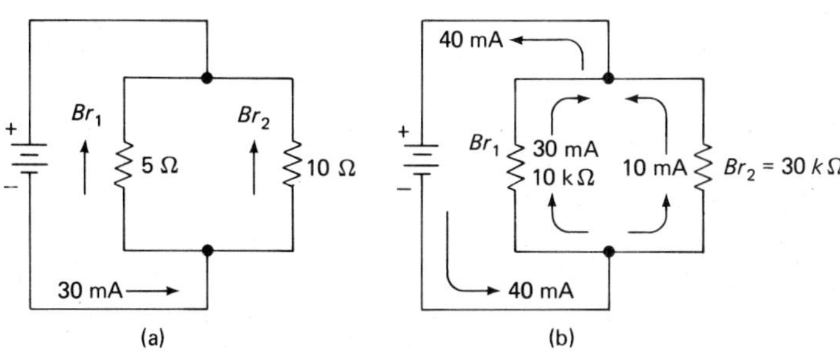

FIGURE 7-12

Note that, in Figure 7-12b, total current equals 40 mA and the branch resistor values are given. Using the formulas above, we find:

$$I_{Br1} = 30 \text{ mA}$$

and

$$I_{Br2} = 10 \text{ mA}$$

With circuits having two branches it is possible to calculate the second-branch current by subtraction. The first-branch current is then subtracted from the total current.

Concerning the use of conductance in determining branch currents, consider Figure 7-13.

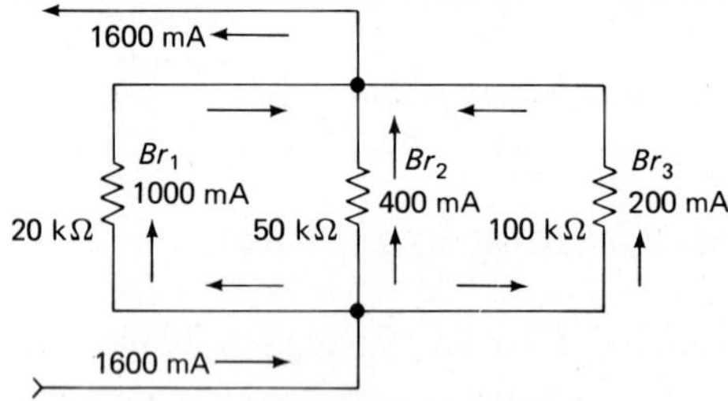

FIGURE 7-13

Again, remember that resistance and conductance are reciprocals of each other. The formula for calculating branch current using G is:

$$I_{Br1} = \frac{G_{Br1}}{G_t} \times I_t$$

where

$$G_{Br1} = \text{conductance of branch 1}$$

and

$$G_t = \text{total conductance of } G_{Br1} + G_{Br2} + G_{Br3}$$

In Figure 7-13, the three-branch conductances are:

$$G_{Br1} = 0.00005$$

$$G_{Br2} = 0.00002$$

$$G_{Br3} = 0.00001$$

$$G_t = 0.00008$$

To find I_{Br1}:

$$G_{Br1} = \frac{0.00005}{0.00008} \times 1600 \text{ mA}$$

$$G_{Br1} = 0.625 \times 1600 \text{ mA}$$

$$G_{Br1} = 1000 \text{ mA}$$

and

$$G_{Br2} = \frac{0.00002}{0.00008} \times 1600 \text{ mA}$$

$$G_{Br2} = 0.25 \times 1600 \text{ mA}$$

$$G_{Br2} = 400 \text{ mA}$$

and

$$G_{Br3} = \frac{0.00001}{0.00008} \times 1600 \text{ mA}$$

$$G_{Br3} = 0.125 \times 1600 \text{ mA}$$

$$G_{Br3} = 200 \text{ mA}$$

then prove using Kirchhoff's current law

$$I_t = I_{Br1} + I_{Br2} + I_{Br3}$$

$$I_t = 1000 \text{ mA} + 400 \text{ mA} + 200 \text{ mA}$$

$$I_t = 1600 \text{ mA}$$

The convenience of using conductance lies in the fact that the branches are stated as conductances rather than resistances. Stating series circuits as conductances, however, leads to the use of a reciprocal formula for their solutions.

Self-Check

Answer each item by indicating whether the statement is true or false.

11. ____ Current flow in a parallel circuit equals the applied voltage.

12. ____ In a parallel circuit, total current flow equals the sum of all branch currents.

13. ____ The current capability of a battery limits the number of branches that can be connected to a given power source.

14. ____ To calculate total voltage in a parallel circuit, add all the separate branch voltages.

15. ____ Kirchhoff's voltage law does not apply to a parallel-resistive circuit.

Voltage Distribution in a Parallel Circuit

From Kirchhoff's voltage law, we know that

The algebraic sum of all the voltages in a closed loop equals zero.

Or, more specifically:

The sum of all voltage drops in a closed loop equals the voltage applied.

We analyze the circuit in Figure 7-14 to illustrate how Kirchhoff's voltage law applies to parallel-resistive circuits.

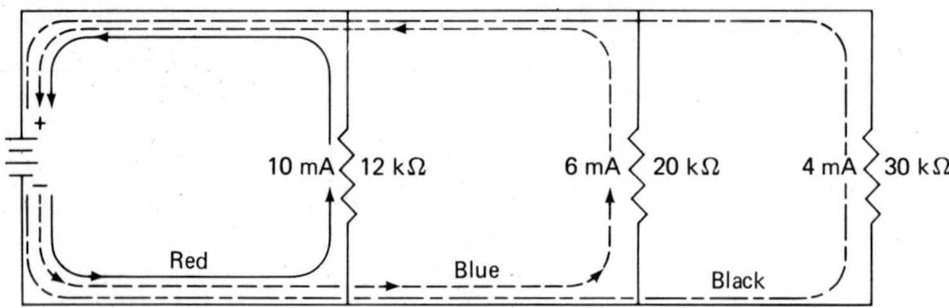

FIGURE 7-14

Note that each separate line forms a complete loop. It is possible for current to flow through this loop even if the other two loops have been removed. With switches in each loop, we can allow current to flow in any or all loops. Conductors that have two or more lines through them are carrying more than one branch current. The solid loop represents branch-1 current, the dashed loop represents branch-2 current, and the dashed loop represents branch-3 current.

When we consider these facts and Kirchhoff's voltage law, we can see that each loop, or branch, voltage drop must equal the applied voltage. We have seen in this chapter that the voltage drop on each branch of a parallel circuit equals the applied voltage:

$$E_t = E_{R1}$$

$$E_t = E_{R2}$$

$$E_t = E_{R3}$$

To prove this, solve for the voltage drop on each branch by using Ohm's law, as follows:

$$E_{R1} = I_{Br1} \times R_1 = 10 \text{ mA} \times 12\text{k} = 120 \text{ V}$$

$$E_{R2} = I_{Br2} \times R_2 = 6 \text{ mA} \times 20\text{k} = 120 \text{ V}$$

$$E_{R3} = I_{Br3} \times R_3 = 4 \text{ mA} \times 30\text{k} = 120 \text{ V}$$

Consider Kirchhoff's voltage law. For each resistor to drop 120 V, it is necessary for the total voltage to be 120 V.

Because of the circuit's design, each branch of a parallel circuit forms a separate closed loop. The calculations just presented prove that the voltage drop of each branch is equal to the applied voltage. This confirms Kirchhoff's voltage law. This solution can be checked as follows. First, find the total resistance.

Calculator solution:

Action	*Entry*	*Display*
ENTER	12,EE,3	12 03
PRESS	1/X	8.33 – 05
PRESS	+	8.33 – 05
ENTER	20,EE,3	20 03
PRESS	1/X	5 – 05
PRESS	+	1.333 – 04
ENTER	30,EE,3	30 03
PRESS	1/X	3.333 – 05
PRESS	=	1.6667 – 04
PRESS	1/X	6 03

6 03 converts to 6 kΩ.

Refer to Figure 7-14. Here, one battery supplies the current for all three branches, which reemphasizes:

$$I_t = 10 \text{ mA} + 6 \text{ mA} + 4 \text{ mA} = 20 \text{ mA}$$

Total voltage can now be confirmed, Ohm's law, I_t, and R_t.

Mathematical Solution	*Action*	*Entry*	*Display*
$E_t = I_t \times R_t$	ENTER	20,EE,+/–,3	20 – 03
$E_t = 20 \text{ mA} \times 6\text{k}\Omega$	PRESS	×	2 – 02
$E_t = 120 \text{ V}$	ENTER	6,EE,3	6 03
	PRESS	=	1.2 02

1.2 02 converts to 120 V.

Rules for Parallel-Circuit Voltage Distribution

1. The voltage drop on each branch of a parallel circuit equals the voltage applied to that portion of the circuit.
2. Each branch of a parallel circuit is part of a separate closed loop.
3. Ohm's and Kirchhoff's voltage laws can be used to solve for voltage distribution in a parallel circuit.

Self-Check

Answer each item by indicating whether the statement is true or false.

16. ____ Kirchhoff's voltage law does not apply to parallel circuits.

17. ____ In a parallel circuit, total voltage is equal to the sum of the branch voltages.

18. ____ Total voltage is dropped by the resistance contained in each branch.

19. ____ In a parallel circuit, the sum of all the voltage drops in a closed loop equals the applied voltage.

20. ____ In a parallel-resistive circuit, the largest resistor drops the most voltage.

Power Distribution in Parallel Circuits

Power dissipations are treated equally, regardless of the circuit. The knowledge you have gained about series circuits can be used to analyze the parallel circuit for power. Figure 7-15 is analyzed to demonstrate the procedures used in analyzing parallel circuits for power distribution.

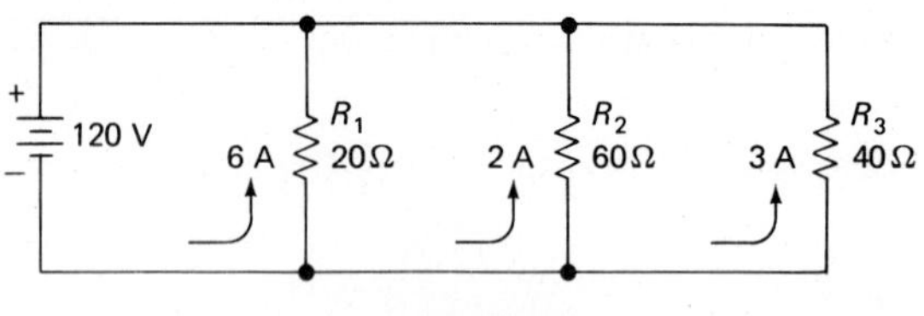

FIGURE 7-15

Voltage and current values are provided in the circuit. They can be used along with the power formula, $P = I \times E$, to solve power values. The power dissipation of R_1 is solved, but that of R_2 and R_3 is merely listed. Proceed as follows:

Mathematical Solution	*Action*	*Entry*	*Display*
$P_{R1} = I_{R1} \times E_{R1}$	ENTER	6	6
$P_{R1} = 6\text{ A} \times 120\text{ V}$	PRESS	×	6
$P_{R1} = 720\text{ W}$	ENTER	120	120
	PRESS	=	720

Solving for P_{R2} and P_{R3}:

$$P_{R2} = 240\text{ W}$$

$$P_{R3} = 360\text{ W}$$

When the power dissipated by each resistor is known, it is possible to calculate total power. An example of each solution, as it would be used with Figure 7-15, is shown. First, we find:

$$I_t = I_{R1} + I_{R2} + I_{R3} = 11\text{ A}$$

Then we solve for total power:

$$P_t = I_t \times E_t$$

$$P_t = 11\text{ A} \times 120\text{ V}$$

$$P_t = 1320\text{ W}$$

or

$$P_t = P_{R1} + P_{R2} + P_{R3}$$

$$P_t = 720 \text{ W} + 240 \text{ W} + 360 \text{ W}$$

$$P_t = 1320 \text{ W}$$

This reinforces a rule learned in Chapter 6.

Total power in a resistive circuit can be calculated by finding the sum of all the separate power dissipations or by finding the product of total current times total voltage.

Other forms of the power formula can also be used to solve parallel circuits. Use of the formula $P = I^2 \times R$ is illustrated in Figure 7-16.

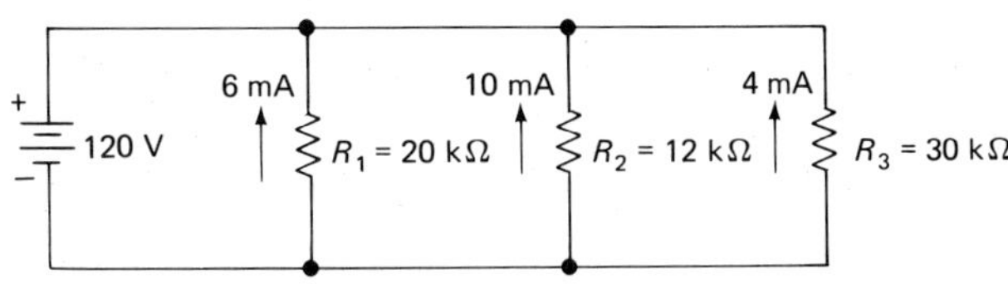

FIGURE 7-16

The circuit provides values for current and resistance, which are applied as follows:

Mathematical Solution	*Action*	*Entry*	*Display*
$P_{R1} = (I_{R1})^2 \times R_1$	ENTER	6,EE,+/−,3	6 − 03
$P_{R1} = (6 \text{ mA})^2 \times 20 \text{ k}\Omega$	PRESS	X^2	3.6 − 05
$P_{R1} = 36\ \mu\text{A} \times 20\text{k}\Omega$	PRESS	×	3.6 − 05
$P_{R1} = 720 \text{ mW}$	ENTER	20,EE.3	20 03
	PRESS	=	7.2 − 01

7.2 − 01 converts to 720 mW.

Current and power for the remainder of the circuit are:

$I_{R1} = 6 \text{ mA}$	$P_{R1} = 720 \text{ mW}$
$I_{R2} = 10 \text{ mA}$	$P_{R2} = 1200 \text{ mW or } 1.2 \text{ W}$
$I_{R3} = 4 \text{ mA}$	$P_{R3} = 480 \text{ mW}$

Use a calculator to confirm the values for R_2 and R_3. Total power can be solved by either form, as follows:

$$P_t = P_{R1} + P_{R2} + P_{R3} = 2.4 \text{ W}$$

$$P_t = I_t \times E_t = 2.4 \text{ W} = 2.4 \text{ W}$$

$$P_t = (I_t)^2 \times R_t = 2.4 \text{ W}$$

$$P_t = \frac{(E_t)^2}{R_t} = 2.4 \text{ W}$$

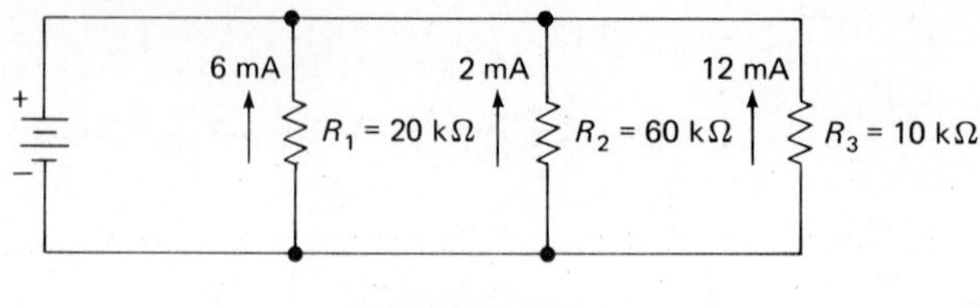

FIGURE 7-17

A third form of the power formula $P = E^2/R$ uses voltage and resistance as known values for solving power. Figure 7-17 illustrates this method. Each branch has current and resistance as its given values, but not voltage; therefore, it is necessary to use Ohm's law to solve for voltage before computing power. This can be done as follows:

$$E_{R1} = I_{Br1} \times R_{Br1}$$

$$E_{R1} = 6 \text{ mA} \times 20 \text{ k}\Omega$$

$$E_{R1} = 120 \text{ V}$$

To solve for total resistance, use the reciprocal formula:

$$\frac{1}{R_t} = \frac{1}{R_1} + \frac{1}{R_2} + \frac{1}{R_3}$$

$$R_t = 6 \text{ k}\Omega$$

Then

Mathematical Solution	*Action*	*Entry*	*Display*
$P_t = \frac{(E_t)^2}{R_t}$	ENTER	120	120
$P_t = \frac{120^2}{6\text{k}}$	PRESS	X^2	14 400
$P_t = \frac{14{,}400}{6\text{k}}$	PRESS	÷	14 400
$P_t = 2.4$ W	ENTER	6,EE,3	6 03
	PRESS	=	2.4 00

2.4 00 converts to 2.4 W.

By using this formula, we can calculate each resistor's dissipation. Use a calculator to verify the following values.

$$P_{R1} = 720 \text{ mW}$$

$$P_{R2} = 240 \text{ mW}$$

$$P_{R3} = 1.44 \text{ W}$$

Six rules have been developed that are useful in analyzing parallel-resistive circuits:

1. The voltage drop across each branch of a parallel circuit equals the voltage applied to that portion of the circuit.
2. The branch with the least resistance has the largest current flow and dissipates more power than the other branches.
3. Total current in a parallel circuit equals the sum of the individual branch currents.
4. The total resistance of a parallel circuit is less than the resistance of the smallest branch.
5. The total power of a parallel circuit equals the sum of all the individual dissipations.
6. The branch with the greatest resistance has the least current flow and dissipates the least power.

Self-Check

Answer each item by indicating whether the statement is true or false.

21. ___ The total power dissipated in a parallel circuit is equal to the sum of all individual dissipations.

22. ___ In a parallel-resistive circuit, the branch with the most resistance dissipates the most power.

23. ___ In a parallel circuit, total power equals total current multiplied by total voltage.

24. ___ In a parallel circuit, the branch having the smallest resistance has the most current.

25. ___ Adding a branch to a parallel-resistive circuit causes total power to increase.

Troubleshooting Parallel-Resistive Circuits

The principles learned in studying series-resistance-circuit troubleshooting also apply to parallel circuits. Although some variation in application occurs, opens and shorts have the same effect. The VOM is an important tool in effective troubleshooting. When the effects of opens and shorts in series circuits are compared to those in parallel circuits, the results are as follows:

1. In a series circuit, the open stops all current. In a parallel circuit, total current may either decrease or stop altogether. We can say that an open in a parallel circuit causes total current to decrease.
2. In a series circuit, the short causes maximum current to flow. In a parallel circuit, we can say only that total current increases. The extent of the increase depends on the circuit design and the location of the short in the circuit.

In Figure 7-18, we analyze the circuit as an illustration of the effect of various troubles that might occur in the circuit.

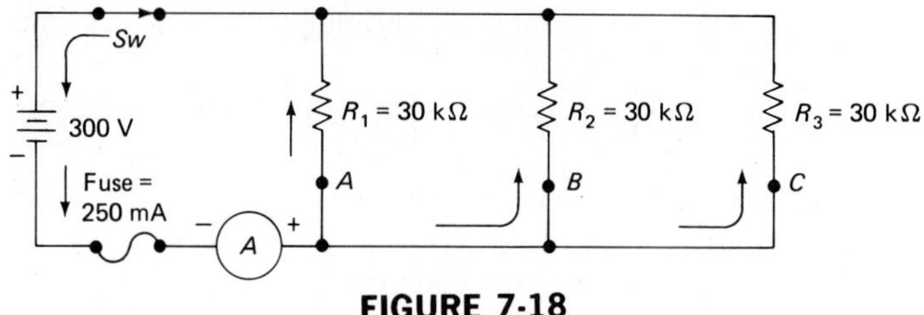

FIGURE 7-18

To identify the source of a problem, we must first know what indications there are when the circuit is operating correctly. Use a calculator to confirm that the conditions are not listed in Figure 7-18 are;

Normal conditions:

$$R_t = R/N = 10\ \text{k}\Omega$$
$$I_t = E_t/R_t = 30\ \text{mA}$$
$$I_{R1} = E_t/R_1 = 10\ \text{mA}$$
$$I_{R2} = E_t/R_2 = 10\ \text{mA}$$
$$I_{R3} = E_t/R_3 = 10\ \text{mA}$$

Now that we know what the normal conditions are, we have a standard against which to compare the results obtained when the same values are solved after a trouble is inserted. Figure 7-19 shows the same circuit, except that R_3 is open, that is, broken in two.

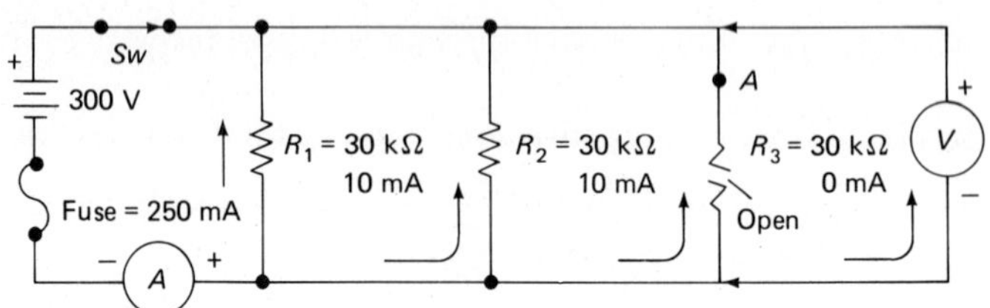

FIGURE 7-19

Both R_1 and R_2 are still connected in complete loops through which current can flow. Each has 300 V applied and 10 mA of current flow. Even though the 300 V is applied to R_3, it is open and no current can flow through it, meaning that it has 0 mA. When the three currents are added, the total current is found to be 20 mA, not 30 mA, as it was when the circuit was normal. The immediate indication of an ammeter check of total current is that $I_t = 20$ mA. The fact that the ammeter reads low confirms that a problem exists, but this low reading does not help us locate the source of the problem. Visually or operationally, the symptom is that the device represented by R_3 is inoperative.

What indication would we get if we connected a voltmeter across R_3, as shown in Figure 7-19? The meter indicates a total voltage of 300 V. Although R_3 is open, the meter is not. The meter bridges the open and current flows through it in sufficient amounts to power the meter's reading. The meter forms a complete loop with the battery and must drop 300 V.

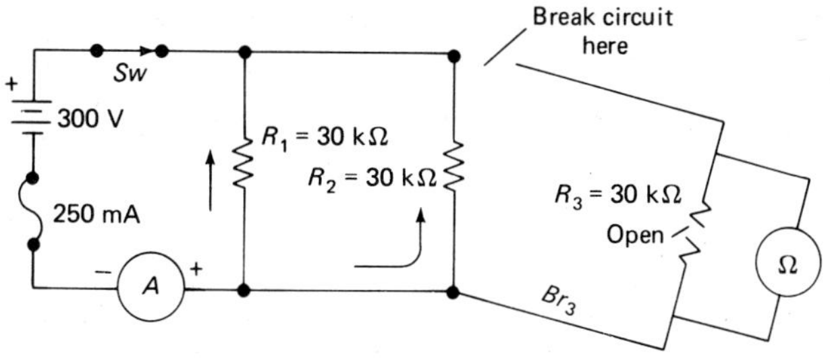

FIGURE 7-20

The fact that total current is low or that the voltmeter reads 300 V does not tell us where the trouble is. What these indications tell us is that somewhere a branch is open. To locate the open component, we must use an ohmmeter and check for loss of continuity. Figure 7-20 illustrates this.

Note that branch 3 of the circuit has been disconnected from the rest of the circuit at one end. This ensures that two precautions are followed: (1) no battery current is allowed to flow through the ohmmeter, which could damage the meter; and (2) the ohmmeter current is contained, so that it can flow only through R_3. Disconnection of branch 3 prevents incorrect indications which would occur when R_1 and R_2 are connected in parallel with R_3. If R_1 and R_2 are left in the circuit, the current supplied by the ohmmeter will have two branches to flow through. That puts R_1 and R_2 in parallel with R_3 and causes the ohmmeter to read incorrectly. By disconnecting branch 3, ohmmeter current is limited to R_3. Limiting ohmmeter current to R_3 ensures that, if the component is open, no current flows. With zero current flow, the meter indicates infinity, which indicates that R_3 is open.

It is possible to use the ammeter when checking for open components, but considerable time is needed to disconnect the circuit at the point shown as an ammeter, insert the meter, make the test, remove the meter, and reconnect the circuit. This procedure must be repeated at each point on the circuit until the branch through which no current flows is found. Thus the ammeter is seldom used in troubleshooting a circuit.

In Figure 7-21 we apply logical troubleshooting procedures to the circuit.

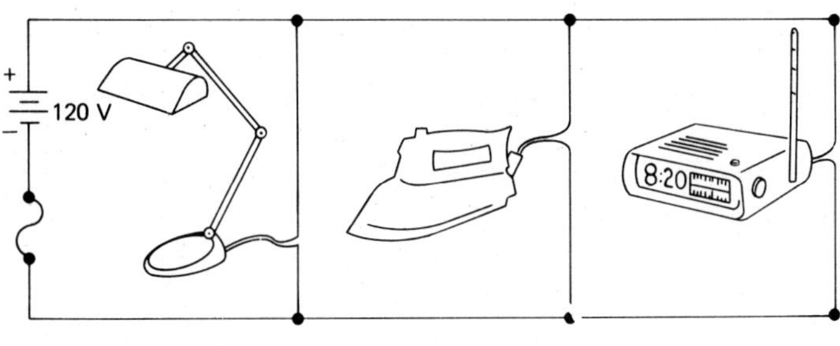

FIGURE 7-21

If the table lamp is open, how would we troubleshoot the circuit? The symptom of a problem is that when the lamp is turned *on*, it does not glow. A check of the other appliances reveals normal operation. Had all the appliances failed to operate, we would check the power source and the conductors that carry total current. Since the lamp is the only problem, we can probably repair it by replacing the bulb. If replacement of the bulb doesn't remedy the problem, we must use the voltmeter to check connections, components, and the switch to find the open. When the open is located, the voltmeter will read total voltage; all other voltage readings will be zero. The following point is illustrated:

When one branch of a parallel circuit is open, the remaining branches continue to operate normally.

Up to now we have been discussing the effect of an open component on a circuit. But what about a *shorted* component? What happens when a short occurs within a branch, or any branch? Recall two rules:

1. A short contains zero ohms of resistance.
2. Total resistance of a parallel circuit is less than the resistance of the smallest branch.

From these statements, we see that when a branch has zero ohms, the entire circuit has zero ohms. Consider Figure 7-21. Any branch that shorts will provide a path for current to flow around the other branches.

Most of the circuit current follows the path of least resistance. All of the current follows a path of zero resistance.

When a branch shorts, maximum current will try to flow through that branch. When current exceeds 10 A, the fuse will open and protect the remaining parts of the circuit from damage. Once the fuse opens, the symptom becomes: None of the devices will operate. Had the fuse not opened, it is possible that the current flow would become so high that the buildup of heat would cause a fire. Unless the circuit opens somewhere, an immediate hazard exists.

With the fuse blown, the ammeter and voltmeter are of little use in troubleshooting the circuit. We still have the ohmmeter, though. To use it, we must isolate (disconnect) each resistor, one at a time, until we find and measure a short (0 Ω). Once we have done that, we replace the resistor, make sure all parts of the circuit are connected, and replace the fuse. The circuit should now operate normally.

When using an ohmmeter, you can expect one of three indications:

An ohmmeter connected across a good resistor indicates the resistor's ohmic value.

An ohmmeter connected across a shorted component indicates zero ohms.

An ohmmeter connected across an open component indicates infinite ohms.

Self-Check

Complete each item by inserting the words and/or numbers needed to make a true and complete statement.

26. An open has ______________ Ω of resistance.

27. A short has ______________ Ω of resistance.

28. Applied voltage is measured across an ______________ resistor.

29. A voltmeter indicates zero volts when connected across a/an ______________ component.

30. An ammeter must be connected in ______________ with the circuit under test.

31. It is necessary to ______________ a component before checking it with an ohmmeter.

Summary

Several rules developed in this chapter are important in your studies. Some of the rules for parallel circuits are:

1. Voltages are equal on all branches of a parallel network.
2. Total current equals the sum of all branch currents.
3. Total resistance is less than the resistance of the smallest branch.
4. Total power equals the sum of all individual power dissipations.
5. An open component in a parallel circuit causes total current to decrease and total resistance to increase.
6. A shorted component causes total current to increase and total resistance to decrease.

Formulas suitable for analyzing parallel circuits were presented and explained.

Review Questions and/or Problems

1. The voltage across any branch of a parallel circuit is equal to
   (a) the total current.
   (b) the total voltage.
   (c) the total resistance.
   (d) less than the smallest branch.

2. A parallel circuit with four branches has equal branch currents when
   (a) the branch voltage drops are equal.
   (b) the resistance of each branch is equal.
   (c) each branch contains two resistors.
   (d) more than two resistors are used.

3. When 10 Ω, 20 Ω, and 30 Ω resistors are connected in parallel, their equivalent resistance is
   (a) more than 10 Ω.
   (b) less than 10 Ω.
   (c) more than 30 Ω.
   (d) 60 Ω.

4. The highest branch current always occurs in the branch that
   (a) contains the greatest resistance.
   (b) contains the least resistance.
   (c) contains the fewest resistors.
   (d) none of the above.

5. Increasing the total applied voltage to a parallel circuit causes total resistance to
   (a) increase.
   (b) decrease.
   (c) remain the same.

6. When a short occurs in a parallel circuit,
   (a) total resistance increases.
   (b) total current decreases.
   (c) total current increases.
   (d) total resistance remains the same.

7. Unlike a series circuit, current continues to flow in a parallel circuit that contains a/an ______________ branch resistor.
   (a) shorted
   (b) open
   (c) only one

8. To calculate branch resistance in a parallel circuit containing one resistor, you must know total voltage and the
   (a) total current.
   (b) total resistance.
   (c) branch voltage drop.
   (d) branch current.

9. When an open occurs in a parallel circuit, the
   (a) total resistance increases.
   (b) total current increases.
   (c) total resistance decreases.
   (d) total resistance does not change.

10. When the voltage applied to a parallel circuit is increased,
    (a) total power increases.
    (b) total resistance remains the same.
    (c) total current increases.
    (d) all the above are correct.

11. A parallel circuit contains 100 branches, each with a 10 kΩ resistor. What is the total resistance of the circuit?
    (a) 1,000,000 Ω
    (b) 100 Ω
    (c) 1000 Ω
    (d) 10,000 Ω

12. When a resistive branch is removed from a parallel circuit, the total resistance ______________ and the total current ______________.
    (a) increases; increases
    (b) increases; decreases
    (c) decreases; decreases
    (d) decreases; increases

13. What voltage is applied to the circuit in Figure 7-22?
    (a) 324 V
    (b) 110 V
    (c) 55 V
    (d) 9 V

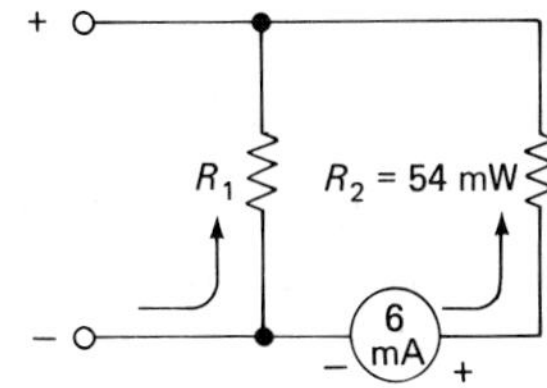

FIGURE 7-22

14. What is the equivalent resistance of the circuit in Figure 7-23?
    (a) 37.5 Ω
    (b) 150 Ω
    (c) 10 Ω
    (d) 5.98 Ω

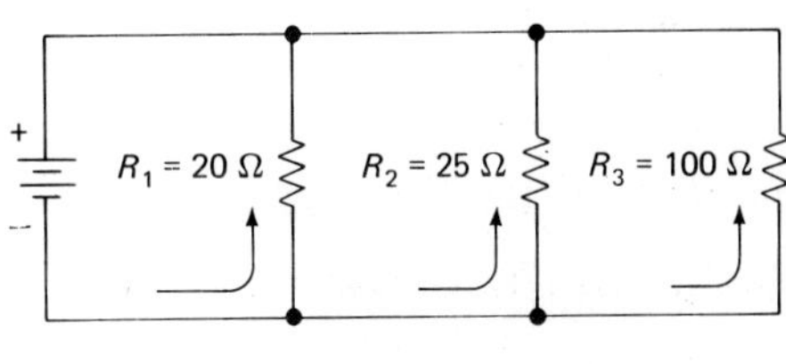

FIGURE 7-23

15. The currents flowing through A_1 and A_2 in Figure 7-24 are:

	A_1	A_2
(a)	22 mA	22 mA
(b)	20 mA	2 mA
(c)	11 mA	11 mA
(d)	2 mA	20 mA

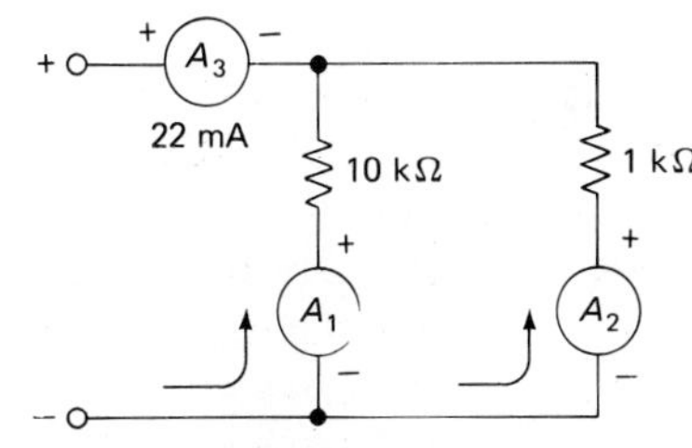

FIGURE 7-24

16. Total current in Figure 7-25 equals
    (a) 2.0 mA.
    (b) 2.2 mA.
    (c) 4.0 mA.
    (d) 6.0 mA.

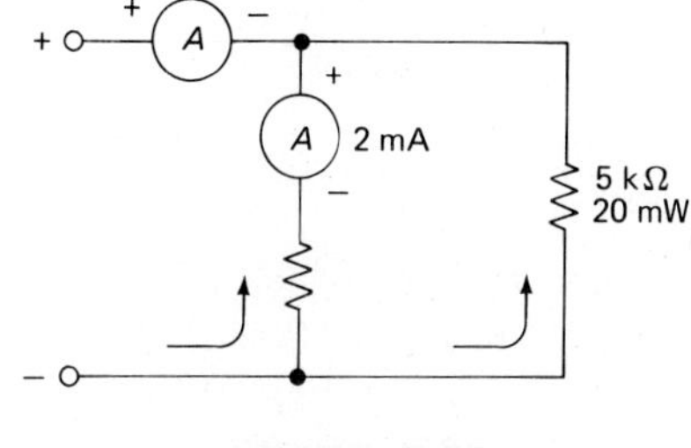

FIGURE 7-25

17. In Figure 7-26, total current equals
    (a) 33 mA.
    (b) 36 mA.
    (c) 34 mA.
    (d) 22 mA.

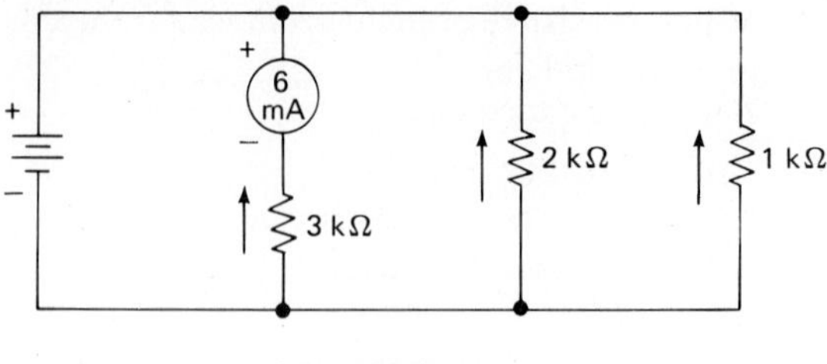

FIGURE 7-26

18. In Figure 7-27, which resistor dissipates the least power?
    (a) R_1
    (b) R_3
    (c) R_2
    (d) R_4

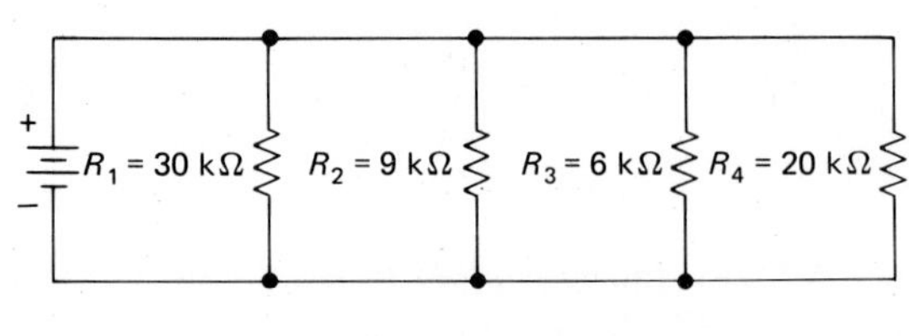

FIGURE 7-27

19. Refer to the circuit in Figure 7-28. If R_3 shorts, the voltage drop on R_1 will
    (a) equal total voltage.
    (b) increase.
    (c) be zero.
    (d) remain the same.

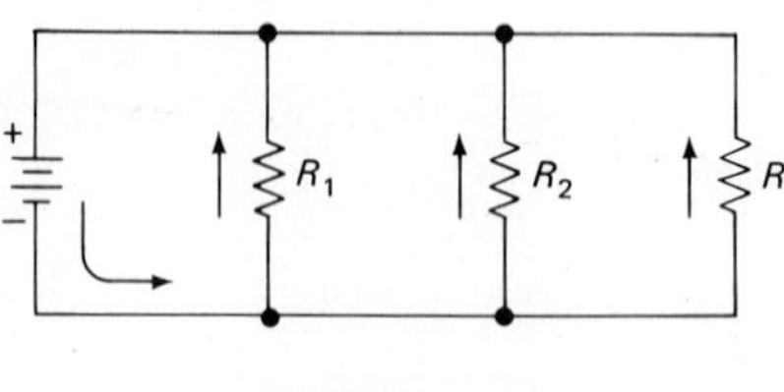

FIGURE 7-28

20. A 24 Ω resistor can be replaced by
    (a) two 12 Ω resistors in parallel.
    (b) two 24 Ω resistors in parallel.
    (c) a 20 Ω and a 40 Ω in series.
    (d) a 40 Ω and a 60 Ω resistor in parallel.

21. Refer to Figure 7-29. What is the ohmic value of R_2?
    (a) 4kΩ
    (b) 6kΩ
    (c) 14kΩ
    (d) 10kΩ

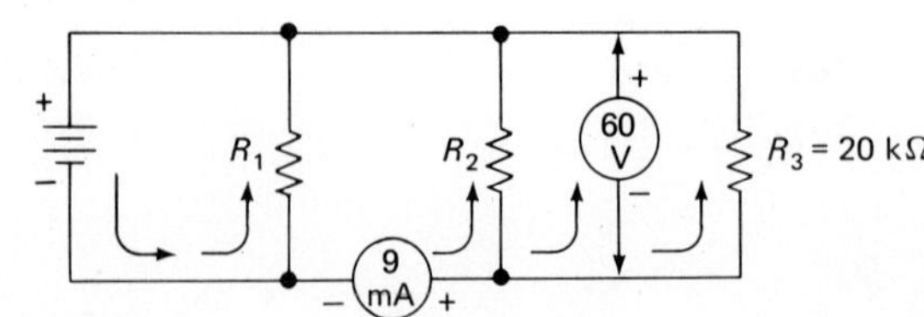

FIGURE 7-29

22. Refer to Figure 7-30. A short across R_1 causes the power dissipated by R_2 to
(a) increase.
(b) decrease.
(c) remain the same.

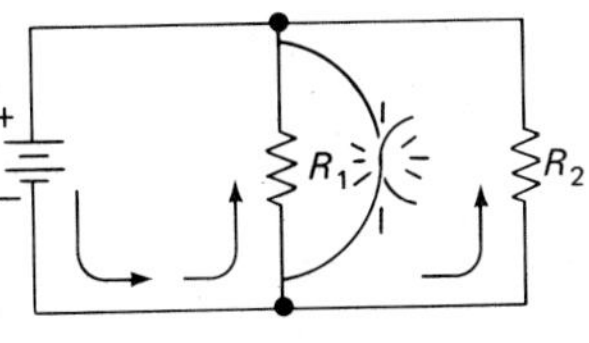

FIGURE 7-30

23. Refer to Figure 7-31. Adding another branch parallel to R_4 causes total power to
(a) increase.
(b) decrease.
(c) remain the same.

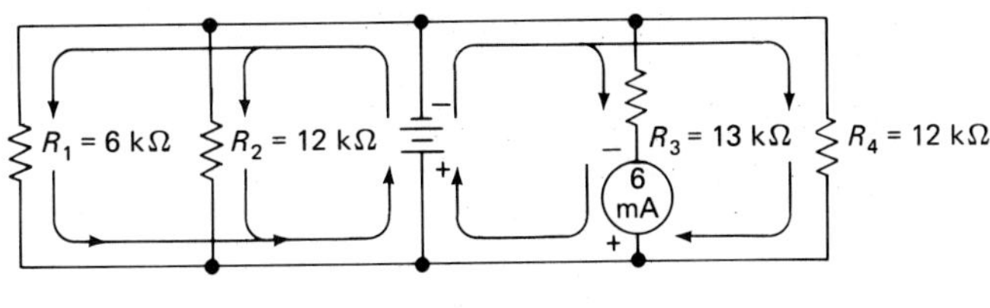

FIGURE 7-31

24. What is the problem in the circuit shown in Figure 7-32?
(a) R_1 is shorted.
(b) R_3 is shorted.
(c) R_1 is open.
(d) R_2 is open.

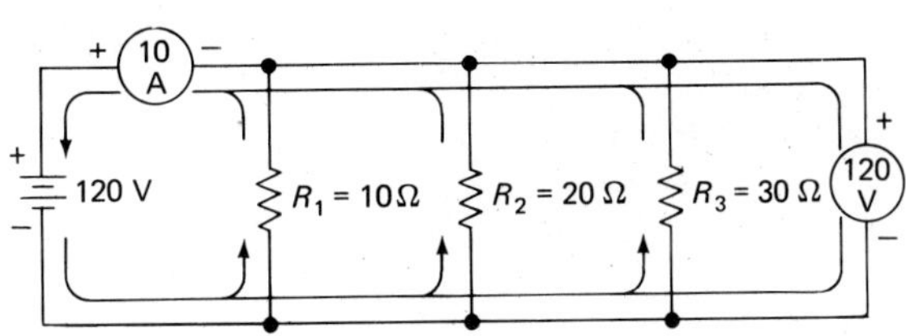

FIGURE 7-32

25. Refer to Figure 7-33. Ammeter A_1 reads 3 mA. If R_4 opens, what will ammeter A_2 read?
(a) 1 mA
(b) 2 mA
(c) 8 mA
(d) 6.5 mA

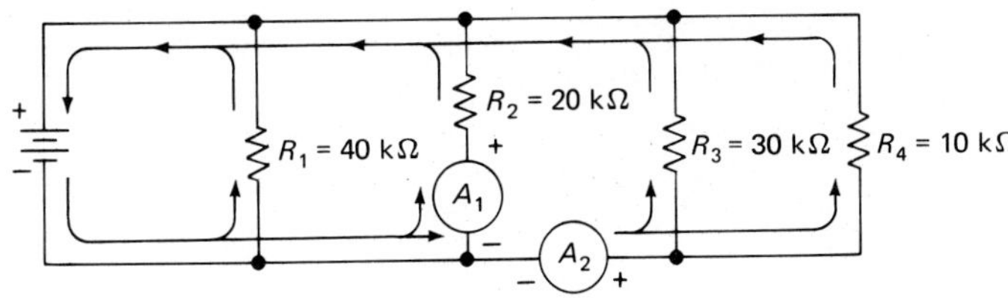

FIGURE 7-33

8

Combination-Resistive Circuits

Introduction

At first glance combination-resistive circuits may appear to be difficult, but in actuality they are combined circuits which include both series and parallel circuits. We have already studied series circuits and parallel circuits; in this section we apply our knowledge of these circuits to combination circuits.

Combination circuits contain a more complex arrangement of series- and parallel-resistor networks. Two types are discussed. First, we cover parallel-series combinations, and later series-parallel combinations.

Objectives

Each student is required to:

1. State the requirements for parallel-series and series-parallel combination circuits.
2. Identify schematic drawings of parallel-series and series-parallel resistive circuits and, if need be, redraw the schematics in simpler form.
3. Given series-parallel and parallel-series resistive circuit schematics:
   (a) Solve for the equivalent resistance of parallel networks and series strings.
   (b) Solve for total resistance of the circuit.
   (c) Calculate total circuit current.
   (d) Draw the equivalent (one-resistor) circuit which represents the entire circuit.
   (e) Solve for voltage, current, and power distribution in combination circuits.
4. Predict the effect on current and voltage distribution in combination circuits when:
   (a) Resistor values change.
   (b) Branches are added to the parallel network.
   (c) Branches are removed from the parallel network.

(d) Resistance is added in series.
(e) Series resistance is removed.

5. Theoretically troubleshoot combination-resistive circuits that contain opens or shorts.

Review of Series and Parallel Circuits

In a series-resistive circuit, components are connected end to end. This forces the current flowing in one resistor to flow through all the other resistors, with the result that applied voltage is divided across all resistors. The percentage of total resistance contained in a series resistor determines the percentage of applied voltage dropped by the resistor. Knowing this, we can make the following statements regarding series-resistive circuits similar to those in Figure 8-1a:

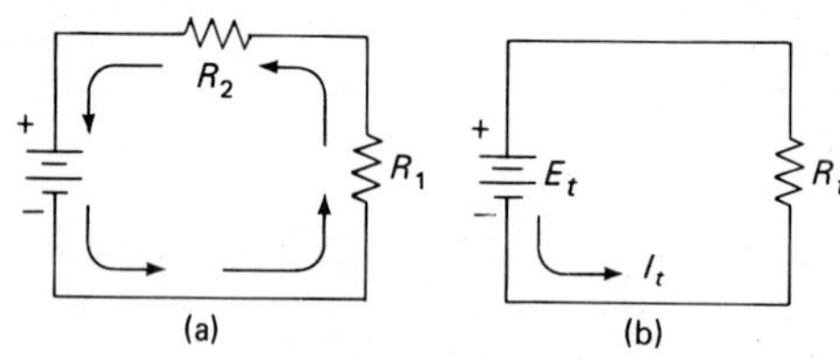

FIGURE 8-1

1. Current is equal at all points in the circuit.
2. Applied voltage is divided between resistors.
3. The largest resistor drops the largest voltage.
4. Total resistance of the circuit equals the sum of all individual resistors, with the result that total resistance is greater than that of either individual resistor.
5. Adding resistors to a circuit causes total resistance to increase and total current to decrease.
6. Series-connected resistors are often called *resistor strings*.
7. Total power in a series circuit equals the sum of all individual power dissipations.
8. Figure 8-1b is the equivalent circuit for Figure 8-1a. Note that it has only one resistor and that the value of that resistor equals R_t in the circuit shown.

Parallel-connected resistors are connected side by side, and the circuit has characteristics significantly different from those of the series string. For parallel circuits, such as that in Figure 8-2a, we can make the following statements:

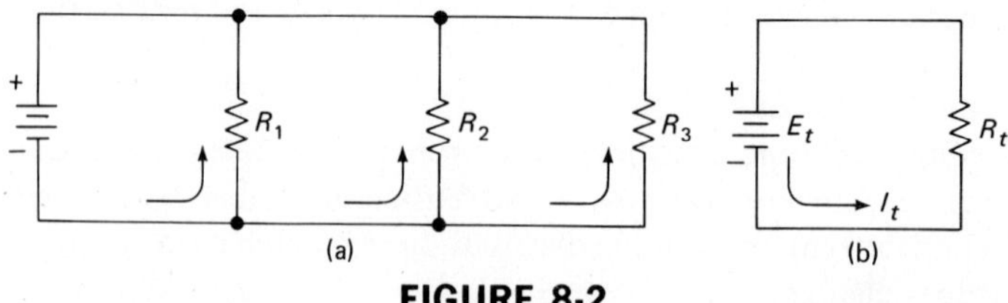

FIGURE 8-2

1. Each branch will have total voltage applied and will therefore drop total voltage.
2. Total current flow in the circuit is divided among the branches.

3. The branch with the smallest resistance has the largest current flow.
4. The total resistance of a parallel circuit is less than the resistance of the smallest branch.
5. As branches are added to a parallel circuit, total resistance decreases and total current increases.
6. Total power in a parallel-resistive circuit equals the sum of all individual power dissipations.
7. Parallel circuits are often called *parallel networks*.
8. Figure 8-2b is the equivalent circuit for Figure 8-2a. Note that its values are E_t, I_t, and R_t.

In combination circuits, the interconnection of series strings and parallel networks can be complex. In this chapter we discuss several combinations of string and network connections.

Parallel-Series Combinations

Branches Containing Two or More Resistors

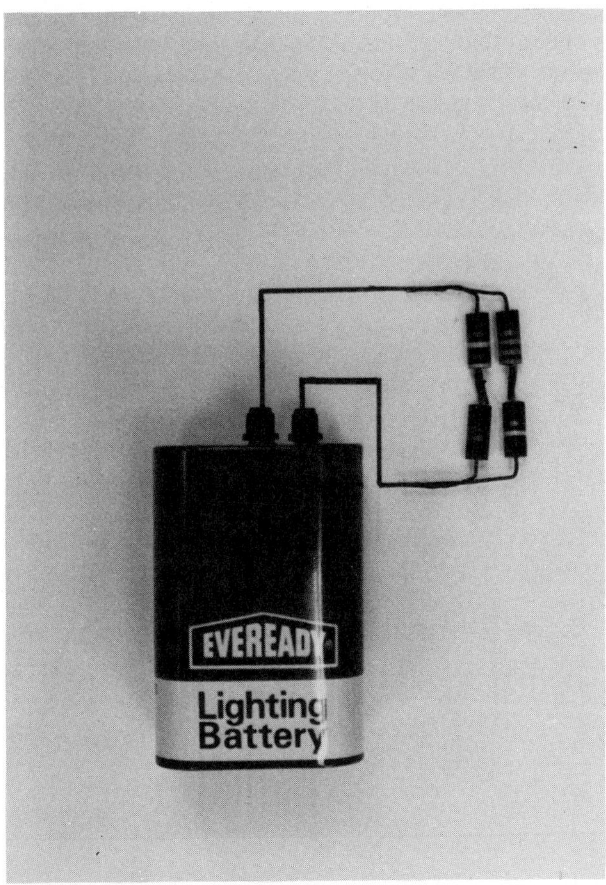

FIGURE 8-3
Parallel-Series Resistive Circuit (Photo by Riddell Allen)

When designing practical circuits it is often necessary to place a string of two or more resistors in a single branch of a parallel circuit (Figure 8-4a).

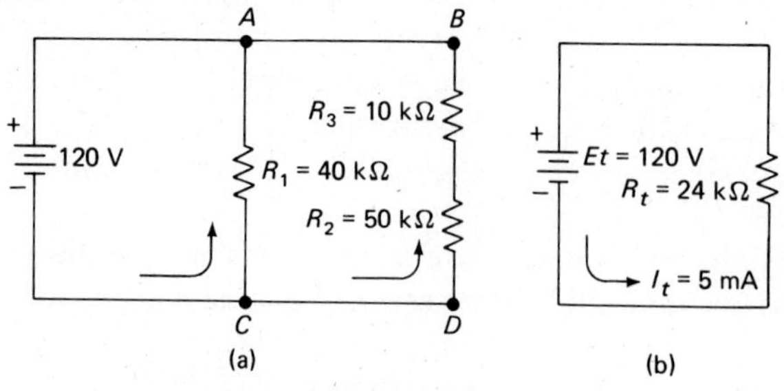

FIGURE 8-4

Branch 1 of the circuit contains one resistor, R_1, connected between points A and C, whereas branch 2 contains a series string, R_2 and R_3, connected between points B and D. This poses a different problem when we are calculating total resistance.

It is necessary to determine the amount of resistance in each branch before we use the formulas discussed for parallel networks. Branch 1 consists of R_1; R_2 and R_3 form the second branch. To find the resistance of each branch, or string, use the series-resistance formula:

$$R_{Br1} = R_1$$

$$R_{Br1} = 40\ \text{k}\Omega$$

and

$$R_{Br2} = R_2 + R_3$$

$$R_{Br2} = 50\text{k} + 10\text{k}$$

$$R_{Br2} = 60\ \text{k}\Omega$$

Once the branch resistances are known, the product-over-sum formula can be used to calculate total resistance:

$$R_t = \frac{R_{Br1} \times R_{Br2}}{R_{Br1} + R_{Br2}}$$

$$R_t = \frac{40\text{k} \times 60\text{k}}{40\text{k} + 60\text{k}}$$

$$R_t = 24\ \text{k}\Omega$$

When we know $E_t = 120$ V and $R_t = 24$ kΩ, we can solve for total current. With I_t known, we can draw the equivalent circuit (Figure 8-4b).

$$I_t = \frac{E_t}{R_t}$$

$$I_t = \frac{120\ \text{V}}{24\text{k}} = 5\ \text{mA}$$

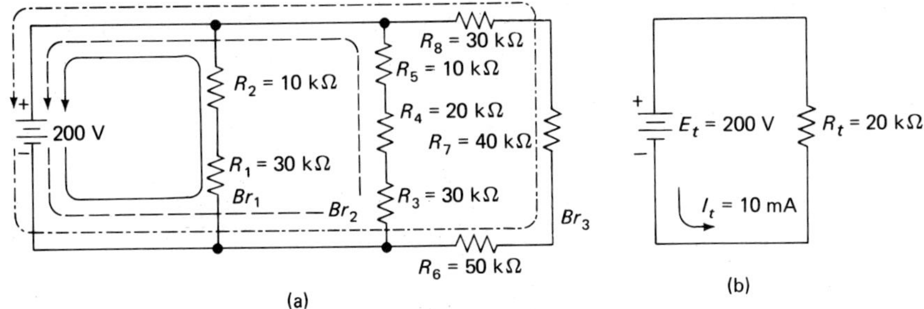

FIGURE 8-5

As mentioned, more complex parallel circuits often have two or more resistors in their branches. In Figure 8-5a, observe that each branch contains more than one resistor. Notice, too, that R_1 and R_2 are contained in one series string; R_3, R_4, and R_5 are in a second string, and R_6, R_7, and R_8 are in a third. These strings are grouped in a three-branch, parallel network whose branches are labeled Br_1, Br_2, and Br_3. Each branch presents an equivalent resistance to the battery which is the same as it would be if it contained only one resistor whose value equaled the sum of all its resistors. Let us examine the circuit one branch at a time. Figure 8-6 illustrates the battery and the closed loop labeled Br_1, which is traced with a solid line.

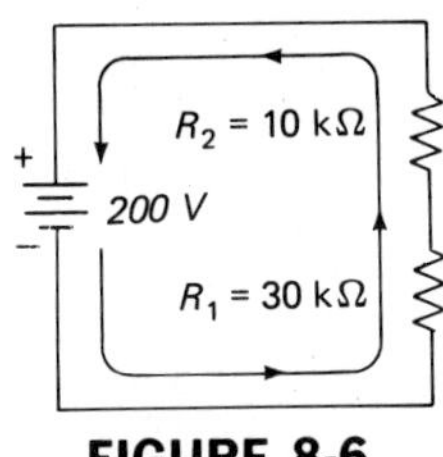

FIGURE 8-6

Analysis of the circuit in Figure 8-6 is identical to that of a series circuit. The formulas used, and the results obtained, are:

$$R_{Br1} = R_1 + R_2 = 40\ \text{k}\Omega$$

$$I_{Br1} = \frac{E_t}{R_{Br1}} = 5\ \text{mA}$$

$$E_{R1} = I_{Br1} \times R_1 = 150\ \text{V}$$

$$E_{R2} = I_{Br1} \times R_2 = 50\ \text{V}$$

$$E_t = E_{R1} + E_{R2} = 200\ \text{V}$$

$$P_{R1} = I_{Br1} \times E_{R1} = 750\ \text{mW}$$

$$P_{R2} = I_{Br1} \times E_{R2} = 250\ \text{mW}$$

$$P_{Br1} = I_{Br1} \times E_t = 1000\ \text{mW}$$

Now we consider branch 2 (dashed loop) as a separate analysis (Figure 8-7). Solutions for this closed loop (branch 2) are:

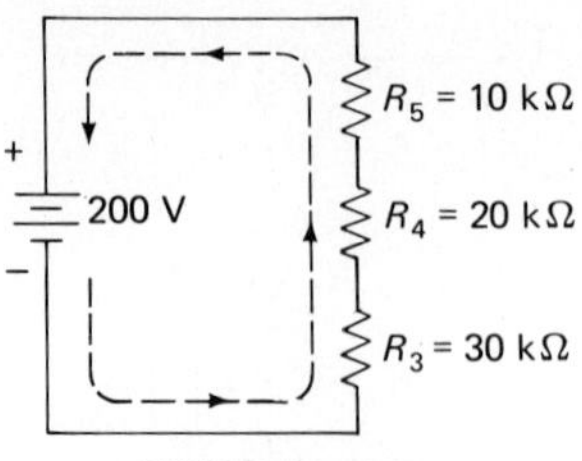

FIGURE 8-7

$$R_{Br2} = R_3 + R_4 + R_5 = 60\text{ k}\Omega$$

$$I_{Br2} = \frac{E_t}{R_{Br2}} = 3.33\text{ mA}$$

$$E_{R3} = I_{Br2} \times R_3 = 100\text{ V}$$

$$E_{R4} = I_{Br2} \times R_4 = 66.7\text{ V}$$

$$E_{R5} = I_{Br2} \times R_5 = 33.3\text{ V}$$

$$E_t = E_{R3} + E_{R4} + E_{R5} = 200\text{ V}$$

$$P_{R3} = I_{Br2} \times E_{R3} = 333\text{ mW}$$

$$P_{R4} = I_{Br2} \times E_{R4} = 222\text{ mW}$$

$$P_{R5} = I_{Br2} \times E_{R5} = 111\text{ mW}$$

$$P_{Br2} = I_{Br2} \times E_t = 666\text{ mW}$$

Consider branch 3 (dashed loop), shown in Figure 8-8 as a separate, closed loop. Branch-3 solutions are:

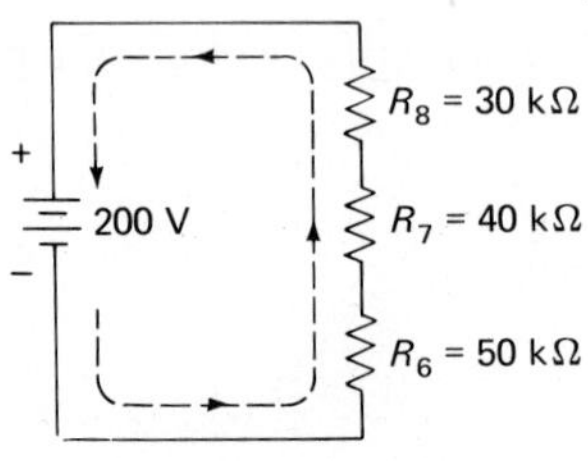

FIGURE 8-8

$$R_{Br3} = R_6 + R_7 + R_8 = 120\text{ k}\Omega$$

$$I_{Br3} = \frac{E_t}{R_{Br3}} = 1.67\text{ mA}$$

$$E_{R6} = I_{Br3} \times R_6 = 83.3\text{ V}$$

$$E_{R7} = I_{Br3} \times R_7 = 66.7\text{ V}$$

$$E_{R8} = I_{Br3} \times R_8 = 50\text{ V}$$

$$E_t = E_{R6} + E_{R7} + E_{R8} = 200\text{ V}$$

$$P_{R6} = I_{Br3} \times E_{R6} = 139.1\text{ mW}$$

$$P_{R7} = I_{Br3} \times E_{R7} = 111.4\text{ mW}$$

$$P_{R8} = I_{Br3} \times E_{R8} = 83.5\text{ mW}$$

$$P_{Br3} = I_{Br3} \times E_t = 334\text{ mW}$$

$$P_t = I_t \times E_t = 2000\text{ mW}$$

Refer to the original circuit in Figure 8-5a. For an analysis of the complete network, it is necessary to find the resistance in each branch as illustrated by analysis of the series loops, which resulted in the following values:

$$R_{Br1} = 40\text{ k}\Omega$$

$$R_{Br2} = 60\text{ k}\Omega$$

$$R_{Br3} = 120\text{ k}\Omega$$

With the branch resistances known, we can calculate total resistance, using the reciprocal formula:

$$\frac{1}{R_t} = \frac{1}{R_{Br1}} + \frac{1}{R_{Br2}} + \frac{1}{R_{Br3}} = 20\text{ k}\Omega$$

$$I_t = \frac{E_t}{R_t} = 10\text{ mA}$$

With E_t, R_t, and I_t known, we can draw the equivalent circuit for Figure 8-5a; Figure 8-5b is this equivalent.

Parallel-Series Circuit Review

Total current can be calculated by addition of the three branch currents or by employing Ohm's law.

The sum of the voltage drops in each loop is equal to applied voltage. Within each branch, the largest resistor drops the most voltage and dissipates the most power. Between branches, the branch with the least branch resistance has the highest current flow and dissipates the most power. Notice here that in the circuit's operation, all the requirements of Kirchhoff's and Ohm's laws are met.

One more item that must be emphasized is power dissipation. In series circuits, the largest resistor dissipates the most power. In parallel circuits, the branch with the smallest resistance dissipates the most power. These two statements are true regarding parallel-series circuits. When it comes to stating which resistor will dissipate the most power, however, we must calculate the power dissipation for each resistor and then select the one that dissipates the most power.

Troubleshooting Parallel-Series Combinations

Each combination of connected components is different. Therefore, each requires varying troubleshooting techniques to meet a particular situation. Troubleshooting the parallel-series combination requires application of both series and parallel-circuit troubleshooting techniques. Before proceeding, let us review these techniques.

Series-Circuit Symptoms and Techniques

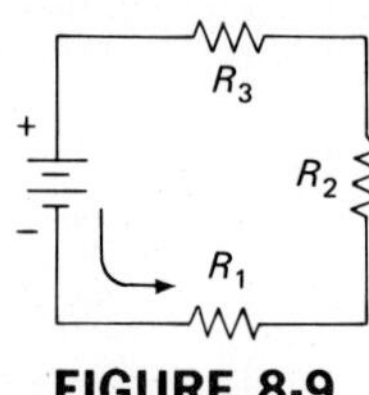

FIGURE 8-9

Figure 8-9 is used to review techniques of series troubleshooting. An open resistor causes:

1. All current flow to stop.
2. Total resistance to be infinity.
3. Total voltage to be dropped across the open resistor.
4. All other resistors to drop zero volts.

A shorted resistor causes:

1. Total current to increase.
2. Total resistance to decrease.
3. Zero volts to be dropped across the shorted resistor.
4. All other resistors to drop more voltage than normal.

Parallel-Circuit Symptoms and Techniques

Figure 8-10 is used to explain troubleshooting parallel networks. An open resistor causes:

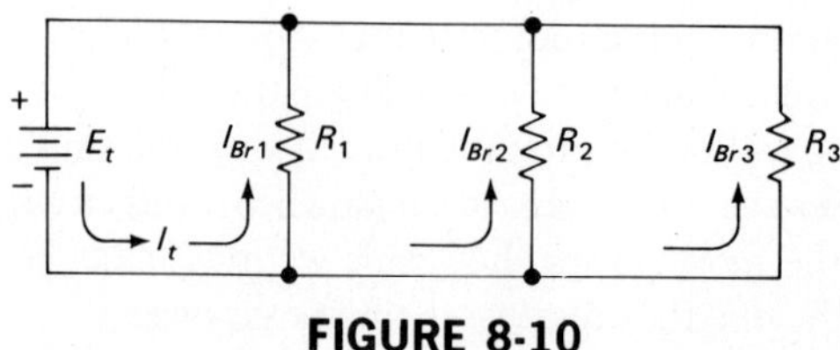

FIGURE 8-10

1. Total current to decrease.
2. Total resistance to increase.
3. All branches to drop E_t whether they are open or not.
4. All devices to continue to work except the one contained in the open branch.

A shorted branch causes:

1. Total current to approach infinity because all current flows through the short and because the current meets no opposition.
2. Total resistance to approach zero ohms because R_t must be less than the resistance of the smallest branch (which is a short).
3. All branches to drop zero volts.
4. All devices to stop working.
5. It is probable that the circuit will be damaged by the current overload.

Applying These Techniques to Parallel-Series Combinations

A simple circuit is illustrated in Figure 8-11. Note that it contains a power source, a fuse, and four resistors. R_1 and R_2 form one string and R_3 and R_4 form another. The strings are then connected in a parallel network. First, let us solve the circuit under *normal* conditions:

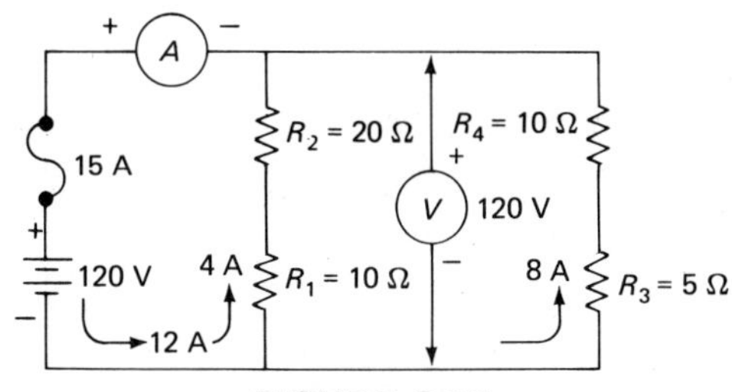

FIGURE 8-11

$$R_{Br1} = R_1 + R_2 = 10\ \Omega + 20\ \Omega = 30\ \Omega$$

$$R_{Br2} = R_3 + R_4 = 5\ \Omega + 10\ \Omega = 15\ \Omega$$

$$R_t = \frac{R_{Br1} \times R_{Br2}}{R_{Br1} + R_{Br2}} = 10\ \Omega$$

$$I_t = \frac{E_t}{R_t} = \frac{120\ \text{V}}{10\ \Omega} = 12\ \text{A}$$

$$I_{Br1} = \frac{E_t}{R_{Br1}} = \frac{120\ \text{V}}{30\ \Omega} = 4\ \text{A}$$

$$E_{R1} = I_{Br1} \times R_1 = 4\ \text{A} \times 10\ \Omega = 40\ \text{V}$$

$$E_{R2} = I_{Br1} \times R_2 = 4\ \text{A} \times 20\ \Omega = 80\ \text{V}$$

$$I_{Br2} = \frac{E_t}{R_{Br2}} = \frac{120\ \text{V}}{15\ \Omega} = 8\ \text{A}$$

$$E_{R3} = I_{Br2} \times R_3 = 8\ \text{A} \times 5\ \Omega = 40\ \text{V}$$

$$E_{R4} = I_{Br2} \times R_4 = 8\ \text{A} \times 10\ \Omega = 80\ \text{V}$$

Now we can determine what happens when a resistor opens or shorts by comparing conditions. First, we consider shorted resistors (Figure 8-12).

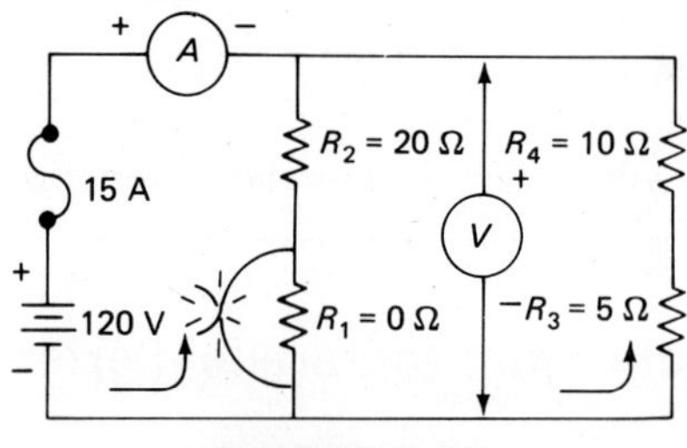

FIGURE 8-12

With R_1 shorted, the following conditions exist:

$$R_{Br1} = R_1 + R_2 = 0\ \Omega + 20\ \Omega = 20\ \Omega$$

$$R_{Br2} = R_3 + R_4 = 5\ \Omega + 10\ \Omega = 15\ \Omega$$

$$I_{Br1} = \frac{E_t}{R_{Br1}} = \frac{120\ \text{V}}{20\ \Omega} = 6\ \text{A}$$

$$I_{Br2} = \frac{E_t}{R_{Br2}} = \frac{120\ \text{V}}{15\ \Omega} = 8\ \text{A}$$

$$I_t = I_{Br1} + I_{Br2} = 6\ \text{A} + 8\ \text{A} = 14\ \text{A}$$

$$E_{R1} = I_{Br1} \times R_1 = 6\ \text{A} \times 0\ \Omega = 0\ \text{V}$$

$$E_{R2} = I_{Br1} \times R_2 = 6\ \text{A} \times 20\ \Omega = 120\ \text{V}$$

$$E_{R3} = I_{Br2} \times R_3 = 8\ \text{A} \times 5\ \Omega = 40\ \text{V}$$

$$E_{R4} = I_{Br2} \times R_4 = 8\ \text{A} \times 10\ \Omega = 80\ \text{V}$$

When these values are compared to normal conditions, we see that I_t, I_{Br1}, E_{R1}, and E_{R2} have changed as a result of R_1 having shorted. If R_1 is good, and one of the other resistors shorts, the following symptoms occur:

R_2 shorts:

1. $R_{Br1} = 10\ \Omega$
2. $I_{Br1} = 12\ \text{A}$

3. $I_t = 20$ A
4. To prevent damage to the rest of the circuit, the fuse opens when I_t exceeds 15 A

R_3 shorts:

1. $R_{Br2} = 10\ \Omega$
2. $I_{Br2} = 12$ A
3. $I_t = 16$ A
4. Fuse blows when $I_t = 15$ A

R_4 shorts:

1. $R_{Br2} = 5\ \Omega$
2. $I_{Br2} = 24$ A
3. $I_t = 28$ A
4. Fuse blows when $I_t = 15$ A

Refer to Figure 8-13. We use it to analyze the result of opening either resistor.

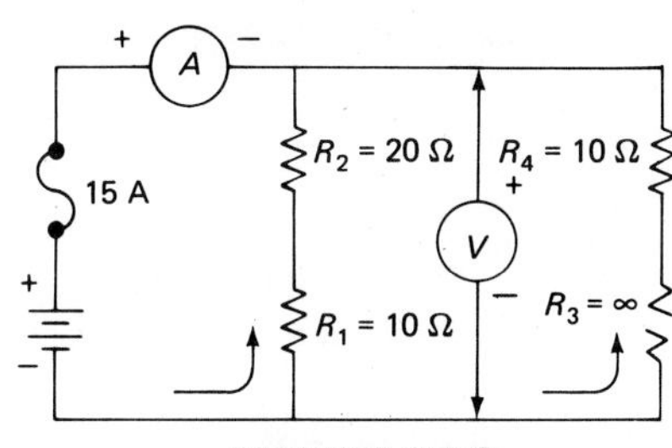

FIGURE 8-13

1. If either R_1 or R_2 opens, R_{Br1} is infinity.
   (a) $I_{Br1} = 0$ A
   (b) R_3 and R_4 are not affected
   (c) I_t decreases
   (d) Devices represented by R_1 and R_2 are inoperative
2. If either R_3 or R_4 opens, R_{Br2} is infinity.
   (a) $I_{Br2} = 0$ A
   (b) R_1 and R_2 are not affected
   (c) I_t decreases
   (d) Devices represented by R_3 and R_4 are inoperative.

Figure 8-14 presents a more complex circuit. Voltmeters and ammeters are connected to the circuit at points where measurements are to be made. R_2 is shorted in branch 1 and R_4 is open in branch 2. Under these conditions, the following indications are presented on the meters as they are connected to the circuit:

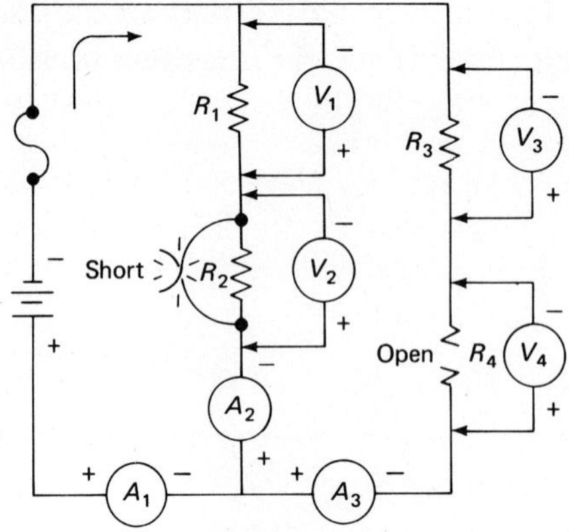

FIGURE 8-14

A_1 indicates some value other than normal.
A_2 indicates higher than normal.
A_3 indicates zero.
V_1 indicates higher than normal and equals E_t.
V_2 indicates zero volts across a short.
V_3 indicates zero volts; no current can reach it.
V_4 indicates approximately the applied voltage because it bridges the open and its high resistance drops most of the applied voltage.

It is now apparent that each combination of connections requires its own analysis. The constants are:

1. A voltmeter indicates zero volts when connected across a short.
2. Voltmeter indications are approximately equal to applied voltage when the voltmeter is connected across an open.
3. An ammeter indicates less than normal when connected in the series portion of a circuit that contains an open component.
4. An ammeter indicates higher than normal when connected in the series portion of a circuit that contains a shorted component.
5. An ammeter connected in parallel branches indicates normal branch current when an open exists in a separate branch, or zero when connected in the branch containing an open.
6. An ammeter indicates branch current higher than normal when connected in a branch that contains a short; it indicates lower than normal when connected in a branch other than the one containing the short.

Summary

We have developed the following points, which should be remembered:

1. Series strings of resistors obey the same rules as series-resistive circuits.
   (a) Applied voltage is divided among individual resistors.
   (b) Current is the same at all points within the string.

   (c) The total resistance of a string equals the sum of all individual resistors.
   (d) An open resistor stops current flow within a string and causes total current to decrease.
   (e) A shorted resistor causes string and total current to increase.
2. Parallel-series resistive networks obey the same rules as parallel-resistive circuits.
   (a) The network functions as though each branch contained only one resistor.
   (b) Total resistance of the circuit is calculated, using parallel analysis.
   (c) Total current of the network equals the sum of all branch currents.
   (d) Applied voltage is dropped across each string.
   (e) An open branch does not affect the operation of any other branch.
   (f) A shorted branch shorts the entire network.
   (g) It is possible for a string to contain a shorted resistor and the circuit continue to operate.
3. Analysis of a parallel-series circuit involves the use of:
   (a) Series analysis to determine string resistance.
   (b) Parallel analysis to determine total resistance.
   (c) Ohm's law to determine total current or voltage.
   (d) Parallel analysis to determine individual branch currents.
   (e) Series analysis to calculate individual voltage drops within the resistive strings.

Self-Check

Answer each item by indicating whether the statement is true or false.

1. ____ The first step in analyzing a parallel-series combination is to solve for R_t, using parallel-circuit techniques.

2. ____ The largest resistor contained in the smallest branch dissipates more power than any other resistor in the circuit.

3. ____ An open resistor in a string stops all current flow.

4. ____ Total current in a parallel-series combination equals the sum of all branch currents.

5. ____ A shorted resistor in a string causes total current to increase.

6. ____ Total current in a parallel-series combination equals the sum of all resistor currents.

Series-Parallel Combination Circuits

Analysis of series-parallel circuits requires that we use formulas learned in the preceding chapters. Although the way formulas are applied varies somewhat, the basic rules are the same. Figure 8-16 contains a simple series-parallel resistive circuit. This circuit is similar to the circuit in the dashboard lights of an automobile. R_1 is a variable resistor whose adjustment varies the brightness of the lights R_2 and R_3. To help us analyze this type of circuit, it is often best to redraw the circuit in a form that allows a better look at the components of the circuits.

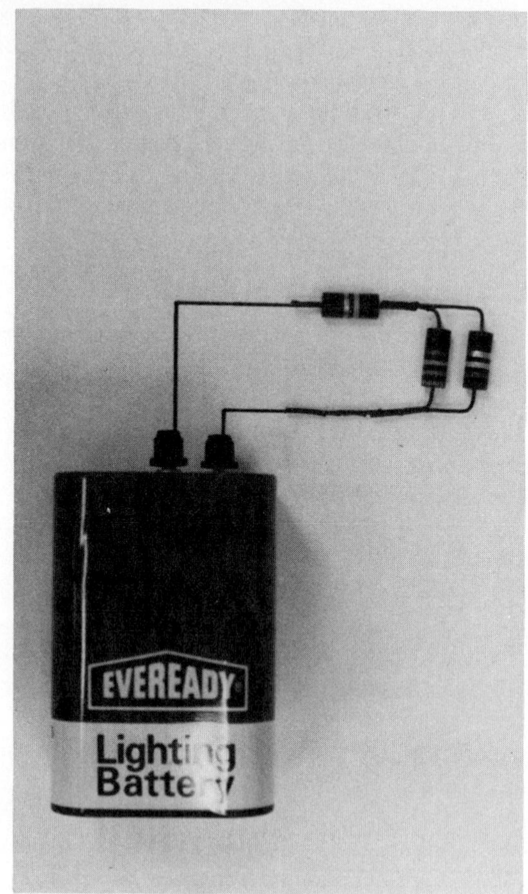

FIGURE 8-15
Series-Parallel Resistive Circuit (Photo by Riddell Allen)

Once the circuit is in recognizable form, we can complete the analysis. A new equivalent circuit results as each step of the analysis is done. An equivalent circuit is one where two or more resistors are combined in an equivalent resistor. In parallel and series circuits, total resistance is the equivalent resistance of all resistors contained in the circuit. To reduce the series-parallel resistive circuit to its equivalent, we must proceed through it in stages. Consider Figure 8-17.

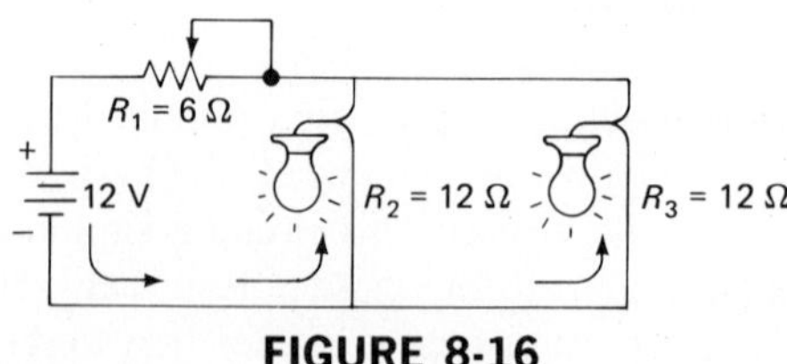

FIGURE 8-16

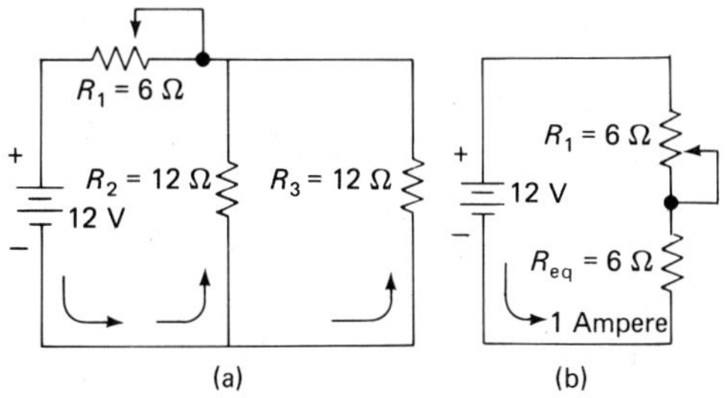

(a) (b)

FIGURE 8-17

Figure 8-17a is identical to Figure 8-16, except that the lamps are represented by resistance symbols. Figure 8-17b represents the same circuit after R_2 and R_3 have been reduced to their equivalent resistance, which can be calculated by using the product-over-sum formula. To solve for R_{eq}, proceed as follows:

$$R_{eq} = \frac{R_2 \times R_3}{R_2 + R_3}$$

$$R_{eq} = \frac{12 \times 12}{12 + 12}$$

$$R_{eq} = \frac{144}{24}$$

$$R_{eq} = 6\ \Omega$$

Once the equivalent has been determined, we can replace the parallel network with its equivalent resistance. By redrawing the circuit, we obtain the circuit shown in Figure 8-17b. To find total resistance from this point requires application of the series formula:

$$R_t = R_1 + R_{eq}$$

$$R_t = 6\ \Omega + 6\ \Omega$$

$$R_t = 12\ \Omega$$

Now that we know total resistance, we can apply Ohm's law and solve for total current, I_t.

$$I_t = \frac{E_t}{R_t}$$

$$I_t = \frac{12\text{ V}}{12\ \Omega}$$

$$I_t = 1\text{ A}$$

We already know that current flow in a series circuit is equal at all points. (Figure 8-17b is a series circuit.) Therefore, we can use I_t to solve E_{R1} and E_{Req}:

$$E_{R1} = I_t \times R_1$$

$$E_{R1} = 1\text{ A} \times 6\ \Omega$$

$$E_{R1} = 6\text{ V}$$

and

$$E_{Req} = I_t \times R_{eq}$$

$$E_{Req} = 1\text{ A} \times 6\ \Omega$$

$$E_{Req} = 6\text{ V}$$

Because R_1 is the series resistor, total current flows through it; but to flow through the two paths of the parallel network, R_2 and R_3, the current must split. Voltage E_{Req} already calculated can be used as a power source for the parallel network (Figure 8-18). This circuit can be analyzed by using parallel analysis identical to that learned in Chapter 7. To solve for branch currents and resistor-power dissipations, we use:

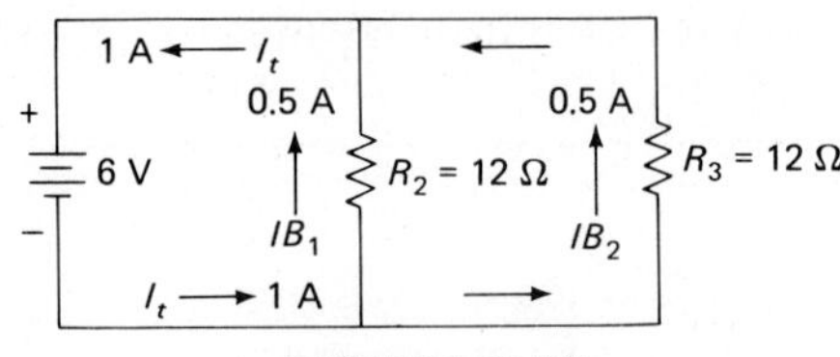

FIGURE 8-18

$$I_{Br1} = \frac{E_{Br1}}{R_2}$$

$$I_{Br1} = \frac{6\text{ V}}{12\ \Omega}$$

$$I_{Br1} = 0.5\text{ A}$$

and

$$I_{Br2} = \frac{E_{Br2}}{R_3}$$

$$I_{Br2} = \frac{6\text{ V}}{12\ \Omega}$$

$$I_{Br2} = 0.5\text{ A}$$

and

$$P_{R2} = I_{Br1}^2 \times R_2$$

$$P_{R2} = 0.5^2 \times 12\ \Omega$$

$$P_{R2} = 3\text{ W}$$

and

$$P_{R3} = I_{Br2}^2 \times R_3$$

$$P_{R3} = 0.5^2 \times 12\ \Omega$$

$$P_{R3} = 3\text{ W}$$

Calculate P_{R1}:

$$P_{R1} = I_t^2 \times R_1$$

$$P_{R1} = 1^2 \times 6\ \Omega$$

$$P_{R1} = 6\ \text{W}$$

Total power then equals:

$$P_t = P_{R1} + P_{R2} + P_{R3}$$

$$P_t = 6\ \text{W} + 3\ \text{W} + 3\ \text{W}$$

$$P_t = 12\ \text{W}$$

The formulas used to reach these solutions could just as easily have been any other formulas suitable for solving the values of this circuit. Other circuits may involve other versions of the formulas, and still others may require different applications. Examine Figure 8-19a as we complete its analysis:

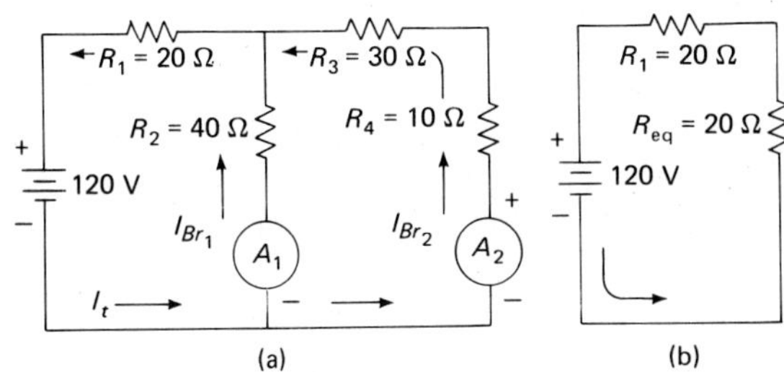

FIGURE 8-19

In this circuit, branch 2 consists of two resistors. To solve parallel circuits, we must calculate the branch resistance; then the equivalent of the parallel networks can be solved. R_{Br2} can be found by:

$$R_{Br2} = R_3 + R_4 = 30\ \Omega + 10\ \Omega = 40\ \Omega$$

We find R_{eq} by:

$$R_{eq} = \frac{R}{N} = \frac{40\ \Omega}{2} = 20\ \Omega$$

When the two-resistor series equivalent of this circuit is drawn, we have Figure 8-19b. Completing the analysis, we get:

$$R_t = R_1 + R_{eq} = 20\ \Omega + 20\ \Omega = 40\ \Omega$$

$$I_t = \frac{E_t}{R_t} = \frac{120\ \text{V}}{40\ \Omega} = 3\ \text{A}$$

and

$$E_{R1} = I_t \times R_1 = 3\ \text{A} \times 20\ \Omega = 60\ \text{V}$$

$$E_{Req} = I_t \times R_{eq} = 3\ \text{A} \times 20\ \Omega = 60\ \text{V}$$

When the parallel network in Figure 8-19a is redrawn, using E_{Req} as its power source, we have Figure 8-20. Solving for the remaining values in the circuit, we get:

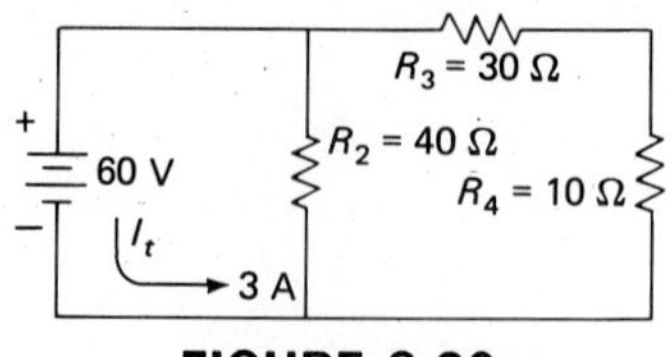

FIGURE 8-20

$$I_{Br1} = \frac{E_{Req}}{R_2} = \frac{60 \text{ V}}{40\ \Omega} = 1.5 \text{ A}$$

$$I_{Br2} = \frac{E_{Req}}{R_{Br2}} = \frac{60 \text{ V}}{40\ \Omega} = 1.5 \text{ A}$$

$$E_{R2} = E_{Req} = 60 \text{ V}$$

$$E_{R3} = I_{Br2} \times R_3 = 1.5 \text{ A} \times 30\ \Omega = 45 \text{ V}$$

$$E_{R4} = I_{Br2} \times R_4 = 1.5 \text{ A} \times 10\ \Omega = 15 \text{ V}$$

$$P_{R2} = I_{Br1} \times E_{Br1} = 1.5 \text{ A} \times 60 \text{ V} = 90 \text{ W}$$

$$P_{R3} = I_{Br2} \times E_{R3} = 1.5 \text{ A} \times 45 \text{ V} = 67.5 \text{ W}$$

$$P_{R4} = I_{Br2} \times E_{R4} = 1.5 \text{ A} \times 15 \text{ V} = 22.5 \text{ W}$$

In solving for P_{R1}, as shown, we use a combination of these solutions to calculate total power. Two examples are:

$$P_{R1} = I_t \times E_{R1} = 3 \text{ A} \times 60 \text{ V} = 180 \text{ W}$$

$$P_t = I_t \times E_t = 3 \text{ A} \times 120 \text{ V} = 360 \text{ W}$$

Then

$$P_t = P_{R1} + P_{R2} + P_{R3} + P_{R4}$$

$$P_t = 180 \text{ W} + 90 \text{ W} + 67.5 \text{ W} + 22.5 \text{ W}$$

$$P_t = 360 \text{ W}$$

When the analysis is complete, we have a circuit that has been reduced to its equivalent values. The values can be used to construct the equivalent circuit for Figure 8-19a. Figure 8-21 is the resulting equivalent circuit.

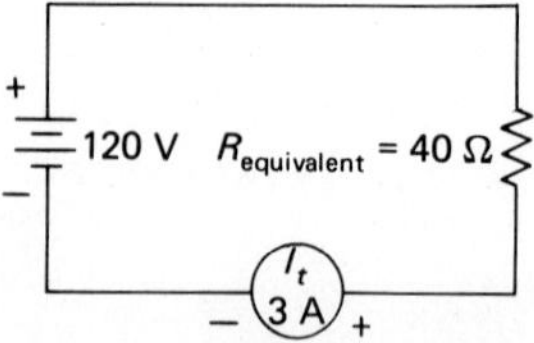

FIGURE 8-21

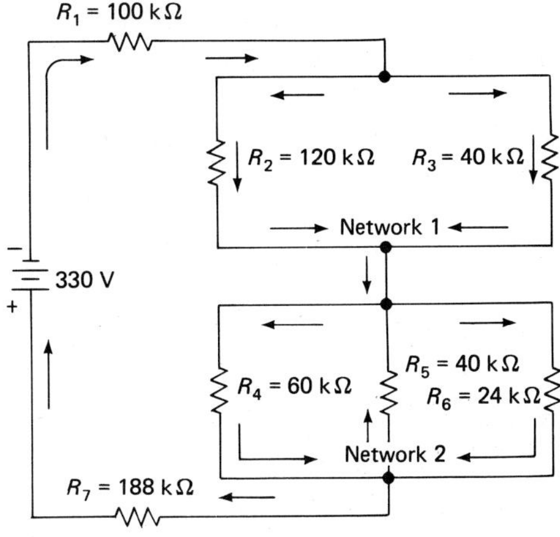

FIGURE 8-22

A series-parallel circuit such as that shown in Figure 8-22 is somewhat more complex. To understand its operation, we must analyze it, using the following procedures. Reduce each parallel network to its R_{eq}. For network 1, we do this by:

$$R_{eq1} = \frac{R_2 \times R_3}{R_2 + R_3} = 30 \text{ k}\Omega$$

and for network 2:

$$\frac{1}{R_{eq2}} = \frac{1}{R_4} + \frac{1}{R_5} + \frac{1}{R_6} = 12 \text{ k}\Omega$$

The circuit can be drawn as the series equivalent in Figure 8-23. Solving Figure 8-23 for the following values, we have:

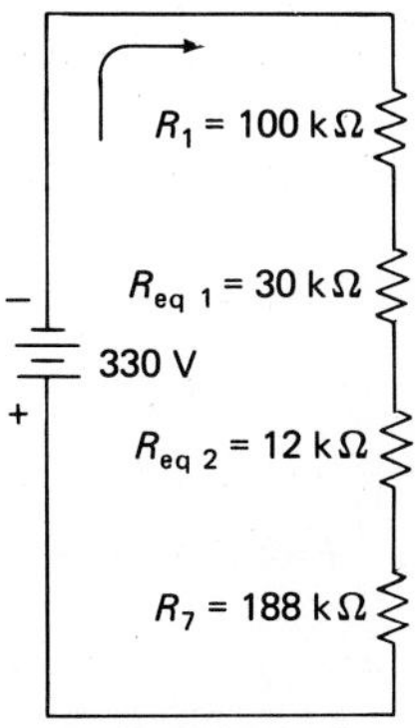

FIGURE 8-23

$$R_t = R_1 + R_{eq1} + R_{eq2} + R_7 = 330 \text{ k}\Omega$$

$$I_t = \frac{E_t}{R_t} = 1 \text{ mA}$$

$$P_t = I_t \times E_t = 330 \text{ mW}$$

$$E_{R1} = I_t \times R_1 = 100 \text{ V}$$

$$P_{R1} = I_t \times E_{R1} = 100 \text{ mW}$$

$$E_{R7} = I_t \times R_7 = 188 \text{ V}$$

$$P_{R7} \times I_t \times E_{R7} = 188 \text{ mW}$$

$$E_{Req1} = I_t \times R_{eq1} = 30 \text{ V}$$

$$E_{Req2} = I_t \times R_{eq2} = 12 \text{ V}$$

Using E_{Req1} as its power source, network 1 can be redrawn as shown in Figure 8-24. Solving the values for Figure 8-24, we get:

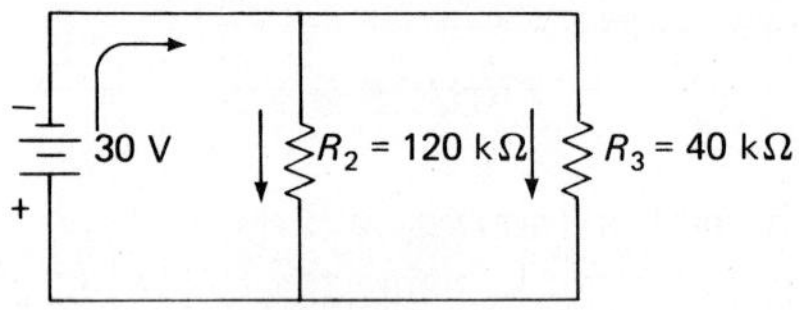

FIGURE 8-24

$$E_{Req1} = E_{R2} = E_{R3}$$

$$I_{R2} = \frac{E_{R2}}{R_2} = 0.25 \text{ mA}$$

$$I_{R3} = \frac{E_{R3}}{R_3} = 0.75 \text{ mA}$$

$$P_{R2} = I_{R2} \times E_{R2} = 7.5 \text{ mW}$$

$$P_{R3} = I_{R3} \times I_{R3} = 22.5 \text{ mW}$$

Using E_{Req2} as the power source, we redraw network 2 as shown in Figure 8-25. Solving for these values, we have:

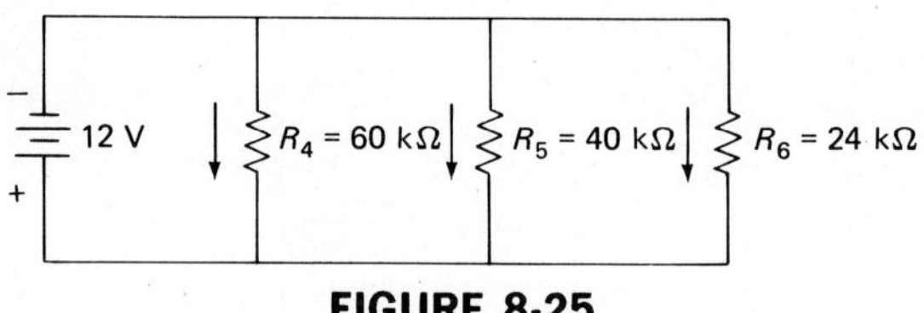

FIGURE 8-25

$$E_{Req2} = E_{R4} = E_{R5} = E_{R6}$$

$$I_{R4} = \frac{E_{R4}}{R_4} = 0.2 \text{ mA}$$

$$I_{R5} = \frac{E_{R5}}{R_5} = 0.3 \text{ mA}$$

$$I_{R6} = \frac{E_{R6}}{R_6} = 0.5 \text{ mA}$$

$$P_{R4} = I_{R4} \times E_{R4} = 2.4 \text{ mW}$$

$$P_{R5} = I_{R5} \times E_{R5} = 3.6 \text{ mW}$$

$$P_{R6} = I_{R6} \times E_{R6} = 6.0 \text{ mW}$$

By way of a check, we can solve for total power as follows:

$$P_t = P_{R1} + P_{R2} + P_{R3} + P_{R4} + P_{R5} + P_{R6} + P_{R7}$$

$$P_t = 100 \text{ mW} + 7.5 \text{ mW} + 22.5 \text{ mW} + 2.4 \text{ mW} + 3.6 \text{ mW} + 6 \text{ mW} + 188 \text{ mW}$$

$$P_t = 330 \text{ mW}$$

Self-Check

Complete each item by inserting the words and/or numbers needed to make a true and complete statement.

7. Refer to Figure 8-23. Resistor ______________ dissipates the most power in this circuit.

8. *Reduce to its equivalent* means to solve the circuit for ______________ ______________, ______________ ______________, and ______________ ______________.

9. Increasing the size of a series resistor in a series-parallel network causes total ______________ to increase and total ______________ to decrease.

10. The ______________ voltage drop must be calculated before we calculate voltage and current distribution in a parallel network that is part of a series-parallel circuit.

11. Adding a branch to the parallel network of a series-parallel circuit causes ______________ ______________ to increase.

12. Removing a branch from the parallel network of a series-parallel circuit causes total resistance to ______________.

Troubleshooting Series-Parallel Resistive Circuits

Troubleshooting series-parallel circuits involves everything we have learned up to now. The problems found will still be those of opens and shorts. The rules already established still apply:

1. Any time an open occurs in a circuit, total resistance increases and total current decreases.

2. A shorted component in a circuit causes total resistance to decrease and total current to increase.
3. An open is represented by infinite resistance and a short by zero resistance.

Figure 8-26 is a series-parallel circuit with R_2 open. Note that R_2 is the series resistor.

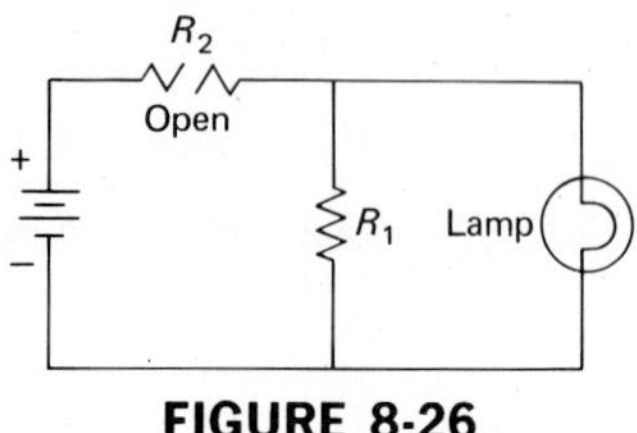

FIGURE 8-26

When a series component is open, all current flow stops.

Since R_2 is the series resistor, the circuit stops operating until the trouble is corrected.

Figure 8-27a illustrates a series-parallel circuit with R_1 open. R_1 is located in the parallel network where it is normally in parallel with the lamp; R_2 is in series with both branches. When R_1 opens, R_2 and the lamp are in series with each other (Figure 8-27a).

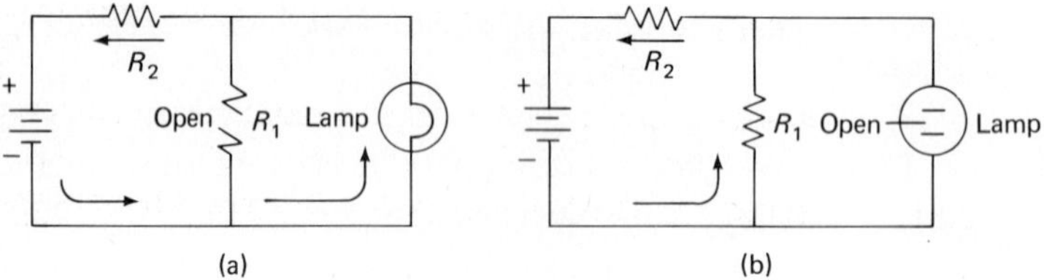

FIGURE 8-27

With R_1 in the circuit open, total resistance of the circuit has increased. This causes total current to decrease, which, in turn, causes E_{R2} to decrease, resulting in the lamp being forced to drop more voltage than normal. When the voltage dropped by the lamp increases, lamp current must increase.

Assume that the lamp is capable of conducting the added current without damage; in that event, it will glow brighter than normal. If it cannot handle the added current, its filament will overheat and will melt in two, thus becoming an open. This open stops current flow in the lamp.

Now let us examine the effect of the open occurring in the other branch. Refer to Figure 8-27b. If an open occurs in the lamp, leaving both R_1 and R_2 normal, the lamp fails to glow. Current can still flow through R_1 and R_2, however, because they form a complete loop with the battery. When this occurs, E_{R1} decreases and E_{R2} increases. Total current decreases while total resistance increases. Even with I_t decreased, R_2 is conducting more current than normal, and under some circumstances, it could be damaged.

How does a short affect the operation of a series-parallel circuit? Figure 8-28 contains a shorted series resistor, R_1. To understand the effect of a shorted series resistor on the circuit, we must first determine what the conditions were before R_1 shorted.

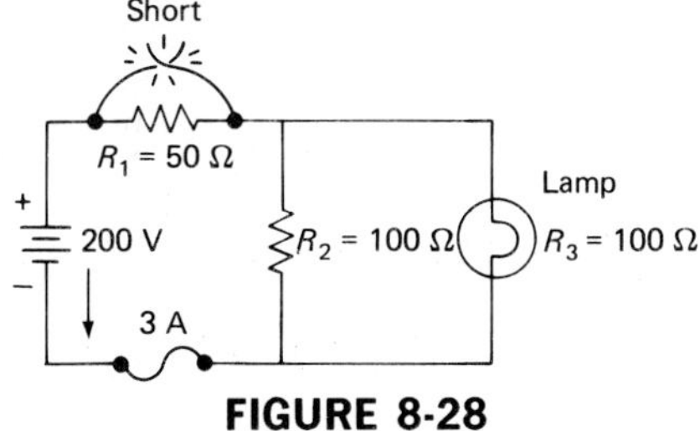

FIGURE 8-28

Normal conditions:

$$R_t = R_1 + \frac{R_2 \times R_3}{R_2 + R_3} = 100\ \Omega$$

$$I_t = \frac{E_t}{R_t} = 2\ \text{A}$$

$$E_{R1} = I_t \times R_1 = 100\ \text{V}$$

$$E_{R2} = E_t - E_{R1} = 100\ \text{V}$$

$$E_{R3} = E_t - E_{R1} = 100\ \text{V}$$

$$I_{R1} = I_t$$

$$I_{R2} = \frac{E_{R2}}{R_2} = 1\ \text{A}$$

$$I_{R3} = \frac{E_{R3}}{R_3} = 1\ \text{A}$$

Conditions with R_1 shorted:

$$R_t = \frac{R_2 \times R_3}{R_2 + R_3} = 50\ \Omega$$

$$I_t = \frac{E_t}{R_t} = 4\ \text{A}$$

$$E_{R1} = 0\ \text{V}\ (0\ \Omega \text{ drops } 0\ \text{V})$$

$$E_{R2} = E_t = 200\ \text{V}$$

$$E_{R3} = E_t = 200\ \text{V}$$

$$I_{R2} = \frac{E_t}{R_2} = 2\ \text{A}$$

$$I_{R3} = \frac{E_t}{R_3} = 2\ \text{A}$$

Examination of the results reveals that the voltage drop on R_2 and R_3 has doubled, as has R_2 and R_3 current. If this condition were allowed to persist, it is probable that R_2 and R_3 would be damaged. But the fuse is rated at 3 A and will open as soon as current exceeds 3 A. The fuse was selected to protect the lamp from a current overload.

When either R_2 or R_3 is shorted, circuit conditions will be similar: Applied voltage would be dropped on R_1, and total current would again rise toward 4 A; but the fuse would open when current reached 3 A.

The voltmeter can be used to troubleshoot the circuit. Refer to Figure 8-29. Let us develop a set of voltmeter indications that would occur should a component open or short.

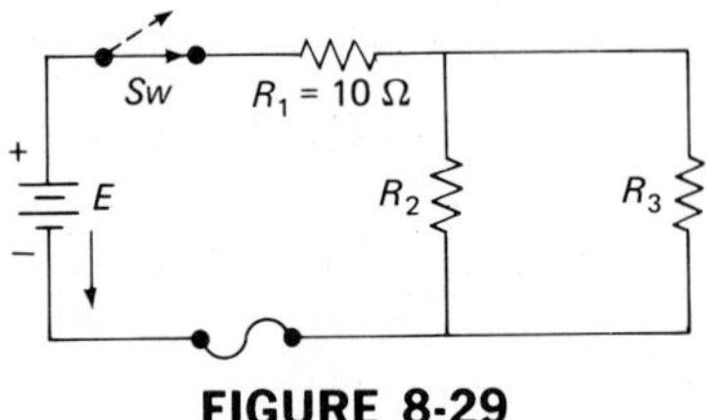

FIGURE 8-29

Table 8-1 lists the voltmeter reading that would occur across any component if that component opened or shorted.

TABLE 8-1

Component	Opened	Shorted	Normal
Switch	E_t	0 V	Closed 0 V, Open E_t
Fuse	E_t	0 V	Extremely small voltage
R_1	E_t	0 V	Part of E_t
R_2	High	0 V	Part of E_t
R_3	High	0 V	Part of E_t

The ohmmeter can be used to troubleshoot this circuit. Remember that the circuit power must be off before you use the ohmmeter. In fact, the component must be isolated (disconnected) from other circuitry before the ohmmeter can make accurate measurements. If it is not isolated, there may be parallel paths for current flow, which can cause erroneous readings. Figure 8-30 illustrates the use of an ohmmeter.

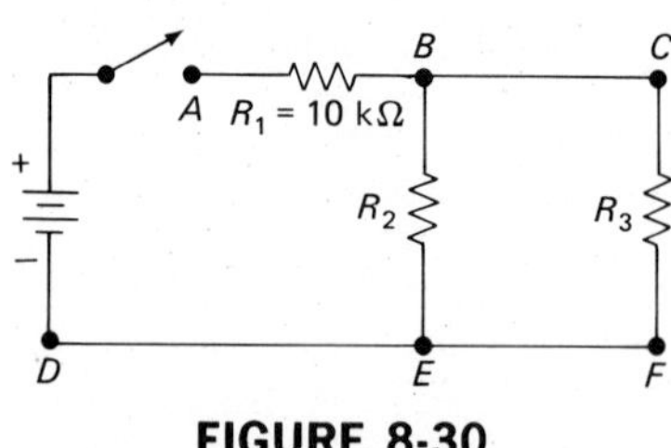

FIGURE 8-30

Opening the switch in the circuit in Figure 8-30 disables the power source. Connecting the ohmmeter between points A and B will read the actual value of R_1 (10k if it is good, zero if shorted, and infinity if open). Connecting the ohmmeter between points B to E or C to F presents a different problem. If both resistors are good, the meter will read their equivalent resistance; if either resistor is open, the meter reads the value of the remaining good resistor. If either resistor is shorted, the meter reads 0 Ω. To obtain accurate measurements it is necessary to disconnect the component at one end and make the check, then reconnect the component.

Self-Check

Complete each item by inserting the words and/or numbers needed to make a true and complete statement.

13. In a series-parallel resistive circuit, an open resistor causes total current to __________.

14. In a series-parallel resistive circuit, a shorted resistor causes total current to __________.

15. In a series-parallel resistive circuit, an open resistor causes total resistance to __________.

16. In a series-parallel resistive circuit, a shorted resistor causes total resistance to __________.

17. An ohmmeter is correctly connected across an open; it will indicate __________ Ω.

18. An ohmmeter is correctly connected across a short; it will indicate __________ Ω.

19. A voltmeter is correctly connected across an open series resistor; it will indicate __________ V.

20. A voltmeter is correctly connected across a shorted series resistor; it will indicate __________ V.

21. A voltmeter is correctly connected across a parallel network that contains an open branch; it will indicate __________ V.

22. A voltmeter is correctly connected across a parallel network that contains a shorted branch; it will indicate __________ V.

Summary

Parallel-Series Circuit

We developed the following points which should be remembered.

1. Series strings of resistors obey the same rules as series-resistive circuits.
   (a) Applied voltage is divided between/among the individual resistors.
   (b) Current is the same at all points in the string.

   (c) The total resistance of a string equals the sum of all individual resistors.
   (d) An open resistor stops current flow within a string and causes total current to decrease.
   (e) A shorted resistor causes string current and total current to increase.
2. Parallel-series resistive networks obey the same rules as parallel-resistive circuits.
   (a) The network functions as though each branch contained only one resistor.
   (b) The total resistance of the network is calculated by parallel analysis.
   (c) The total current of the network equals the sum of all branch currents.
   (d) Applied voltage is dropped across each string.
   (e) One open branch does not affect the operation of any other branch.
   (f) A shorted branch shorts the entire network.
   (g) It is possible for one string to contain a shorted resistor and the circuit continue to operate.
3. Analysis of a parallel-series circuit involves using:
   (a) Series analysis to determine string resistance.
   (b) Parallel analysis to determine total resistance.
   (c) Ohm's law to determine total current or voltage.
   (d) Parallel analysis to determine individual branch currents.
   (e) Series analysis to calculate individual voltage drops within resistive strings.

Series-Parallel Combinations

A series-parallel circuit contains one or more series resistors connected in series with one or more parallel networks.

A series resistor in a series-parallel circuit has the current for two or more parallel branches flowing through it.

An open in a series-parallel circuit causes total resistance to increase.

A shorted component in a series-parallel circuit causes total resistance to be lower than normal.

Total power dissipated in a series-parallel circuit equals the sum of all individual power dissipations.

Review Questions and/or Problems

1. If a voltmeter is connected across a good series resistor in a series-parallel circuit, the meter will indicate
   (a) total voltage.
   (b) part of the applied voltage.
   (c) zero volts.

2. When the series resistor shorts in a series-parallel circuit, it causes total current to
   (a) increase.
   (b) decrease.
   (c) remain the same.
   (d) become zero.

3. When an open occurs in the parallel portion of a series-parallel circuit, the voltage drop on the series resistor
   (a) increases.
   (b) decreases.
   (c) remains the same.
   (d) becomes zero volts.

4. When an open occurs in the parallel portion of a series-parallel circuit, the total resistance
   (a) increases.
   (b) decreases.
   (c) remains the same.
   (d) becomes zero ohms.

5. If total voltage applied to a series-parallel circuit is increased while total resistance remains unchanged, total current
   (a) increases.
   (b) decreases.
   (c) remains the same.
   (d) becomes zero amperes.

6. One definition of a series-parallel circuit is:
   (a) a group of resistors connected in series with other resistors.
   (b) a group of series resistors connected in series with other resistors.
   (c) a group of parallel resistors connected in series with other resistors.
   (d) a group of resistors connected in parallel with other resistors.

7. If a parallel network consists of branch resistors of unequal size,
   (a) they will dissipate equal amounts of power.
   (b) the largest resistor will dissipate the most power.
   (c) the smallest resistor will dissipate the most power.
   (d) the total power dissipated by the network will be the difference between the two.

8. In the circuit shown in Figure 8-31, what is the total resistance?
   (a) 2 kΩ
   (b) 6 kΩ
   (c) 8 kΩ
   (d) 33 kΩ

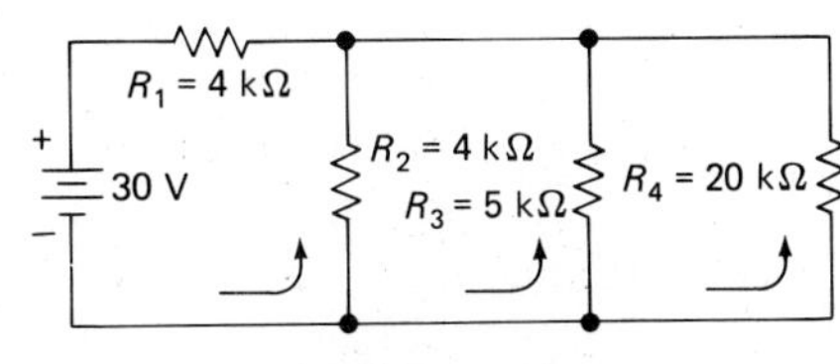

FIGURE 8-31

9. In the circuit in Figure 8-32, all the resistors within the dotted line can be replaced by one ______________ resistor.
   (a) 100 kΩ
   (b) 25 kΩ
   (c) 20 kΩ
   (d) 150 kΩ

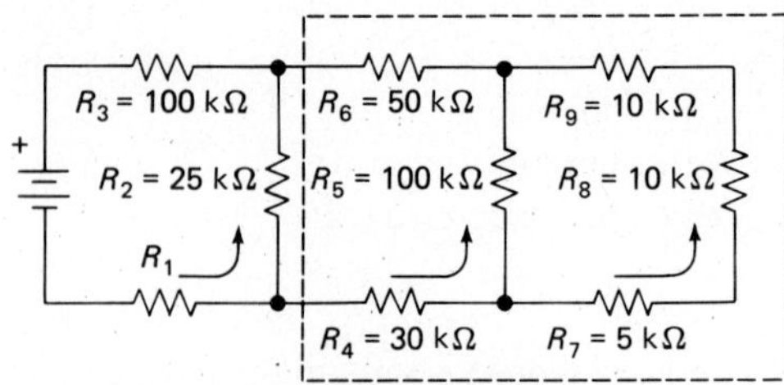

FIGURE 8-32

10. In the circuit in Figure 8-33, R_3 is shorted. What will the voltage across R_2 be?
   (a) total voltage
   (b) zero volts
   (c) infinity
   (d) part of the applied voltage

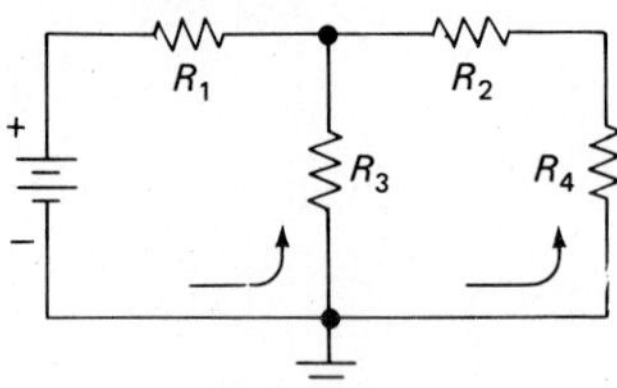

FIGURE 8-33

11. When Sw_1 is closed (Figure 8-34), the voltage drop across R_1 will
   (a) increase.
   (b) decrease.
   (c) remain the same.
   (d) be zero volts.

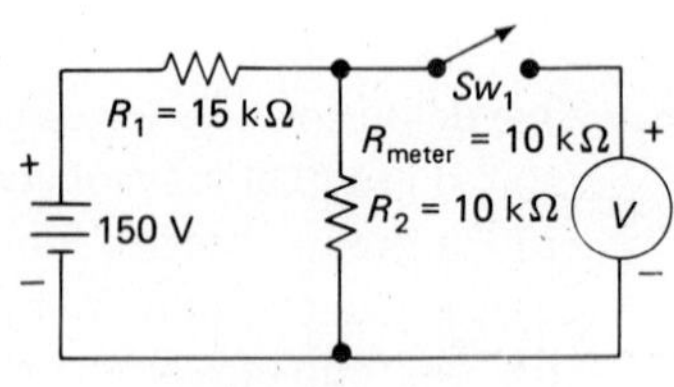

FIGURE 8-34

12. In the circuit in Figure 8-35, what is the total current?
   (a) 0.25 mA
   (b) 0.50 mA
   (c) 1.5 mA
   (d) 2.0 mA

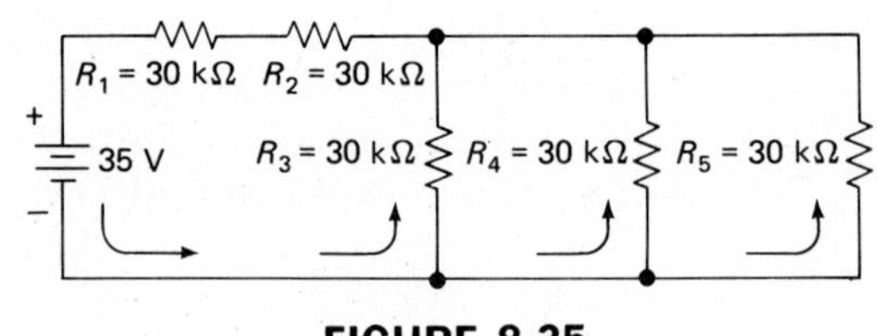

FIGURE 8-35

13. If R_2 opens in the circuit shown in Figure 8-36 which statement is true?
   (a) R_t decreases
   (b) I_t decreases
   (c) E_{R1} increases
   (d) E_{R2} decreases

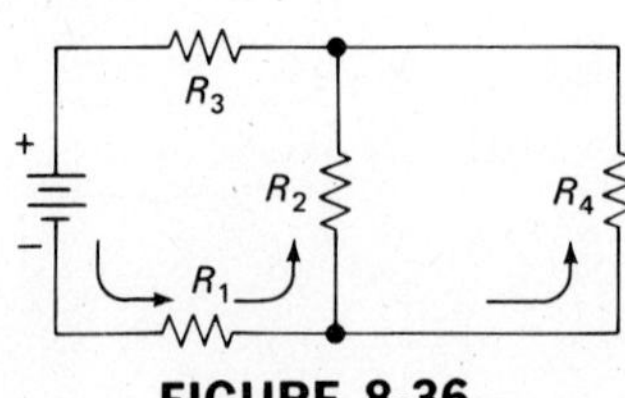

FIGURE 8-36

14. Which of the resistors shown in Figure 8-37 will dissipate the least power?
    (a) R_1
    (b) R_2
    (c) R_4
    (d) R_5

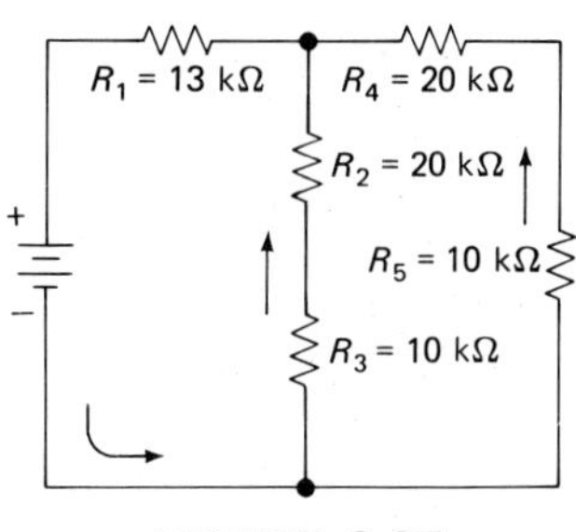

FIGURE 8-37

15. In the circuit in Figure 8-38, if R_1 shorts, what will the ammeter read?
    (a) 20 μA
    (b) 10 mA
    (c) 20 mA
    (d) 30 mA

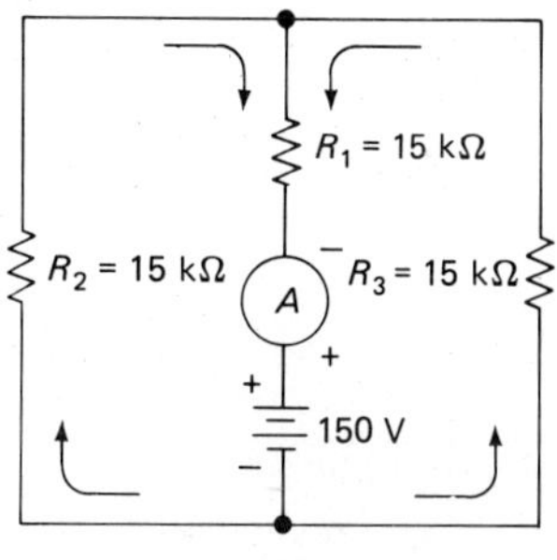

FIGURE 8-38

16. In the circuit in Figure 8-39, if R_4 opens, the current flowing through
    (a) R_1 will increase.
    (b) R_1 will decrease.
    (c) R_2 will decrease.
    (d) R_2 will remain the same.

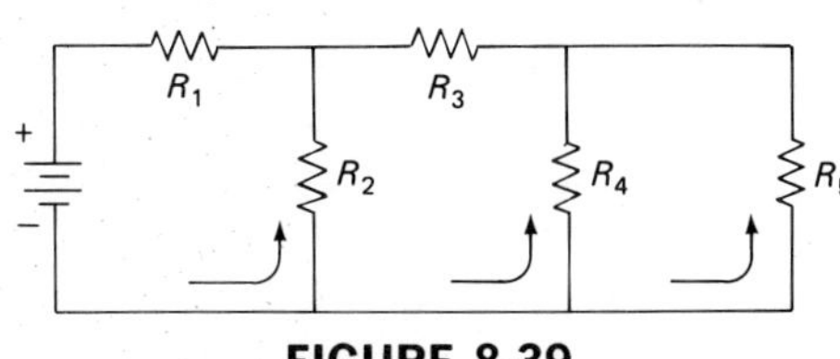

FIGURE 8-39

17. In the circuit in Figure 8-40, the ammeter reads 5 mA. What is wrong?
    (a) R_5 shorted
    (b) R_2 opened
    (c) R_5 opened
    (d) R_1 shorted

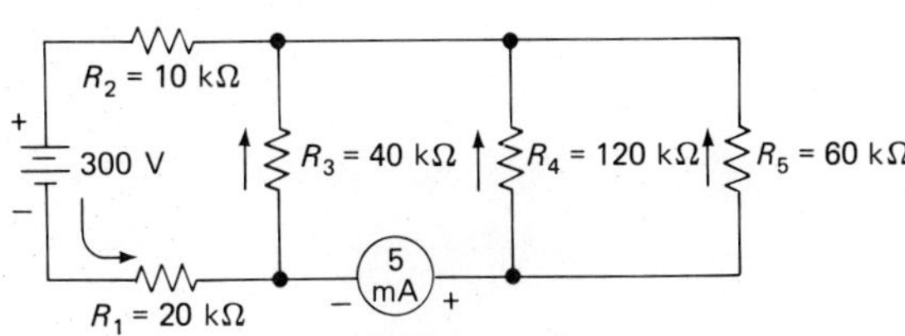

FIGURE 8-40

18. In the circuit in Figure 8-41, if R_5 opens, the current flow through
   (a) R_1 increases.
   (b) R_2 decreases.
   (c) R_4 increases.
   (d) R_4 remains the same.

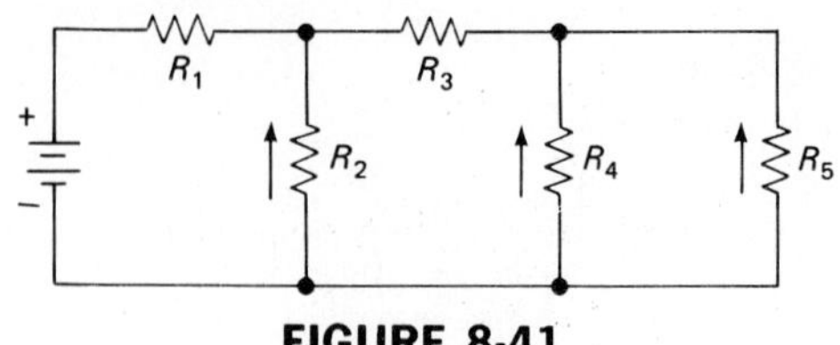

FIGURE 8-41

19. In Figure 8-42, what would cause lamp$_1$ (L_1) to become dimmer while lamp$_2$ (L_2) becomes brighter?
   (a) R_1 shorting
   (b) R_2 opening
   (c) R_1 opening
   (d) R_4 shorting

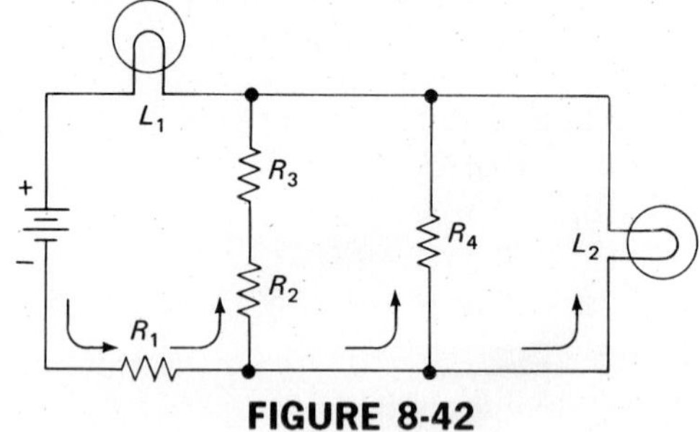

FIGURE 8-42

20. In Figure 8-43, what would cause both lamps to become brighter?
   (a) R_1 shorting
   (b) R_2 opening
   (c) R_3 shorting
   (d) R_4 opening

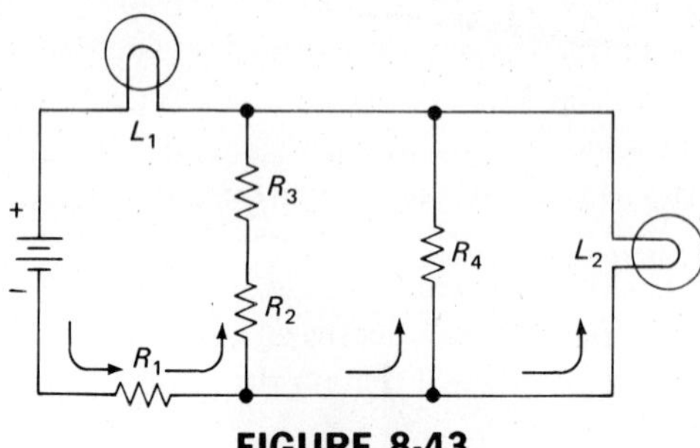

FIGURE 8-43

9

Variable Resistors and Voltage Dividers

Introduction

It is often necessary to provide the user with control of a circuit. Examples are a radio's volume control and the brightness adjustment on a car's dashboard lights. One way of providing operator control is to insert a variable resistor in a circuit.

Variable resistors are classified in two general categories, rheostat and potentiometer. As its name suggests, a potentiometer is used to vary voltage; the voltage selected by a potentiometer can be applied to another circuit.

At other times, the operator must be able to control the current flow in a circuit, and a rheostat is installed. By using a rheostat, the operator can select the correct heat level for a heating element in an industrial oven or set the light level in a room or a theater. In each case, the user controls the amount of current flow in a circuit.

A potentiometer is installed when the user requires control of voltage level. A stereo's volume control is one example. Varying the position of the potentiometer's wiper changes the amount of voltage selected for input to the stereo amplifier. As the amount of voltage is changed, so is the volume.

Other circuits discussed in this chapter allow two or more voltages to be obtained from one power source. These circuits are called *voltage dividers.* They can be designed to meet specific power requirements of a circuit which, in turn, requires a voltage less than that of the power source available. When the circuit is providing current to the second circuit, it is said to be *loaded.* When the second circuit is not connected, or is turned off, the voltage divider is said to be *unloaded.*

Objectives

Each student is required to:

1. Define *potentiometer* and *rheostat* from the viewpoint of construction, connection, and application.

(a)

(b)

FIGURE 9-1
(a) Potentiometer and (b) Assorted Rheostats
(Courtesy of Allen-Bradley)

2. Identify the schematic symbols for a potentiometer and a rheostat.
3. Draw circuit diagrams that contain the symbols for potentiometers and rheostats.
4. State the purpose and use of a voltage divider.
5. Analyze unloaded and loaded voltage divider circuits for distribution of current and voltage.
6. Define the term *load* as it applies to electronics.
7. Explain what the phrase *loaded voltage divider* means.
8. Design a simple voltage divider.

Variable Resistors

Theoretically, potentiometers and rheostats have different configurations. They also operate differently. Today, it is more accurate to say that the main difference between potentiometers and rheostats is the way they connect into a circuit and the purpose that each serves. In most applications, both resistors consist of a circular carbon wafer. Each has a movable contact mounted on a shaft so it can move along the wafer and select a specific amount of resistance for each point passed. The resistance values available vary from zero to the maximum value of the wafer.

Traditionally, the rheostat has been a high-current device; but in the circuits used today, high currents are usually not present. Lower currents allow a cheaper design of rheostat to be used.

Originally, a rheostat was made of resistive wire wound around an insulating frame. Today, a carbon wafer is commonly used as the resistance element. The potentiometers presented here are low-current devices made of carbon wafers. Figure 9-2 illustrates the visual features of each type.

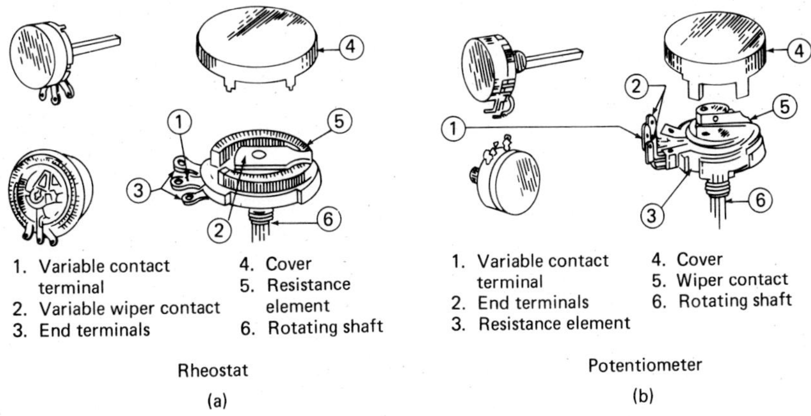

FIGURE 9-2

A rheostat is any variable resistor connected so that it can be used to vary the amount of current flow in a circuit. Figure 9-3 illustrates a circuit that contains a rheostat connected in series with a lamp or similar device. In this illustration, R_2 is a lamp. R_1 is used to vary the circuit resistance. Ohm's law is based on the fact that resistance and current are inversely proportional. Therefore, circuit current will be affected by any change in resistance. As resistance increases, circuit current decreases, and the lamp will dim. By decreasing resistance we increase current, and the lamp glows more brightly.

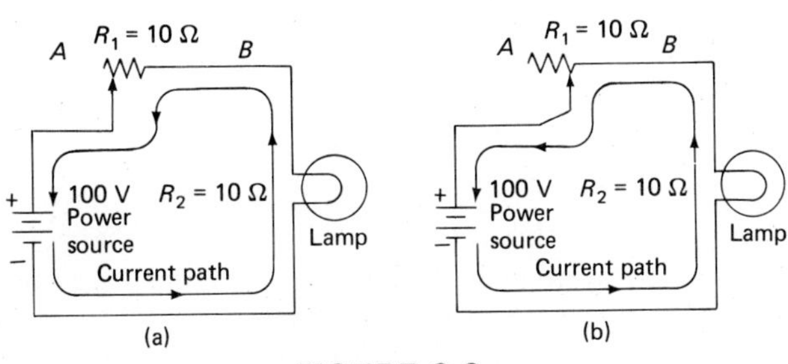

FIGURE 9-3

To better understand what happens, let us examine both circuits shown in Figure 9-3. As connected, the rheostat has only two electrical contacts. The connection marked by the arrowhead is a movable contact, or wiper. As it moves along the resistance of R_1, the ohmic value of R_1 is changed. This changes total resistance and total current in the circuit. In Figure 9-3a, R_1 is entirely within the circuit. In Figure 9-3b, none of R_1 is within the circuit. To better understand how this affects the lamp, compare the following statements:

Figure 9-3a:

$$R_t = R_1 + R_2 = 10\ \Omega + 10\ \Omega = 20\ \Omega$$

$$I_t = \frac{E_t}{R_t} = \frac{100\text{ V}}{20\ \Omega} = 5\text{A}$$

$$E_{R2} = I_t \times R_2 = 5\text{ A} \times 10\ \Omega = 50\text{ V}$$

Figure 9-3b:

$$R_t = R_1 + R_2 = 0\ \Omega + 10\ \Omega = 10\ \Omega$$

$$I_t = \frac{E_t}{R_t} = \frac{100\text{ V}}{10\ \Omega} = 10\text{ A}$$

$$E_{R2} = I_t \times R_2 = 10\text{ A} \times 10\ \Omega = 100\text{ V}$$

In Figure 9-3b, voltage and current at the lamp have both doubled when they are compared to the values in Figure 9-3a. This causes the lamp to glow much more brightly. As the wiper is moved to the right, the lamp glows brighter; when the wiper is moved to the left, the lamp dims.

A potentiometer's connection differs from that of the rheostat. It has three electrical connections. See the example in Figure 9-4.

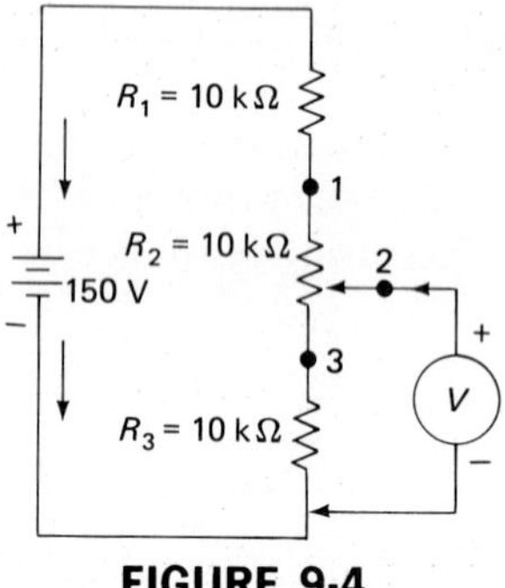

FIGURE 9-4

In this circuit, total current flows through all parts of R_2's resistance wafer, regardless of the wiper's position. Points 1, 2, and 3 are connected separately, but point 2 is connected as the input to a separate part of the circuitry. An example is the volume control used to adjust the level of input to the final amplifier of a stereo system. Using the values in Figure 9-4, let us solve the circuit.

$$R_t = R_1 + R_2 + R_3 = 10\text{k} + 10\text{k} + 10\text{k} = 30\text{ k}\Omega$$

$$I_t = \frac{E_t}{R_t} = \frac{150\text{ V}}{30\text{ k}\Omega} = 5\text{ mA}$$

$$E_{R1} = I_t \times R_1 = 5\text{ mA} \times 10\text{ k}\Omega = 50\text{ V}$$

$$E_{R2} = I_t \times R_2 = 5\text{ mA} \times 10\text{ k}\Omega = 50\text{ V}$$

$$E_{R3} = I_t \times R_3 = 5\text{ mA} \times 10\text{ k}\Omega = 50\text{ V}$$

Analyze this circuit from the standpoint of the voltmeter's reading. Remember that any voltmeter indicates the sum of the voltages dropped by all the resistors connected between its leads. With the wiper set to point 3, the meter indication equals E_{R3} (50 V). As the wiper moves toward point 1, more of R_2 is included between the leads of V_1, and the voltage indicated on V_1 increases. When the wiper is at point 1, the meter reads $E_{R2} + E_{R3}$ (100 V). In other words, R_2 can select any voltage between 50 V and 100 V. In this application, the setting on R_2 serves to "divide" the voltages of R_2 as desired.

The resistance wafer of a variable resistor can be designed to meet the needs of any circuit. Resistance value, current capability, and power dissipation can be high or low as needed for a specific circuit; resistance within the wafer can be linear or tapered. Tapered resistors have variations in the amount of resistance contained on different parts of the wafer. The resistance can be distributed in a linear, logarithmic, or nonlinear pattern. Tapers are used to customize a circuit, thus enabling the control to compensate for other things. The volume control on a stereo is tapered to compensate for the nonlinearity of the human ear.

The variable resistors just discussed are represented by the schematic symbols shown in Figure 9-5. Note that the rheostat is represented by two different symbols. A potentiometer can be modified to act as a rheostat. This modification is made by shorting the wiper to either pin 1 or pin 3. This places the two shorted tabs at the same electrical potential, and the device operates as though it had only two contacts.

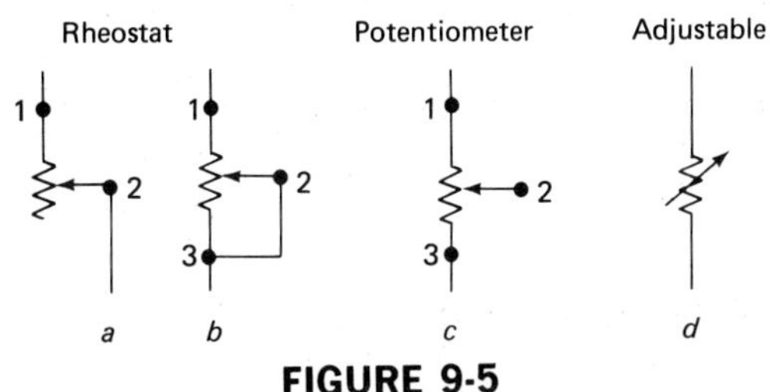

FIGURE 9-5

In many pieces of equipment, variable resistors are used that are not user-controlled. Often called *adjustable* resistors, these resistors are normally adjusted during manufacture and do not require further adjustment except during maintenance. Adjustable resistors can be either potentiometers or rheostats, and their schematic symbols may be any of those shown in Figure 9-5. Many adjustable resistors have devices attached that allow their wipers to be locked in position after the resistive adjustment has been made.

Self-Check

Complete each item by inserting the words and/or numbers needed to make a true and complete statement.

1. Two types of variable resistors are the ______________ and the ______________.

2. A potentiometer has ______________ electrical connections.

3. The rheostat is used to control circuit ______________.

4. The potentiometer is used to select the ______________ applied to a subsequent ______________.

5. The terms *linear* and *tapered* refer to the resistance distribution of a ______________.

6. A rheostat has ______________ electrical connections.

Unloaded Voltage Dividers

A voltage divider can be used to supply voltage to a circuit that is less than the voltage of a battery or power supply. To do this, the voltage divider must contain two or more resistors connected in series across the power source. Figure 9-6 illustrates a three-resistor (three-voltage), unloaded voltage divider.

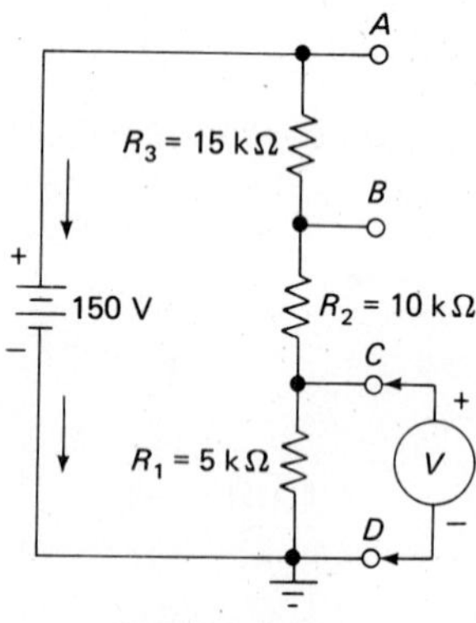

FIGURE 9-6

As you can see, this circuit is nothing more than a series circuit. All voltage dividers have a series circuit as the basic part of their design. Notice that the circuit has taps at points labeled A, B, C, and D. Analysis of the circuit reveals:

$$R_t = R_1 + R_2 + R_3 = 30\ \text{k}\Omega$$

$$I_t = \frac{E_t}{R_t} = 5\ \text{mA}$$

$$E_{R1} = I_t \times R_1 = 25\ \text{V}$$

$$E_{R2} = I_t \times R_2 = 50\ \text{V}$$

$$E_{R3} = I_t \times R_3 = 75\ \text{V}$$

By connecting the negative lead of a voltmeter at reference point D, we can measure the voltages between D and the other taps:

Point C = 25 V

Point B = 75 V

Point A = 150 V

This makes three different voltages available. In other words, the battery voltage has been divided into three different voltages: 25 V, 75 V, and 150 V.

To designate point D or any other point as the reference point, a ground symbol is placed at that point in the schematic. The ground represents the point from which all voltages are referenced. The ground is more than a mere schematic symbol; it represents a common connection point, for example, the chassis of an automobile. In most cars, the negative pole of the battery is connected to the chassis. At each light, horn, or radio, a wire is connected between the chassis and the device. From the other side of the device, the circuit continues through a switch and fuse, back to the positive pole of the battery. Using the chassis as a common connection allows all the current being used in the car to be conducted through the chassis, which is acting as one large conductor. This reduces vehicle weight and cost. This also has the advantage of placing the reference point nearby, which is useful when measuring voltage. When the negative pole of the battery is connected to the ground, we say the system has a *negative ground;* when the positive pole is connected to the ground, the system has a *positive ground.* When no ground symbol is indicated on the schematic, the negative pole of the battery is used as ground.

Some circuits require placement of the ground at a point other than the positive or negative pole of the power source (Figure 9-7).

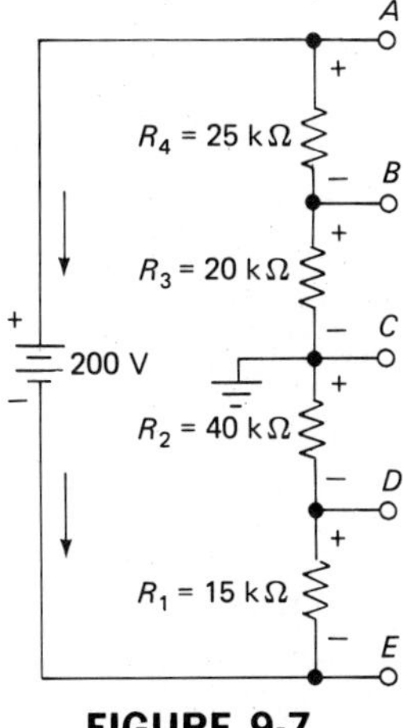

FIGURE 9-7

Note that the ground is at point C in this circuit. When that is the case, voltages within the circuit need further description. Two rules are useful for understanding these voltages:

When current flows from the negative side of a power source, through a resistor, and toward ground, the voltage dropped by that resistor is said to be* negative *voltage.

***When current flows away from ground, through a resistor, and to the positive pole of a power source, the voltage is said to be* positive.**

These rules apply when the negative lead of the voltmeter is connected to the ground and all voltages are measured with respect to it. Because the positive lead is used to check voltages, any voltage measured between the ground and the positive pole are positive; those measured between the ground and the negative pole are negative. To see how this applies, we analyze Figure 9-7. These values are:

$$R_t = R_1 + R_2 + R_3 + R_4 = 100\text{ k}\Omega$$

$$I_t = \frac{E_t}{R_t} = 2\text{ mA}$$

$$E_{R1} = I_t \times R_1 = 30\text{ V}$$

$$E_{R2} = I_t \times R_2 = 80\text{ V}$$

$$E_{R3} = I_t \times R_3 = 40\text{ V}$$

$$E_{R4} = I_t \times R_4 = 50\text{ V}$$

Connecting the negative lead of the voltmeter to the ground, and measuring the four voltages, results in:

Point D with respect to ground $= -80\text{ V }(E_{R2})$

Point E with respect to ground $= -110\text{ V }(E_{R2} + E_{R1})$

Point B with respect to ground $= +40\text{ V }(E_{R3})$

Point A with respect to ground $= +90\text{ V }(E_{R3} + E_{R4})$

Trace the current path in Figure 9-7 and mark each resistor with the polarities, as shown in the drawing. This also makes it easy to determine whether a voltage is positive or negative. On Figure 9-7, note that the ends of each resistor are marked by polarities. The end of each resistor farthest from the ground is labeled with the polarity that will be assigned to the voltage measured at that point.

Self-Check

Complete each item by inserting the words and/or numbers needed to make a true and complete statement.

7. Voltages developed by current flowing toward ground are ______________ in polarity.

8. A ______________ ______________ circuit is used as an unloaded voltage divider.

9. Two purposes for ground in a circuit are.
   (a) ______________.
   (b) ______________.

10. Voltages that are dropped by current flowing away from ground are ______________ in polarity.

11. The voltage divider is used to ______________ ______________.

12. Refer to Figure 9-7. Move ground to point B. What four voltages are available?
(a) ______________ V
(b) ______________ V
(c) ______________ V
(d) ______________ V

13. In a circuit, the ground provides a ______________ ______________ point.

Loaded Voltage Dividers

A loaded voltage divider is a complex series-parallel circuit used to provide correct voltages and current to one or more separate circuits. The additional circuits perform useful functions. The divider serves only to supply voltage and current. *Load* refers to current demand, not the device through which current flows. The term *loaded* comes from the fact that when a power source is supplying current, it is said to be "loaded."

Any time a circuit is connected in parallel with the power source, the load on the source is increased; the same thing happens in the case of the loaded voltage divider. A circuit connected at any tap of a voltage divider increases the load on the power supply. Power supplies providing small amounts of current are said to have a *low load;* those supplying heavy current are said to have a *high load.*

Batteries have a maximum current capability. Because all power supplies have a maximum load which they can supply, care must be exercised when adding circuits to a power supply to prevent overloading the source.

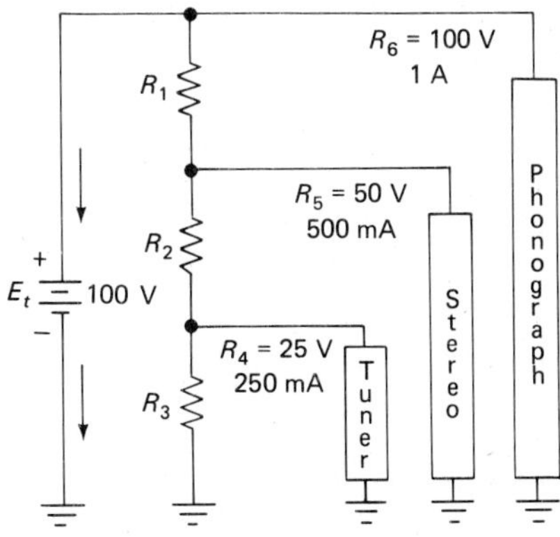

FIGURE 9-8

Figure 9-8 represents a loaded-voltage-divider circuit. The circuit is used to provide three separate voltages and currents:

1. 100 V at 1 A to a phonograph
2. 50 V at 500 mA to a stereo amplifier
3. 25 V at 250 mA to an FM tuner

When we substitute current and voltage requirements for each load device, Figure 9-9 results. In analyzing this circuit, we use the formulas and methods already learned. The sequence in which they are applied, however, is quite different.

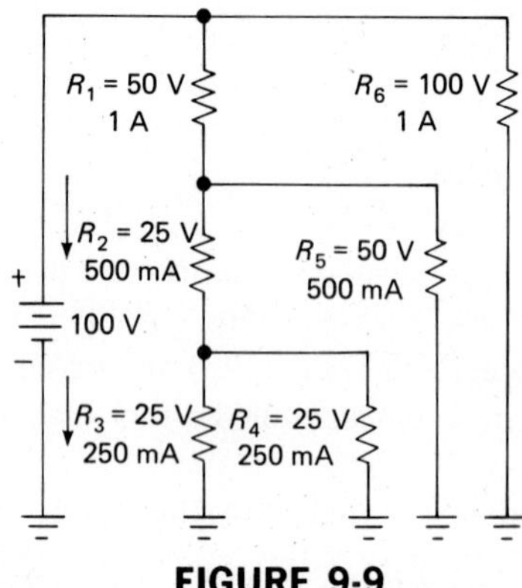

FIGURE 9-9

The voltages and currents listed in Figure 9-9 can be used with Ohm's law to calculate the resistance of each device in the circuit. R_1, R_2, and R_3 are fixed resistors whose values must be selected when the circuit is designed; tuners, stereos, and phonographs are manufactured with resistance values tailored to their specific functions. Using the data supplied, we find the resistance of each device as follows:

$$R_1 = \frac{E_{R1}}{I_{R1}} = \frac{50\text{ V}}{1\text{ A}} = 50\ \Omega$$

$$R_2 = \frac{E_{R2}}{I_{R2}} = \frac{25\text{ V}}{500\text{ mA}} = 50\ \Omega$$

$$R_3 = \frac{E_{R3}}{I_{R3}} = \frac{25\text{ V}}{250\text{ mA}} = 100\ \Omega$$

$$R_4 = \frac{E_{R4}}{I_{R4}} = \frac{25\text{ V}}{250\text{ mA}} = 100\ \Omega$$

$$R_5 = \frac{E_{R5}}{I_{R5}} = \frac{50\text{ V}}{500\text{ mA}} = 100\ \Omega$$

$$R_6 = \frac{E_{R6}}{I_{R6}} = \frac{100\text{ V}}{1\text{ A}} = 100\ \Omega$$

These values can be used to solve for total resistance of the circuit. To solve for total resistance, we start at the lower left, R_3, and work upward and outward. First, we use the product-over-sum formula to solve equivalent resistance of R_3 and R_4:

$$R_{eq1} = \frac{R_3 \times R_4}{R_3 + R_4} = 50\ \Omega$$

Note that R_3 and R_4 are *in parallel.*

Not all of this information is provided when you design your own voltage dividers. With the finding of resistor values, most of the data regarding this circuit is known. To illustrate the method used to calculate voltage and current for the different parts of a voltage divider, we complete the calculations. This also proves the data listed above.

Using each result as part of the next solution, proceed upward and outward, solving each step as follows:

$$R_{eq2} = R_{eq1} + R_2 = 100\ \Omega$$

This is in parallel with R_5.

$$R_{eq3} = \frac{R_{eq2} \times R_5}{R_{eq2} + R_5} = 50\ \Omega$$

This is in series with R_1.

$$R_{eq4} = R_{eq3} + R_1 = 100\ \Omega$$

This is in parallel with R_6.

$$R_t = \frac{R_{eq4} \times R_6}{R_{eq4} + R_6} = 50\ \Omega$$

$$I_t = I_{R3} + I_{R4} + I_{R5} + I_{R6} = 2\ \text{A}$$

$$R_t = \frac{E_t}{I_t} = 50\ \Omega$$

R_6 is in parallel with E_t, so

$$E_{R6} = E_t = 100\ \text{V}$$

then

$$I_{R6} = \frac{E_{R6}}{R_6} = 1\ \text{A}$$

and

$$I_{R1} = I_t - I_{R6} = 1\ \text{A}$$

then

$$E_{R1} = I_{R1} \times R_1 = 50\ \text{V}$$

R_5 is in parallel with the bottom of R_1.

$$E_{R5} = E_t - E_{R1} = 50\ \text{V}$$

and

$$I_{R5} = \frac{E_{R5}}{R_5} = 500 \text{ mA}$$

and then

$$I_{R2} = I_{R1} - I_{R5} = 500 \text{ mA}$$

then

$$E_{R2} = I_{R2} \times R_2 = 25 \text{ V}$$

and

$$E_{R3} = E_{R5} - E_{R2} = 25 \text{ V}$$

Then, since R_3 and R_4 are in parallel,

$$E_{R4} = E_{R3} = 25 \text{ V}$$

Then

$$I_{R3} = \frac{E_{R3}}{R_3} = 250 \text{ mA}$$

and

$$I_{R4} = \frac{E_{R4}}{R_4} = 250 \text{ mA}$$

These results confirm the accuracy of our circuit design in Figure 9-9. The result is three separate power sources which supply:

100 V at 1 A to the phonograph, R_6
50 V at 500 mA to the stereo amplifier, R_5
25 V at 250 mA to the FM tuner, R_4

This results in a load on the power source that equals

$$I_{R3} + I_{R4} + I_{R5} + I_{R6} = 2 \text{ A}$$

These procedures can be reversed for designing simple voltage dividers that can be used to operate the devices chosen. The only considerations are that the voltage required by the load be less than the available power-supply voltage, and that the total load placed on the power source not exceed its rating. To illustrate the design of a simple voltage divider, consider the following*:

Situation:

You have a 12-V battery with sufficient current capability for use as a power source. You also have a 6-V radio with an internal resistance of 24 Ω, which requires 250 mA to operate.

*Material following is used by permission of *School Shop Magazine*. John E. Lackey, *School Shop*, **41**:4, November 1981, Prakken Publications.

Problem:

Design a voltage divider suitable for operating this radio from the available 12-V battery.

The circuit to be used appears in Figure 9-10.

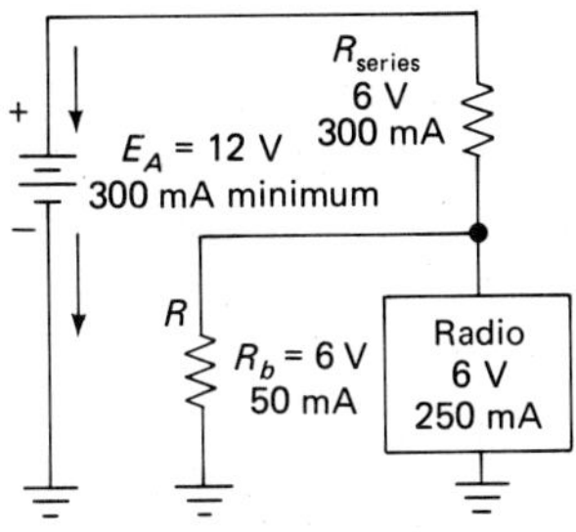

FIGURE 9-10

A "bleeder" resistor is used in the circuit; it serves two purposes: (1) minimizing voltage fluctuations should other circuits that share the same power source be turned on or off; and (2) providing a safety feature by discharging any circuits in the power supply that might be a safety hazard. Because R_b is of little use operationally, we keep its current as small as possible. A bleeder current of 50 mA is assumed. R_b is in parallel with the radio, and must drop the same voltage as the radio (6 V).

Now we can solve for the ohmic value of R_b:

$$R_b = \frac{E_b}{I_b} = \frac{6\text{ V}}{50\text{ mA}} = 120\ \Omega$$

Current arriving at the bottom of the series resistor, R_s, is the sum of the bleeder and radio currents, which is

$$I_{Rs} = I_b + I_{radio} = 50\text{ mA} + 250\text{ mA} = 300\text{ mA}$$

To calculate the voltage drop for R_s, use Kirchhoff's law:

$$E_{Rs} = E_t - E_{Rb} = 12\text{ V} - 6\text{ V} = 6\text{ V}$$

To calculate the ohmic value of R_s, use Ohm's law:

$$R_s = \frac{E_{Rs}}{I_{Rs}} = \frac{6\text{ V}}{300\text{ mA}} = 20\ \Omega$$

These results tell us the value of each resistor and its current flow. We have already learned, however, that a resistor's power rating must be considered in designing a circuit. To determine each resistor's power rating, do the following calculations:

$$P_{Rb} = I_{Rb} \times E_{Rb} = 50\text{ mA} \times 6\text{ V} = 300\text{ mW}$$

$$P_{Rs} = I_{Rs} \times E_{Rs} = 300\text{ mA} \times 6\text{ V} = 1.8\text{ W}$$

To build this circuit, select a 0.5 W or higher, 120 Ω resistor for R_b; and a 2 W or higher, 20 Ω resistor for R_s. When the circuit is complete, and the radio is connected, it should work well.

Note: Troubleshooting these circuits is identical to troubleshooting the series-parallel circuits discussed in Chapter 8; for this reason, troubleshooting is not included here.

Self-Check

Answer each item by indicating whether the statement is true or false.

14. ___ Adding a parallel branch to a series voltage divider converts the circuit to a loaded voltage divider.

15. ___ A series resistor contained in a loaded voltage divider conducts total current.

16. ___ Removing a branch from a loaded voltage divider causes total current to increase.

17. ___ A loaded voltage divider can be used to supply current to an unlimited number of devices.

18. ___ As additional branch circuits are added to a voltage divider, the *load* on the power supply increases.

Summary

Variable Resistors

1. Variable resistors allow resistance values to be varied as needed in a circuit.
2. Two types of variable resistors are:
   (a) Rheostats used to vary current.
   (b) Potentiometers used to vary voltage.
3. Adjustable resistors may be of either type.

Voltage Dividers

Unloaded-voltage-divider analysis consists of:

1. Solving for each resistor's voltage drop.
2. Locating the ground in the circuit.
3. Assigning polarities to each voltage available.
4. And involves only series analysis.

In loaded-voltage-divider analysis,

1. To solve for R_t, start at the lower left corner of the resistor network and work up and out.
2. With R_t known, solve for I_t, using Ohm's law.
3. Use Ohm's law to calculate the voltage drop for any series resistors present.
4. Solve for voltage and current distribution by starting at the upper right and working down and in toward the bottom left.
5. Both series and parallel analysis are involved.

Review Questions and/or Problems

1. What is meant by the term *loaded voltage divider?*
   (a) A tapped resistor is used in series to form a divider.
   (b) A load device is placed in series with a tapped resistor.
   (c) An electronic component or circuit that draws current from the source is placed across a divider.
   (d) An electronic component or circuit that does not draw current is placed across a divider.

2. A loaded voltage divider has
   (a) two or more paths for current flow.
   (b) only one path for current flow.
   (c) one or more paths for current flow.
   (d) none of the above.

3. The term *load* refers to
   (a) voltage.
   (b) resistance.
   (c) current.
   (d) power.

4. What is the polarity of the ground in a voltage divider?
   (a) positive
   (b) negative
   (c) neutral
   (d) none of these

5. Placing a load device across a voltage divider causes total resistance to
   (a) increase.
   (b) decrease.
   (c) remain the same.
   (d) become zero.

6. In the the circuit shown in Figure 9-11, closing Sw_1 causes
   (a) voltage between point A and ground to increase.
   (b) current through R_3 to decrease.
   (c) current through R_2 and R_3 to increase.
   (d) voltage between point A and ground to decrease.

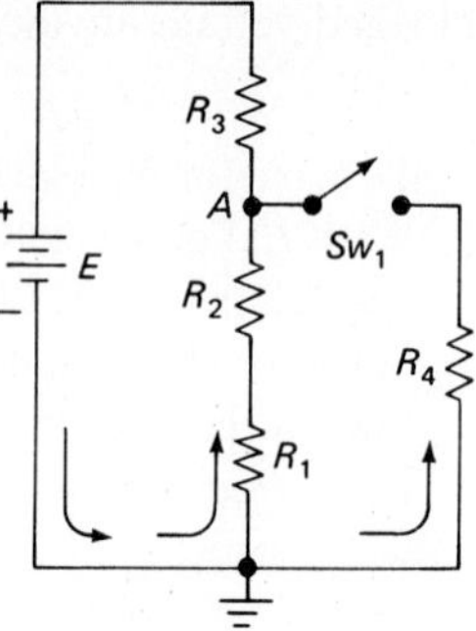

FIGURE 9-11

7. When Sw_1 is closed, the voltmeter's reading in Figure 9-12
   (a) increases.
   (b) decreases.
   (c) remains the same.
   (d) decreases to 0 V.

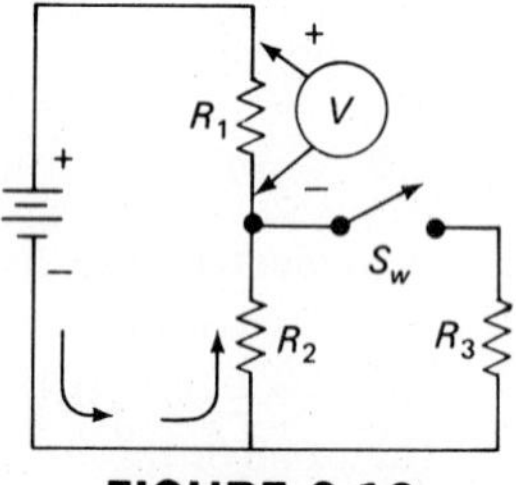

FIGURE 9-12

8. Total resistance of the circuit in Figure 9-13 is
   (a) 19 kΩ.
   (b) 9 kΩ.
   (c) 20 kΩ.
   (d) 28 kΩ.

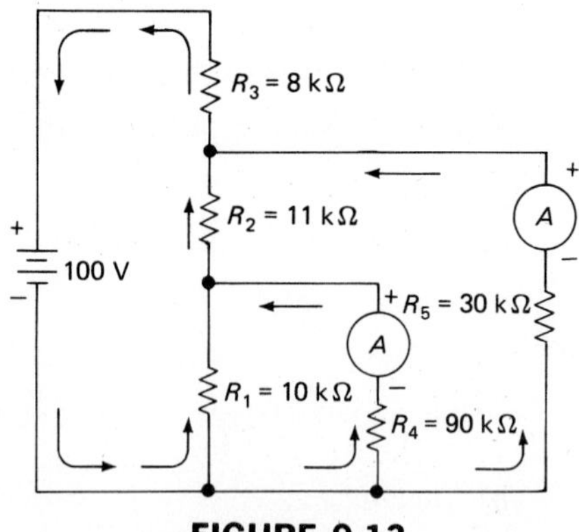

FIGURE 9-13

9. What is the ohmic value of R_3 in the circuit in Figure 9-14?
   (a) 2.67 kΩ
   (b) 2 kΩ
   (c) 5 kΩ
   (d) 20 kΩ

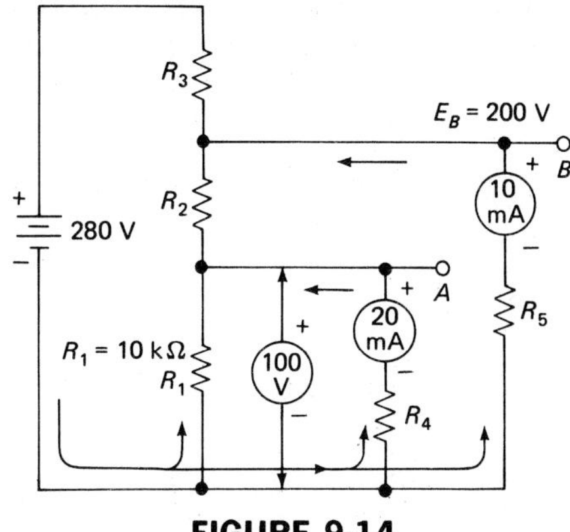

FIGURE 9-14

10. The voltage at point C, with respect to ground in the circuit in Figure 9-15, is
   (a) −75 V.
   (b) −45 V.
   (c) −56 V.
   (d) −10 V.

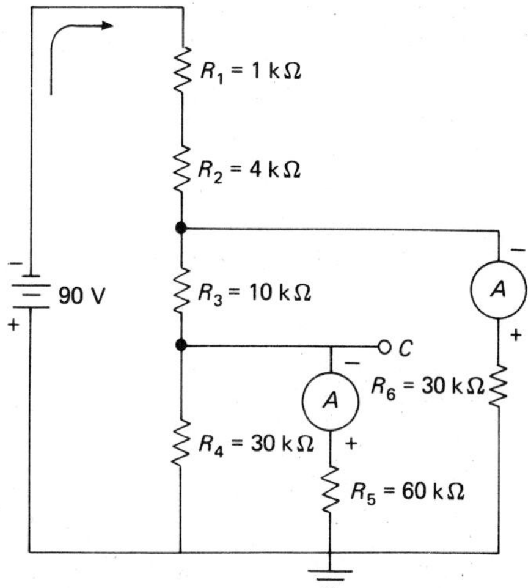

FIGURE 9-15

11. In Figure 9-16 what is the minimum voltage obtainable at point A with respect to ground?
   (a) 112 V
   (b) 120 V
   (c) 67.2 V
   (d) 52.8 V

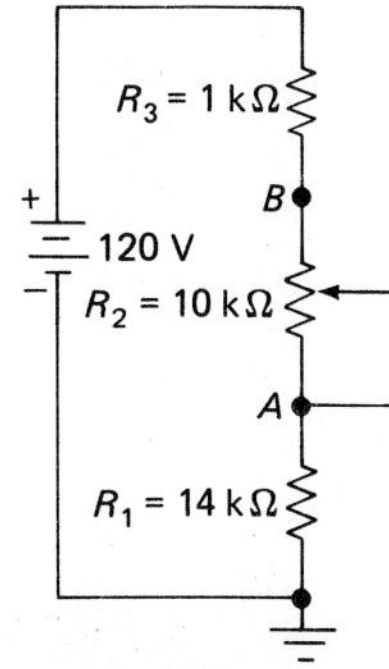

FIGURE 9-16

12. What are the minimum and maximum voltages between point E_{out} and ground in Figure 9-17?
    (a) 24 to 30 V
    (b) 20 to 24 V
    (c) −24 to −30 V
    (d) −20 to −24 V

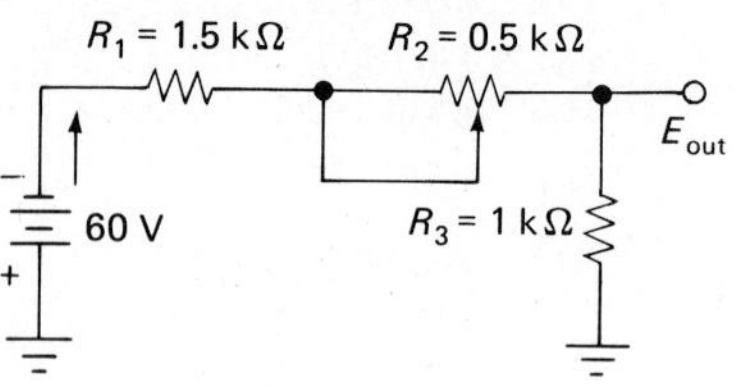

FIGURE 9-17

13. A rheostat is used to vary
    (a) current.
    (b) voltage.
    (c) power.
    (d) none of these.

14. Which of the symbols in Figure 9-18 is the schematic symbol for a potentiometer?

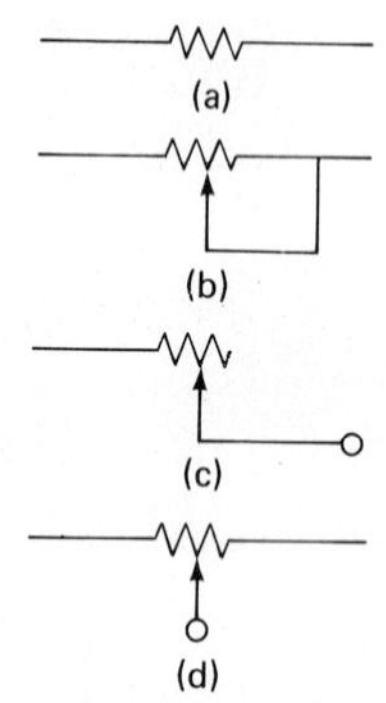

FIGURE 9-18

15. How many connections does a potentiometer have?
    (a) one
    (b) two
    (c) three
    (d) four

16. The ground symbol used in a voltage divider denotes a
    (a) voltage source.
    (b) current source.
    (c) reference point.
    (d) none of these

17. How many electrical connections does a rheostat have?
    (a) one
    (b) two
    (c) three
    (d) four

18. The volume control of a stereo system is an example of
    (a) a potentiometer.
    (b) a rheostat.
    (c) an adjustable resistor.
    (d) none of these

19. In Figure 9-19, which is the symbol for a rheostat?

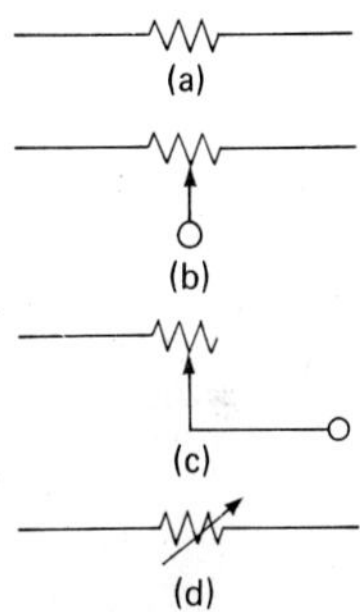

FIGURE 9-19

20. Connecting the battery to the chassis in an automobile allows the manufacturer to omit one conductor of the circuit.
    (a) true
    (b) false

10

Resistive-Bridge Circuits

Introduction

Frequently, one circuit is used to sense a change in operation that might occur in another circuit. Once a change is sensed, still other circuits can be activated, to compensate for the change.

One circuit used to sense changes is the resistive-bridge circuit. Some bridge circuits are made of fixed resistors, while others may contain one or more variable resistors. Use of variable resistors makes it possible for the Wheatstone bridge to be a highly accurate ohmmeter. Bridge circuits that contain devices other than resistors are discussed later. In this chapter, we concentrate on the DC resistive bridge circuit and its uses.

Objectives

Each student is required to:

1. Identify schematic diagrams for resistive-bridge circuits.
2. State the ratio of resistors in a resistive-bridge circuit.
3. Use given quantities (current, voltage, and resistance values) to calculate the voltage conditions in a bridge circuit.
4. Given three known resistance values and the ratio from objective 2, solve for the value of the unknown resistance required to balance a bridge.
5. Predict voltage values for all parts of the bridge circuit after having analyzed and solved the circuit's values.
6. Determine whether a bridge is balanced or unbalanced.
7. Explain the design and operation of a galvanometer.
8. Predict the direction of current flowing through the galvanometer in an unbalanced resistive bridge circuit.
9. Explain the use of a resistive bridge circuit as an ohmmeter.

Balanced Bridge Circuits

Figure 10-1 is used to review the parallel-series-resistive circuit. The circuit has two current paths, with each path containing two resistors. Points A and B are located in the circuit as shown.

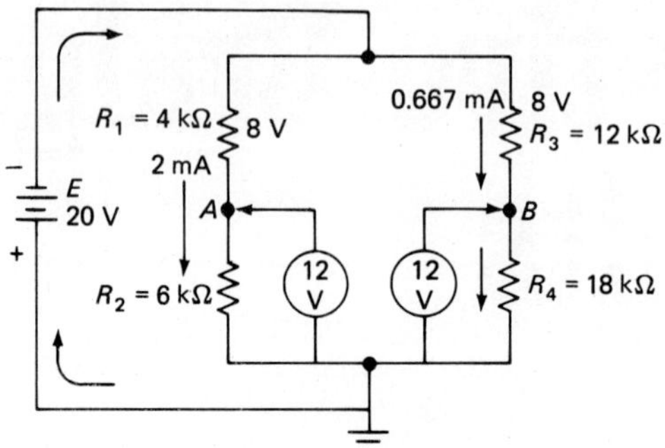

FIGURE 10-1

Using the values assigned in the schematic, we can calculate branch currents, individual voltage drops, and the difference in potential between points A and B. The formulas and procedures are:

Mathematical Solution	Calculator Solution Action	Entry	Display
$R_{Br1} = R_1 + R_2$	ENTER	4,EE,3	4 03
$R_{Br1} = 4\text{ k}\Omega + 6\text{ k}\Omega$	PRESS	+	4 03
$R_{Br1} = 10\text{ k}\Omega$	ENTER	6,EE,3	6 03
	PRESS	=	1 04

1 04 converts to 10 kΩ.

Mathematical Solution	Calculator Solution Action	Entry	Display
$I_{Br1} = \dfrac{E_t}{R_{Br1}}$	ENTER	20	20
$I_{Br1} = \dfrac{20\text{ V}}{10\text{ k}\Omega}$	PRESS	÷	20
$I_{Br1} = 2\text{ mA}$	ENTER	10,EE,3	10 03
	PRESS	=	2 – 03

2 – 03 converts to 2 mA.

Mathematical Solution	Calculator Solution Action	Entry	Display
$E_{R2} = I_{Br1} \times R_2$	ENTER	2,EE,+/–,3	2 – 03
$E_{R2} = 2\text{ mA} \times 6\text{ k}\Omega$	PRESS	×	2 – 03
$E_{R2} = 12\text{ V}$	ENTER	6,EE,3	6 03
	PRESS	=	1.2 01

1.2 01 converts to 12 V.

Looking ahead to the analysis of bridge circuits there, we are interested in the voltage of point A. This voltage equals the voltage drop of R_2, or 12 V. Therefore:

Voltage at point A = 12 V to ground

To solve branch 2 and for the voltage at point B:

Mathematical Solution	Calculator Solution Action	Entry	Display
$R_{Br2} = R_3 + R_4$	ENTER	12,EE,3	12 03
$R_{Br2} = 12\ \text{k}\Omega + 18\ \text{k}\Omega$	PRESS	+	1.2 04
$R_{Br2} = 30\ \text{k}\Omega$	ENTER	18,EE,3	18 03
	PRESS	=	3 04

3 04 converts to 30 kΩ.

Mathematical Solution	Calculator Solution Action	Entry	Display
$I_{Br2} = \dfrac{E_t}{R_{Br2}}$	ENTER	20	20
$I_{Br2} = \dfrac{20\ \text{V}}{30\ \text{k}\Omega}$	PRESS	÷	20
$I_{Br2} = 0.6667\ \text{mA}$	ENTER	30,EE,3	30 03
	PRESS	=	6.667 – 04

6.667 – 04 converts to 0.6667 mA.

Mathematical Solution	Calculator Solution Action	Entry	Display
$E_{R4} = I_{Br2} \times R_4$	ENTER	.6667,EE,+/–,3	.6667 – 03
$E_{R4} = 0.6667\ \text{mA} \times 18\ \text{k}\Omega$	PRESS	×	6.67 – 04
$E_{R4} = 12\ \text{V}$	ENTER	18,EE,3	18 03
	PRESS	=	1.2 01

1.2 01 converts to 12 V. Rounding off the mathematical calculations using a calculator can cause small errors.

Voltage at point B = E_{R4} = 12 V

Comparison of the voltages present at points A and B reveals that the voltages at A and B are equal. Stated another way, there is no difference in potential between A and B. Resistor R_5 can be connected between points A and B, and no current flows through it, because it has no applied voltage.

Figure 10-2 contains the same component values as Figure 10-1. Parallel branches are arranged differently, and a load resistor, R_L has been inserted. The voltages at points A and B are equal, thus placing 0 V of potential across R_L.

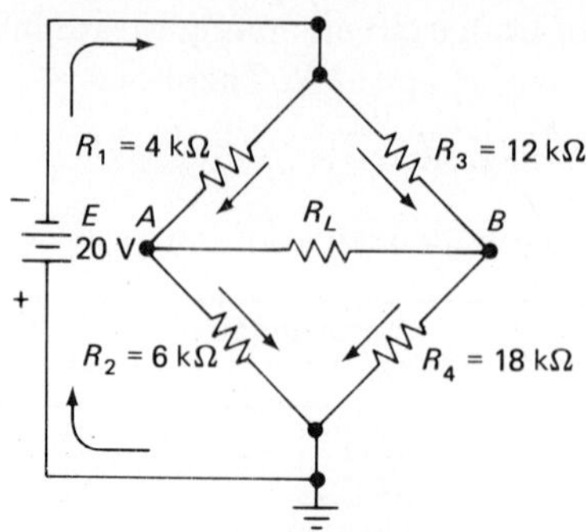

FIGURE 10-2

By inserting R_L between points A and B, we form a *bridge*. The relationship in the circuit should be evident. When a difference in potential occurs between points A and B, current flows through the load device, R_L. The direction current flows through R_L is determined by which points, A or B, is most positive. If A is most positive, current flows from B to A; if B is most positive, current flows from A to B.

A galvanometer can be used in test equipment as the load device. A highly sensitive ammeter with zero at the center of the scale, the galvanometer can measure the current flow in either direction. Figure 10-3 shows a galvanometer scale and presents its schematic symbol.

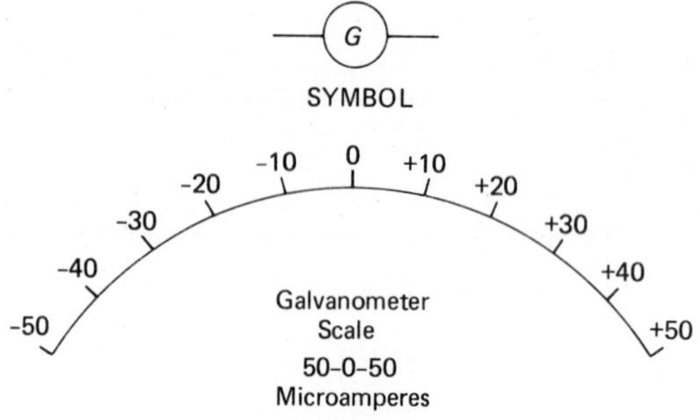

FIGURE 10-3

The pointer on a galvanometer deflects either left or right. Because it behaves this way, and because the galvanometer is highly sensitive to small changes in current, it is possible to adjust a bridge very accurately for a zero-current indication. When the current shown on the galvanometer is zero, the bridge is said to be balanced.

Figure 10-4 illustrates a bridge circuit with a galvanometer connected. If no current flows through the meter, its pointer rests at midscale (a reading of zero). A reading of zero means that zero potential is applied. The balanced condition of a bridge circuit can be checked mathematically, using the ratio

$$\frac{R_1}{R_2} = \frac{R_3}{R_4}$$

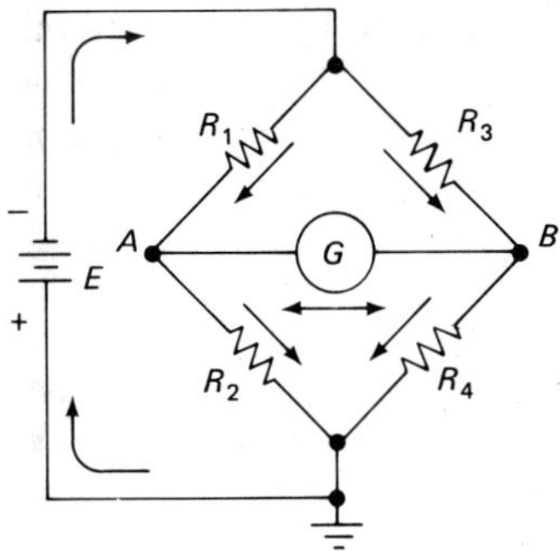

FIGURE 10-4

Resistor values are given in Figure 10-5. Using the values and the ratio, we can calculate whether the bridge is balanced:

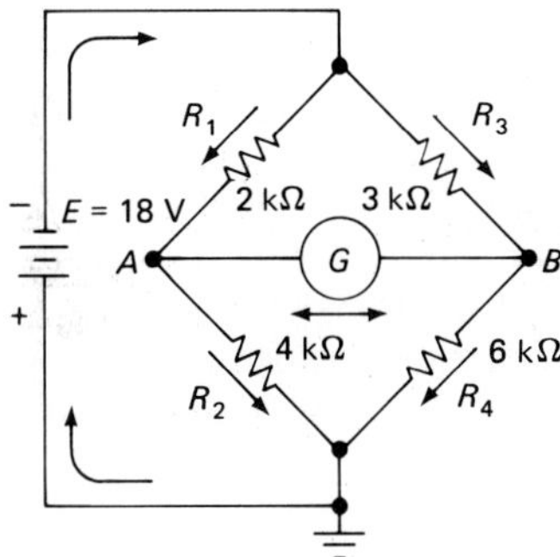

FIGURE 10-5

$$\frac{R_1}{R_2} = \frac{R_3}{R_4}$$

$$\frac{2\ \Omega}{4\ \Omega} = \frac{3\ \Omega}{6\ \Omega}$$

resulting in

$$0.5 = 0.5$$

Another method can be used with the ratio, to establish whether a bridge is balanced or unbalanced:

$$\frac{R_1}{R_2} \gtrless \frac{R_3}{R_4}$$

By cross-multiplying the terms connected by the arrows, we get:

$$R_1 \times R_4 = R_2 \times R_3$$

Substituting the resistor values in this ratio yields:

$$2\ \Omega \times 6\ \Omega = 3\ \Omega \times 4\ \Omega$$

Completing the multiplication, we have:

$$12\ \Omega = 12\ \Omega$$

These ratios can be stated in several different forms:

$$\frac{R_1}{R_2} = \frac{R_3}{R_4} \quad \text{or} \quad R_1 \times R_4 = R_2 \times R_3$$

$$\frac{R_1}{R_3} = \frac{R_2}{R_4} \quad \text{or} \quad R_1 \times R_4 = R_3 \times R_2$$

$$\frac{R_2}{R_1} = \frac{R_4}{R_3} \quad \text{or} \quad R_2 \times R_3 = R_1 \times R_4$$

$$\frac{R_3}{R_1} = \frac{R_4}{R_2} \quad \text{or} \quad R_3 \times R_2 = R_1 \times R_4$$

When application of either ratio results in equal quantities, the bridge is balanced and no current flows through the galvanometer. Applying any of the four ratios results in equal quantities when solving Figure 10-5. This indicates that the bridge is balanced.

Using any form of the ratio and three known resistor values, we can calculate the value of a fourth resistor, which is suitable for balancing the bridge (Figure 10-6).

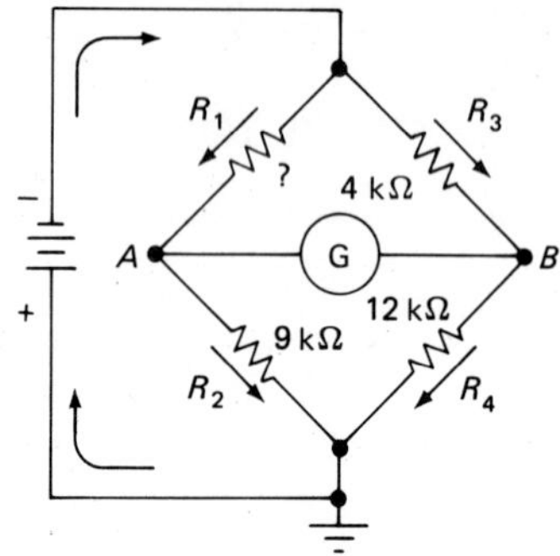

FIGURE 10-6

To determine what value R_1 must have to balance the bridge, substitute the ohmic values of R_2, R_3, and R_4 in the ratio:

$$R_1 \times R_4 = R_2 \times R_3$$

$$R_1 \times 12 = 9 \times 4$$

$$12R_1 = 36$$

The completed solution results in

$$R_1 = 3\ \text{k}\Omega$$

Inserting a 3 kΩ resistor in place of R_1 in Figure 10-6 results in a balanced-bridge circuit. If the solving of a ratio results in quantities that are not precisely equal, the bridge is unbalanced.

Unbalanced Bridge Circuits

An unbalanced bridge has a difference in potential existing across the galvanometer. Current flows through the meter, and the meter's deflection indicates the amount and direction of current flow. You can familiarize yourself with the unbalanced bridge circuit by analyzing Figure 10-7. You may be able to determine that the bridge is unbalanced by comparing the size of the resistors in each branch. To confirm that the bridge is unbalanced, let us analyze the circuit by using the ratio,

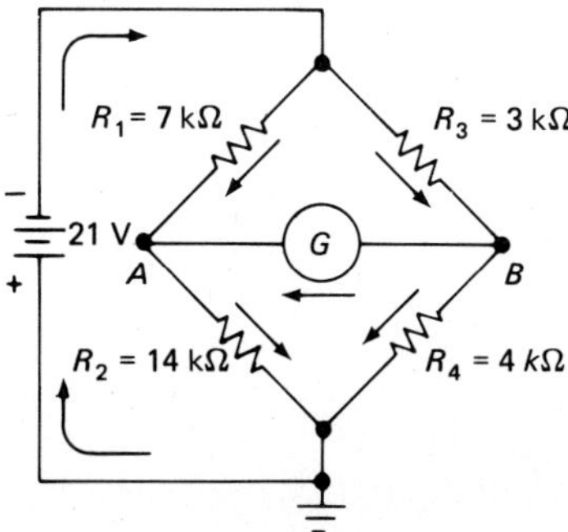

FIGURE 10-7

$$R_1 \times R_4 = R_2 \times R_3$$

$$7\text{k} \times 4\text{k} = 3\text{k} \times 14\text{k}$$

$$28\text{M} = 42\text{M}$$

Evidently, 28M does not equal 42M; therefore, the bridge is not balanced. When the bridge is unbalanced, a difference in potential exists across the galvanometer, and current flows through the meter. To determine the potential difference and current, we solve, for point A:

$$R_{Br1} = R_1 + R_2 = 21 \text{ k}\Omega$$

$$I_{Br1} = \frac{E_t}{R_{Br1}} = 1 \text{ mA}$$

$$E_{\text{point A}} = I_{Br1} \times R_2 = -14 \text{ V}$$

and for point B:

$$R_{Br2} = R_3 + R_4 = 7 \text{ k}\Omega$$

$$I_{Br2} = \frac{E_t}{R_{Br2}} = 3 \text{ mA}$$

$$E_{\text{point B}} = I_{Br2} \times R_4 = -12 \text{ V}$$

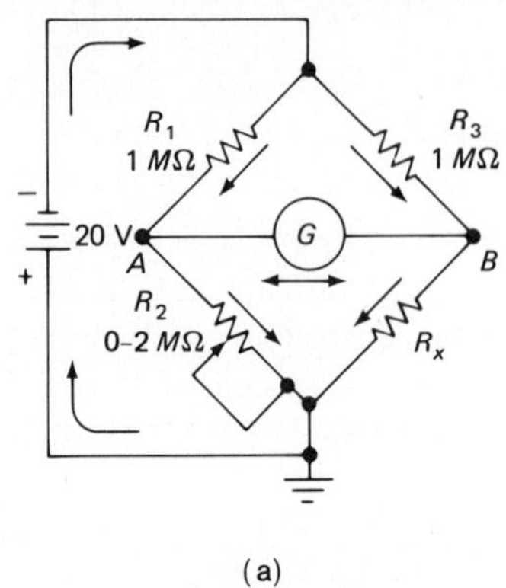

(a)

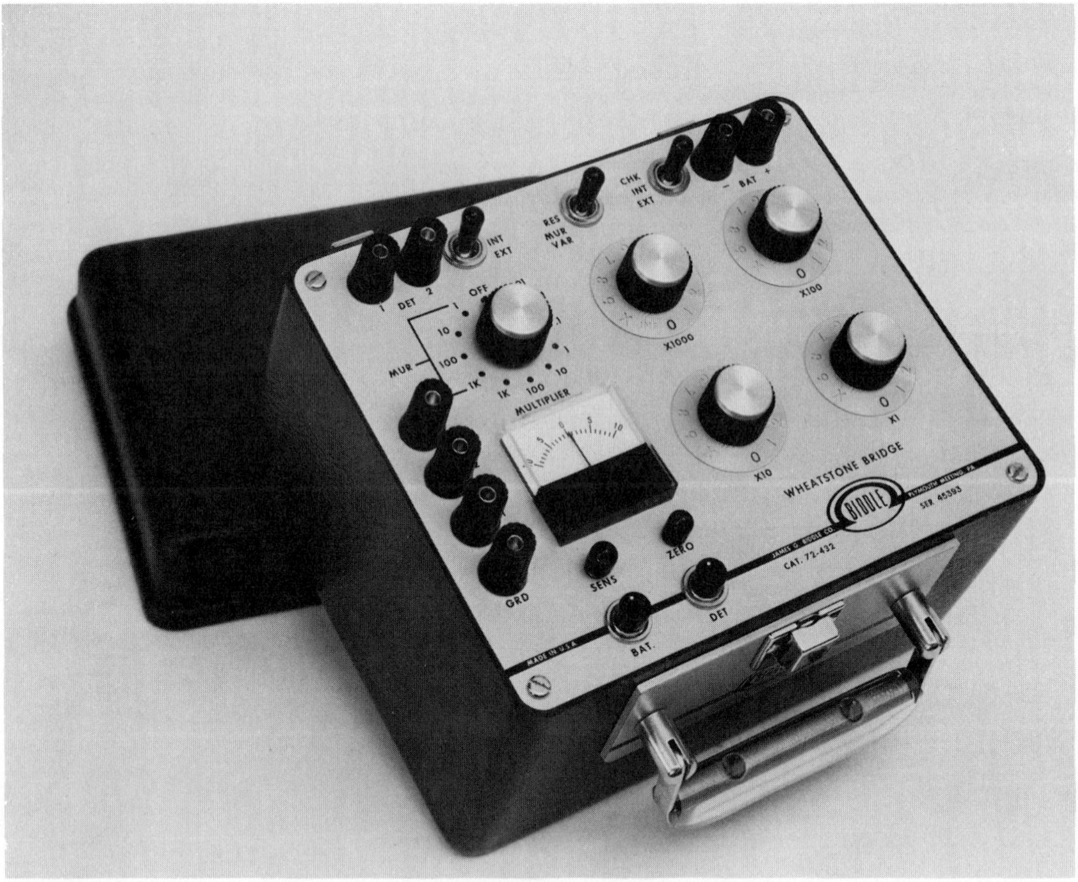

(b)

FIGURE 10-8
(b) Wheatstone Bridge (Photo Courtesy of Biddle Instruments)

To find the difference in potential between points A and B:

$$-14 \text{ V} - (-12 \text{ V}) = -2 \text{ V (difference in potential)}$$

The difference in potential is -2 V, with point B less negative than point A. A voltmeter connected between points A and B would read 2 V. Because a potential is present, the connection of a galvanometer between A and B results in current flow through the meter. To determine the direction of current flow, we use the following:

1. Current flows from a negative potential to a positive potential.
2. Current flows from a positive potential to a more positive potential.
3. Current flows from a negative potential to a less negative potential.

The third rule applies to this circuit, with galvanometer current flowing from point A to point B. The galvanometer's pointer deflects to indicate this current flow. Replacing the meter with a DC motor results in the motor rotating at a set speed as long as the circuit is operating. Ideally, it is better to be able to vary the speed of the motor to meet different requirements. This can be done using the Wheatstone bridge circuit shown in Figure 10-8.

In this circuit, R_2 has been replaced by a variable resistor connected as a rheostat. The circuit can be used to vary the speed of a motor connected in place of the galvanometer. Picture the motor in a slot-car racer or an electric train. If R_2 has been replaced by a variable 2 MΩ resistor, it is possible to change the direction of the car or train as well as vary the speed in either direction. The key is reversing and changing current between points A and B.

The Wheatstone bridge circuit can be used as an ohmmeter. R_1 and R_3 are both 1 MΩ; R_2 is a variable 0–2 MΩ resistor. When a resistor of unknown ohms is connected in place of R_x, R_2 can be adjusted so that the bridge is balanced. In order for the bridge to be balanced, R_2 must equal R_4. If R_2 is marked with a calibrated scale, its adjustment will indicate its ohmic value by moving a pointer along the scale. Because $R_x = R_2$, this value equals R_x. Values of R_x between 0 and 2 MΩ can be measured using this arrangement. Resistors exceeding 2 MΩ cannot be measured using this circuit; for small resistors, greater accuracy is required. Better accuracy is achieved by making the changes shown in Figure 10-9.

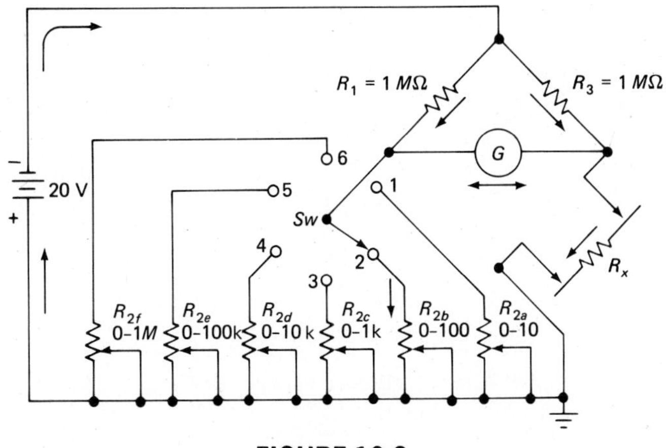

FIGURE 10-9

When checking resistors of 0 to 10 Ω, the switch is set to position 1; for resistors 0 to 1 kΩ, the switch is set to position 3; and so on. In some cases, you may have to try more than one range in order to balance the bridge and get a correct reading.

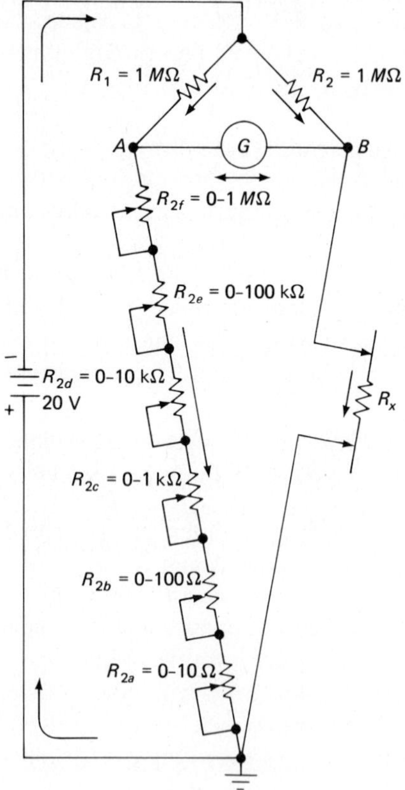

FIGURE 10-10

Figure 10-10 illustrates another way to wire a Wheatstone bridge. In this circuit, the wiper of each resistor, $R_{2a\text{-}f}$, is connected to a switch with calibrated positions 0–9. When R_x is connected, each switch is rotated until the combination of all switch settings balances the bridge. Reading the value of each switch setting, and starting with the R_{2f} position, gives us a direct readout of the unknown resistor value. This circuit can measure any resistance from 0 to 1,111,110 Ω. To extend the upper limit of measurement, it is necessary to install a larger (10 MΩ) resistor as R_{2g}. The accuracy of these measurements is only as accurate as the tolerance of the resistors used as R_1, R_2, and R_3 in the circuit.

Self-Check

Complete each item by inserting the words and/or numbers needed to make a true and complete statement.

1. One type of bridge circuit can be used as an ohmmeter; it is called a ______________ bridge.

2. A galvanometer measures ______________.

3. When zero current flows through the load in a bridge circuit, the bridge is ______________.

4. The variable resistor in a resistive bridge circuit allows us to ______________ the circuit.

5. A bridge circuit can be used to provide highly accurate control of a motor's ______________.

Summary

Resistive-bridge circuits are designed for special uses. They can provide exact control of a motor's speed, vary the current through a load device, or be used as a test instrument, among many other uses.

Bridge circuits are said to be balanced when no current flows through the load device, and unbalanced when current flows in the load device. By inserting a galvanometer as the load device, a visual indication is available, which indicates the balanced or unbalanced condition of the circuit. A balanced bridge causes the meter to indicate zero; an unbalanced bridge causes it to indicate the direction and amount of current flow in the load device. The balance of a resistive bridge can be checked by the ratio $R_1/R_2 = R_3/R_4$, as it is applied in Figure 10-11.

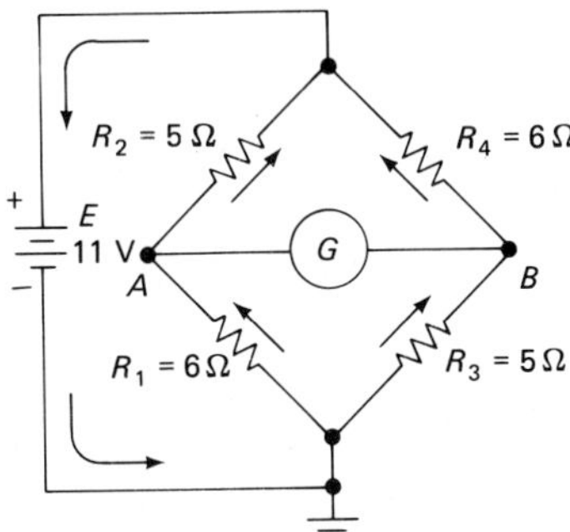

FIGURE 10-11

Inserting a rheostat for one of the fixed resistors will allow conditions in the bridge to be varied. Inclusion of a rheostat results in a Wheatstone bridge circuit. If the rheostat is selected correctly, current flow in the load device can be varied and current direction reversed.

A Wheatstone bridge is used as a highly accurate test instrument. The ohmmeter was discussed; other versions of the Wheatstone bridge are used for similar measurements in later studies. To design an ohmmeter, we will make R_1 and R_3 of equal value. The rheostat inserted as R_2 determined the maximum size of an unknown resistor, R_x, that could be measured. Varying the setting of R_2 until the meter indicates a balance resulted in the value of R_2 equaling that of R_4. If we know the value of R_2, we also know the value of R_4.

Review Questions and/or Problems

1. Identify the conditions that exist within a balanced bridge circuit.
   (a) Current flows through the galvanometer; no difference in potential exists between A and B.
   (b) There is no current flow in the galvanometer, and no difference in potential exists between A and B.
   (c) There is no current flow in the galvanometer, but a difference in potential exists between A and B.
   (d) Current flows in the galvanometer; a difference in potential exists between A and B.

2. The bridge in Figure 10-12 can be used as
   (a) an ammeter.
   (b) a voltmeter.
   (c) an ohmmeter.
   (d) an oscilloscope.

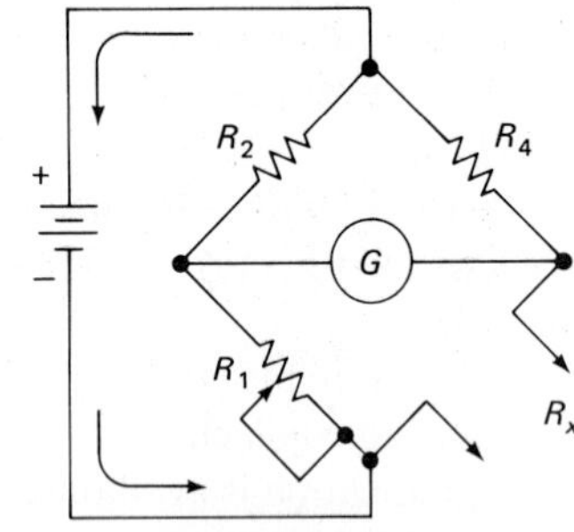

FIGURE 10-12

3. Decreasing the voltage applied to the circuit in Figure 10-13
   (a) increases current from A to B.
   (b) increases current from B to A.
   (c) decreases current from A to B.
   (d) has no effect on the galvanometer reading.

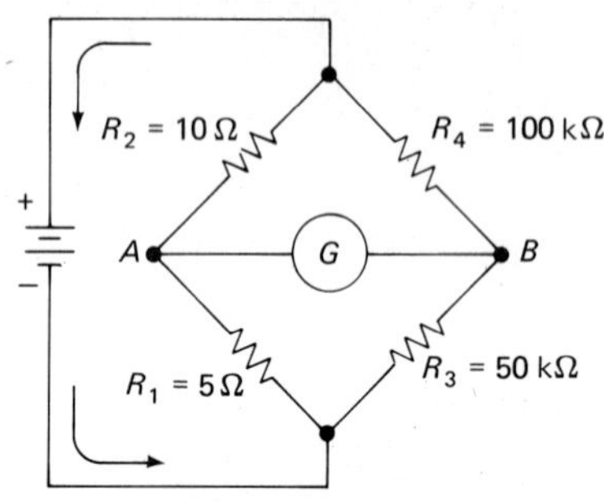

FIGURE 10-13

4. If the voltage applied to a bridge circuit is increased, the current through the galvanometer
   (a) increases if the bridge is balanced.
   (b) decreases if the bridge is balanced.
   (c) decreases if the bridge is unbalanced.
   (d) increases if the bridge is unbalanced.

5. To use the bridge for measuring unknown resistors, the circuit must at least contain
   (a) no variable resistors.
   (b) one variable resistor.
   (c) two variable resistors.
   (d) three variable resistors.

6. Current flow in the circuit in Figure 10-14 is from
   (a) B to A.
   (b) A to B.

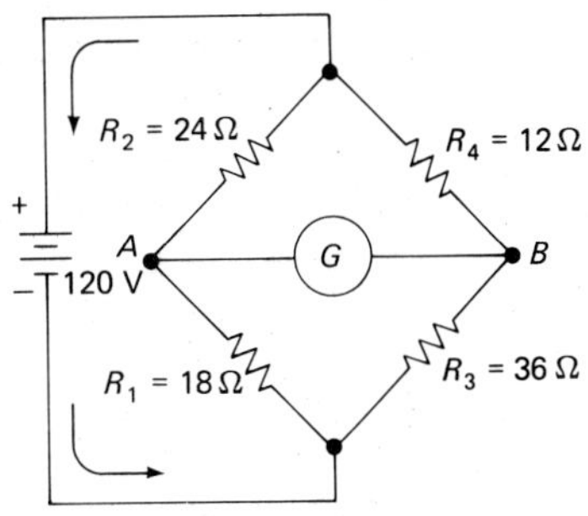

FIGURE 10-14

7. The bridge in Figure 10-15 is balanced. What size is R_x?
   (a) 10 kΩ
   (b) 20 kΩ
   (c) 30 kΩ
   (d) 45 kΩ

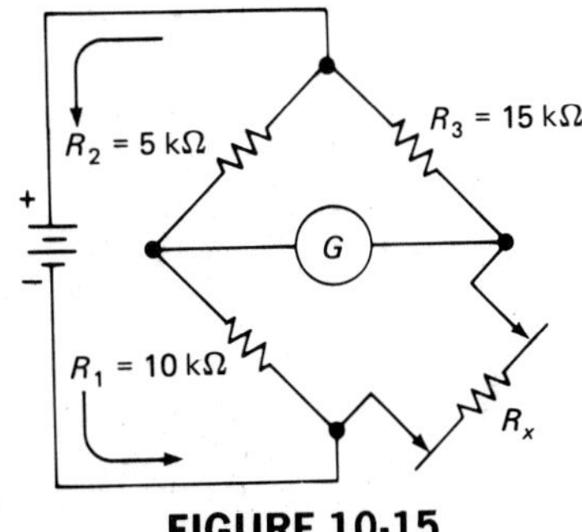

FIGURE 10-15

8. The galvanometer current in Figure 10-16 is zero. What is the size of R_1?
   (a) 3.3 kΩ
   (b) 6.7 kΩ
   (c) 13.3 kΩ
   (d) 20 kΩ

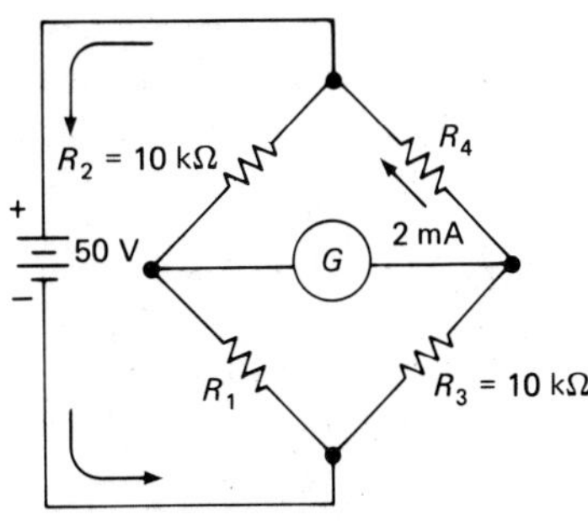

FIGURE 10-16

9. If R_3 shorts in the circuit shown in Figure 10-17, what will be the value of R_1 when the bridge is balanced?
   (a) 2.5 kΩ
   (b) 6 kΩ
   (c) 12 kΩ
   (d) 1.0 kΩ

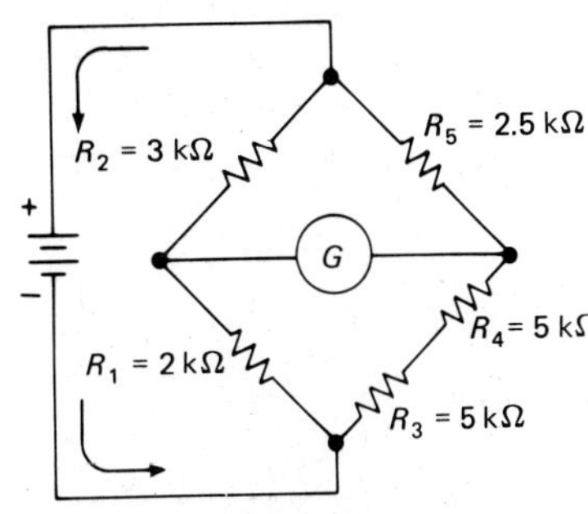

FIGURE 10-17

10. How much voltage will R_2 drop in the circuit in Figure 10-18?
   (a) 37.5 V
   (b) 15 V
   (c) 75 V
   (d) 60 V

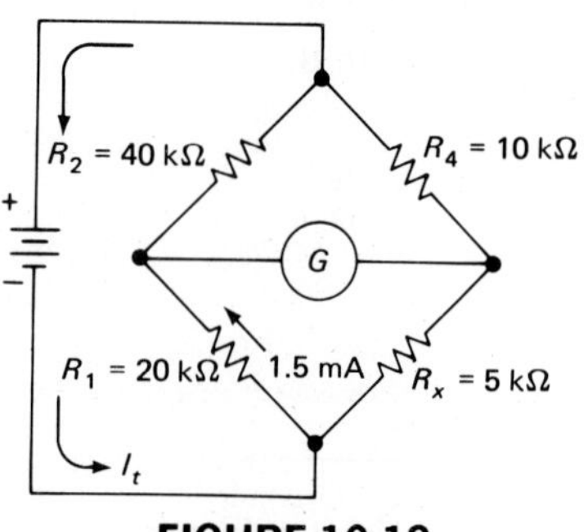

FIGURE 10-18

11

Network Analysis

Introduction

Often it is necessary to analyze a circuit where known quantities do not fit Ohm's law. In those cases, we rely on other methods to complete the analysis.

In some circuits it is possible for one component to be influenced by two or more power sources. When that happens, the voltage and current effect for any component is the algebraic sum of the effects produced by each source acting separately.

For other circuits, such as those with a power source, it is good to know the basic characteristics of the circuit's output. That allows us to predict the effect when a circuit is connected across the output. Thevenin's and Norton's theorems provide us with this knowledge.

Objectives

Each student is required to:

1. Apply Kirchhoff's laws to circuits that supply insufficient information for analysis by Ohm's law.
2. Solve problems with one or two sources, either series-aiding or series-opposing, by use of the superposition theorem.
3. Reduce circuits to their Thevenin's equivalent circuit by use of Thevenin's theorem.
4. Use Norton's theorem to reduce circuits to their Norton's equivalent.
5. Convert Norton's equivalent circuits to Thevenin's equivalent circuits.
6. Convert Thevenin's equivalent circuits to Norton's equivalent circuits.
7. Use simultaneous equations to solve circuits containing three unknowns.
8. Analyze resistive-bridge circuits, using Thevenin's theorem.
9. State the maximum-power-transfer theorem and explain the effect that power transfer has on circuit operation.

Analysis Using Kirchhoff's Laws

Before proceeding to an analysis of circuits using Kirchhoff's laws, let us first review them.

Kirchhoff's voltage law states:

The algebraic sum of all the voltages around any closed path equals zero.

In application, we found that the law means, when all voltage drops in a closed loop are added, their sum equals the total voltage. We have used this law to analyze series, parallel, and series-parallel resistive circuits. Kirchhoff's current law states:

The algebraic sum of the currents into any point of a circuit must equal the algebraic sum of the currents out of that point.

In application, we found that this law means, the sum of all currents flowing away from any point in a circuit must equal the current that flows into that point. In series circuits, current is the same at all points within the circuit; but in a parallel network, total current equals the sum of all branch currents.

Application of these laws allows us to analyze circuits which, at first glance, might appear to be unsolvable.

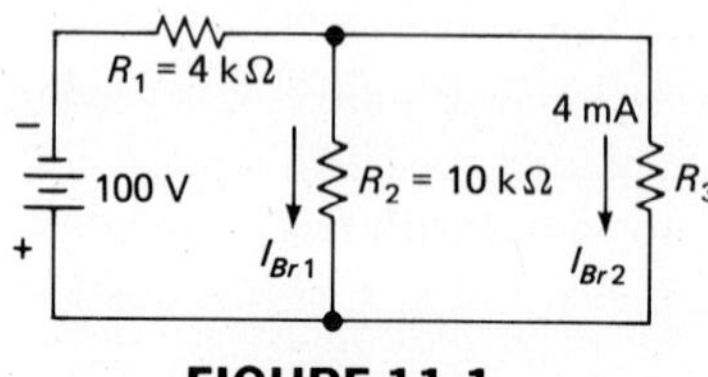

FIGURE 11-1

Examination of the schematic in Figure 11-1 fails to locate a point where two known quantities are available. Without two known quantities, it is impossible to use Ohm's law. We can, however, use Kirchhoff's laws, and start the analysis.

To analyze the circuit in Figure 11-1, we first apply Kirchhoff's voltage law. In this circuit, R_1 and R_2 form a complete loop for current flow. From Kirchhoff's law, we know that the sum of their voltage drops, $E_{R1} + E_{R2}$, must equal the total voltage, E_t. These circuit conditions are known:

$$E_t = E_{R1} + E_{R2} = 100\ \text{V}$$

$$I_t = I_{Br1} + I_{Br2}$$

$$E_{R1} = I_t \times R_1$$

$$E_{R2} = I_{Br1} \times R_2$$

Substituting $I_t \times R_1$ for E_{R1} and $I_{Br1} \times R_2$ for E_{R2} in the first formula, we have:

$$I_t \times R_1 + I_{Br1} \times R_2 = E_t$$

Substituting *total current* in the equation, we have:

$$(I_{Br1} + I_{Br2}) \times R_1 + I_{Br1} \times R_2 = E_t$$

After multiplying quantities:

$$I_{Br1}(4\text{ k}\Omega) + (4\text{ k}\Omega)(4\text{ mA}) + I_{Br1}(10\text{ k}\Omega) = 100\text{ V}$$

Completing the solution, we find:

$$I_{Br1}(14\text{ k}\Omega) + 16\text{ V} = 100\text{ V}$$

$$I_{Br1}(14\text{ k}\Omega) = 100\text{ V} - 16\text{ V}$$

$$I_{Br1} = \frac{84\text{ V}}{14\text{ k}\Omega}$$

$$I_{Br1} = 6\text{ mA}$$

With I_{Br1} known, we can complete the circuit analysis, using Ohm's law:

$$I_t = I_{Br1} + I_{Br2} = 6\text{ mA} + 4\text{ mA} = 10\text{ mA}$$

$$E_{R1} = I_t \times R_1 = 10\text{ mA} \times 4\text{k} = 40\text{ V}$$

$$E_{R2} = I_{Br1} \times R_2 = 6\text{ mA} \times 10\text{k} = 60\text{ V}$$

$$E_{R2} = E_{R3} = 60\text{ V}$$

$$R_3 = \frac{E_{R3}}{I_{Br2}} = \frac{60\text{ V}}{4\text{ mA}} = 15\text{ k}\Omega$$

From these calculations, we can see that a thorough understanding of all the laws and theorems studied is necessary, especially of Kirchhoff's laws. Kirchhoff's laws are applied to other circuits later in this chapter.

Self-Check

Answer each item by indicating whether the statement is true or false.

1. ____ Kirchhoff's voltage law tells us that the sum of all voltage drops in a closed loop equals the applied voltage.

2. ____ Kirchhoff's current law tells us that the sum of all currents flowing in a circuit equals total current.

3. ____ It is not necessary to use Ohm's law when analyzing a circuit using Kirchhoff's law.

4. ____ To use Kirchhoff's law, we must be given two knowns pertaining to some point in the circuit.

5. ____ To solve for current using Kirchhoff's law, all results must have positive quantities.

Superposition Theorem

For circuits containing two or more power supplies, the superposition theorem can be used to determine the power source with the superior position. When two or more sources are connected across the same component, this theorem can be used to determine circuit conditions. The superposition theorem is stated thus:

> ***In a network with two or more sources, the current or voltage for any component is the algebraic sum of the effects produced by each source acting separately.***

Take special note of the following statement.

> ***When using the superposition theorem for analysis, you must be very careful to observe all voltage polarities.***

By way of illustration, let us analyze the circuit in Figure 11-2. The purpose of this analysis is to determine the voltage reading of V_1, which is connected between point A and ground. To analyze the effect of each source on the circuit, it is necessary to kill (short) all other sources. Next, we analyze the circuit for the effect of each source. Finally, by algebraic addition, we combine the result of all sources.

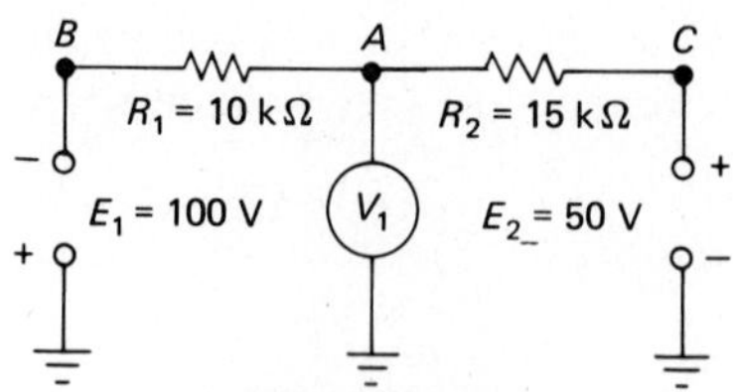

FIGURE 11-2

The first source to be analyzed is $E_1(-100\text{ V})$. For this, the 50 V source must be killed, which results in the circuit shown in Figure 11-3. An analysis of this circuit reveals:

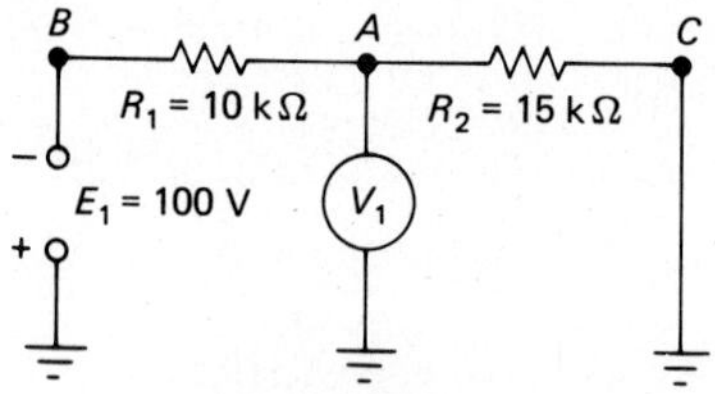

FIGURE 11-3

$$R_t = R_1 + R_2 = 25\text{ k}\Omega$$

$$I_t = \frac{E_t}{R_t} = 4\text{ mA}$$

$$E_{R1} = I_t \times R_1 = 40\text{ V}$$

$$E_{R2} = I_t \times R_2 = 60 \text{ V}$$

$$E_t = E_{R1} + E_{R2} = 100 \text{ V}$$

The voltmeter in Figure 11-3 is connected between point A and ground, which is in parallel with E_{R2}. Therefore the voltage read by V_1 equals -60 V.

We use Figure 11-4 to analyze for the effect of E_2. E_1 has been replaced by a short. The analysis reveals:

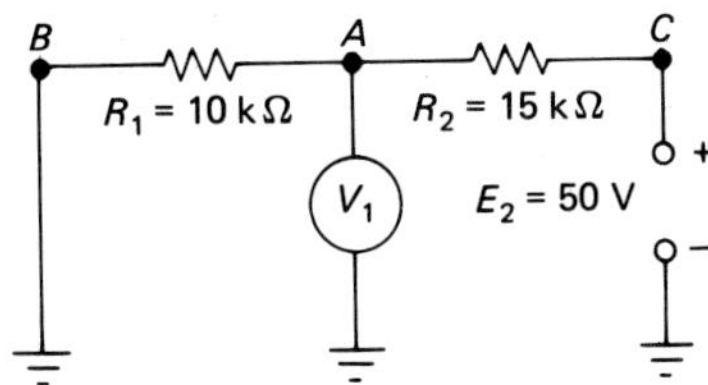

FIGURE 11-4

$$R_t = R_1 + R_2 = 25 \text{ k}\Omega$$

$$I_t = \frac{E_t}{R_t} = 2 \text{ mA}$$

$$E_{R1} = I_t \times R_1 = 20 \text{ V}$$

$$E_{R2} = I_t \times R_2 = 30 \text{ V}$$

$$E_t = E_{R1} + E_{R2} = 50 \text{ V}$$

In the circuit, V_1 is connected between point A and ground, which is in parallel with E_{R1}. Therefore, the voltage read by V_1 equals 20 V.

To calculate the reading for V_1 in Figure 11-2, we combine the two values already identified by using algebraic addition:

Figure 11-3 point A to ground $= -60$ V

Figure 11-4 point A to ground $= +20$ V

Voltmeter, V_1, reads: $-60 \text{ V} + (+20 \text{ V}) = -40 \text{ V}$

Up to now, our discussion has centered on a circuit containing series-aiding power sources. For a circuit containing series-opposing sources, we analyze Figure 11-5. The analysis procedures are the same. When the 50 V source is killed, Figure 11-6 represents the 100 V circuit. Analyzing the circuit, we get:

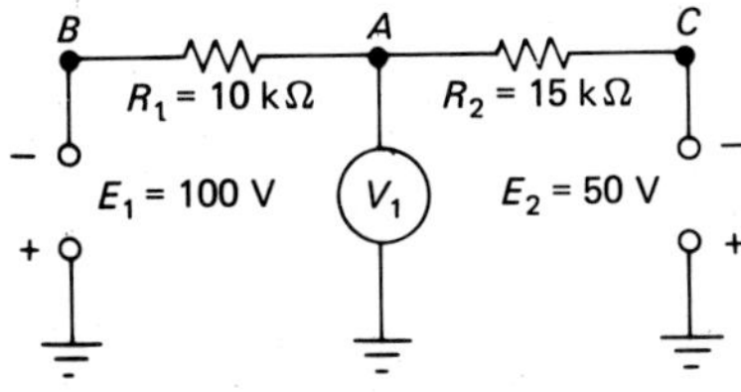

FIGURE 11-5

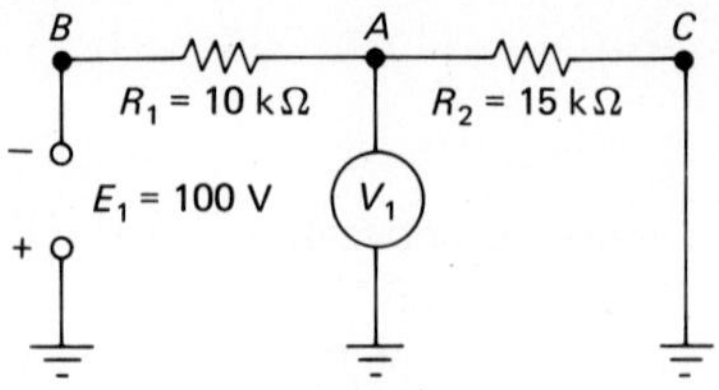

FIGURE 11-6

$$R_t = R_1 + R_2 = 25 \text{ k}\Omega$$

$$I_t = \frac{E_t}{R_t} = 4 \text{ mA}$$

$$E_{R1} = I_t \times R_1 = 40 \text{ V}$$

$$E_{R2} = I_t \times R_2 = 60 \text{ V}$$

$$V_1 = E_{R2} = -60 \text{ V}$$

Figure 11-7 represents the 50-V source, with 100 V killed.

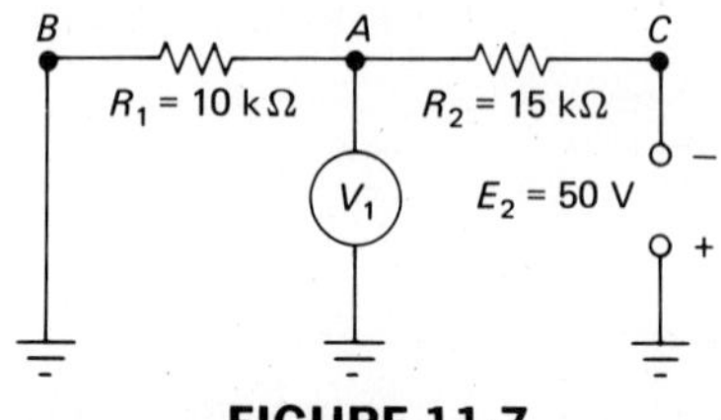

FIGURE 11-7

Analyzing the circuit, we find:

$$R_t = R_1 + R_2 = 25 \text{ k}\Omega$$

$$I_t = \frac{E_t}{R_t} = 2 \text{ mA}$$

$$E_{R1} = I_t \times R_1 = 20 \text{ V}$$

$$E_{R2} = I_t \times R_2 = 30 \text{ V}$$

$$V_1 = E_{R1} = -20 \text{ V}$$

Superimposing the two voltages, we get:

$$-60 \text{ V} + (-20 \text{ V}) = -80 \text{ V}$$

With both power sources connected, V_1 reads -80 V.

Applying Kirchhoff's Laws

To confirm our analysis of these circuits, we apply Kirchhoff's laws. Figure 11-8 contains a circuit with two series-aiding power sources. In this circuit, current flows from the negative pole of E_1 to the positive pole of E_2. This means they are connected series-aiding. Analysis reveals:

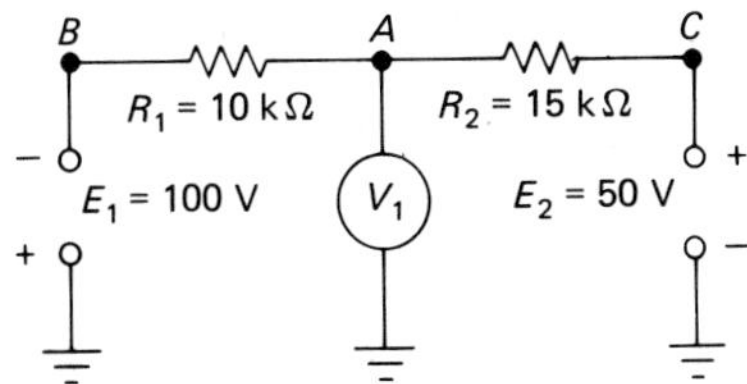

FIGURE 11-8

$$E_t = E_1 + E_2 = 150 \text{ V}$$

$$R_t = R_1 + R_2 = 25 \text{ k}\Omega$$

$$I_t = \frac{E_t}{R_t} = 6 \text{ mA}$$

$$E_{R1} = I_t \times R_1 = 60 \text{ V}$$

$$E_{R2} = I_t \times R_2 = 90 \text{ V}$$

To determine the reading of the voltmeter, V_1, redraw the circuit and label all components with their polarity and voltage, as shown in Figure 11-9.

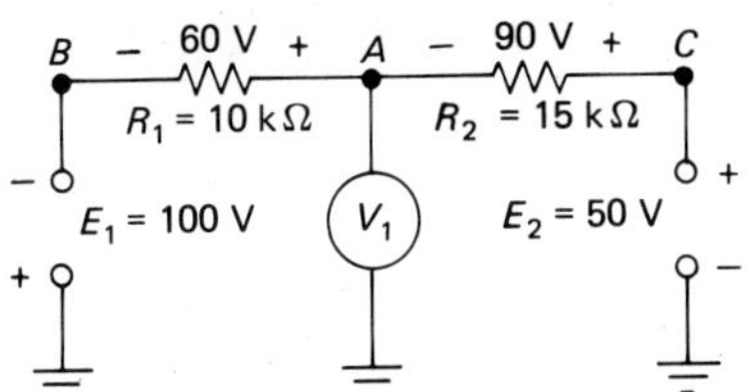

FIGURE 11-9

Leaving point A and following the circuit in either direction, we pass through a resistor, and a power source, proceeding to ground. Either route can be used to determine the reading on V_1. The solutions are:

$$E_{R1} + E_1 = +60 \text{ V} + (-100 \text{ V}) = -40 \text{ V}$$

$$E_{R2} + E_2 = -90 \text{ V} + (+50 \text{ V}) = -40 \text{ V}$$

This proves that voltage from point A to ground equals -40 V. Note that these values are the same as those obtained using the superposition theorem and Figure 11-2.

Figure 11-10 represents a circuit with two opposing power sources. Performing the analysis, we get:

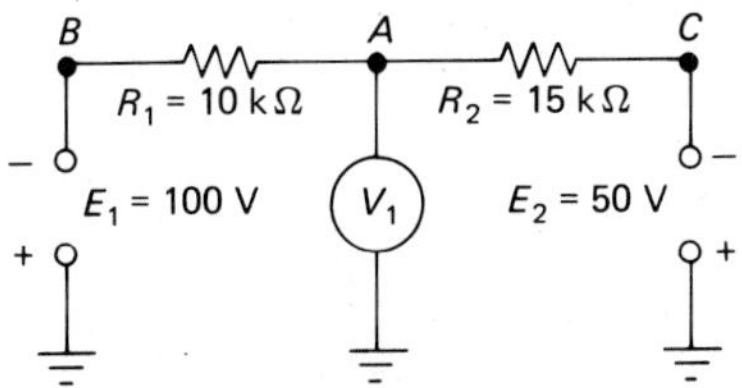

FIGURE 11-10

$$R_t = R_1 + R_2 = 25\ \text{k}\Omega$$

$$E_t = 100\ \text{V} - 50\ \text{V} = 50\ \text{V}$$

$$I_t = \frac{E_t}{R_t} = 2\ \text{mA}$$

$$E_{R1} = I_t \times R_1 = 20\ \text{V}$$

$$E_{R2} = I_t \times R_2 = 30\ \text{V}$$

Current flow from the larger source (100 V) overcomes the smaller source (50 V). This causes current to flow from the negative pole of the 100-V source to the negative pole of the 50-V source. Redrawing the circuit with polarities and voltages inserted results in the circuit shown in Figure 11-11. Solving both branches gives us:

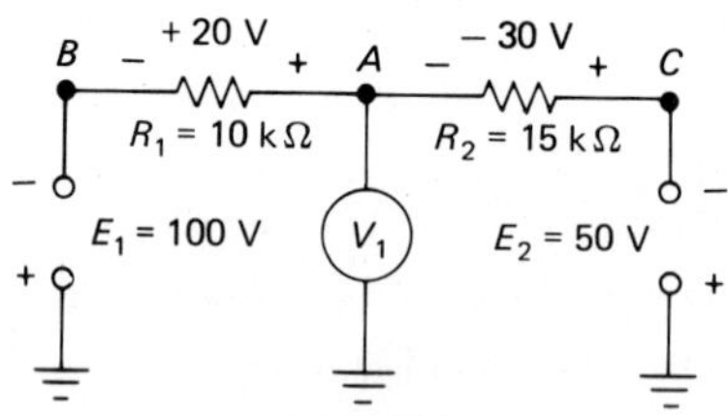

FIGURE 11-11

$$+20\ \text{V} + (-100\ \text{V}) = -80\ \text{V}$$

$$-30\ \text{V} + (-50\ \text{V}) = -80\ \text{V}$$

In either case, V_1 reads -80 V.

Using either Kirchhoff's laws or the superposition theorem, we get the same values for the readings that we would have obtained if the voltmeter, V_1, were connected between point A and ground.

Self-Check

Answer each item by indicating whether the statement is true or false.

6. ____ The superposition theorem determines which power source has the greatest effect on operation of the circuit.

7. ____ To use the superposition theorem for analysis, all power sources are opened.

8. ____ Analysis by Kirchhoff's laws and the superposition theorem does not always result in the same values.

9. ____ It is not necessary to consider voltage polarities when using the superposition theorem.

10. ____ The voltage drop between point A and ground in the Kirchhoff and the superposition solutions is the voltage that would be applied to a load device connected to this circuit.

Introduction to Theorem Use

Important notes for solutions involving Thevenin's and Norton's theorems are:

Thevenin's equivalent circuits operate from a constant voltage source similar to those already discussed.

Norton's equivalent circuits operate from a constant-current source and require different analysis techniques.

To kill a* current source, *we must remove (open) the branch containing the source.

To kill a* voltage source, *the branch containing the source must be shorted.

Thevenin's Theorem

Thevenin's theorem is useful in simplifying the voltage analysis of complex circuits. Using this theorem, we can reduce any circuit to its open-terminal voltage, E_{th}, and open-terminal resistance, R_{th}. A network can consist of any number of components and power sources (Figure 11-12a), but when it is reduced to its most basic equivalent, it can be represented by the component and power source shown in Figure 11-12b.

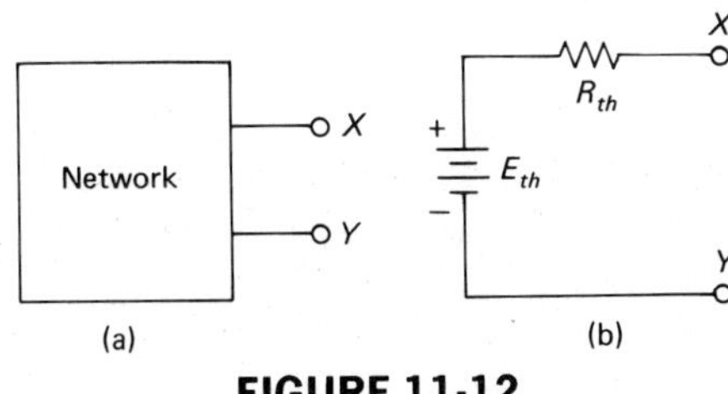

FIGURE 11-12

Thevenin's theorem is stated as follows:

The entire circuit between points X and Y of Figure 11-12a can be replaced by one voltage source, E_{th}, connected in series with one equivalent resistance, R_{th}, as in Figure 11-12b.

E_{th} is the open-terminal (zero current) voltage present between points X and Y. R_{th} is the open-terminal resistance measured by an ohmmeter connected across points X and Y, with a short replacing the voltage source.

Reducing a Circuit to Its Thevenin's Equivalent

Refer to Figure 11-13. We solve Figure 11-13a for the voltage and current present at a load resistor connected between points X and Y. The steps involved in the use of Thevenin's theorem are:

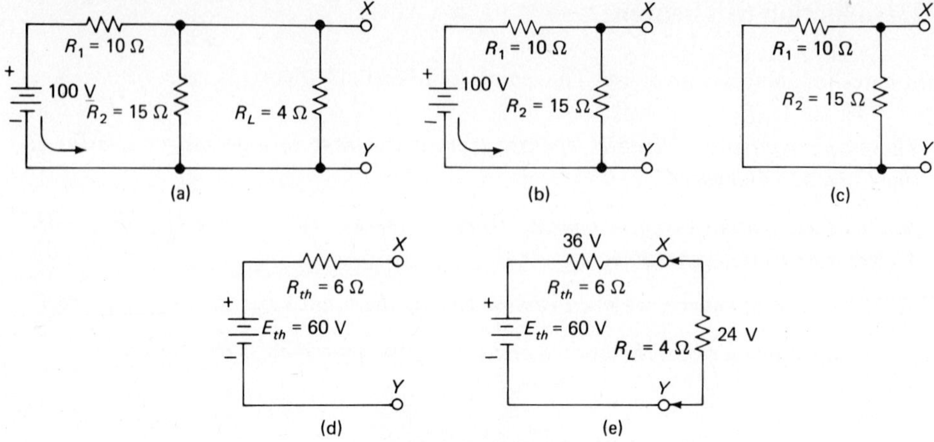

FIGURE 11-13

1. Open the circuit between points X and Y by removing R_L as shown in Figure 11-13b.
2. Short the power source. This results in Figure 11-13c.
3. Calculate the total resistance in the circuit as it would appear to an ohmmeter connected across points X and Y of Figure 11-13c. Calculated mathematically, total resistance is

$$R_t = \frac{R_1 \times R_2}{R_1 + R_2} = 6\ \Omega$$

   This 6 Ω is R_{th} for the equivalent circuit shown in Figure 11-13d.
4. With the open between points X and Y, as in step 1, solve for E_{th}, as follows:
   (a) Remove the short from the power source that was connected in step #2 above.
   (b) Calculate the voltage between points X and Y and find E_{th}.
   With an open between X and Y, a series circuit results. R_2 is in parallel with points X and Y, so the difference in potential between X and Y equals E_{R2}.
   To calculate E_{th} for Figure 11-13c, we must solve for:

$$R_t = R_1 + R_2 = 25\ \Omega$$

$$I_t = \frac{E_t}{R_t} = 4\ \text{A}$$

$$E_{R2} = I_t \times R_2 = 60\ \text{V}$$

The 60 V becomes E_{th} and the 6 Ω, R_{th}. Figure 11-13d represents the Thevenin's equivalent of Figure 11-13a.

To determine the voltage and current present at R_L in Figure 11-13a, we connect the 4-Ω resistor between points X and Y of Thevenin's equivalent (Figure 11-13e). Solving for I_{RL} and E_{RL} results in values for load current and voltage drop.

Use series-circuit rules as follows:

$$R_t = R_{th} + R_L = 10\ \Omega$$

$$I_{RL} = \frac{E_{th}}{R_t} = 6 \text{ A}$$

$$E_{RL} = I_{RL} \times R_L = 24 \text{ V}$$

With a 4-Ω load device connected, the difference in potential between X and Y is 24 V. Load current is 6 A.

Thevenin's Equivalent of a Two-Source Circuit

The techniques we have been using can be applied to more complex circuits. Figure 11-14a has two power sources. Thevenin's theorem will be used to reduce the circuit to its Thevenin's equivalent.

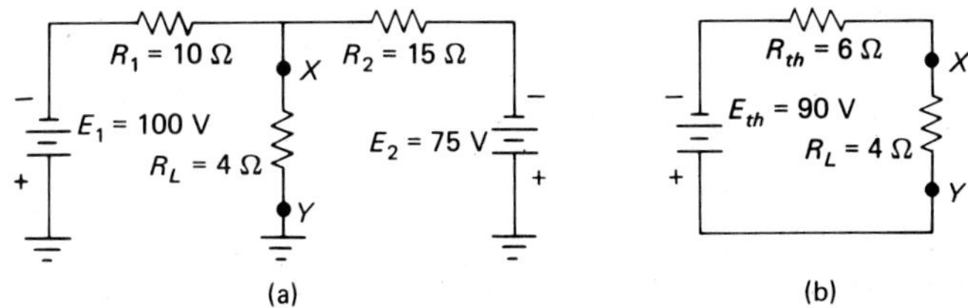

FIGURE 11-14

Solution involves the same steps:

1. Remove R_L, leaving an open between points X and Y.
2. Place a short across both power sources.
3. Solve for the parallel resistance of R_1 and R_2. In this case, R_t is used as R_{th}:

$$R_t = \frac{R_1 \times R_2}{R_1 + R_2} = 6 \, \Omega$$

4. Remove both shorts from the power sources.
5. Use the superposition theorem to solve for the voltage between points X and Y. This results in the following:

$$-30 \text{ V} + (-60 \text{ V}) = -90 \text{ V}$$

6. Redraw the Thevenin's equivalent circuit shown in Figure 11-14b. Use 90 V as E_{th} and 6 Ω as R_{th}. Note that the load resistor, R_L, has been inserted in the circuit.
7. Solve for E_{RL} and I_{RL} as follows:

$$R_t = R_{th} + R_L = 10 \, \Omega$$

$$I_{RL} = \frac{E_{th}}{R_t} = 9 \text{ A}$$

$$E_{RL} = I_{RL} \times R_L = 36 \text{ V}$$

8. Therefore, the voltage between points X and Y is 36 V.

Thevenin's Equivalent of a Bridge Circuit

Using Thevenin's theorem, we can analyze resistive-bridge circuits for the voltage and current present at the load. Figure 11-15a shows a bridge circuit that has a 2 Ω R_L connected. The steps used in the solution are:

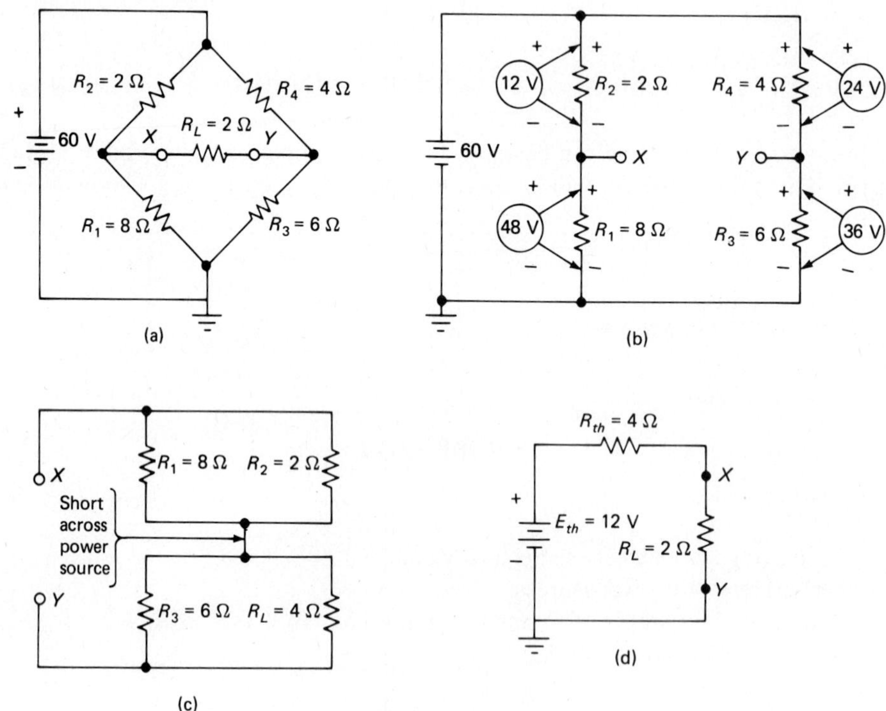

FIGURE 11-15

1. Remove R_L, leaving an open between points X and Y.
2. Examine the circuit. You will find that it consists of two parallel voltage dividers connected as shown in Figure 11-15b.
3. Use the circuit in Figure 11-15b, and solve for the voltages present at points X and Y, as follows:

$$R_{Br1} = R_1 + R_2 = 10\ \Omega$$

$$I_{Br1} = \frac{E_t}{R_{Br1}} = 6\ \text{A}$$

$$E_{R1} = I_{Br1} \times R_1 = 48\ \text{V}$$

$$E_{R2} = I_{Br1} \times R_2 = 12\ \text{V}$$

$$E_{R2} = \text{voltage at point X} = 48\ \text{V}$$

and

$$R_{Br2} = R_3 + R_4 = 10\ \Omega$$

$$I_{Br2} = \frac{E_t}{R_{Br2}} = 6\ \text{A}$$

$$E_{R3} = I_{Br2} \times R_3 = 36\ \text{V}$$

$$E_{R4} = I_{Br2} \times R_4 = 24\ \text{V}$$

$$E_{R4} = \text{voltage at point Y} = 36\ \text{V}$$

With polarities and voltage drops for each component labeled in Figure 11-15b, we can determine that when voltage is read between point X and ground, the voltage is 48 V. Between point Y and ground the potential is 36 V. To combine these algebraically, we use the equation,

$$+48\ \text{V} - (+36\ \text{V}) = 12\ \text{V for } E_{th}$$

Point X is 12 V positive with respect to point Y. Because current flows from a positive point to a more positive point, the current flows through the load from point Y to point X.

4. To solve for R_{th}, we short the 60-V power source.
5. Examine Figure 11-15c. There, two parallel networks exist that are connected in series by the short that replaced the battery.
6. Find the equivalent resistance of each network.

$$R_{eq1} = \frac{R_1 \times R_2}{R_1 + R_2} = 1.6\ \Omega$$

$$R_{eq2} = \frac{R_3 \times R_4}{R_3 + R_4} = 2.4\ \Omega$$

7. Find the total resistance:

$$R_t = R_{eq1} + R_{eq2} = 4\ \Omega$$

8. Draw the Thevenin's equivalent circuit, using $E_{th} = 12$ V, and $R_{th} = 4\ \Omega$, with R_L inserted, as in Figure 11-15d.
9. Solve for voltage and current between points X and Y:

$$R_t = R_{th} + R_L = 6\ \Omega$$

$$I_t = \frac{E_{th}}{R_t} = 2\ \text{A}$$

$$E_{RL} = I_t \times R_L = 4\ \text{V}$$

The current flow between points X and Y is 2 A, flowing from point Y to point X. The voltage drop across R_L is 4 V; since R_L is connected between the two points, the voltage between X and Y must be 4 V.

Self-Check

Answer each item by indicating whether the statement is true or false.

11. ___ In using Thevenin's theorem, we reduce a circuit to an equivalent parallel circuit.

12. ___ Thevenin's equivalent voltage, E_{th}, is open-terminal voltage.

13. ___ Thevenin's theorem cannot be used to analyze resistive-bridge circuits.

14. ___ Thevenin's equivalent resistance is found by shorting the power source.

15. ___ Thevenin's theorem can be used to analyze circuits with more than one power source.

Norton's Theorem

Norton's theorem is used for the same purpose as Thevenin's; both attain the same results, but they take different approaches. To understand the difference between the two, you should know:

Thevenin's theorem is used to analyze a circuit from the standpoint of voltage. This analysis results in a constant voltage being available at an open terminal. The voltage divides between R_{th} and R_L when a load is connected.

Norton's theorem is used to analyze the current aspects of a circuit. The analysis results in a constant current output from a power source. This current divides between the circuit's equivalent resistance, R_N, and R_L when the source is loaded.

Norton's theorem is stated thus:

Any circuit connected to two points (X and Y) can be reduced to a constant current source, I_N, in parallel with an equivalent resistance, R_N.

To demonstrate Norton's theorem, we analyze the circuit shown in Figure 11-16a. This circuit is identical to that in Figure 11-13a used with Thevenin's theorem. The results obtained here are the same as those using Thevenin's theorem.

To analyze the circuit, we:

1. Use the procedures set forth in Thevenin's analysis for finding equivalent resistance. R_{th} and R_N are always equal; therefore, the same procedures are used to solve for their values. By way of review, the procedure is restated here. Examine Figure 11-16b. Find R_N by disconnecting R_L so that points X and Y are open; short out the power source; and treat the circuit as a parallel circuit

having R_1 parallel with R_2. This analysis results in:

$$R_N = \frac{R_1 \times R_2}{R_1 + R_2} = 6\ \Omega$$

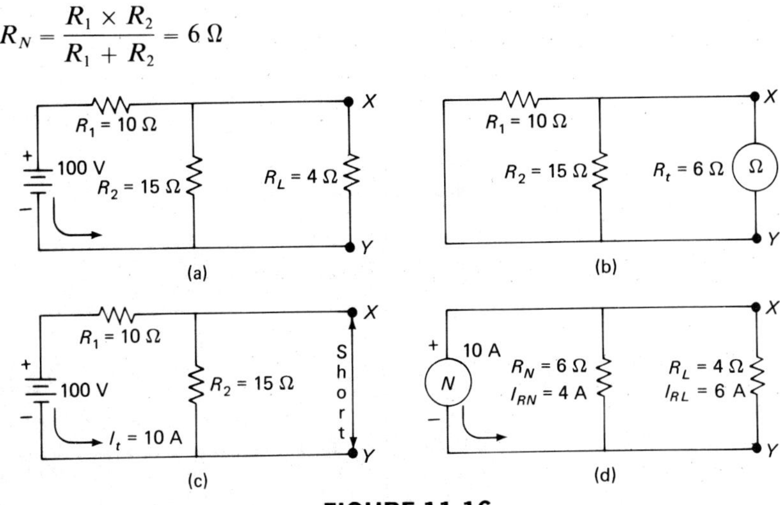

FIGURE 11-16

2. Short points X and Y together (Figure 11-16c).
3. Calculate R_t and I_N as follows:

$$R_t = R_1 = 10\ \Omega$$

$$I_N = \frac{E_t}{R_L} = 10\ \text{A}$$

4. Drawing Norton's equivalent circuit results in Figure 11-16d. *Note:* The source is a current source, not voltage.
5. Solve for the current division between R_N and R_L:

$$R_t = \frac{R_N \times R_L}{R_N + R_L} = 2.4\ \Omega$$

$$E_t = I_N \times R_t = 24\ \text{V}$$

$$E_t = E_{RL} = 24\ \text{V}$$

$$I_{RL} = \frac{E_t}{R_L} = 6\ \text{A}$$

This results in $E_{RL} = 24$ V and $I_{RL} = 6$ A. These values are the same as those found for this circuit, using Thevenin's theorem for analyzing Figure 11-13.

Norton's Equivalent of a Complex Circuit

Figure 11-17a is identical to Figure 11-14a, which was solved using Thevenin's theorem. It is used for analysis of Norton's theorem as well; the results obtained are the same. The steps are:

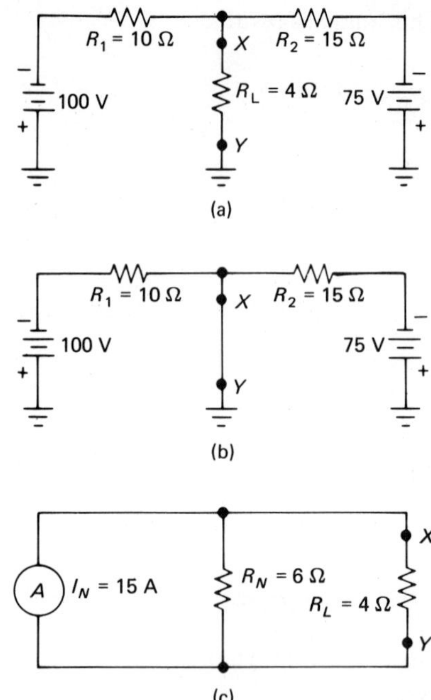

FIGURE 11-17

1. Use the procedures set forth in Thevenin's analysis, for finding equivalent resistance. R_{th} and R_N are always equal; therefore, the same procedures are used to solve for their values. In this case, with points X and Y open and the source shorted, $R_N = 6\ \Omega$.
2. Short points X and Y together. This places ground at point X.
3. Observe Figure 11-17b. With ground at X, two series circuits are present:
   (a) R_1 is in series with 100 V and has a current flow of 10 A.
   (b) R_2 is in series with 75 V and has a current flow of 5 A.
4. Total current, I_N, can be calculated, using parallel analysis:

$$I_N = I_{R1} + I_{R2} = 15\ \text{A}$$

5. Figure 11-17c represents the Norton's-equivalent circuit. The equations used in the circuit are:

$$I_N = 15\ \text{A, and } R_N = 6\ \Omega$$

6. To solve for E_{RL} and I_{RL}, we must find:

$$R_t = \frac{R_N \times R_L}{R_N + R_L} = 2.4\ \Omega$$

$$E_t = I_N \times R_t = 36\ \text{V}$$

$$I_{RL} = \frac{E_t}{R_L} = 9\ \text{A}$$

This results in $E_{RL} = 36$ V and $I_{RL} = 9$ A. These values are the same as those found when we used Thevenin's theorem to analyze the circuit.

Converting Norton's Theorem to Thevenin's Theorem

To convert the results of a Norton's solution to Thevenin's, we must apply Thevenin's analysis to Norton's equivalent. Figure 11-18 results from an analysis of Norton's theorem. Now we convert it to Thevenin's equivalent. The steps required are:

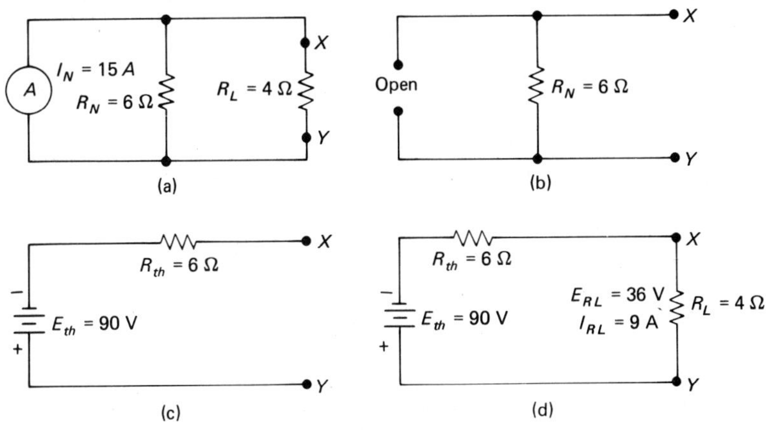

FIGURE 11-18

1. Remove R_L. This opens the circuit between points X and Y.
2. Kill the branch containing the current source. Calculate the equivalent resistance of the remaining circuit. With the source open, resistor R_N is the only resistor in a series circuit with points X and Y open (Figure 11-18b). Therefore, R_{th} equals R_N, which is 6 Ω.
3. R_{th} equals 6 Ω.
4. To find E_{th}, we must:
   (a) Make sure an open exists between points X and Y. With a constant current of 15 A flowing in the circuit, the entire 15 A will be flowing through R_N.
   (b) This means that $E_{th} = I_N \times R_N = 90$ V.
   (c) Redraw the circuit showing E_{th} and R_{th} equivalents, and you have the circuit in Figure 11-18c.
5. Insert R_L between points X and Y, and you have the circuit shown in Figure 11-18d.
6. Solve for E_{RL} and I_{RL} as follows:

$$R_t = R_{th} + R_L = 10\ \Omega$$

$$I_{RL} = \frac{E_{th}}{R_L} = 9\ \text{A}$$

$$E_{RL} = I_{RL} \times R_L = 36\ \text{V}$$

These values are the same as the values already obtained in solutions of this circuit.

A Thevenin's equivalent circuit can be converted to a Norton's equivalent circuit by doing the opposite. Observe the analysis in Figure 11-18d. The steps are:

1. To find R_N, open the circuit between points X and Y. Kill the power source and calculate the total resistance (6 Ω) between X and Y. Remember: R_N equals R_{th}; therefore, R_{th} equals 6 Ω.

2. To find I_N—short points X and Y, and with the power source connected to the circuit—solve for I_t:

$$I_N = I_t = \frac{E_{th}}{R_{th}} = 15 \text{ A}$$

3. Draw Norton's equivalent with a constant current source of 15 A as I_N parallel to an R_N of 6 Ω. With this diagram we have converted Figure 11-18d back into that of Figure 11-18a.

Self-Check

Answer each item by indicating whether the statement is true or false.

16. ____ Thevenin's analysis reduces a circuit to its equivalent values as they would exist with a constant-voltage source.

17. ____ Insertion of a load resistor in a Thevenin's equivalent circuit results in a parallel circuit.

18. ____ A Norton's equivalent circuit is powered by a constant-current source.

19. ____ Using either system, the equivalent resistance, R_{th}, or R_N, is always the same.

20. ____ Connecting R_L to a Norton's equivalent causes the total current to divide in proportion to the size of R_L and R_N.

21. ____ Converting Norton's equivalent circuit to Thevenin's equivalent circuit results in the same values as would have resulted had Ohm's law been used.

Solutions Involving Three Unknowns

It is sometimes convenient to use simultaneous equations in solving more complex circuits. Use of this technique requires mixing Ohm's law, Kirchhoff's laws, and algebra, to reach a solution.

Figure 11-19 is the same circuit as that solved with Figures 11-14 and 11-15. We will do our analysis by using "three unknowns," and we will confirm the values already solved.

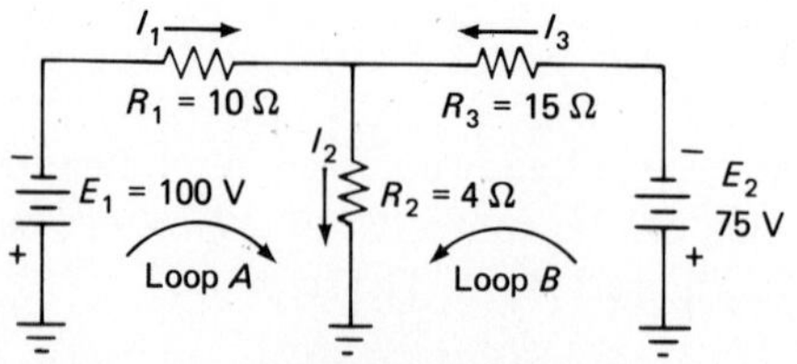

FIGURE 11-19

First, we assume that there is a path through which total current flows. It makes no difference what path we pick; analysis results in the same values. Therefore, we choose the path marked I_2 through R_2. If the wrong path is selected for total current, one of the current values solved will be negative. Once you select a path for total current,

You must use the same path in all stages of the analysis.

To establish the equations needed in using this method, we:

1. Assume that total current flows through path I_2. If this assumption is correct, then R_1 and R_3 are parallel and $I_1 + I_3 = I_t$. Current flowing through I_2 divides and flows through I_1 and I_3. This can be stated by the formula:

   $I_1 + I_3 - I_2 = 0$ (equation 1)

2. To set up equation 2 we use Kirchhoff's voltage law.
   (a) In loop A, $E_{R1} + E_{R2} = E_t(100 \text{ V})$
   (b) Ohm's law tells us that

   $E_{R1} = I_1 \times R_1$

   and

   $E_{R2} = I_2 \times R_2$

   We can use these quantities as substitutes for E_{R1} and E_{R2} in Kirchhoff's law, as follows:

   $I_1 \times R_1 + I_2 \times R_2 = 100 \text{ V}$

   But the equations we are working with must have three unknowns. We can include E_{R3} in the equation, by allowing it to have zero value, it has no affect on the voltage drops of loop A. Adding E_{R3} value to the equation gives us equation 2:

   $I_1 R_1 + I_2 R_2 + 0R_3 = 100 \text{ V}$ (equation 2)

3. Applying Kirchhoff's and Ohm's laws to loop B, then including $0R_1$, we have an equation of three unknowns:

   $0R_1 + I_2 R_2 + I_3 R_3 = 75 \text{ V}$ (equation 3)

4. Grouping the equations, we have:

   $I_1 + I_3 - I_2 = 0$

   $I_1 R_1 + 0R_3 + I_2 R_2 = 100 \text{ V}$

   $0R_l + I_3 R_3 + I_2 R_2 = 75 \text{ V}$

5. We can substitute known values from the circuit in these formulas, with the result:

   $I_1 + I_3 - I_2 = 0$ (equation 1)

   $10(I_1) + 0(I_3) + 4(I_2) = 100 \text{ V}$ (equation 2)

   $0(I_1) + 15(I_3) + 4(I_2) = 75 \text{ V}$ (equation 3)

6. By applying the algebraic law of elimination, we can reduce the equations by eliminating those that exhibit zero quantities. To compare equations 1 and 2 for elimination, we multiply equation 1 by 4. This results in equations 4, 5, and 6. (Note: $-4I_2$ cancels $+4I_2$)

$$4(I_1) + 4(I_3) - 4(I_2) = 0 \text{ V} \qquad \text{(equation 4)}$$

$$10(I_1) + 0(I_3) + 4(I_2) = 100 \text{ V} \qquad \text{(equation 5)}$$

$$14(I_1) + 4(I_3) + \quad = 100 \text{ V} \qquad \text{(equation 6)}$$

Note that equation 6 has been reduced to two unknowns.

7. To eliminate R_2 from equation 3, we compare equations 1 and 3. Multiplying equation 1 by 4 gives us equations 7, 8, and 9: (Note: $-4I_2$ cancels $+4I_2$)

$$4(I_1) + 4(I_3) - 4(I_2) = 0 \qquad \text{(equation 7)}$$

$$0(I_1) + 15(I_3) + 4(I_2) = 75 \qquad \text{(equation 8)}$$

$$4(I_1) + 19(I_3) + \quad = 75 \qquad \text{(equation 9)}$$

Note that equation 9 has been reduced to the same two unknowns as equation 6.

8. Now we can combine and eliminate one quantity from equations 6 and 9. We could eliminate either quantity, but we choose to eliminate the I_1 quantity. To do this, we multiply equation 6 by 4 and equation 9 by 14, and subtract 11 from 10. This results in equations 10, 11, and 12:

$$56\,(I_1) + 266\,(I_3) = 1050 \qquad \text{(equation 10)}$$

$$56\,(I_1) + 16\,(I_3) = 400 \qquad \text{(equation 11)}$$

$$00\,(I_1) + 250\,(I_3) = 650$$

$$I_3 = \frac{650}{250}$$

$$I_3 = 2.6 \text{ A} \qquad \text{(equation 12)}$$

9. The value of I_3 can be substituted in equation 9, to solve for I_1, resulting in:

$$4\,(I_1) + 19\,(I_3) = 75 \qquad \text{(equation 9)}$$

$$4\,(I_1) + 19\,(2.6) = 75$$

$$4\,(I_1) + 49.4 = 75$$

$$4\,(I_1) = 75 - 49.4$$

$$4\,(I_1) = 25.6$$

$$I_1 = \frac{25.6}{4}$$

$$I_1 = 6.4 \text{ A}$$

10. Substitute the values for I_1 and I_3 in equation 1, and solve for I_2:

$$I_1 + I_3 - I_2 = 0 \qquad \text{(equation 1)}$$

$$6.4\text{ A} + 2.6\text{ A} - I_2 = 0$$

$$9.0\text{ A} - I_2 = 0$$

$$-I_2 = -9\text{ A}$$

$$I_2 = 9\text{ A}$$

The answer obtained for I_2 is the same as that obtained by solving this circuit with Thevenin's and Norton's theorems. The major difference is that by using this system we can solve for all three currents. To complete the analysis of this circuit, solve for:

$$E_{R1} = I_1 \times R_1 = 64\text{ V}$$

$$E_{R2} = I_2 \times R_2 = 36\text{ V}$$

$$E_{R3} = I_3 \times R_3 = 39\text{ V}$$

Kirchhoff's voltage law states that the sum of the voltage drops in any closed loop equals the applied voltage of that loop. Loop A has 100 V applied and contains R_1 and R_2. When E_{R1} and E_{R2} are added, their sum is 64 V + 36 V = 100 V. Loop B has 75 V applied and contains R_2 and R_3. When E_{R2} and E_{R3} are added, their sum is 39 V + 36 V = 75 V. Both loops conform to Kirchhoff's voltage law; therefore, the results of this analysis must be correct.

Self-Check

Answer each item by indicating whether the statement is true or false.

22. ___ An advantage gained by using the three-unknown method is that all currents can be calculated.

23. ___ Zero values are inserted in the equations for components with no effect on that particular loop.

24. ___ Selection of the wrong resistor as the one carrying total current results in a negative current in one answer.

25. ___ The three-unknown method provides the same results as using Thevenin's theorem.

26. ___ It is necessary to multiply equations with numbers that result in like quantities being equal, but opposite when removing one quantity from a result in a three-unknown analysis.

Maximum Power Transfer Theorem

In working with DC circuits, it is important to understand the principles of power transfer. Mechanically, power transfer can be illustrated as follows. An automobile engine can run at high speed, but the car will not move until some of that power is applied to the drive wheels. In electronics, a stereo system can develop large amounts of power, but the power is unusable until delivered to the speaker. Regardless of the amount of power involved, only power that is coupled to the work-producing device is usable power.

In DC circuits, usable power is coupled according to the theorem,

The maximum power transfer that can be delivered to a load occurs only when the internal resistance of the power source equals the resistance of the load.

This theorem can be stated mathematically:

$$R_{int} = R_L$$

or

$$R_i = R_L$$

Consider Figure 11-20 and the solutions that follow. These solutions develop data that are used to prove this theorem. (The data are given in Table 11-1.) Note that R_L is small (100 Ω) compared to R_i (1000 Ω). In the development of these data, R_i remains 1000 Ω, but R_L varies upward to 10 kΩ. The formulas used are:

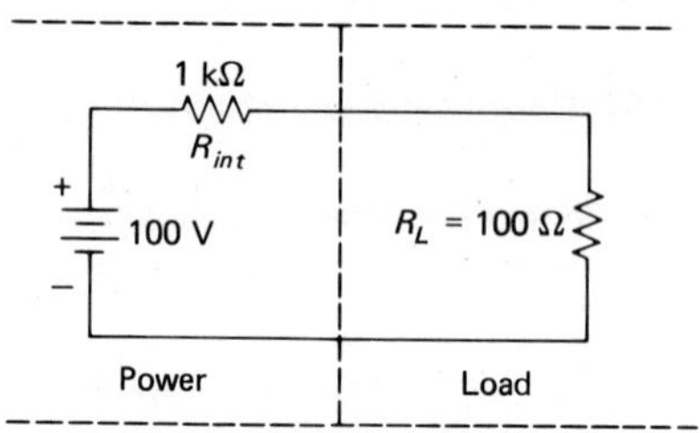

FIGURE 11-20

$$R_t = R_i + R_L$$

$$I_t = \frac{E_t}{R_t}$$

$$P_t = I_t \times E_t$$

$$P_{RL} = (I_{RL})^2 \times R_L$$

$$P_{Ri} = (I_{Ri})^2 \times R_i$$

Each formula is repeated for each value of R_L. The data resulting from these solutions are given in Table 11-1.

TABLE 11-1

Graph Point	R_L Ω	R_t Ω	I_t mA	P_t W	P_{RL} W	P_{Ri} W
1	100	1100	90.91	9.090	0.826	8.264
2	200	1200	83.33	8.333	1.389	6.944
3	300	1300	79.92	7.992	1.775	5.916
4	400	1400	71.43	7.143	2.041	5.102
5	500	1500	66.67	6.667	2.222	4.445
6	600	1600	62.50	6.250	2.344	3.906
7	700	1700	58.82	5.882	2.422	3.460
8	800	1800	55.56	5.556	2.469	3.087
9	900	1900	52.63	5.263	2.493	2.770
10	1000	2000	50.00	5.000	2.500	2.500
11	1100	2100	47.62	4.762	2.494	2.268
12	1200	2200	45.45	4.545	2.479	2.066
13	1500	2500	40.00	4.000	2.400	1.600
14	2000	3000	33.33	3.333	2.221	1.112
15	5000	6000	16.67	1.667	1.389	0.278
16	10000	11000	9.090	0.909	0.826	0.164

Figure 11-21 presents a graphical display of the affect on power transfer that occurs with changes in R_L with respect to R_i. Note the conditions that prevail in each equation:

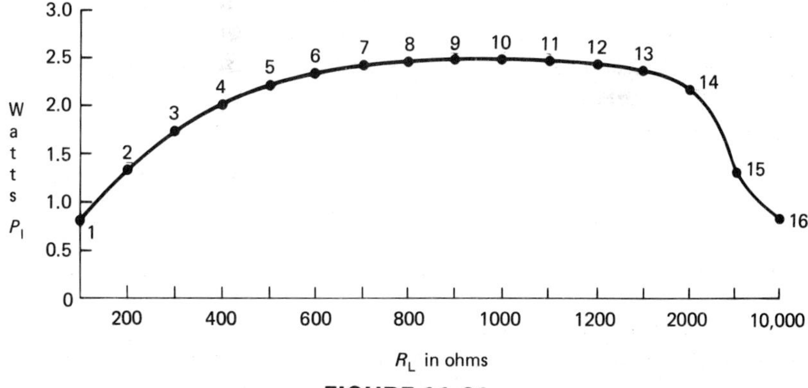

FIGURE 11-21

$R_L < R_i$

$R_L = R_i$

$R_L > R_i$

Note that as R_L approaches R_i from either direction, the power transferred to R_1 increases. Only when R_L equals R_i is maximum power transferred. The curve plotted on the graph confirms that maximum power transfer occurs when both resistances are 1000 Ω.

Self-Check

Answer each item by indicating whether the statement is true or false.

27. ___ Usable power is power that is coupled to the working portion of a circuit.

28. ___ Maximum power transfer occurs when $R_i = R_L$.

29. ___ Usable power equals total power of the circuit.

30. ___ Power dissipated in the power source serves no useful purpose.

Summary

Many complex circuits cannot be solved by using Ohm's law; the information available does not support use of the law. For these situations, other methods must be used. By using Kirchhoff's laws, the superposition theorem, Thevenin's theorem, Norton's theorem, and the three-unknown method, we can analyze many of these circuits.

Kirchhoff's voltage and current laws establish circuit standards that can be stated for any circuit. Once these standards are expressed in Kirchhoff form, Ohm's-law values can be substituted in the formulas, to aid solution. For example, since $E_{R1} = I_{R1} \times R_1$, either value can be substituted for the other.

The superposition theorem allows us to analyze circuits containing two or more power sources. When a component is influenced by more than one source, its total voltage and current equal the algebraic sum of the two sources acting separately. Using this theorem we can determine the effect of each power source. Then the total effect is found by algebraic addition.

Thevenin's and Norton's theorems were used to reduce complex circuits to their equivalent values of resistance and voltage (or current). Using these procedures, we solved for Thevenin's or Norton's resistance. In the case of Thevenin's theorem, we determined the open-terminal voltage characteristic of a constant voltage source. In the case of Norton's theorem, we solved for the open-terminal current characteristic of a constant current source.

The use of simultaneous equations containing three unknowns is another method for analyzing complex circuits. To use this method, we established equations that represent the assumed total current at any point in the original circuit, and wrote a Kirchhoff's current law equation for that point. We then wrote a current equation for each power source. Comparison of the three equations, and elimination of unknowns, resulted in a complete analysis.

The power-transfer theorem was introduced and explained. Calculations and graphical analysis were included which proved that maximum power is transferred to a load only when internal resistance is equal to load resistance.

Review Questions and/or Problems

1. Refer to Figure 11-22. Solve for E_{R1}, using Kirchhoff's analysis.
   (a) 60 V
   (b) 16 V
   (c) 40 V
   (d) 84 V

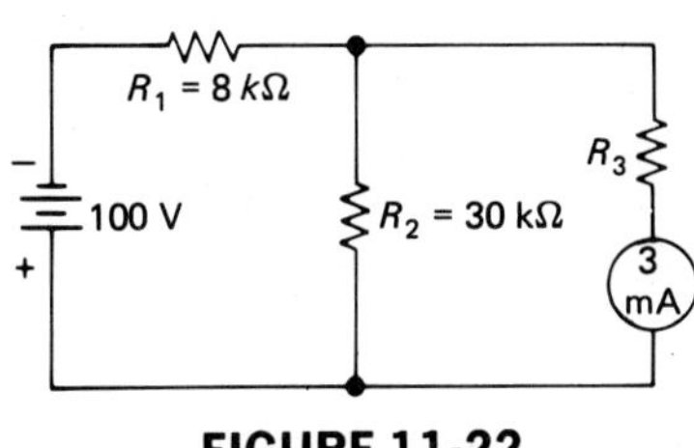

FIGURE 11-22

2. In Figure 11-23, use the superposition theorem to determine the reading for V_1.
   (a) −32 V
   (b) −14 V
   (c) −18 V
   (d) −46 V

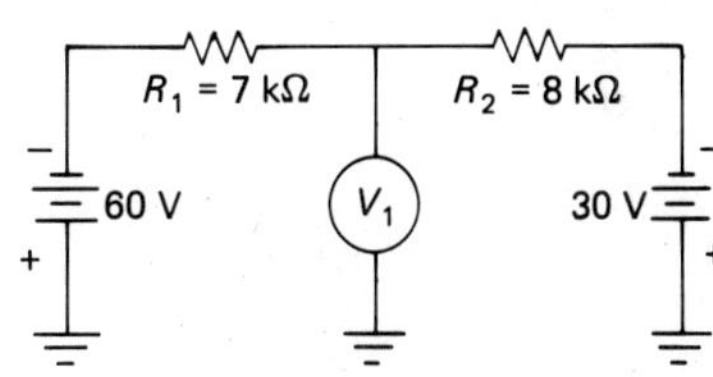

FIGURE 11-23

3. Refer to Figure 11-24. Use Kirchhoff's law to determine the voltage reading on V_1.
   (a) −32 V
   (b) +14 V
   (c) −46 V
   (d) −18 V

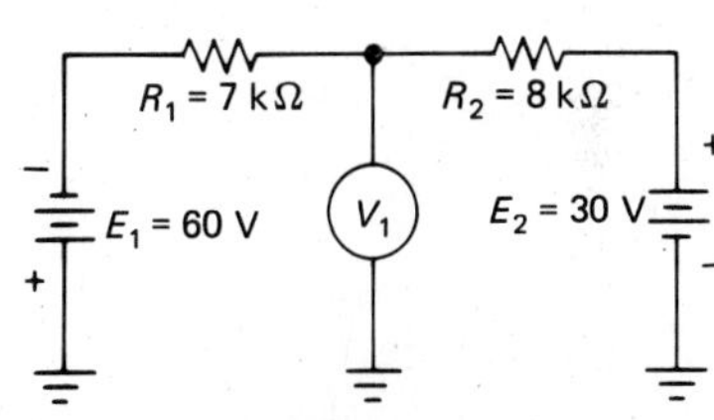

FIGURE 11-24

4. Which diagram in Figure 11-26 is the correct Thevenin's equivalent circuit for Figure 11-25?

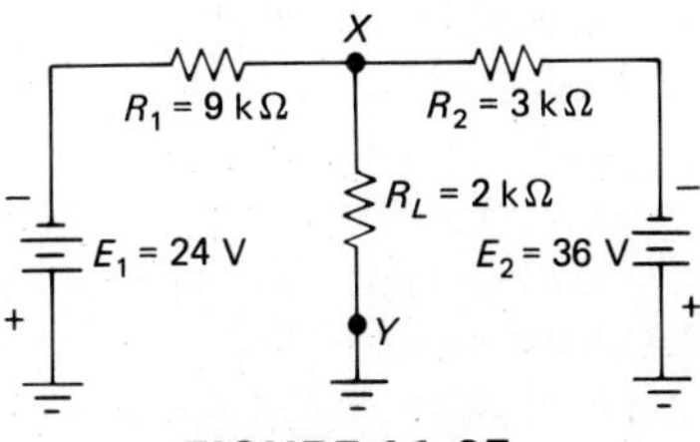

FIGURE 11-25

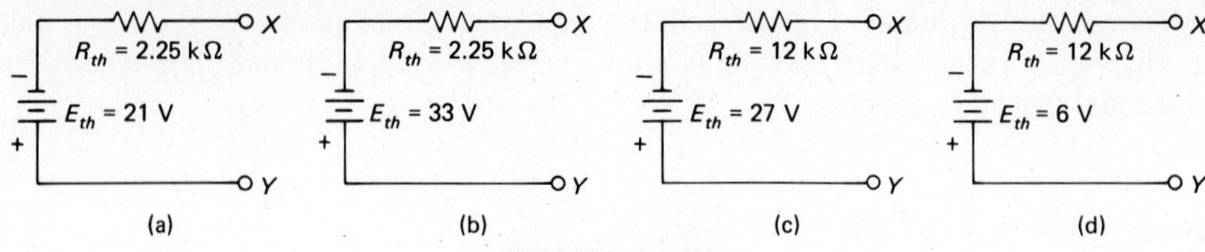

FIGURE 11-26

5. Which diagram in Figure 11-28 is the correct Norton's equivalent circuit for Figure 11-27?

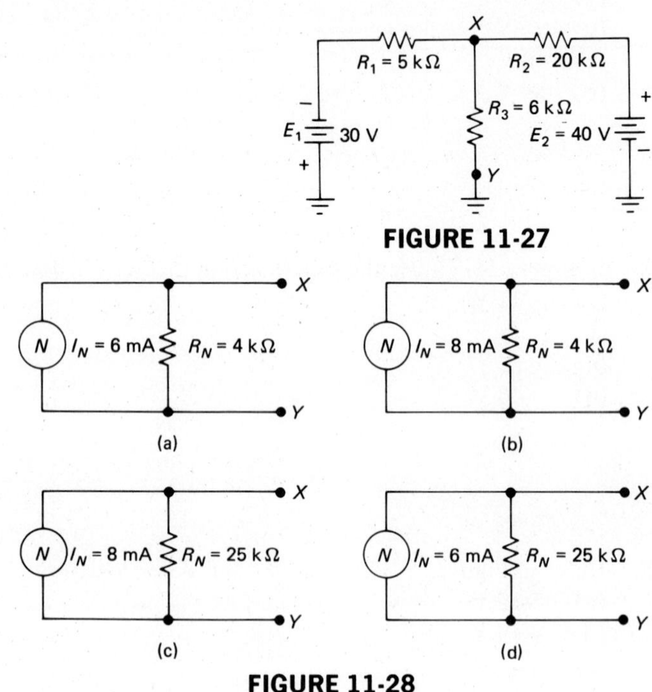

FIGURE 11-28

6. Refer to Figure 11-29, and use the three-unknown method to solve for I_1, I_2, and I_3 (in milliamperes).

	I_1	I_2	I_3
(a)	3.5	0.5	4.0
(b)	0.5	4.0	3.5
(c)	4.0	0.5	3.5
(d)	5.0	0.5	3.5

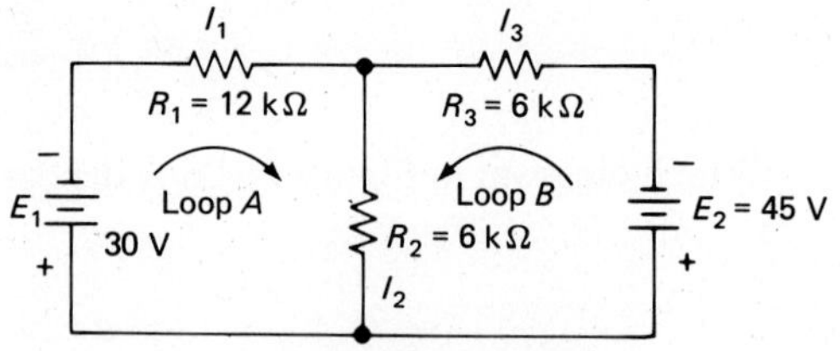

FIGURE 11-29

7. The superposition theorem is used to determine the effect of two power sources on a single component.
   (a) true
   (b) false

8. Thevenin's theorem is used to analyze a circuit for its voltage characteristics.
   (a) true
   (b) false

9. Reducing a bridge circuit to its Thevenin's equivalent is the same procedure used to analyze bridge circuits in Chapter 10.
   (a) true
   (b) false

10. Norton's theorem is the same as Thevenin's theorem.
    (a) true
    (b) false

11. Maximum power is transferred when internal resistance and load resistance are equal.
    (a) true
    (b) false

12. Given a circuit having $E_{th} = 20$ V and $R_{th} = 100\ \Omega$, which of the following resistors can be used as the load device for obtaining maximum power transfer?
    (a) 10 Ω
    (b) 100 Ω
    (c) 1000 Ω
    (d) 200 Ω

12

Magnetism and Relays

Introduction

Magnetism, like electricity, is another invisible force that has been known for centuries. As early as 2600 BC, the Chinese were aware of magnetism and its use as a rudimentary compass. The ancient Greeks experimented with magnetism, followed closely by experiments with electricity. Not until 1819, however, did Hans Christian Oersted discover the close relationship between electricity and magnetism. Today, electricity and magnetism are recognized as the basic ingredients of modern communications. Magnetism and/or the magnetic effect are found in almost every modern electrical circuit. The uses of magnetic effect vary from a doorbell to an electromagnet. The subjects of magnetism and electronics are recognized as inseparable. To become a good technician, you must understand basic magnetism.

Objectives

Each student is required to:

1. Define
   (a) Magnetism.
   (b) Magnet.
   (c) Permeability.
   (d) Retentivity.
   (e) Flux, flux density, and flux intensity.
   (f) Reluctance.
   (g) Residual magnetism.
   (h) Magnetic materials.
   (i) Nonmagnetic materials.
   (j) Relay.

   (k) Weber.
   (l) Ampere-turn.
2. State the difference between
   (a) Natural and artificial magnets.
   (b) Permanent and temporary magnets.
3. Describe the characteristics of the magnetic lines of force that exist around a magnet.
4. State the magnetic law of attraction and repulsion.
5. Describe the molecular arrangement that exists in
   (a) Nonmagnetized materials.
   (b) Magnetized materials.
6. State the purpose of a relay.
7. List and identify the parts of a simple relay.
8. Identify the schematic drawings for:
   (a) Holding relays.
   (b) Vibrating relays.
   (c) Time-delay relays.
   (d) Overload, or latch, relays.
9. Use test equipment to troubleshoot relay circuits.
10. Describe common problems found in relay circuits.

Magnetism

Any metallic material that is capable of attracting other metals is called a *magnet*. Any material that can be attracted by a magnet is a *magnetic material*. Iron and steel are magnetic materials. The Chinese found that when a magnet is suspended horizontally and allowed to rotate freely, the same end of the magnet always points to the earth's north pole. Early compasses suitable for navigation were based on this finding. Magnetic materials taken from the ground are sometimes called lodestones, from the Chinese "leading stone." These magnets are classified as *natural magnets*.

Magnets can be made by rubbing iron or steel with a natural magnet or by the use of an electrical current. Magnets thus created are said to be *artificial magnets*. Artificial magnets can be either permanent or temporary. When a permanent magnet becomes magnetized it remains magnetized for a long time. Temporary magnets remain magnetized only as long as the magnetizing force is present; when the force is removed, the material is no longer magnetized.

Hard steel, which is used to make permanent magnets, is difficult to magnetize. The molecules of hard steel are so closely spaced that the magnetic lines of force cannot easily distribute themselves throughout the material. The opposition a material presents to magnetic lines of force is called *reluctance* ($\mathcal{R}$). All permanent magnets result from the magnetizing of materials that have a high reluctance. When compared to electrical units, reluctance and resistance are similar.

Materials that have low reluctance, such as soft iron, are easy to magnetize because they offer little opposition to the distribution of lines of force. Once the magnetizing force is removed, soft iron retains only a small part of its magnetism. Hard steel can be used to make a permanent magnet and soft iron to make a temporary one. A temporary magnet is used, with a heavy crane, in a junkyard to pick up and move old automobiles. A permanent magnet is used in a loudspeaker.

The amount of magnetism left in a material once the magnetizing force has been removed is known as *residual magnetism*. The ability of a material to remain magnetized after removal of the magnetizing force is referred to as its *retentivity*.

The difference between a permanent magnet and a temporary magnet results from the reluctance of the materials used in their manufacture. These magnets are also classified according to their *permeability* (μ), the ability of a material to conduct magnetic lines of force; it compares with conductance in an electrical circuit. Those used in permanent magnets have low permeability and high reluctance, whereas materials used in temporary magnets have high permeability and low reluctance.

Magnets are made in many different shapes. Bar magnets may be round or flat bars of metal. Taking a bar magnet and bending it into the shape of a horseshoe results in a horseshoe magnet. Bending the bar magnet to the point that the two ends are joined results in a circular magnet.

Magnetic lines of force are known collectively as *flux*. Magnetic flux is comparable to electrical current, but it is not flow in the sense that current is. Flux is measured in the number of magnetic lines passing through a magnetic circuit. The unit of flux is the *weber* (Wb; symbol, the Greek letter phi ϕ). The degree to which a magnet is magnetized is expressed as the magnet's *flux density* or *magnetizing potential* (mmf). Flux density refers to the number of magnetic lines per square centimeter. The unit of flux density is the *tesla* (T), where $1 \text{ T} = \text{Wb/m}^2$, where m = meters. The symbol for flux density is the Greek letter beta, β.

Magnetic Poles

The magnetic force that surrounds a magnet is not uniformly distributed. Magnetic force is heavily concentrated at the ends of a magnet and is weakly concentrated at the middle (Figure 12-1).

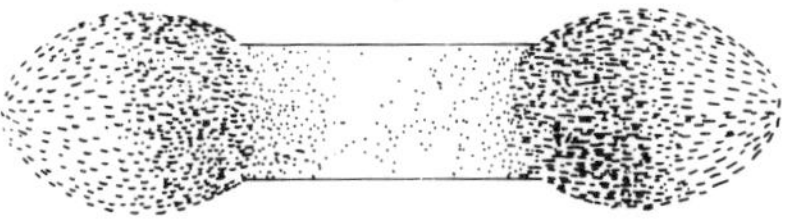

FIGURE 12-1

Law of Magnetic Poles

When iron filings are sprinkled on a magnet, the pattern shown in Figure 12-1 results. Notice that most of the filings are near the two ends, with fewer filings near the middle. This indicates that the force attracting the filings is much greater at the ends (called *magnetic poles*); also, one end of the magnet has a north-seeking capability. The pole that seeks the earth's north pole is called the *north* pole, and the pole that seeks the earth's south pole is the *south* pole.

Theories of Magnetism

Weber's Theory

Weber's theory (named for the German physicist Wilhelm Eduard Weber, 1804–91) is based on the molecular construction of matter. The theory assumes that, since all material consists of many molecules, and since these molecules are small bits of the same material, anything said about the material can also be said about a single molecule; therefore, each molecule must be a small magnet. In nonmagnetic materials, magnets are randomly aligned and adjacent magnets cancel out. In a magnetized material, all magnets (molecules) are aligned such that all north poles point one way and all south poles the other. Thus aligned, all molecules help each other form one large magnet. Figure 12-2 illustrates the use of a permanent magnet to change an unmagnetized metal bar into a bar magnet. The polarity of the new magnet depends on which pole is used to do the stroking and on the direction in which the stroking magnet is moved.

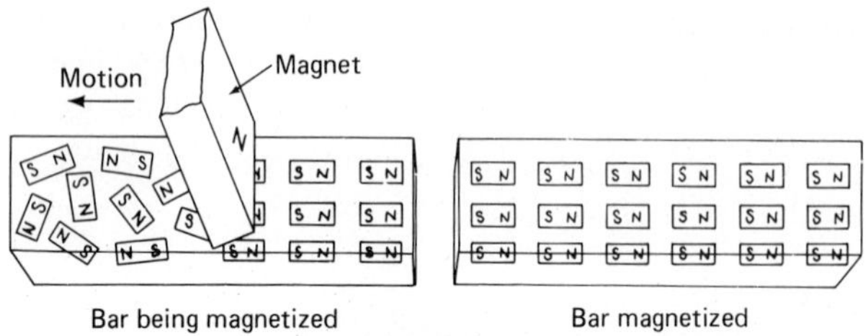

FIGURE 12-2

Weber's theory is supported by the fact that when a bar magnet is broken in two, two new magnets are formed. The poles of each new magnet lie in the same direction as those of the original magnet. As the bar magnet is broken further, new magnets will have the same polarity. Weber surmised that this occurs down to the last molecule (Figure 12-3).

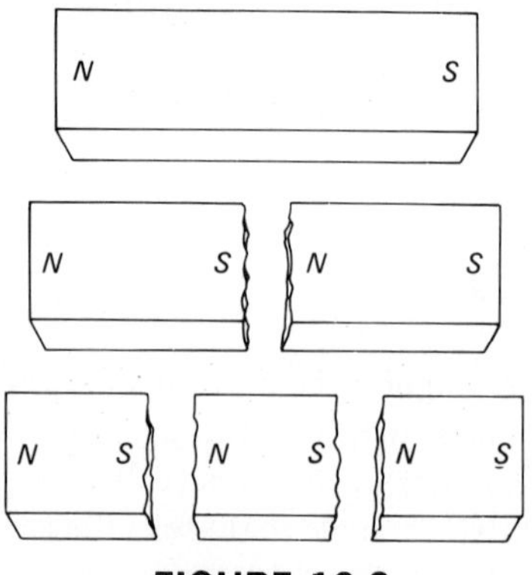

FIGURE 12-3

Weber's theory is further supported by the fact that when a magnet is held out of alignment with the earth's magnetic field, and is continually heated or jarred, its magnetism is disarranged and it becomes demagnetized. Electronic test equipment can be affected by this fact.

Domain Theory

The domain theory is based on the principle of electron spin. As we have noted, the electron theory holds that electrons spin on their axis, much as the earth does. Electrons, however, spin in either direction; some spin one way, and others the opposite way. The electron has a negative electrical charge, but it also has a magnetic field. The domain theory holds that as electrons spin, the total effect of their separate magnetic fields results in magnetic and nonmagnetic materials.

Atoms with equal numbers of electrons spinning in each direction are nonmagnetic because the fields exerted by the electrons cancel each other. An atom with more electrons spinning in one direction than the other is a magnetic atom; the greater the imbalance, the stronger the magnet. The atoms then bond into "domains" of intense magnetic fields. About 10 million domains can exist in 1 mm^3, with the result that any material contains a great many domains. The random alignment of these domains regulates the material's degree of magnetization. When brought under the influence of an external magnet, however, all domains align with the external magnet, thus creating a magnet with a strength equal to that of the domains it includes.

Law of Magnetic Force

The magnetic strength of a magnetic pole is measured in *unit poles*. A unit pole has a strength such that when placed 1 cm from a pole of equal strength (within air or in a vacuum), a force of 1 dyne is exerted. This force can be either repelling or attracting, depending on the polarity of the poles.

The intensity of attraction and repulsion between two poles can be related to Coulomb's law of charged bodies: The force between two poles is directly proportional to the product of the pole strengths and inversely proportional to the square of the distance between the poles. For magnets in air, the law can be stated:

$$F = \frac{M_1 \times M_2}{d^2}$$

where

F = force, in dynes, between two poles separated by air

M_1 = magnetic strength of the first pole in unit poles

M_2 = magnetic strength of the second magnet in unit poles.

d^2 = distance between the two poles, in centimeters squared

Magnetic Fields

The area around a magnet, governed by magnetic lines of force, is called the *magnetic field.* The influence of a magnetic field can be observed by taking a compass and moving it completely around a magnet. As the compass moves from point to point, its needle also moves. By moving the compass completely around the magnet the needle is caused to rotate one full rotation. (See Figure 12-4.)

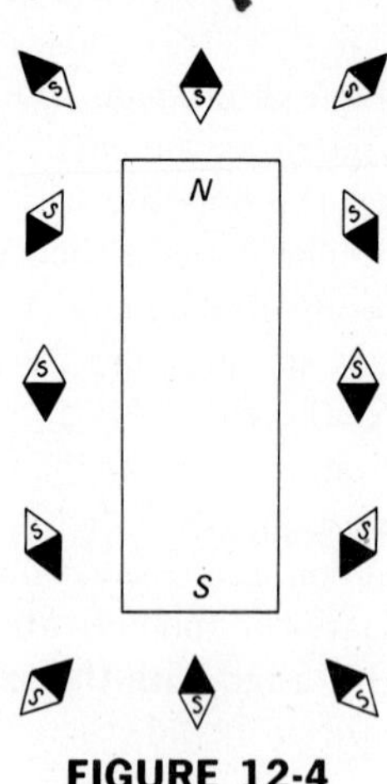

FIGURE 12-4

By placing a bar magnet on a flat surface, then placing a piece of glass or paper over it, and sprinkling iron filings on the cover, we can observe the effect of the lines of force around a magnet (Figure 12-5).

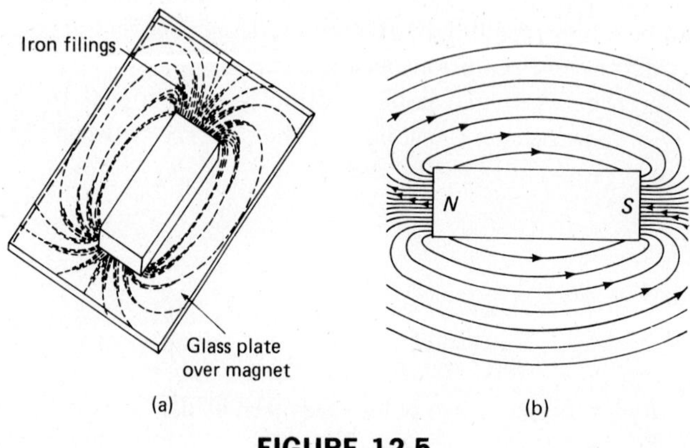

FIGURE 12-5

Figure 12-5a represents the pattern assumed by the filings. In Figure 12-5b, the filings have been replaced by representative lines; these are *lines of flux,* which are measured in webers. Magnetic lines of force are assumed to have certain characteristics. All lines of force:

1. Start at the north pole and end at the south pole.
2. Are continuous and always form complete loops.
3. *Never* cross.
4. Tend to shorten themselves; therefore, the magnetic lines between two unlike poles cause the poles to be pulled closer together.
5. Enter and leave a magnetic material at right angles to the surface.
6. Pass through all materials, magnetic and nonmagnetic. Stated another way, there is no insulator for magnetic lines of force.

This explanation gives the indication that opposite poles have opposite charges, whereas, when two magnets are brought near each other,

Unlike poles attract, and like poles repel.

Figure 12-6 illustrates this effect. Figure 12-6a shows the effect when two like poles are placed next to each other. Figure 12-6b shows the effect of two unlike poles.

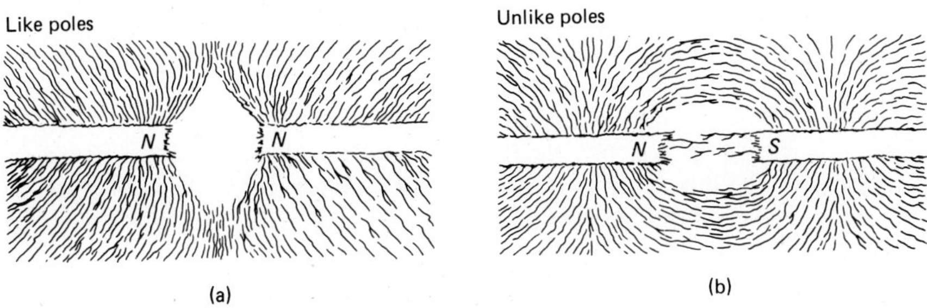

FIGURE 12-6

Magnetizing of Magnetic Materials

Any material that can be attracted by a magnet can be magnetized; any material that can be attracted by a magnet can become a magnet. When brought under the influence of another magnet, a magnetic material becomes a separate magnet. By now, you should be able to understand the affect of magnetic lines on magnetic materials placed in the magnet's field (Figure 12-7a).

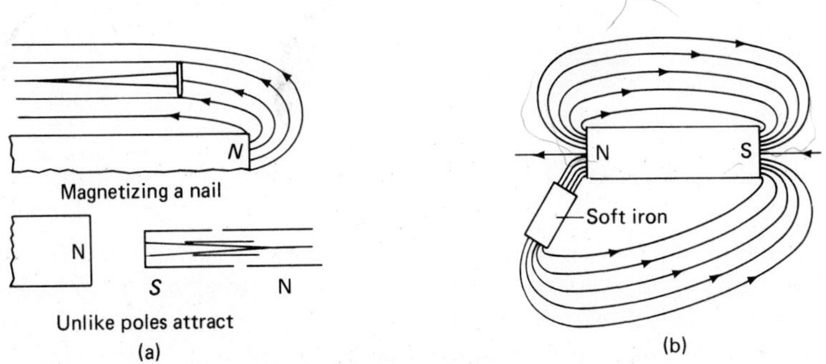

FIGURE 12-7

Such magnetic materials as a nail have a much lower reluctance than air. Therefore, as the nail is brought under the influence of a bar magnet's field, some flux lines will flow through the nail. In fact, lines of force will bend out of shape to flow through low-reluctance materials (Figure 12-7b). Lines leaving the north pole of the bar magnet enter the nail and pass through it, emerging from the opposite end. Within a magnet, the lines of force are from south to north. This results in the nail becoming a magnet whose south pole is nearest the bar magnet's north pole and whose north pole is nearest the bar magnet's south pole. The strength of the new magnet depends on the number of lines of force flowing through it and on the magnet's magnetic characteristics.

When the bar magnet is removed, the magnetic properties of the nail virtually disappear. This occurs because the nail is composed of soft iron, which has low retentivity. Had the nail been made of hard steel, it would hold much of its magnetism and become a permanent magnet.

The force that creates a magnet from unmagnetized metal is called a *magnetizing force* or *magnetizing potential* (symbol, mmf). Because most electrical use of magnets is in the form of electromagnets, the measure for mmf is the ampere-turn. Before the measures were standardized, this was called *magnetomotive force,* from which the symbol mmf was derived.

Magnetic Shielding

One characteristic of magnetic flux is its ability to pass through any material; there is no known insulator. That is a problem in some electronic circuitry; excess magnetism can cause erratic operation. In that case, the flux must be rerouted around the circuit we wish to protect. This can be done by providing a low-reluctance path that conducts lines through paths where they have no effect. Remember that lines bend out of their normal patterns and follow a low-reluctance path. In effect, we remove the flux from the area by giving it a more desirable path. This is called *magnetic shielding* and is illustrated in Figure 12-8.

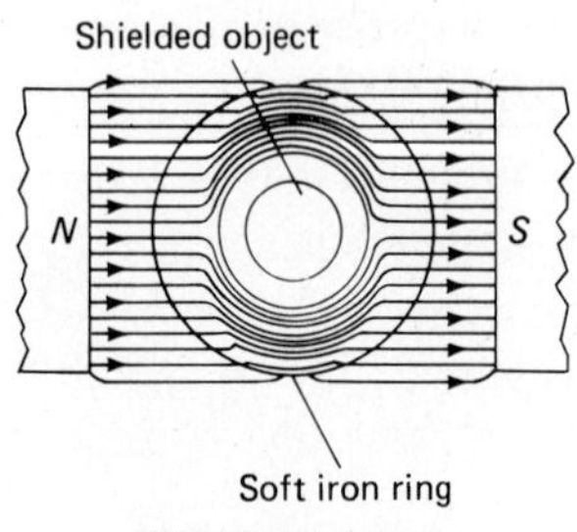

FIGURE 12-8

International Standard of Units

As has been mentioned, the tendency in recent years has been to standardize units of measure on an international basis. Today, practically all areas of magnetism and electricity have been standardized. An agreement made in 1960 sets standards for measuring quantities associated with magnetism and electronics. Some electronic quantities that have been studied and standardized are ampere, volt, coulomb, and ohm. The system established is called *mks units;* "mks" stands for meter (distance), kilogram (weight), seconds (time). The standards were established under the Système International, abbreviated SI. The quantities and units for magnetic materials are given in Table 12-1.

Do you remember resistance and conductance in electricity? They are reciprocals; that is, they are inversely proportional to each other. Magnetic conduction is also governed by two quantities, reluctance and permeance. They, too, are reciprocals. In the case of electricity, we said that the unit of measure is the Siemens, but that the symbol mho ℧ is a designation commonly used for expressing conductance.

We saw that magnetism involves a practical approach. The names and symbols used in the discussion are identical to those in the SI system (Table 12-1). Table 12-1 also correlates each quantity to its SI unit.

TABLE 12-1
mks Units (SI) for Magnetism

Quantity	*Symbol*	*Unit*
Flux	ϕ	Weber (Wb)
Flux density	β	Weber/m^2 = tesla (T)
Potential	mmf	Ampere-turn (A)
Field intensity	$\mathcal{H}$	Ampere-turn per meter (A/M)
Reluctance	$\mathcal{R}$	Ampere-turn per weber (A/Wb)
Permeance	P = 1/$\mathcal{R}$	Weber per ampere-turn (Wb/A)
Relative μ	μ^r or K^m	None; measured by numbers
Permeability	See note*	B/H = tesla (T)/A/M

**Permeability:* $(\mu) = \mu^r \times 1.26 \times 10^{-6}$

Self-Check

Answer each item by indicating whether the statement is true or false.

1. T Magnetism has been known for centuries.

2. T Magnets are classified as natural or artificial.

3. F The ability of a material to remain magnetized is called *retentivity.*

4. F Like magnetic poles attract each other.

5. T Magnetic lines of force form complete loops.

6. T Permeability is a measure of the ease with which magnetic lines of force pass through a material.

7. T If a bar magnet is broken in half, the poles on either side of the break repel each other.

8. T One magnetic line of force is called an oersted.

9. F A weber is the unit of measure for field intensity.

10. T Employing the laws of attraction and repulsion, the north-seeking pole of a compass is actually a magnetic south pole.

11. ____ Magnetic induction is used to protect circuits from undesirable magnetic lines of force.

12. ____ Materials can be magnetized by magnetic induction.

Electromagnetism

Electromagnetism is an important part of electronics. As was mentioned, Oersted discovered the relationship between magnetism and electronics. He found that a current-carrying conductor has the following characteristics:

1. The conductor is surrounded by a magnetic field.
2. The magnetic lines form complete circles around the conductor and are continuous for the length of the conductor.
3. A compass indicates the direction in which magnetic lines rotate around a conductor.
4. A magnetic field is perpendicular to the plane of the conductor.
5. Increasing current flow in a conductor causes the magnetic field of the conductor to increase. When the current stops, the field disappears; when the current returns, the field reappears.
6. Reversing the direction of current flow causes the magnetic lines of force to reverse direction.
7. The strength of an electrically excited magnetic field is the product of the number of turns (N) around a coil and the current (I) flowing through the coil. This is rated in ampere-turns and is assigned the symbol, mmf.

Magnetic Lines Around a Straight Conductor

We can repeat Oersted's experiment by moving a compass around a current-carrying conductor. As the compass moves, the compass needle moves such that it remains aligned with the conductor's magnetic field. One complete rotation around the conductor causes the needle of the compass to make one complete rotation.

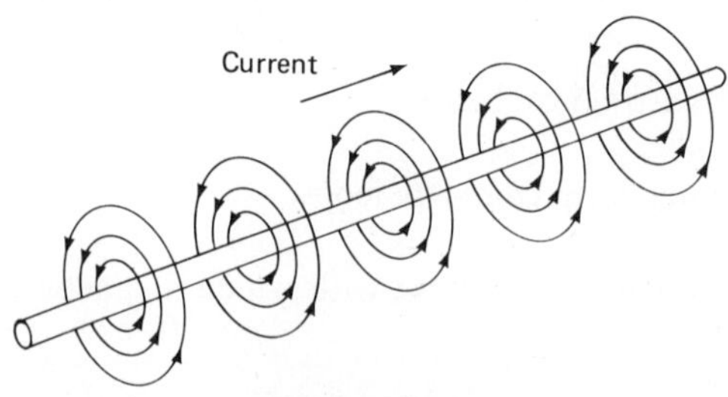

FIGURE 12-9

When the conductor is a straight wire, its magnetic field resembles the field shown in Figure 12-9. Several points along the wire were selected in order to show field characteristics; actually, the field is continuous, and no gaps exist.

In future studies, it will be necessary for you to determine the direction of movement of magnetic lines around a current-carrying conductor. The easiest way to do this is to use the *left-hand rule* (Figure 12-10). To use this rule,

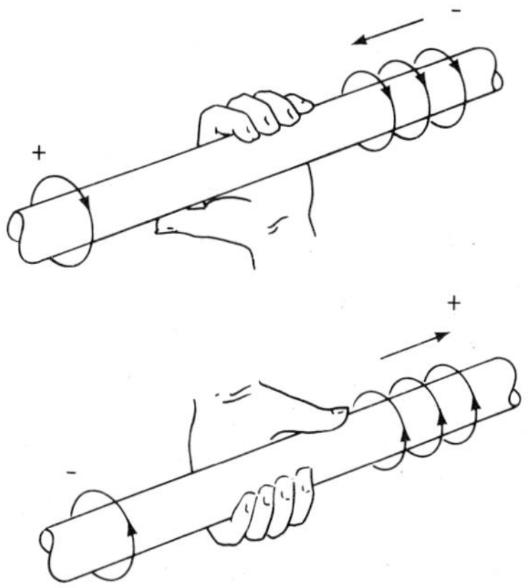

FIGURE 12-10

1. Point your *left* thumb in the direction of current flow in the conductor.
2. Grasp the conductor in you *left* hand.
3. *Curl* your fingers around the conductor.

Your fingers will point in the direction of the magnetic field.

Figure 12-11 contains another way to determine the direction of the magnetic field. Note that Figure 12-11a has a dot (·) in its center and that Figure 12-11b has a cross (×) in its center. These symbols are used to indicate the direction of current flow. The dot represents current flowing toward you (out of the page); the cross represents current flowing away from you (into the page). Some people find this easier to remember if they associate the symbols with an arrow. The dot represents the end of an arrow as it approaches you, and the cross represents the flight of an arrow as it leaves you.

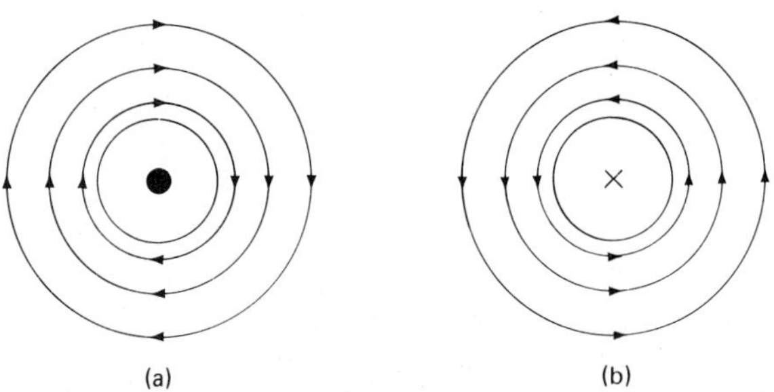

FIGURE 12-11

In each figure, the left-hand rule has been applied. Pointing your left thumb in the direction of current flow (*out* of the page in a, and *into* the page in b), and closing your fingers causes the fingers to point in the direction of magnetic line movement. In Figure 12-11a, the field rotates *clockwise* and in 12-11b, *counterclockwise*.

Magnetic Field Around a Coil of Wire

When we use the current-carrying conductor to make the loop shown in Figure 12-12, the circuit takes on characteristics that are even more magnetic. Note that all lines of flux enter the coil from one side and leave from the other. The end where the lines enter the coil is the *south* pole and the end where they leave is the *north* pole. This is also true of a bar magnet. We have created a short electromagnet with the same characteristics as a short bar magnet.

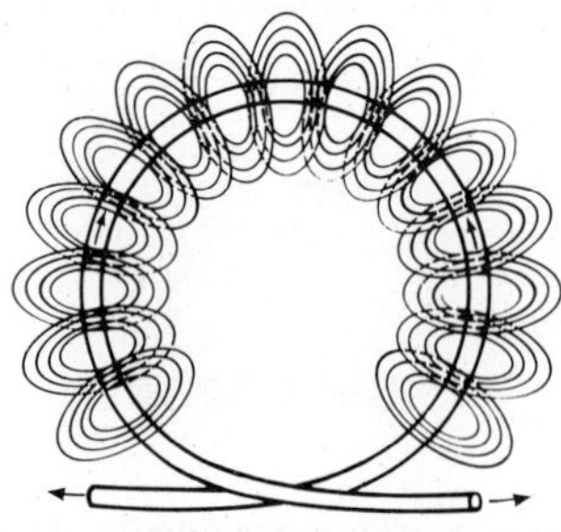

FIGURE 12-12

Winding several turns on the coil (Figure 12-13) results in a much stronger bar magnet. Lines of force are concentrated in the center of the coil, but outside the coil, they fan out in all directions. This coil acts similarly to a bar magnet; its effect, fields, and uses are the same. When two coils are placed near each other, they obey the laws of attraction and repulsion. In fact, if the coil is suspended so it can rotate, the coil will align with the earth's magnetic field, as will the bar magnet. This coil is an *electromagnet*. The best way to determine its north and south poles is to use another left-hand rule:

> ***Grasp the coil with your left hand so that your fingers point in the direction of current flow across the face of the coil. Your thumb points to the north pole.***

In Figure 12-13, lines of magnetic force leave the top of the coil and reenter the bottom, thus indicating that the top is north and the bottom is south. Since we cannot see lines of force, we cannot make this deduction. By using the left-hand rule, however, we can locate the north pole. The left-hand rule is applied as follows:

1. Determine the direction of current flow through the part of the coil nearest you (the coil's face). Note that, on the diagram, the arrows point to the left.
2. Place your *left* hand across the face of the coil, with your fingers pointing in the direction of current flow. Note that, in this diagram, your fingers point left.

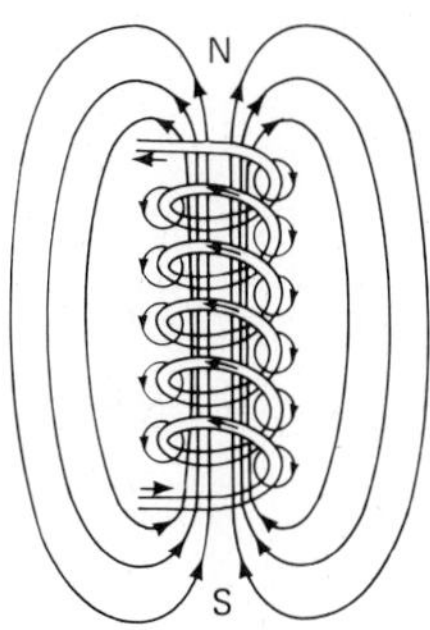

FIGURE 12-13

Your *left* thumb points to the north pole, which is at the top of the coil. This is the same polarity as that marked on the diagram: north at the top and south at the bottom.

Strong magnets can be made by winding many turns of wire around a piece of soft iron. The soft iron used as a magnet's center is called a *core.* (Because reluctance of the core is much lower than that of air, many more lines of force can flow through the center of the magnet before the magnet reaches saturation. (Magnetic saturation occurs when an increase in current flow through the coil results in no appreciable change in the number of lines of flux.) In the case of the soft-iron core, it is possible to have higher current flow in the coil without the coil reaching saturation. The additional current causes greater flux, and the magnet becomes stronger. Another way to increase the strength of the magnet is to wind more turns around the coil; additional coils produce more lines of force and thus greater magnetic strength.

The strength of an electromagnet is a product of the turns-to-current ratio, called the ampere-turns ratio. It is possible to achieve the same magnetic strength as long as the product of current and turns is the same. Take, for example, a magnet with 2 A flowing and 5000 turns. When these are multiplied, the ampere-turns ratio is found to be 10,000. Another magnet has 20 A and 500 turns; the ampere-turns ratio is also 10,000. The two magnets have the same strength, even though they are constructed differently.

Two coils having the same ampere-turns ratio have fields of equal strength.

The Electromagnetic Effect

When a bar of soft iron is placed in the magnetic field of a current-carrying coil, the bar becomes magnetized. The bar has much less reluctance than air, and magnetic lines tend to shorten themselves in order to flow through it. This results in more lines flowing through the area than flowed there before the bar was inserted. If the field is strong enough, the tendency of lines to shorten themselves will cause the bar to be attracted into the center of the coil. The bar will be pulled toward the coil until its center is near the center of the coil, where the field is strongest. The action is the same regardless of which end of the coil is nearest the bar. The bar is always magnetized in such a way that it is attracted to the coil.

Self-Check

Answer each item by indicating whether the statement is true or false.

13. ___ As the distance from an electromagnet increases, its magnetic field increases.

14. ___ The magnetic field surrounding a conductor can be varied by varying the current in the conductor.

15. ___ To determine the direction of flow for magnetic lines around a conductor, use the right-hand rule.

16. ___ When current flows through a coil, a magnet is formed.

17. ___ An electromagnet is a permanent magnet.

18. ___ The position of the north pole of an electromagnet is determined by the direction of current flow in the coil.

19. ___ When applying the left-hand rule to a coil, your fingers should be pointing toward the north pole.

20. ___ The cross (+) is used to indicate current flow into the page.

21. ___ In an electromagnet, lines of flux enter the core at the south pole and exit at the north pole.

22. ___ Magnetic lines of force tend to shorten themselves.

Electromagnetic Relays

In some situations it is not practical to operate a switch manually to control a circuit's operation. Here, we can use a switch, called a relay, that is operated by electromagnetism.

A relay is a remotely controlled switch.

Relays come in many shapes and have numerous applications. Regardless of their shape, appearance, or application, all operate similarly. Relays operate on the principle that an electromagnet attracts a piece of soft iron (Figure 12-14).

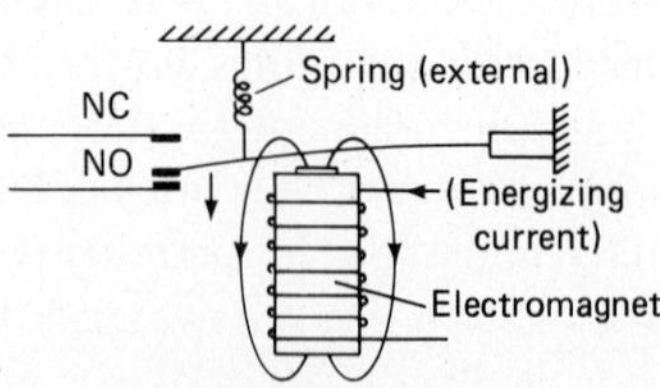

FIGURE 12-14

A basic relay consists of a coil, an iron core, and a movable magnetic material, usually an iron bar. When current flows through the coil, a magnetic field builds around the coil. If the field is strong enough, it will attract the metal bar. If the bar is used to close the contacts of a switch, a circuit can be turned on or off as needed.

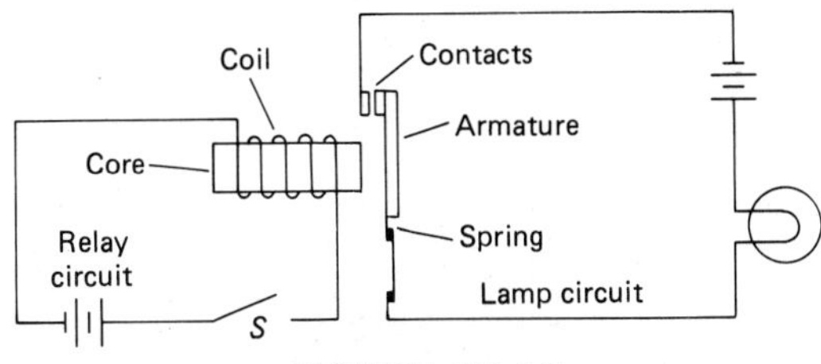

FIGURE 12-15

A basic relay is shown in Figure 12-15. It consists of a coil, a core, movable contacts, an armature, and a tension spring. When the switch is closed, current flow in the coil creates a magnetic field which pulls the contacts together. When the contacts *make,* the circuit of the lamp is complete and the lamp glows. To turn the lamp off, we open the switch. This stops the flow of current in the coil, which, in turn, causes the magnetic field to decay. As soon as the field is weak enough, the spring takes over and *breaks,* or opens, the circuit, thus turning the lamp off. When the relay turns the circuit *on,* we say it has been *energized.* With the relay turned *off,* we say it has been *deenergized.* Contact operation in a relay can be designed to turn a circuit on or off, as desired.

Any switch we have discussed can be designed into the contacts of a relay. A relay that is deenergized is said to be in the *normal* position. All contacts are identified as either normally open (NO) or normally closed (NC). Some contact designs available are shown in Figure 12-16. Any combination of these or other contacts is available as an off-the-shelf relay. In a schematic diagram, the relay is indicated by the abbreviation K, with subsequent relays labeled K_1, K_2, and so on, as noted in the diagram.

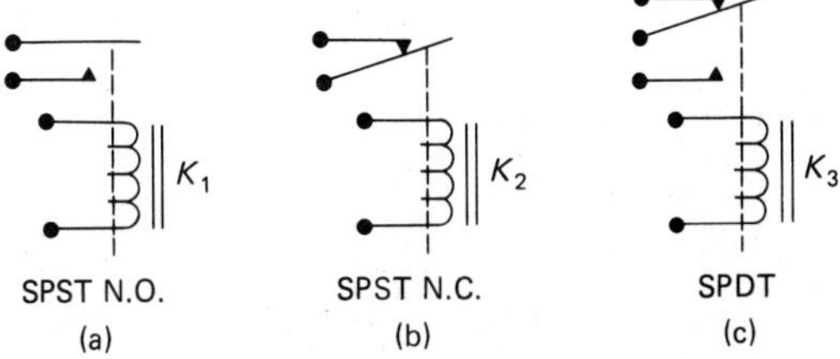

FIGURE 12-16

Each time you start your car, a relay (the starter solenoid) is used. The schematic of an automobile starting circuit is shown in Figure 12-17. This system uses a low-current circuit to energize the relay that controls the heavy-current (up to 400 A) circuit needed to drive the starter motor. When the ignition is turned to *start,* the relay energizes and closes its contacts. These contacts complete the high-current circuit, allowing current to flow in the starter motor. This action continues until the switch is released and the relay is deenergized.

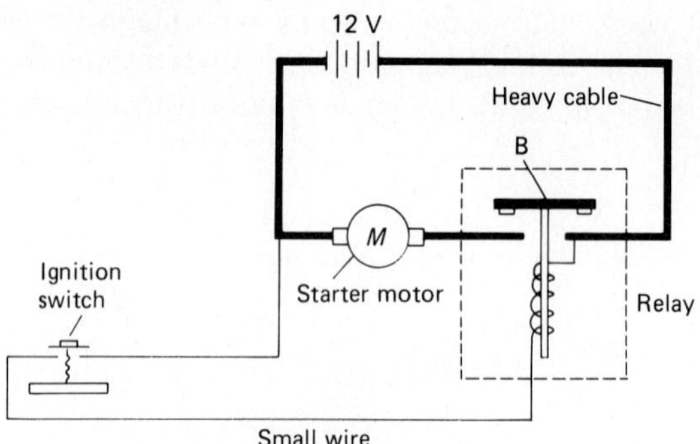

FIGURE 12-17

Holding Relays

A type of relay that is often used is the *holding relay*. Its schematic diagram is shown in Figure 12-18. This type of relay has separate on/off switches and will operate as follows. Switch 1 is a pushbutton switch (Sw_1) that is normally closed; switch 2 (Sw_2) is a pushbutton switch that is normally open. When Sw_2 is closed, current flows through the coil, energizing K_1 and causing contact A to close. Because Sw_2 is a momentary on pushbutton switch, contact is made only when the switch is depressed. As soon as the switch is released, its contacts return to the normally open position. Contact A, though, allows current to flow in the coil, thus keeping K_1 energized. The relay's magnetic field *holds* contact A in the *make* position, and any circuit controlled by K_1 can continue to operate. To turn the circuit off, we must press Sw_1. When that happens, current flow in K_1 stops, the field decays, contacts *break*, and contact A returns to its NO position; the circuit is turned off. This circuit is used as the start/stop circuit on such heavy machines as table saws and drill presses.

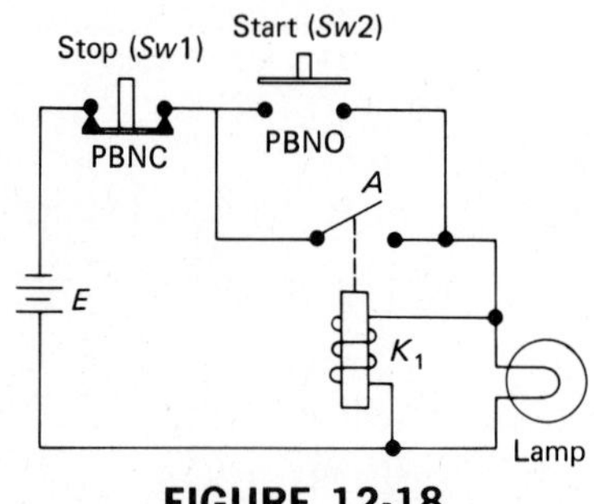

FIGURE 12-18

Vibrating Relays

Before the advent of transistors, most automobile radios used a *vibrator* as part of its power supply. A vibrator converts DC into pulsating DC (PDC) (Figure 12-19).

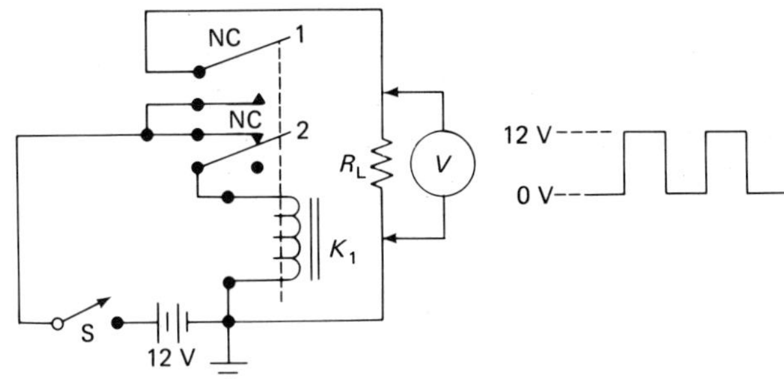

FIGURE 12-19

The vibrator operates as follows: When the switch is closed, current starts to flow in the coil of K_1. A magnetic field begins to expand around K_1; as soon as it becomes strong enough, contacts 1 and 2 move to their opposite positions. This starts current flow in R_L, causing the voltmeter to read 12 V. At the same time, this movement stops current flow in the coil. Field decay is not immediate; considerable time will elapse before K_1 deenergizes. With K_1 deenergized, current stops in the load, the voltmeter reads 0 V, and current starts in K_1. As soon as the field is strong enough, K_1 energizes; the voltmeter now reads 12 V. The relay continues to operate in this manner as long as the switch is closed. The voltmeter reading will pulsate between 0 and 12 V DC. The pulsating DC output can be represented by the waveshape shown in Figure 12-19.

A more common use of the vibrating relay is the alarm circuit. A bell that signals the start or end of class is an alarm circuit (Figure 12-20). Because the armature of the circuit is spring-loaded, it is held in the NC position. When the switch is closed, the flow of current in the coil causes the relay to energize, which, in turn, causes the armature to move away from the other contact and interrupt the flow. When the field decays, the armature returns to the closed position, and the action starts over. As long as the switch remains closed, the armature will vibrate. If a hammer is connected to the relay solenoid so that when the contacts open the hammer strikes a bell, we have an alarm circuit.

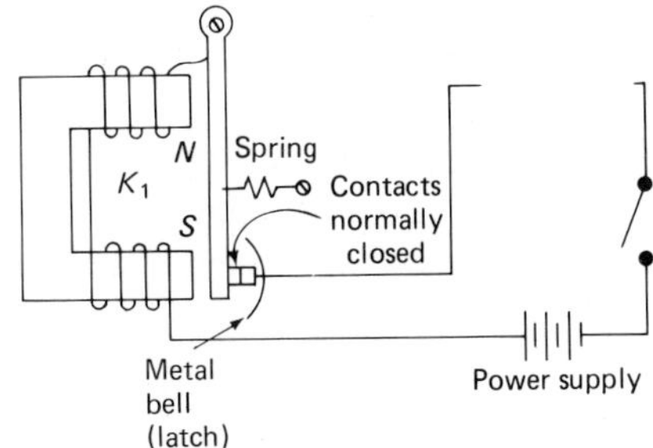

FIGURE 12-20

Overload (Latch) Relays

In some circuits, relays replace fuses, to provide protection from current overload. These relays are classified as *overload* or *latch* relays. An overload relay does not energize until the current reaches a predetermined level. When current exceeds that level, the relay energizes, stopping all current to the load device. Usually, this relay must be reset manually. Figure 12-21 illustrates the latch relay.

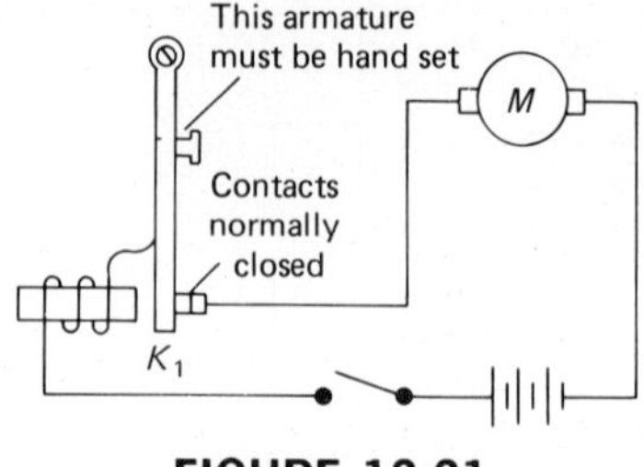

FIGURE 12-21

Assume that K_1 has been designed to energize at 15 A. If the load is normally 10 A, the relay will not energize until a problem occurs in the load circuit. If current in the load increases to 15 A, K_1 energizes, stopping the current through R_L. To restart the circuit, the latch must be released physically.

Time-Delay Relays

In some cases, expensive equipment can be damaged by application of a high voltage as soon as the switch is closed. If that happens, a *time-delay* relay can be used to delay application of the voltage to part of the circuit. Such relays work according to several principles, one of which is the heat principle. A small piece of metal is connected so that, when a switch is turned on, current flows through it and causes it to heat up. As the metal is heated, it expands and at some predictable time makes contact with another conductor, causing another circuit to operate. Time-delay relays are available that provide accurate control over the duration of delays.

Problems in Relays

Most problems that occur in relays are mechanical ones—for example, spring tension, sticking contacts, and relay chatter. Careful inspection and care of relays can reduce these problems.

Electrical problems result from either of two troubles, open coils or partially shorted coils. An open coil stops the operation of relays; it can be detected by an ohmmeter check. Chatter results from part of the windings shorting through their insulation. An ohmmeter can be used to locate the problem, but you must know the coil's normal resistance. The present ohms reading can then be compared to a normal reading.

Self-Check

Answer each item by indicating whether the statement is true or false.

23. ___ A relay is operated by a permanent magnet.

24. ___ A relay can be used to protect against overload.

25. ____ Vibrating relays convert direct current into pulsating direct current.

26. ____ Circuits with separate start-stop switches must have two relays.

27. ____ Overload relays must be reset manually.

28. ____ The contact in a relay is made of aluminum.

29. ____ A tension spring is used to return relay contacts to the normal position.

30. ____ A deenergized relay has current flow in its relay coil.

31. ____ The expansion and decay of a magnetic field occur immediately upon current starting and stopping in the coil.

32. ____ A deenergized relay can have current flowing through its contacts.

Summary

Magnetism and electricity are so closely related that it is impossible to have electricity without magnetism. The existence of magnetism was known centuries before this relationship was recognized. Magnetic ores can be mined from the earth, but they are so weak that they are of little use today.

All magnets have characteristics in common. They all have north and south poles, as well as magnetic fields that surround them. A magnetic field is made up of magnetic lines of force with properties that allow prediction of their reaction in given situations. All magnets obey the rule, unlike poles attract, and like poles repel.

Current-carrying conductors are surrounded by a magnetic field which varies in strength as current in the conductor varies. Coiling the conductor into turns causes the magnetic properties of the coil to be identical to those of a bar magnet. The strength of a magnet depends on the ampere-turns ratio. By using this ratio and suitable core materials, we can construct magnets of desired strength. This type of magnet is called an electromagnet.

By the induction process, we can use strong electromagnets to magnetize soft iron for use as a temporary magnet. For instance, a piece of hard steel can be magnetized for use as a permanent magnet. Magnets thus built are classified as artificial magnets; they are much more suitable than natural magnets for our needs.

Another use for an electromagnet is to operate the contacts of a switch located in a position with limited access. Here, it is called a relay. Various types of relays are available. Holding and latch types are in wide use.

Review Questions and/or Problems

1. In Figure 12-22, where are the magnet's poles located?
   (a) 2 and 3 north
   (b) 1 and 4 north
   (c) 1 and 3 south
   (d) 2 and 4 opposite

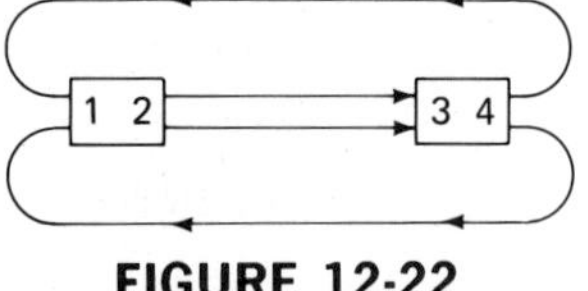

FIGURE 12-22

2. In the diagram in Figure 12-23, where are the north poles?
   (a) A and D
   (b) B and C
   (c) C and A
   (d) D and B

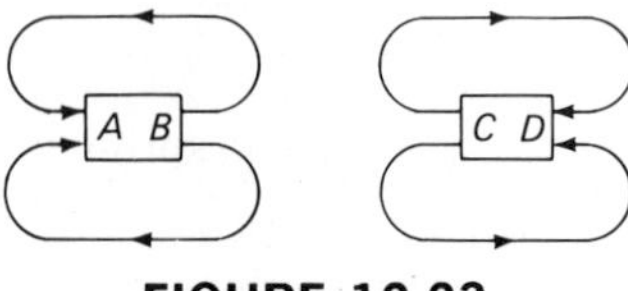

FIGURE 12-23

3. Which term is used to describe the ease with which magnetic lines of force pass through a material?
   (a) Permeability
   (b) Retentivity
   (c) Reluctance
   (d) Force

4. Residual magnetism is defined as
   (a) the ease with which flux flows in a permanent magnet.
   (b) opposition to magnetic lines of flux.
   (c) a small amount of magnetism in a nonmagnetic material.
   (d) magnetism remaining in a magnetized piece of material after removal of the magnetizing force.

5. What are the laws of magnetism?
   (a) Like poles repel, unlike poles attract.
   (b) Like poles repel, unlike poles repel.
   (c) Like poles attract, unlike poles repel.
   (d) Like poles attract, unlike poles attract.

6. The weber is a unit of
   (a) field intensity.
   (b) pole strength.
   (c) flux density.
   (d) reluctance.

7. The force between two poles is
   (a) inversely proportional to the distance between them.
   (b) directly proportional to the distance between them.
   (c) inversely proportional to the square of the distance between them.
   (d) directly proportional to the square of the distance between them.

8. Lines of force moving in the same direction are in parallel, and they will ________ each other.
   (a) cross
   (b) repel
   (c) attract
   (d) unite with

9. What theory of magnetism assumes that all magnetic substances are composed of small molecular magnets?
   (a) Coulomb's
   (b) Weber's
   (c) Domain
   (d) Oersted's

10. In Figure 12-24, the magnetic lines of force rotate
   (a) clockwise.
   (b) counterclockwise.

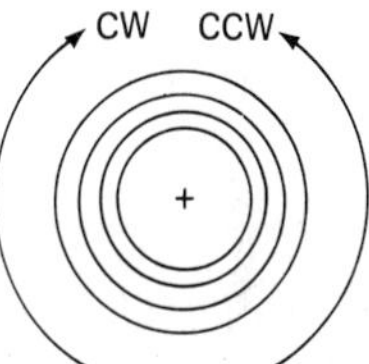

FIGURE 12-24

11. How many parts does a basic relay have?
   (a) Three
   (b) Four
   (c) Five
   (d) Six

12. In Figure 12-25, where is the north pole?
   (a) Point A
   (b) Point B

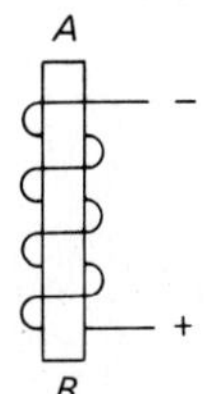

FIGURE 12-25

13. What is a remotely operated switch called?
   (a) A core
   (b) A relay
   (c) A contact
   (d) An armature

14. The product of the number of turns in a coil and amperes of current is called the
    (a) ampere-turns ratio.
    (b) turns ratio.
    (c) ampere capacity.
    (d) saturation.

15. The core material used in an electromagnet has a low
    (a) reluctance.
    (b) permeability.
    (c) retentivity.
    (d) residual magnetism.

13

Direct Current Meters

Introduction

Throughout this book, we have been using meters to measure current, voltage, and resistance. The modern technician must be thoroughly familiar with the use of test equipment. Two types of meters are available: analog and digital. An analog meter indicates values by the angular movement of a pointer along a graduated scale; pointer movement results from the interaction of magnetic fields. A digital meter is much more complex; it presents values in numerical form.

Because digital circuitry is more advanced, we must wait until later to discuss it. In this chapter we concentrate on the analog meter. Pointer positioning on this type of meter is accomplished by a *meter movement.* Meter movements take numerous forms and serve numerous functions. One type of analog meter is the VU meter on a stereo amplifier; the fuel gauge on a car is another.

In class, you have probably been using a multimeter capable of measuring voltage, current, and resistance. These are the three electrical units used in all calculations involving Ohm's law. Meter readings are used to confirm Ohm's-law values. Each meter (whether amp, volt, or ohm) has circuitry and measuring scales specifically designed for such use.

Objectives

Each student is required to:

1. Define the following:
   (a) Thermocouple.
   (b) Hot-wire meter.
   (c) Moving-coil meter.
2. Identify the electrical effect used by each type of meter listed in number 1.
3. Describe linear, inverse, and square-law scales and state how each is used.

4. Identify schematic diagrams for
   (a) An ammeter.
   (b) A voltmeter.
   (c) An ohmmeter.
5. Explain *ammeter sensitivity* and tell why it is important.
6. Compute values for ammeter *shunts.*
7. Define *voltmeter sensitivity* and state how it affects voltmeter use.
8. Compute the value of voltmeter *multipliers.*
9. Explain what is meant by the term *voltmeter loading.*
10. Describe the process involved in *zeroing* an ohmmeter.
11. Calculate the value of *limiter* resistors for use in an ohmmeter.
12. Identify the volt, ohm, and ammeter sections of a VOM schematic.
13. List the precautions that must be observed in using different meters.
14. Make a list of the step-by-step procedures involved in the correct use of an analog ohmmeter.

Types of Meters

To understand meters and measurement, you must first understand the effects of electricity. There are five separate effects which can be used to detect the presence of electricity: heat-sensitive, electromagnetic, electrostatic, chemical, and physiological. The last, physiological, is the result of an electrical shock to a living body, and is not suitable for use in measurement. Chemical and electrostatic effects are suitable for laboratory use only. That leaves heat-sensitive and electromagnetic types for everyday use.

Heat-Sensitive

Heat-sensitive meters are mainly of two types, hot-wire and thermocouple. Both depend on reaction to heat for their measurements. Heat-sensitive meters are used primarily for measuring heavy alternating currents. Since that is not normally part of the technician's job, the explanation of this type of meter is short.

As a resistive device is heated, its power dissipation occurs in direct proportion to its current squared—I^2. This allows heat to be translated into a proportional amount of electric current. As heat varies in a piece of resistive wire, the wire expands and contracts, the amount of expansion or contraction is directly proportional to changes in current. Figure 13-1 is a pictorial representation of a *hot-wire meter* and its scale. The resistive wire connected between points A and B is heat-sensitive; it expands and contracts with changes in heat. Connected to the heat-sensitive wire is a piece of bronze wire. The bronze wire is wrapped around a pulley and connected to a tension spring. When the heat-sensitive wire is heated, it expands, and the bronze wire is pulled downward by the spring, thus maintaining tension. As the bronze wire moves down, its rotation around the pulley causes the pulley to turn. This moves the pointer across the scale indicating the current flow. As the wire cools, the sequence is reversed, and the pointer moves back to zero.

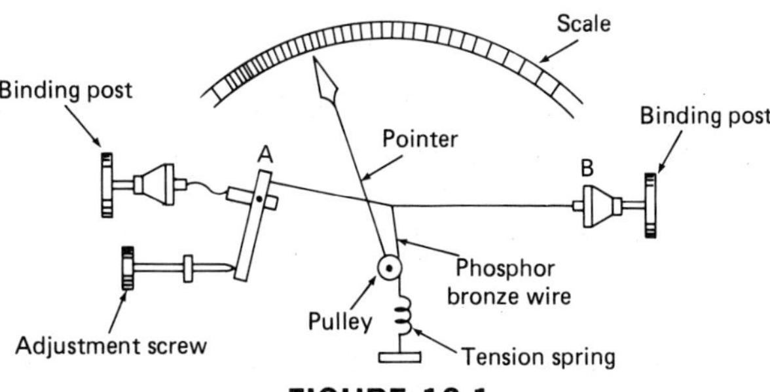

FIGURE 13-1

An unevenly graduated square-law scale is a result of the fact that heat is directly proportional to current squared. This type of scale is shown in Figure 13-1. It is difficult to get accurate readings from this scale, however. Another disadvantage of the hot-wire meter is the time required for heat changes to stabilize in the wire. Finally, the hot-wire meter lacks sensitivity and is inefficient. Because of these disadvantages, the hot-wire meter is seldom used, and then only in applications where other measuring devices are not available.

The *thermocouple* in Figure 13-2 is not a meter; but when it is used with a separate meter movement, it is a valuable piece of test equipment. The thermocouple operates on the principle that when two dissimilar metals are joined and heated, a difference of potential exists between the unjoined ends. The alloys that provide the highest EMF, and that are used in many thermocouples, are constantan, bismuth-antimony, and nickel-copper. When a meter movement is connected between the two unjoined ends, current flow is in direct proportion to the heat applied at the opposite end.

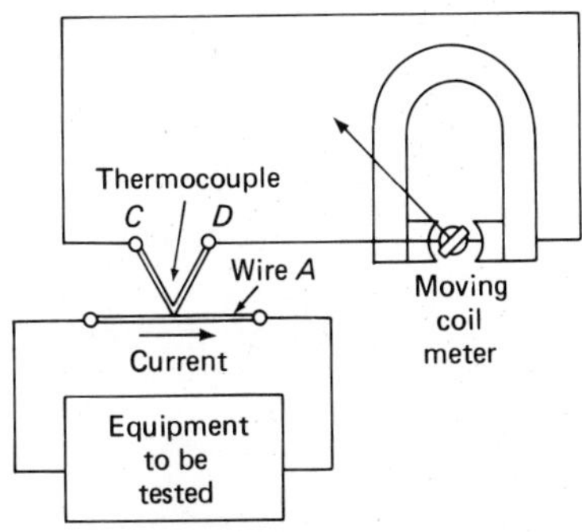

FIGURE 13-2

Many disadvantages of the hot-wire meter are also present in the thermocouple. Because of the importance of the thermocouple in such heated areas as industrial furnaces and rocket engines, its use must be mentioned, and the technician must be aware of it as an option.

Self-Check

Complete each item by inserting the words and/or numbers needed to make a true and complete statement.

1. List the five effects of electricity.
   (a)
   (b)
   (c)
   (d)
   (e)
2. The ______________ type meter presents its indication by use of a pointer and a calibrated scale.
3. List two types of meters that use heat for measurement.
   (a)
   (b)
4. A meter that displays its indications numerically is called a/an ______________ meter.
5. List two advantages of heat-sensitive meters.
   (a)
   (b)

Electromagnetic

The three classes of meters based on the electromagnetic effect are: moving coil, moving iron, and dynamometer. All operate on the principle of magnetic repulsion and attraction. When current flows through a coil, a magnetic field builds around the coil that is directly proportional to the current flow in the coil. The strength of the field can be used to represent the amount of current flow through the coil. Linear scales are used to measure current and voltage (Figure 13-3). Note that all divisions along the scale are equal, giving it linear characteristics. Linear scales can be calibrated to indicate current, voltage, or any other unit that is proportional to current. The moving-coil meter is the most popular type and will be discussed first and fullest. The *moving-coil meter* (Figure 13-4) is the most widely used movement. This movement is based on the interaction of an electromagnetic field and the field exerted by a permanent magnet.

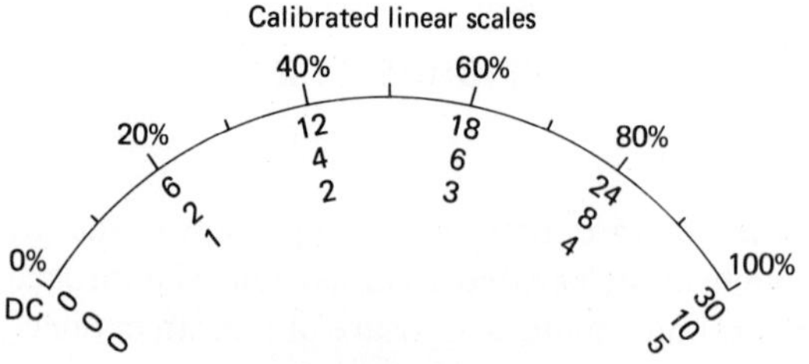

FIGURE 13-3

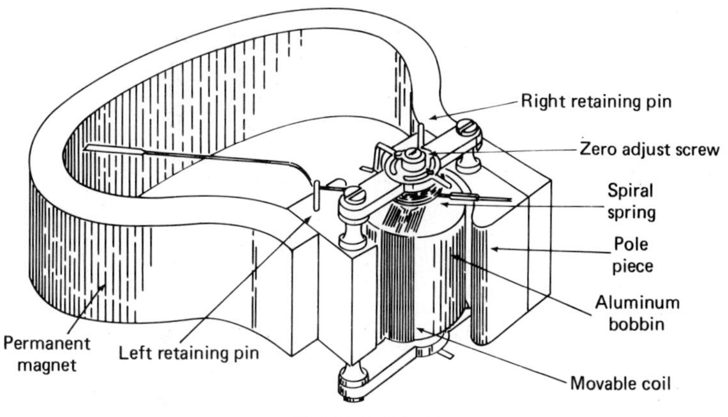

FIGURE 13-4

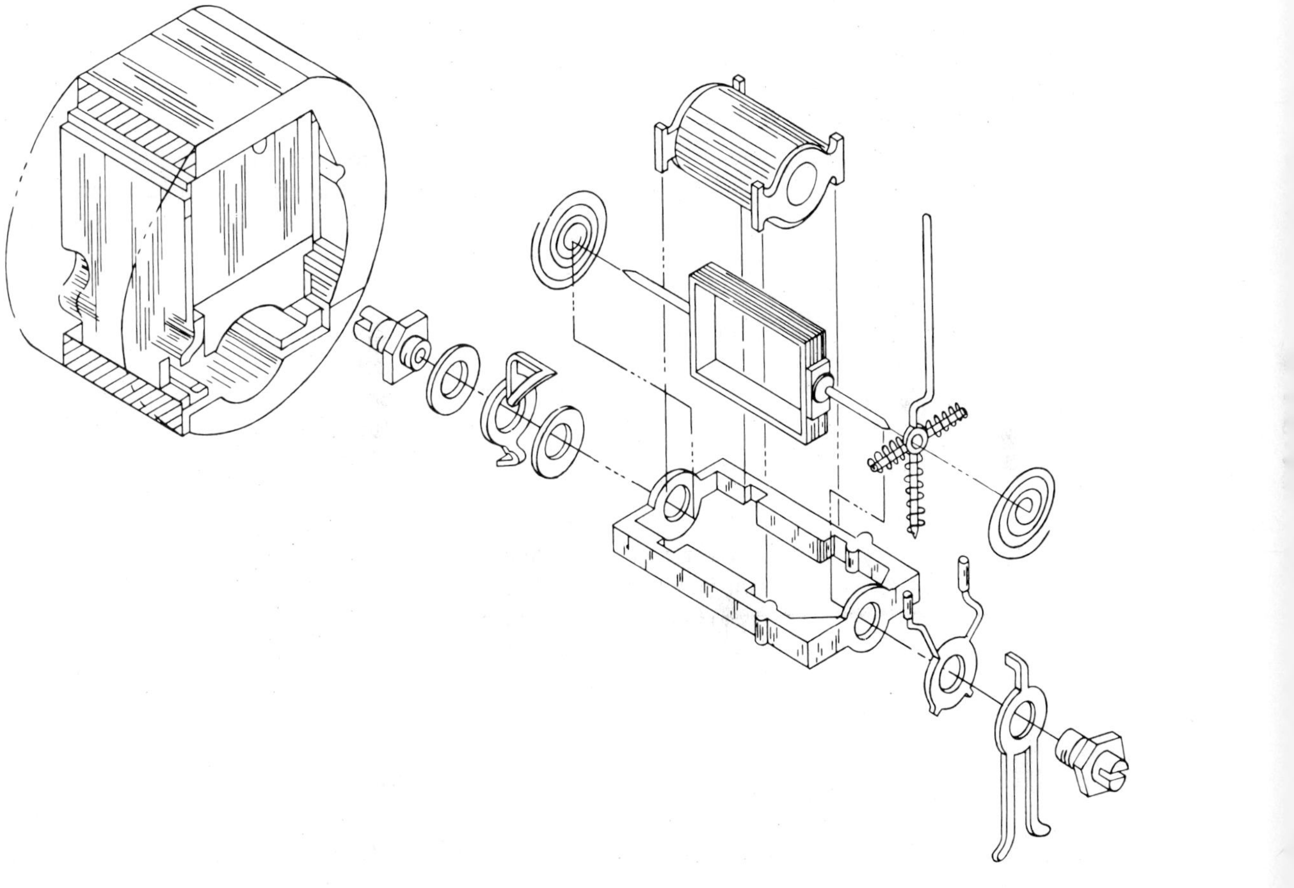

FIGURE 13-5
D'Arsonval Meter Movement (Drawing Courtesy of Triplett Corporation)

The scale, most often used with a moving-coil movement, is linear (Figure 13-3). The movement of the moving-coil meter was invented in 1882 by Jacques Arsène d'Arsonval (1851–1940). D'Arsonval used the DC-motor (moving-coil) principle to design a highly sensitive, accurate galvanometer. This instrument, however, is suitable only for lab use. It has fine wire springs that support the bobbin and is too delicate for ordinary use.

In 1886, Edward Weston modified d'Arsonval's movement by supporting the bobbin with rigid, jeweled bearings and securing the coiled springs more firmly to make it rugged and portable. Today, the Weston movement is used almost exclusively, but the movement is still called d'Arsonval. The galvanometer studied with the Wheatstone bridge is a version of the Weston movement. Other analog meters are designed so that scale zero is at either end of the scale.

Figure 13-5 illustrates a moving-coil movement. The permanent magnet exerts a magnetic field between its two pole pieces. A coil is wound around an aluminum bobbin. When current flows through the coil, a magnetic field builds around the coil and interacts with the permanent field, causing the bobbin to rotate in a direction that indicates the amount and direction of current flow in the coil. As the bobbin rotates, the pointer moves across the scale. Because this movement is directly proportional to the current flow, the points on the scale can be calibrated to indicate amounts of current flow. Some other design considerations are involved, but these are the most important for our purposes.

The moving-coil meter is polarity-sensitive. To ensure proper polarity, we must connect the negative lead nearest the negative pole of the battery and the positive lead nearest the positive pole. Connecting the meter in the circuit to reverse polarity causes the pointer to deflect off the scale in the opposite direction. When it is connected correctly, the meter movement requires a specific amount of current flow to deflect full scale; this is called *full-scale deflection,* or FSD. The amount of current required for FSD is a measure of *meter sensitivity.* The smaller the amount of current required for FSD, the greater the sensitivity. Meter sensitivity is stated in milliamperes or in microamperes. Two popular movements are those with sensitivities of 1 mA and 50 μA. Once a movement has been manufactured, its sensitivity is permanent and cannot be changed. The schematic symbol for a moving coil is shown in Figure 13-6a, the common way of representing the meter in a circuit is shown in Figure 13-6b.

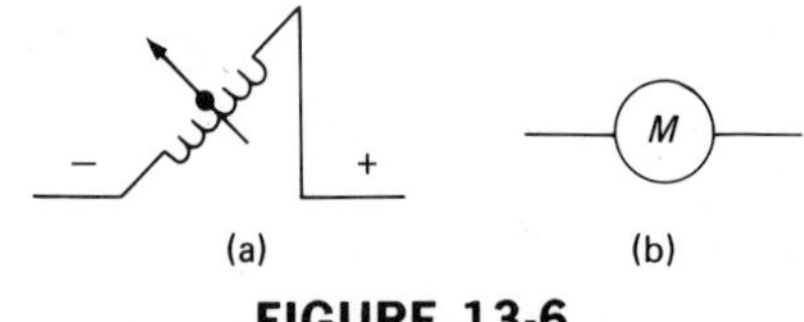

FIGURE 13-6

A single-meter movement can be used in any number of measuring devices. The most common uses are volt, ohm, and ammeter, with each function operating in one or more ranges.

Self-Check

Answer each item by indicating whether the statement is true or false.

6. ____ A moving-coil meter presents its indication by means of a rotating bobbin and pointer.

7. ____ The bobbin's rotation results from electromagnetism within the movement.

8. ____ *Meter sensitivity* is the ability of a meter to measure small amounts of voltage or current.

9. ____ The d'Arsonval movement is a delicate laboratory instrument.

10. ____ The Weston meter is a rugged version of the d'Arsonval movement.

The Ammeter

Electrically, the ammeter circuit is identical to the parallel circuit shown in Figure 13-7. If this circuit has a total current of 10 mA, and R_1 can conduct only 1 mA, then R_2 must conduct 9 mA. From Kirchhoff's current law, we know that the total current in a parallel circuit equals the sum of the branch currents. Therefore,

$$I_t = I_{R1} + I_{R2}$$

$$I_t = 1 \text{ mA} + 9 \text{ mA}$$

$$I_t = 10 \text{ mA}$$

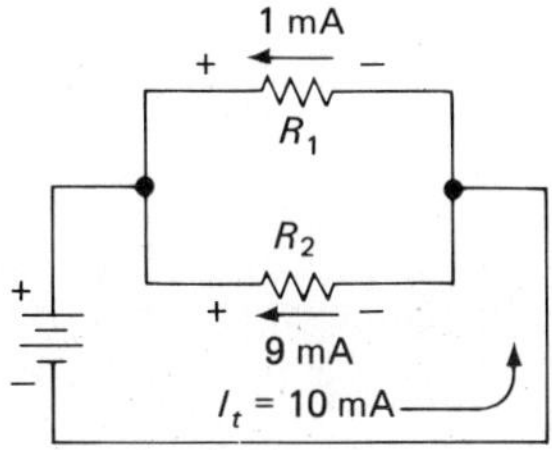

FIGURE 13-7

If the current flowing in the circuit is less than 10 mA, that current will divide, with one-tenth flowing through R_1 and the remaining nine-tenths flowing through R_2. If R_1 is constructed so that any current flow in excess of 1 mA destroys it, we have a special problem. For total current to exceed 10 mA, the value of R_2 must be changed to a value capable of carrying all current in excess of 1 mA. The parallel path occupied by R_2 is called a *shunt path,* and the resistor connected in the path, a *shunt resistor.*

Let's extend this line of thought to include an ammeter circuit. The moving-coil movement, designed for maximum FSD current, replaces R_1 in the circuit. This current is the sensitivity rating for that meter movement. Figure 13-8a represents a movement having a sensitivity of 1 mA and a resistance of 100 Ω.

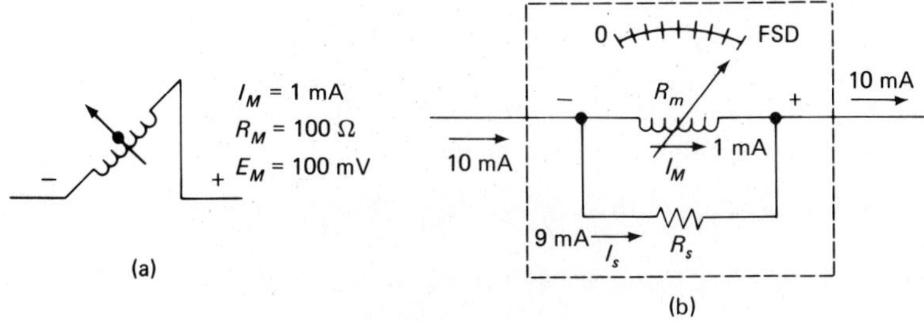

FIGURE 13-8

The movement in Figure 13-8 can conduct a maximum of 1 mA. Exceeding that limit causes the pointer to deflect past full scale, which could damage the movement. With the current and resistance values noted, we can use Ohm's law to determine the meter's voltage drop when 1 mA is flowing:

$$E_M = I_M \times R_M$$

$$E_M = 1\text{ mA} \times 100\ \Omega$$

$$E_M = 100\text{ mV}$$

where

I_M = meter FSD current
E_M = meter voltage drop

In order to use this meter to measure currents larger than 1 mA, we must shunt the additional current through a parallel branch (Figure 13-8b). If we wish to use this movement in a circuit that can measure 10 mA, we calculate the shunt value as follows:

1. Determine shunt current, I_s:

$$I_s = I_t - I_M$$

$$I_s = 10\text{ mA} - 1\text{ mA}$$

$$I_s = 9\text{ mA}$$

2. Use I_s and E_M to calculate the shunt resistor value, R_s:

$$R_s = \frac{E_M}{I_s}$$

$$R_s = \frac{100\text{ mV}}{9\text{ mA}}$$

$$R_s = 11.11\ \Omega$$

Connecting an 11.11 Ω resistor in the shunt position of Figure 13-8b results in a circuit where the meter movement reaches FSD when the total current entering the circuit reaches 10 mA. The scale used with this meter is calibrated 0–10 mA on the linear scale. Any current less than 10 mA causes a proportionate amount of deflection. The amount of deflection is directly proportional to current flow in the circuit; that is, a deflection of one-half scale indicates a current flow of 5 mA.

To use a meter movement as the indicating device for several different ranges of current, we design the circuit as shown in Figure 13-9.

The movement used in Figure 13-9 is a 50-μA movement. Its resistance is $R_M = 1\text{ k}\Omega$ and its voltage is $E_M = 50\text{ mV}$. We know that this movement acting without a shunt can measure 0 to 50 μA, we will add 1 mA, 10 mA, and 150 mA ranges to the circuit. To do this, we install a range switch and proceed as follows. Solve for the ohmic value of the shunt R_1:

$$I_{R1} = I_t - I_M = 1\text{ mA} - 50\ \mu\text{A} = 0.95\text{ mA}$$

$$R_1 = \frac{E_M}{I_{R1}} = \frac{50\text{ mV}}{0.95\text{ mA}} = 52.63\ \Omega$$

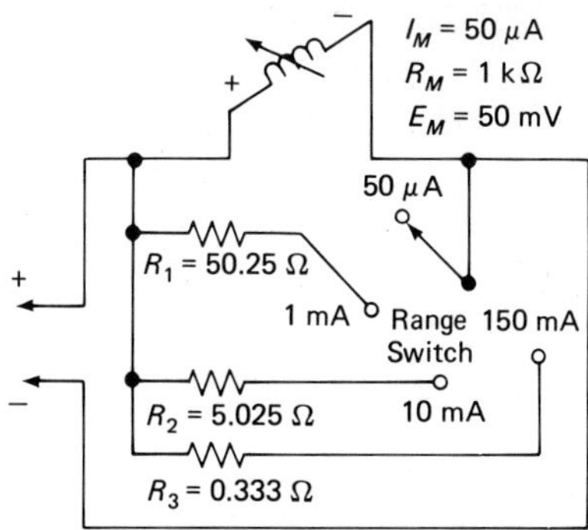

FIGURE 13-9

Solve for R_2:

$$I_{R2} = I_t - I_M = 10 \text{ mA} - 50\ \mu\text{A} = 9.95 \text{ mA}$$

$$R_2 = \frac{E_M}{I_{R2}} = \frac{50 \text{ mV}}{9.95 \text{ mA}} = 5.025\ \Omega$$

Solve for R_3:

$$I_{R3} = I_t - I_M = 150 \text{ mA} - 50\ \mu\text{A} = 149.95 \text{ mA}$$

$$R_3 = \frac{E_M}{I_{R3}} = \frac{50 \text{ mV}}{149.95 \text{ mA}} = 0.333\ \Omega$$

The shunts used in designing an ammeter must be extremely accurate. The values calculated here are not unrealistic, since shunts can be purchased that have values this precise.

An ammeter circuit must have a very small resistance value. Any time the ammeter is connected in a circuit, its resistance is added in series with the load, causing current to decrease. The mere fact that the meter was connected caused an error. To keep this error as small as possible, the ammeter resistance is kept as low as possible.

Some ammeters use a resistive network as the shunt. The movement and selected resistors are connected in a closed loop. One point of this network is connected permanently; the other connection is selected by the range switch. All resistors are used in all ranges, and different combinations of series and shunt resistors exist for each range. These networks are called universal shunts. Regardless of the type of shunt used, all analog ammeters operate similarly, and the accuracy of an ammeter depends on the accuracy of the shunts used in its design.

In using an ammeter, some precautions are necessary, to protect the meter and the person using it:

1. Do not work on "hot" circuits unless it is absolutely necessary.
2. Never connect an ammeter in parallel with a component.
3. Make sure the meter is connected in correct polarity in the circuit.
4. Never use a DC meter to measure AC values.

Self-Check

Answer each item by indicating whether the statement is true or false.

11. ____ A parallel-resistive circuit is similar to an ammeter circuit.

12. ____ A linear scale is calibrated so that the length of each division on its scale is different.

13. ____ An ammeter shunt is used to modify an ammeter circuit so it can measure larger amounts of current.

14. ____ A 100-μA meter movement is more sensitive than a 50-μA movement.

15. ____ When modifying an ammeter circuit for measurement of higher currents, you should add a larger shunt resistor.

The Voltmeter

A voltmeter can be compared to a series-resistive circuit. Figure 13-10 is used to analyze a series circuit that has similarities to the voltmeter circuit.In Figure 13-10a, R_1 drops 1 V when total current equals 1 mA. If R_1 is designed so that its maximum current cannot exceed 1 mA, and we wish to increase the voltage applied to the circuit, we must insert a series resistor capable of dropping all voltage in excess of 1 V. Figure 13-10b is typical. With R_2 = 19 kΩ connected in series, the voltage applied could reach 20 V before 1 mA of current flows. At that point, E_{R1} = 1 V and E_{R2} = 19 V. Varying the size of R_2 enables us to select the combination of resistors needed to drop any applied voltage and not cause E_{R1} to exceed 1 V. In each case, E_{R2} must equal the applied voltage, less E_{R1}.

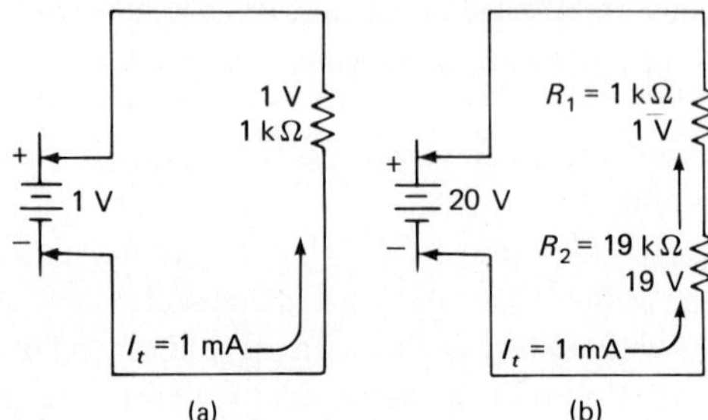

FIGURE 13-10

This analysis can be related to the voltmeter's circuit if we realize that a moving-coil meter movement is manufactured with a maximum current capability and a fixed resistance. Substituting the meter movement for R_1 would cause the circuit to have a maximum total current equal to the current rating of the meter movement; R_2 would be required to drop all voltage applied to the circuit in excess of that dropped by the meter, E_M. For ammeters, we used a movement with I_M = 50 μA, R_M = 1 kΩ, and E_M = 50 mV. If we use this movement, the circuit for a voltmeter would resemble Figure 13-11.

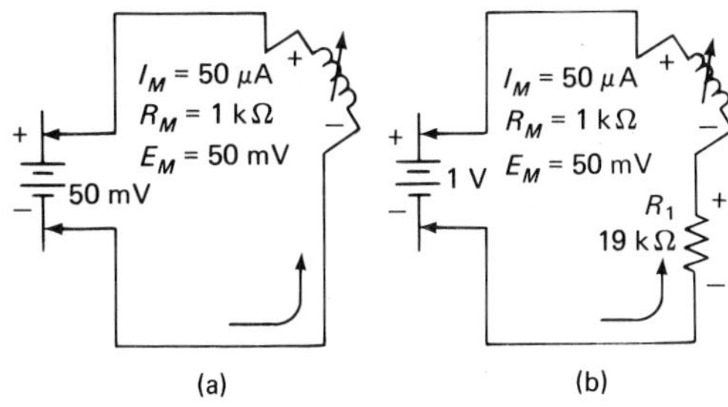

FIGURE 13-11

Figure 13-11a shows only the movement. When $I_M = 50\ \mu A$, $E_M = 50$ mV as calculated by Ohm's law. In order to use this movement for measuring voltages larger than 50 mV, we must connect a series (multiplier) resistor to the circuit, as shown in Figure 13-11b. If R_1 is a 19-kΩ resistor, we can determine the amount of voltage applied to the circuit when 50 μA of current is flowing by:

1. $E_M = I_M \times R_M = 50\ \mu A \times 1\ k\Omega = 50\ mV$
2. $E_{R1} = I_M \times R_1 = 50\ \mu A \times 19\ k\Omega = 950\ mV$
3. $E_t = E_M + E_{R1} = 50\ mV + 950\ mV = 1000\ mV$, or 1 V

With R_1 connected as the multiplier, we extend the meter's range to read 0-1 V. For other voltages, we must find the value of the multiplier for each range we wish to provide. To use the movement to measure 0-25 V, we could design and build the circuit in Figure 13-12, which illustrates a four-range voltmeter.

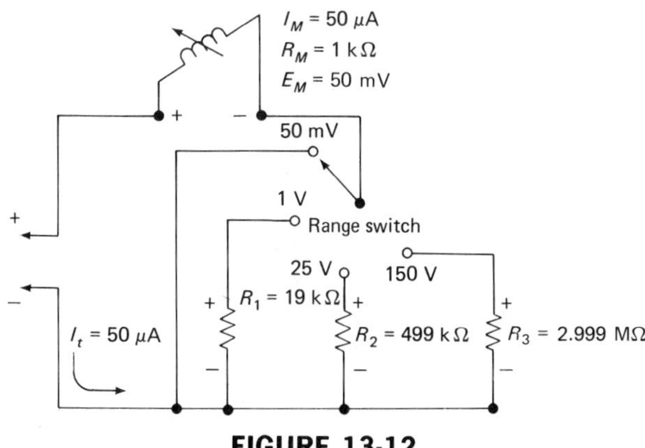

FIGURE 13-12

To determine the size of the multipliers required to complete this circuit, we solve:

$$E_{R2} = E_t - E_M = 25\ V - 50\ mV = 24.95\ V$$

$$R_2 = \frac{E_{R2}}{I_M} = \frac{24.95\ V}{50\ \mu A} = 499\ k\Omega$$

and

$$E_{R3} = E_t - E_M = 150 \text{ V} - 50 \text{ mV} = 149.95 \text{ V}$$

$$R_3 = \frac{E_{R3}}{I_M} = \frac{149.95 \text{ V}}{50\ \mu\text{A}} = 2.999 \text{ M}\Omega$$

By installing these resistors as R_2 and R_3, we add two branches to those already discussed. The meter can now measure voltages over four ranges: 0-50 mV, 0-1 V, 0-25 V, and 0-150 V. Ranges can be selected by use of the range switch. To add other ranges, multiplier values are calculated, the resistors are installed, and a switch is installed that allows the added ranges to be selected.

Voltmeters and Ammeters

Both voltmeters and ammeters must be connected in the circuit using correct polarity. The negative lead of the meter is connected to the negative potential in the circuit, and the positive lead is connected to the positive potential.

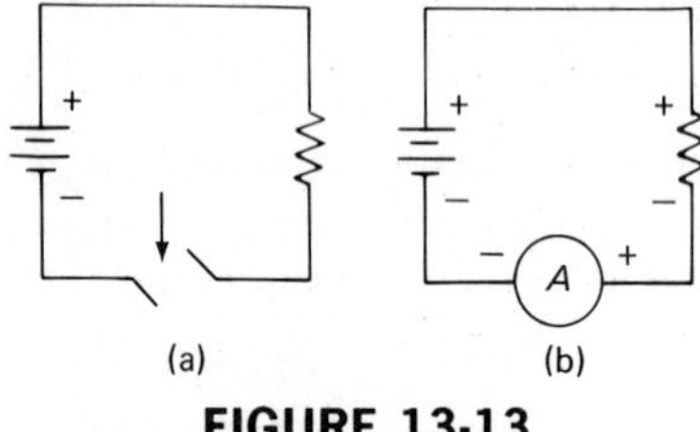

FIGURE 13-13

Figure 13-13 represents the ammeter connection. Diagram a illustrates how the circuit must be broken; diagram b shows how the meter is connected into the break. Note the connection of leads to the circuit. By inserting the ammeter in this manner, we add the ammeter's resistance to the path of the current, which, in turn, causes the current to decrease, resulting in an incorrect reading. To keep the error as small as possible, the resistance of an ammeter must be very small.

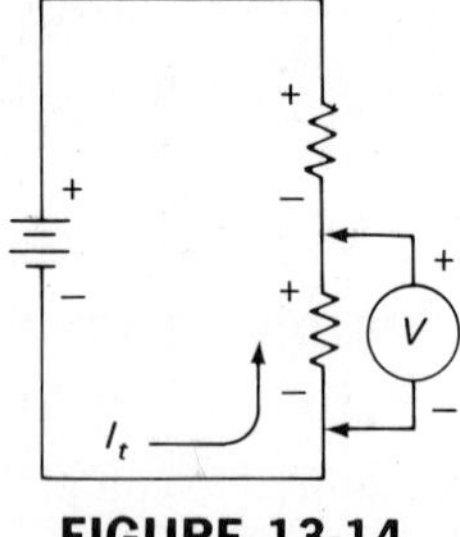

FIGURE 13-14

Figure 13-14 illustrates the correct procedure for connecting a voltmeter to a circuit. Polarity has been observed. After connection of the voltmeter, the following conditions exist:

1. Adding an extra branch (the voltmeter) to the circuit causes total current to increase.
2. Equivalent resistance of the resistor and meter is less than for the resistor alone, because the equivalent must be less than the resistance of the smallest branch.
3. This results in a voltage reading less than that expected.

The voltage read by a voltmeter is less than was actually present before the meter was connected. To keep this error as small as possible, the voltmeter must have high internal resistance. Note that the *ohms/volt* ratio indicates voltmeter sensitivity. Meter movements with the lowest current sensitivity have the best voltage sensitivity.

Let's consider the following meters and their use as voltmeters: meter number 1, a 1-mA meter movement; and meter number 2, a 50-μA meter movement. To determine each meter's sensitivity, we use Ohm's law and divide meter current into 1 V. This results in:

$$\text{Meter 1 sensitivity} = \frac{1\text{ V}}{1\text{ mA}} = 1\text{ k}\Omega/\text{V}$$

$$\text{Meter 2 sensitivity} = \frac{1\text{V}}{50\ \mu\text{A}} = 20\text{ k}\Omega/\text{V}$$

As an illustration of the effect of these calculations on the voltage readings taken by each meter, we examine Figure 13-15. Each movement is used to measure the voltage while the range switch is set to a range of 20 V FSD. With it set to the 20-V position, the internal resistance of each meter is:

$$\text{meter 1: } 20\text{ V} \times 1\text{ k}\Omega/\text{V} = 20\text{ k}\Omega$$

$$\text{meter 2: } 20\text{ V} \times 20\text{ k}\Omega/\text{V} = 400\text{ k}\Omega$$

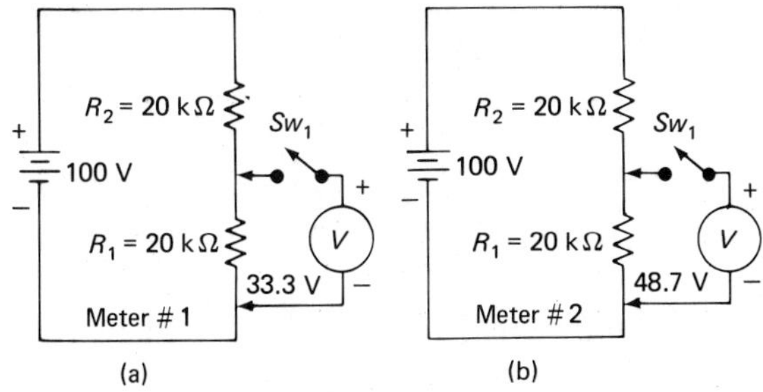

FIGURE 13-15

Refer to Figure 13-15a. The effect of each meter's internal resistance on the voltage reading is as follows. Using meter 1, and with the switch open, each resistor drops

$$E_{R1} = 50\text{ V}$$

$$E_{R2} = 50\text{ V}$$

When the switch is closed:

$$R_{eq} = \frac{R_1 \times R_M}{R_1 + R_M} = 10\text{ k}\Omega$$

$$R_t = R_2 + R_{eq} = 30\ \text{k}\Omega$$

$$I_t = \frac{E_t}{R_t} = 3.33\ \text{mA}$$

$$E_{R1} = I_t \times E_{Req} = 33.3\ \text{V}$$

$$E_{R2} = I_t \times R_2 = 66.7\ \text{V}$$

Connecting meter 1 causes an error of 16.7 V.

Using Figure 13-15b and meter 2, we calculate what happens when this particular meter is used. When the switch is open, the voltage drops are the same:

$$E_{R1} = 50\ \text{V}$$

$$E_{R2} = 50\ \text{V}$$

When the switch is closed:

$$R_{eq} = \frac{R_1 \times R_M}{R_1 + R_M} = 19\ \text{k}\Omega$$

$$R_t = R_2 + R_{eq} = 39\ \text{k}\Omega$$

$$I_t = \frac{E_t}{R_t} = 2.56\ \text{mA}$$

$$E_{R1} = I_t \times E_{Req} = 48.7\ \text{V}$$

$$E_{R2} = I_t \times R_2 = 51.3\ \text{V}$$

The measurement using meter 2 is 48.7 V, an error of only 1.3 V. Obviously, use of meter 2 results in a much more accurate reading. The error caused by connecting a voltmeter is known as *voltmeter loading*.

Both the voltmeter and the ammeter use the current flow in a circuit to make their measurements. Power must be on for the readings to be taken. This results in a hazardous condition. Care must be taken to avoid electrical shock. When inserted, both meters cause circuit loading, which results in slight errors of measurement.

When making measurements in a DC circuit, be sure to use a DC meter. Do not attempt to use a DC meter to read AC, or it may be damaged. Always observe polarity with meters that are polarity-sensitive. It is a good practice to always use the highest range available when making current or voltage readings; this provides meter protection when high voltages or currents are measured. If the deflectional obtained is small, turn the range switch until an accurate reading is obtained.

In recent years many advances have been made in the design of meters. The addition of special transistors and other devices has made them more reliable and accurate and has reduced the damage caused by current overloads.

Self-Check

Complete each item by inserting the words and/or numbers needed to make a true and complete statement.

16. The voltmeter can be compared to the ______________ resistive circuits already discussed.

17. A 50-μA meter movement has ______________ meter sensitivity than a 100-μA movement.

18. When a voltmeter is connected to a circuit, it causes total resistance to ______________.

19. When a voltmeter is connected to a circuit, it causes total current to ______________.

20. Circuit power must be ______________ when making voltage and ampere readings.

The Ohmmeter

The ohmmeter is discussed separately because it uses a different method for making measurements. The ohmmeter has its own power source; it does not depend on the circuit for current. In fact, for the ohmmeter to make accurate measurements, the component to be checked must be isolated from all other paths through which current might flow.

Ohmmeter design uses a series-resistive circuit. Again, a meter movement replaces a resistor in the circuit; other resistors are then inserted. These allow the meter to be used accurately and for different ranges. A battery or other power source is an integral part of the ohmmeter. To use this meter for measuring ohms, it is necessary to calibrate it before each measurement. Calibration compensates for changes in the circuit that might develop over time. To calibrate the ohmmeter circuit shown in Figure 13-16,

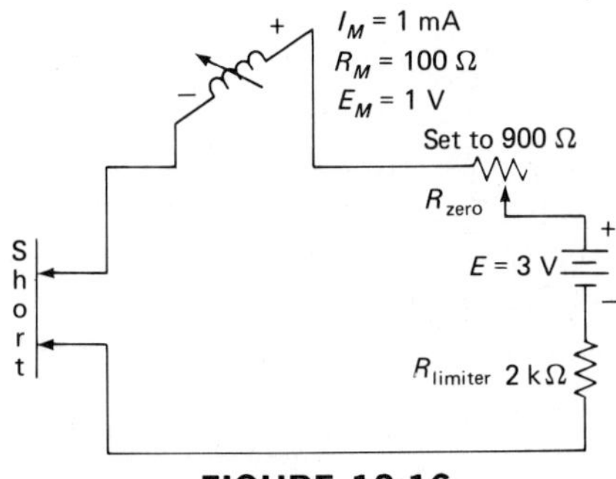

FIGURE 13-16

1. Short the test leads, as shown in the schematic.
2. Observe the pointer position on the scale.
3. Adjust the R_{zero} for pointer position at zero.

When the ohmmeter is calibrated (zeroed), total current is set to the meter's FSD current value. Because current is zeroed for maximum, scale zero is at the right side (FSD) of the scale. Inserting a resistor between the test leads results in a reduced current flow. An inverse scale is used for an ohmmeter (Figure 13-17).

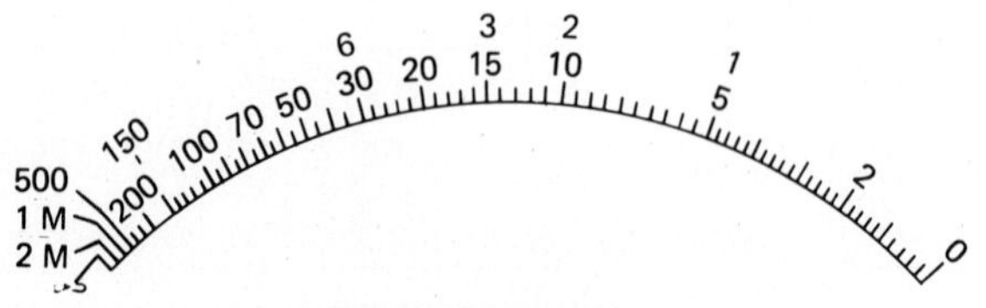

FIGURE 13-17

Note that the scale is highly nonlinear. Because of this nonlinearity, the most accurate readings are made on the right side of the scale. The left side of this scale represents infinite ohms (an open), the reading shown when the short is removed from the test leads.

An ohmmeter circuit's construction is displayed in Figure 13-18. If the leads were shorted, current would flow as indicated by the arrows. This meter has a 1-mA sensitivity and an internal resistance of 100 Ω; it will drop 100 mV. A full-scale deflection on the meter indicates that 1 mA is flowing.

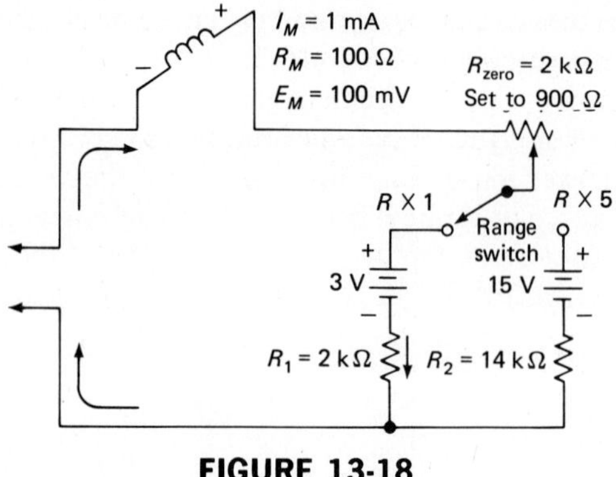

FIGURE 13-18

In order for R_{zero} to zero the meter, R_1 must drop some voltage. R_1 is inserted in the circuit to limit the amount of current flow; this is called a *limiter* resistor. Setting R_{zero} to 900 Ω allows us to calculate the ohmic value of the limiter:

$$R_t = \frac{E_{source}}{I_M} = \frac{3\text{ V}}{1\text{ mA}} = 3\text{ k}\Omega$$

$$R_1 = R_t - R_{zero} - R_M = 2\text{ k}\Omega$$

Selecting a 2-kΩ rheostat for R_{zero} allows us to set R_{zero} to 900 Ω and have approximately the same number of ohms on either side of this setting, for future adjustments. Then we connect the 2-kΩ limiter in series with R_{zero}, the battery, and the meter.

The circuit in Figure 13-18 has two ranges that can be used for checking resistance and continuity. The 3-V range has already been solved, but now we calculate the value of limiter R_2.

$$R_t = \frac{E_t}{I_M} = 15\ \text{k}\Omega$$

$$R_2 = R_t - R_{\text{zero}} - R_M = 14\ \text{k}\Omega$$

In calibrating the scale of an ohmmeter, it helps to understand the effect of resistance on current. The use of Ohm's law implies that resistance and current are inversely proportional. Connecting a resistor between the test leads causes total resistance to increase and total current to decrease. With less current flow, the pointer deflects only partway over the scale. After the scale has been zeroed, and if resistance is connected between the test leads, we can predict the scale-calibration value. The 3-V circuit of the ohmmeter, already discussed, has an internal resistance of 3 kΩ. If we connect a 3-kΩ resistor between the leads,

$$R_t = R_x + R_M + R_{\text{zero}} + R_1 = 6\ \text{k}\Omega$$

$$I_t = \frac{E_{\text{source}}}{R_t} = 0.5\ \text{mA}$$

The current is exactly half the FSD current, which causes the pointer to move to midscale. Therefore, the half-scale position is calibrated as 3 kΩ. By inserting other known resistance values between the leads, we can calibrate the pointer position on the scale with that resistor's ohmic value.

Because the ohmmeter contains its own battery source, it is impossible to connect it in reversed polarity. When making a check, you must make sure the component being checked is isolated from all other current paths.

Self-Check

Complete each item by inserting the words and/or numbers needed to make a true and complete statement.

21. An ohmmeter is calibrated by adjusting the ______________ adjust.

22. An ohmmeter scale is a/an ______________ scale.

23. The value of a third limiter that could be added in Figure 13-18, which would be powered by a 25-V battery, is ______________ Ω.

24. When zeroing the ohmmeter, you must ______________ the test leads.

25. Center scale on the ohmmeter is calibrated to a resistance value that equals the ______________ ______________ of the meter.

Multimeters and VOM

FIGURE 13-19
Volt-Ohm-Milliammeter (Courtesy Simpson Electric Company)

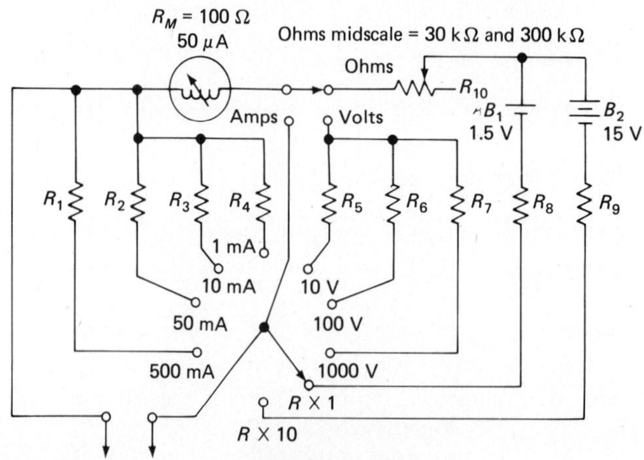

FIGURE 13-20

To make a more useful test instrument from a relatively expensive meter movement, it is common to use one movement to measure at least three electrical units. Meters designed to do this are called either multimeters or VOMs. Figure 13-19 shows a VOM that is available commercially. This meter is discussed in Chapter 4. It is designed so the operator can measure voltage, current, and resistance by selecting the function and range needed. Range is selected by a switch at the center of the meter. The function is selected by a +DC, −DC, AC switch at the left. A typical VOM circuit is illustrated in Figure 13-20.

Multimeters are similar, though they usually include extra circuits for measuring AC volts and AC amps.

Self-Check

Complete each item by inserting the words and/or numbers needed to make a true and complete statement.

26. A ______________ contains several different meters within one instrument.

27. The function switch on a multimeter is used to select settings for measuring ______________, ______________, or ______________.

28. The VOM often has two switches, called ______________ and ______________.

29. The circuit illustrated in Figure 13-20 has ______________ functions.

30. The meter illustrated in Figure 13-19 has ______________ voltmeter ranges available.

The Moving-Iron Meter

An iron-plunger meter movement is shown in Figure 13-21. This movement is based on the principle of magnetic attraction of a soft-iron core resulting from the flow of current through a coil that causes a magnetic field. The distance the coil is moved is directly proportional to the amount of current flow and the strength of the field. The plunger-type meter movement was an early invention used primarily to measure alternating currents. It is seldom used today.

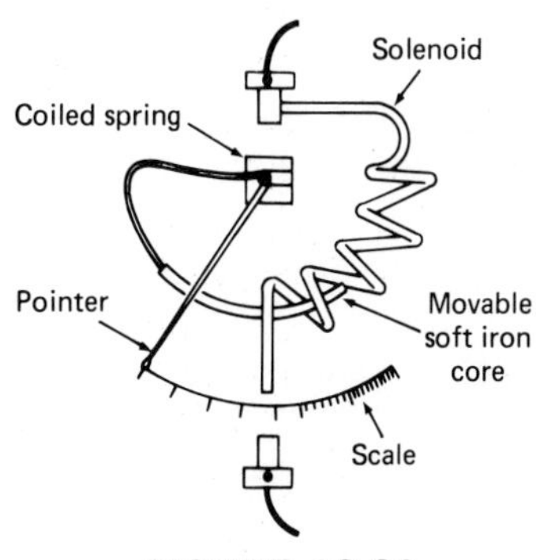

FIGURE 13-21

The Electrodynamometer

The dynamometer consists of two stationary coils and one movable coil (Figure 13-22a). The movable coil supports and positions the pointer. When the meter is connected correctly to a circuit, current flows through L_1, L_2, and L_3, making each an electromagnet. Because all are magnetized with the same polarity, the movable coil, L_3, aligns itself with L_1 and L_2, causing the pointer to move along the scale in proportion to current flow.

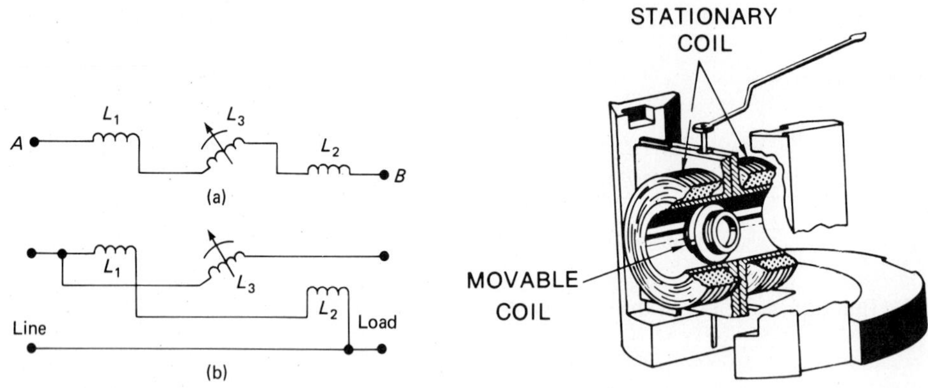

FIGURE 13-22
Dynanometer Movement (Drawing (c) courtesy of Triplett Corporation)

The principle of the electrodynamometer can be used to measure AC or DC voltage and current, but more often it is used as a power meter to measure watts. When the meter is used as a wattmeter, the fixed coils L_1 and L_2 are connected to each other, in series, but in parallel with the power source (Figure 13-22b). The movable coil, L_3, is placed in series with the load to be measured. Currents through the meter now become part of the circuit current and voltage network, which allows the dynamometer to measure power accurately. The meter becomes a wattmeter whose scale is calibrated in watts.

Self-Check

Complete each item by inserting the words and/or numbers needed to make a true and complete statement.

31. The moving-iron meter operates on the principle of ______________.

32. The moving-iron meter is most effective when used as a ______________ measuring device.

33. In a dynamometer, the ______________ serves the same purpose as the permanent magnet in a moving-coil movement.

Summary

The five effects of electricity and their use in detecting electrical current were explained. Measuring devices based on the effects of heat and electromagnetism serve for most purposes. This chapter concentrates on meters designed to take advantage of these two effects.

The heat-sensitive meter movement was explained, the hot-wire and thermocouple meters were discussed and their operation explained; of the two, the thermocouple is used the most widely.

The moving-coil movement was introduced. This movement, discovered by d'Arsonval, was later modified by Weston, making it portable and more durable. Today, the movement is used for most electrical measurements. The d'Arsonval movement presents readings along a scale designed for specific measurements. Because of the angular movement of a pointer, the movement is classified as an *analog* device. More recently, the *digital* meter has been developed; but because of cost and other factors, the analog is still used extensively.

The adaption of the moving-coil movement to measurements of DC voltage, DC current, and resistance was discussed. The schematic symbol for each type was presented, along with the analysis for designing specific ranges in each measuring device. The scales used with each meter were discussed.

Multimeters and VOMs were discussed; these are multipurpose meters which include two or more functions, for example, volts, amperes, and ohms. By the use of switches, we can select a specific function and range for use.

Review Questions and/or Problems

1. The most common effect of current that is used for measurement is
   (a) heating effect.
   (b) electromagnetic effect.
   (c) physiological effect.
   (d) chemical effect.

2. The most common type of meter movement is the
   (a) thermocouple.
   (b) dynamometer.
   (c) moving coil.
   (d) moving iron.

3. One advantage of the d'Arsonval movement is that it uses a
   (a) nonlinear scale.
   (b) logarithmic scale.
   (c) square-law scale.
   (d) linear scale.

4. In a typical ammeter circuit, shunt resistors are connected
   (a) in series with the meter movement.
   (b) parallel to the meter movement.
   (c) in series with and parallel to the movement.

5. Which of the following is *not* an operational precaution to be observed when measuring direct current?
   (a) Observe polarity.
   (b) Make sure a DC meter is used.
   (c) Begin a measurement with the highest range.
   (d) Take all measurements with the power off.

6. In a voltmeter circuit, multipliers are connected
   (a) in series with the meter movement.
   (b) parallel with the meter movement.
   (c) in series with and parallel to the movement.

7. Greater sensitivity indicates
   (a) less circuit loading.
   (b) more circuit loading.
   (c) a narrower range of measurement.
   (d) greater meter currents.

8. Voltmeter loading produces
   (a) readings larger than normal.
   (b) readings smaller than normal.
   (c) more accurate readings.
   (d) no effect on meter readings.

9. Which of the following is the same in an ohmmeter and a voltmeter?
   (a) The type of scale used.
   (b) Circuit power is turned on.
   (c) A device for zeroing is present.
   (d) A d'Arsonval movement can be used.

10. Ohmmeter readings should be taken on
   (a) the right one-third of the scale.
   (b) the center one-third of the scale.
   (c) the left one-third of the scale.
   (d) any part of the scale.

11. Which of the following are *not* found in a conventional multimeter?
   (a) Voltmeter
   (b) Ammeter
   (c) Wattmeter
   (d) Ohmmeter

14

Alternating Current: Characteristics, Frequency, and Wavelength

Introduction

In our discussion thus far, we have been concerned primarily with direct-current technology. In fact, with the exception of magnetism, relays, and meter circuits, all the resistive circuits work equally well, whether powered by DC or AC. What is DC? What is AC? What are the differences between them? Some definitions are in order:

A circuit operating with direct current has current flow in the same direction at all times.

A circuit operating with alternating current has periodic reversals in the direction of current flow.

The definition of *alternating current* is:

Alternating current has a continually changing amplitude, and will have periodic changes in direction.

Stated otherwise, direct current is "unidirectional" and alternating current is "bidirectional." An AC power source causes current to flow in one direction for a period of time. During this period, current starts at zero, increases to the maximum, and decreases to zero. At the end of the period, current flow reverses and current flows in the opposite direction for a period of time. During this period, current again starts at zero, increases to the maximum, and decreases to zero. This series of events continues as long as the AC circuit is in operation.

Objectives

Each student is required to:

1. Define
   (a) Alternating current
   (b) Cycle
   (c) Negative alternation and positive alternation
   (d) Negative half-cycle and positive half-cycle
   (e) Period
   (f) Duration
   (g) Time
   (h) Frequency
   (i) Hertz
   (j) Kilohertz
   (k) Megahertz
   (l) Millisecond and microsecond
   (m) Sine wave
   (n) Negative peak and positive peak
   (o) Peak-to-peak voltage
   (p) Peak voltage
   (q) Effective voltage
   (r) RMS voltage
   (s) Direct current
   (t) Audio frequencies
   (u) Radio frequencies
   (v) Wavelength
   (w) Velocity of an electromagnetic wave
2. Identify the following items when given a drawing of a sine wave:
   (a) Positive peak and negative peak.
   (b) Positive amplitude and negative amplitude.
   (c) Positive alternation and negative alternation.
   (d) Positive half-cycle and negative half-cycle.
   (e) Peak-to-peak.
   (f) Cycle, period, duration.

3. Use any voltage value supplied to convert between the following:
   (a) Peak voltage.
   (b) Effective (RMS) voltage.
   (c) Peak-to-peak voltage.
4. Determine the number of current reversals that will occur for a given frequency.
5. Explain the relationship of frequency, time, and wavelength.
6. Use given values to calculate
   (a) Wavelength.
   (b) Frequency
   (c) Time.

Waveforms

To illustrate the changes that can occur in voltage or current over time, we use drawings called *waveforms*. Figure 14-1 illustrates five waveforms common in electronics. They have three characteristics:

1. The length of the horizontal axis of each wave represents a period of time.
2. The height (amplitude) of a waveform represents change in either voltage or current over the period of time represented by the horizontal axis.
3. All AC waveforms have both positive- and negative-amplitude components.

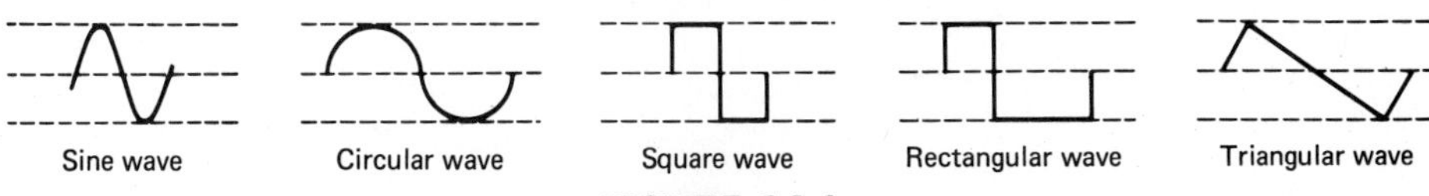

FIGURE 14-1

To be alternating current, current flowing in a circuit must have all three characteristics. For current to reverse, voltage polarity must also reverse; such reversals are noted on the waveforms as amplitude above or amplitude below a center reference. Voltage or current changes are stated with reference to the center (zero) axis. A graphical representation of this, presented in Chapter 3, is repeated here as Figure 14-2.

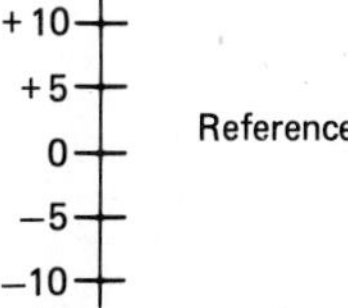

FIGURE 14-2

In the figure, note that all points above the center (zero) are labeled positive (+), and those below zero are labeled negative (−). These notations are stated with reference to the center of the diagram, which represents zero. AC waveshapes use the same form of reference.

Except for polarity, waveforms are symmetrical if their first and second halves are identical. In Figure 14-1 the circular and sine waves are symmetrical. The sine wave is the most common AC waveform; it is used as the symbol for alternating current. The alternating current used in a house has sine-wave characteristics. Figure 14-3 shows an AC generator capable of supplying a sine-wave AC output.

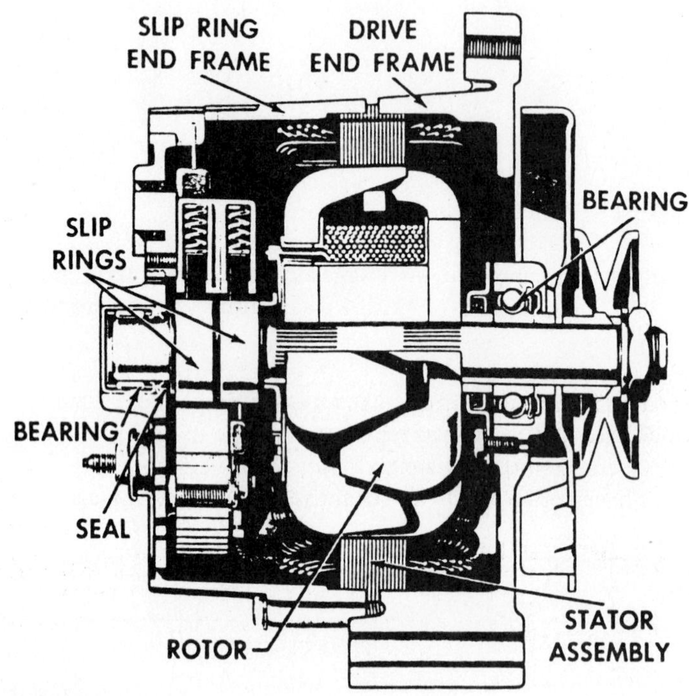

FIGURE 14-3
An AC Generator (Courtesy Delco-Remy, Division of General Motors)

The Sine Wave

The shape of a sine wave results from plotting all sine values for each point generated by rotation of the radius of a circle through 360 degrees (Figure 14-4).

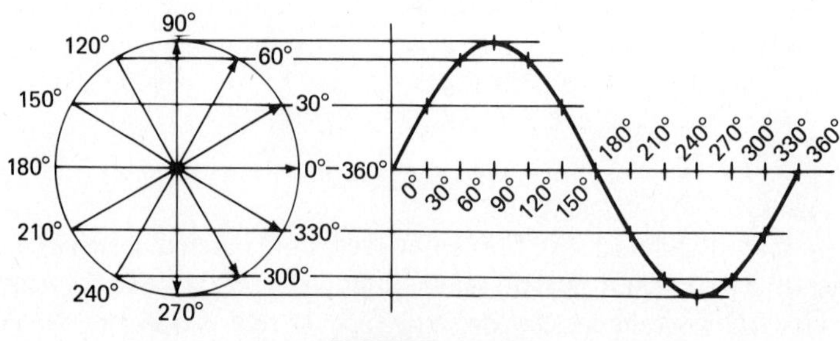

FIGURE 14-4

Plotting the sine values for each angle through which the tip of the radius passes results in the wave shape shown; therefore, it is called a *sine wave.* A sine wave can be used to represent either AC current or voltage.

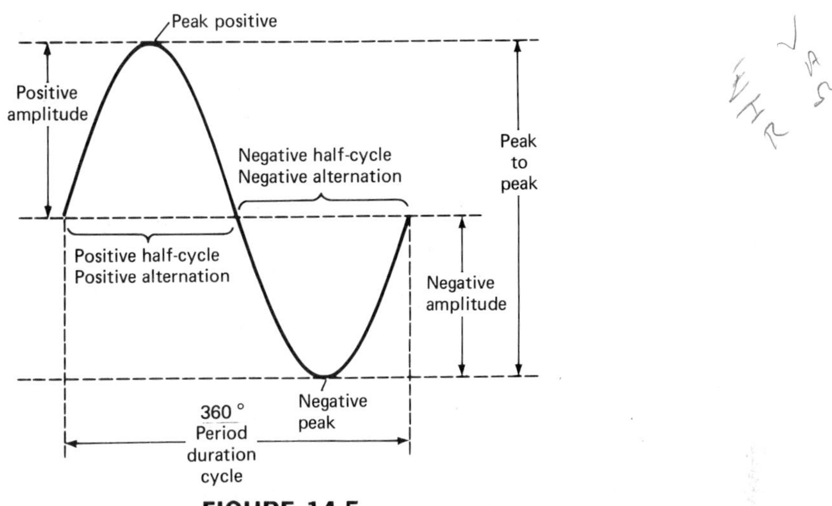

FIGURE 14-5

Figure 14-5 is a representation of a sine wave. Several important points are noted. Notice that the dashed horizontal line divides the wave in two equal parts (positive and negative). The part above the line represents positive voltage, or current; and the lower part represents negative voltage, or current. The total angular change is 360 degrees. A half-wave equals 180 degrees. The positive 180 degrees is called the *positive half-cycle,* or *positive alternation.* The negative 180 degrees is the *negative half-cycle,* or *negative alternation.* The most positive point (90 degrees) is the *peak positive* and the most negative point (270 degrees), the *peak negative.* The amount of change between zero and peak is the *amplitude. Peak-to-peak* value is the total voltage change between positive and negative peaks; this means peak-to-peak voltage is twice that of one peak.

Peak and peak-to-peak values are used for some electronic analysis, but the most valuable unit of AC voltage or current is the *effective* value. Let us analyze a sine wave in terms of voltage. Because DC was known and understood years before the advent of AC, many standards used for DC were later adopted for AC. *Effective voltage* or current is the amount of AC needed to provide the same heating effect as an equal amount of DC voltage or current.

Heating effect is directly proportional to the square of current. By using this standard, we can calculate the effective value of an AC current or voltage. Take an infinite number of voltages along the wave; square each, find the average of all squares, and extract the square root of the average. The voltage that results is the *root mean square* voltage, or RMS.

Effective voltage and RMS voltage are equal.

Unless otherwise identified, AC voltage and current values are stated in effective values. The voltage connected to a house is 120-V effective (RMS). Converting an AC voltage with 1 V peak to effective (RMS) voltage results in 0.707 V. With the ratio (0.707) as the standard, we can use the following formulas to convert *peak* voltage to *effective* voltage:

$$E_{eff} = 0.707 \times E_{\text{peak}}$$

$$E_{RMS} = 0.707 \times E_{\text{peak}}$$

At other times it may be necessary to convert *peak-to-peak voltage* to *effective voltage:*

$$E_{eff} = 0.3535 \times E_{pk\text{-}pk}$$

$$E_{RMS} = 0.3535 \times E_{pk\text{-}pk}$$

To convert effective voltages to peak and peak-to-peak values:

$$E_{\text{peak}} = 1.414 \times E_{eff}$$

$$E_{pk\text{-}pk} = 2.828 \times E_{eff}$$

These conversions are important in future studies of AC. To complete several of the lessons, you must know and be able to use these formulas.

As we have seen, alternating current reverses its direction periodically. Two reversals of current occur for each sine wave. A complete sine wave equals one *cycle;* a cycle consists of a positive alternation and a negative alternation. Alternations are often referred to as *half-cycles,* the positive alternation as the *positive* half-cycle and the negative alternation as the *negative* half-cycle. During each half-cycle, voltage starts at zero, increases to a peak, then decreases to zero. During the positive half-cycle, current flows in one direction; during the negative half-cycle, it flows in the opposite direction. During each half-cycle, however, current starts at zero, increases to a peak, and decreases to zero. The time required for current to go through a complete cycle is the *period;* the time required for a cycle is the *duration.* (An example is household current, in which there are 60 cycles per second [s].) The duration of a cycle is 1/60 s; the period of a cycle is 1/60 s; the time for completing a cycle is 1/60 s; a cycle is 360 degrees of angular movement. The terms *time, duration, cycle,* and *360 degrees* are used to describe completion of two half-cycles (alternations) of alternating current.

Sine waves that represent an AC voltage or current can be of any duration. Sixty cycles happen to be convenient. Figure 14-6 relates various AC currents to their duration and cycles. The standard for identifying AC waves is the *hertz* (Hz).

1 hertz equals 1 cycle per second.

Figure 14-6a (A) represents 1 Hz of an AC that has 1 cycle per second. When we compare Figure 14-6a (A) to Figure 14-6a (B), we see that two cycles occur during each second; in this case, we say that 2 Hz are present. The term *hertz* is used as a statement of *frequency*.

Frequency ***is the number of*** **hertz** ***(cycles) per second.***

Household current, for example, has 60 cycles per second and is referred to as 60 Hz. The frequency in Figure 14-6a (B) is 2 Hz.

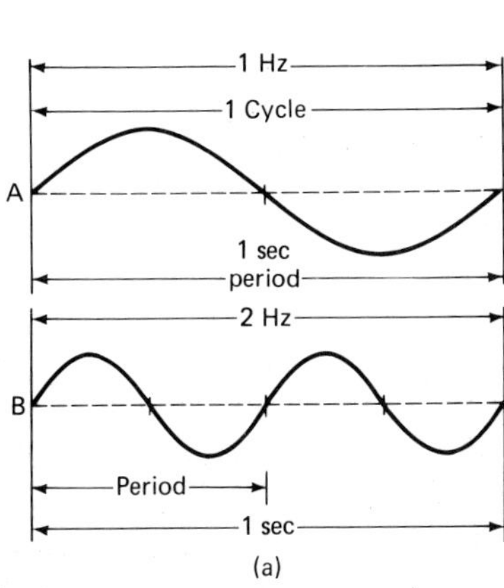

(a)

(b)

FIGURE 14-6
(b) Vibrating Reed Frequency Meter with Pro·visions for either 50-Hz or 60-Hz Operation (Photo Courtesy of J-B-T Instruments)

Notice that in Figure 14-6, period, time, and hertz are related. For 2 Hz, the period is half the period of 1 Hz. The same is true of other frequencies. For frequencies over 1000 Hz, we use the term *kilohertz* (kHz), and for those over 1 million, we use *megahertz* (MHz). For a frequency of 10 kHz, the period is 1/10,000 s. The unit 10 kHz tells us that the frequency is 10,000 cycles per second and that the current reverses 20,000 times per second.

As frequency increases, the time required for a cycle decreases and the period becomes shorter. To state the time and period components for higher frequencies, we use *millisecond* (ms) and *microsecond* (μs).

An AC sine wave represents pictorially AC or voltage. Just as the wave has two half-cycles, it also has two current reversals. For 300 Hz, current reverses 600 times per second. For all alternating current,

There are two current reversals for each cycle (Hz).

To find the number of current reversals for a 500-Hz source, we use:

$$\text{number of reversals} = 2 \times \text{frequency}$$

$$\text{number of reversals} = 2 \times 500 \text{ Hz}$$

$$\text{number of reversals} = 1000$$

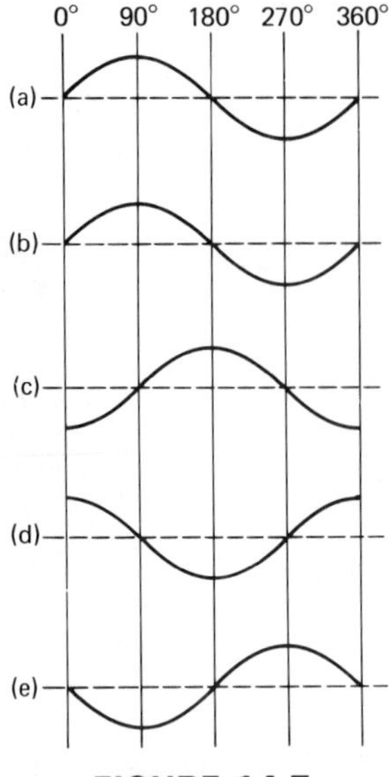

FIGURE 14-7

We have established that for one cycle and its reversals to occur, time must pass. The time relationship of sine waves is referred to as the *phase relationship.* In Figure 14-7, note that 14-7a and 14-7b are identical. Every point is identical; as one goes positive, so does the other. These two waves are said to be *in phase.* Even if their amplitudes had not been equal, they would still be in phase, because their timing is the same. In DC, as voltage increases, so does current; and they are always in phase. When AC is applied to resistive circuits, voltage and current are always in phase. With some AC circuits, however, this is not true; sometimes, voltage and current are *out of phase.* In still other circuits we can compare two voltages or currents as to their phase relationship. Compare Figure 14-7c to Figures 14-7a and 14-7b. Note that every point on sine wave c is 90 degrees behind a and b. This wave "lags" the other two by 90 degrees. Figure 14-7d always leads 14-7a and 14-7b by 90 degrees; it is said to "lead" by 90 degrees. In Figure 14-7e the wave is exactly 180 degrees out of phase with a and b. By using phase, we can compare current and voltage throughout the circuit, to understand what occurs at any given time.

Self-Check

Answer each item by indicating whether the statement is true or false.

1. ____ There is one positive peak and one negative peak for each hertz of frequency.

2. ____ The positive half-cycle must always occur first in an AC circuit.

3. ____ 10 MHz describes a frequency whose period equals 1/10,000 s.

4. ____ In a resistive AC circuit, voltage and current are always in phase.

5. ___ *Peak voltage* of an AC wave refers to the difference between peak-positive voltage and peak-negative voltage.

6. ___ One alternation of AC equals 360 degrees of rotation.

7. ___ Alternating current must have equal periods of current in each direction.

8. ___ Effective voltage equals 1.414 × E_{peak}.

9. ___ Peak-to-peak voltage equals the total voltage change that occurs between positive peak voltage and negative peak voltage.

10. ___ *Negative alternation* and *negative half-cycle* refer to the same part of a sine wave.

Types of Frequency

The range, or spectrum, of frequencies extends from DC (0 Hz) to infinity. All sound waves, radio signals, TV signals, X rays, sunlight, and so on are transmitted by a frequency or group of frequencies within this spectrum. Light frequencies fall within the visible spectrum; sound falls within the audible spectrum. The frequencies involved in electronics usually are neither visible nor audible. For our purposes, we can divide the spectrum into three divisions: DC, audio, and radio. Later, we may divide them further; but for our purposes here, the divisions are:

DC—direct current (0 Hz)
Power frequencies—50 Hz, 60 Hz, 400 Hz
AF—audio frequency (20 Hz to 20 kHz)
RF—radio frequency (20 kHz to 300 GHz: GHz)

Let us discuss each division in greater depth.

Direct Current (DC)

As we have learned, direct current has no frequency. It has zero cycles per second, or 0 Hz. Current flows in the same direction at all times. Frequencies slightly above zero, however, have more DC characteristics than they do AC characteristics. Frequencies from 0 Hz to approximately 20 Hz fall in this category and are treated much like direct current.

Power Frequencies

The transmission of electrical power by alternating current is affected by both distance and frequency. If the frequency is too low, current variations take such a long time that light filaments cool and dim (the "flicker" effect). Cooling between alternations can become a problem in areas where constant heat is required. If the frequency is too high, transmission distance is affected. Three frequencies—50 Hz, 60 Hz, and 400 Hz—are used to transmit nearly all alternating current worldwide. In Japan and Europe, commercial power systems operate on 50 Hz, a frequency so low that some people complain that they can always detect flicker. In the United States, we use 60 Hz, which has few disadvantages, as far as heating effect is concerned. For special applications, such as in airplanes and submarines, a 400-Hz system is sometimes used. Note that all the power frequencies are within the AF range.

Audio Frequencies (AF)

Starting at approximately 20 Hz and extending to 16 kHz–20 kHz is a band of frequencies which, when converted to mechanical frequency, can be heard by the human ear. Frequencies that can be heard by the human ear are called *audio frequencies*. Because the hearing range of individuals varies, the extremes of this range are arbitrarily selected as 20 Hz to 20 kHz.

Radio Frequencies (RF)

The transmission of electromagnetic waves through the air is accomplished by using frequencies above 20 kHz. Frequencies from 20 kHz to approximately 300 GHz are classified as *radio frequencies* and are used for that purpose. Within this band, smaller groups of frequencies are assigned for specific uses:

540 kHz to 1600 kHz—AM radio
88 MHz to 108 MHz—FM radio
54 MHz to 72 MHz—TV channels 2, 3, 4
76 MHz to 88 MHz—TV channels 5, 6
174 MHz to 216 MHz—TV channels 7-13
470 MHz to 890 MHz—TV channels 14-83

In our study of communications, we consider these and other frequency bands.

Self-Check

Answer each item by indicating whether the statement is true or false.

11. ____ Household power in the United States is transmitted using 50 Hz.

12. ____ The power frequencies used are 50 Hz, 60 Hz, and 400 Hz.

13. ____ Most people can hear a mechanical frequency of 20 kHz.

14. ____ Power transmission is affected by distance and frequency.

15. ____ Power frequencies are separate from audio frequencies.

16. ____ Radio frequencies are divided into use bands.

17. ____ The radio-frequency band is 20 kHz to 300 GHz.

18. ____ To be heard, audio frequencies must be mechanical vibrations.

19. ____ Frequencies below 20 Hz have much in common with direct current.

20. ____ Light is transmitted by a visible frequency.

Frequency Spectrum

Alternating current and the sine wave are representative of each other. All frequencies can be represented by a sine wave. The period, or time, of a sine wave can be represented as a characteristic of a frequency. Another characteristic of frequency and the sine wave is *wavelength,* which is a measure of the physical length of a wave.

When we speak of a wave, we visualize movement—for example, a wave moving across water. When a wave travels from one point to another, time elapses. In an electromagnetic wave, we consider the distance that energy travels during the period of one sine wave of a particular frequency. Because each frequency has a different period, that frequency also has a different wavelength. Electrical energy has a velocity of 300 million meters per second (186,000 miles per second). If the duration of a wave is 1 s, its initial energy will have traveled 300 million meters during that period.

Two factors are involved when considering wavelength: velocity and frequency. Velocity is stated as 3×10^8 meters per second. Frequency depends on the band and method of operation. In electronics,

Wavelength is the distance an electromagnetic wave can travel in the time of 1 cycle.

Wavelength is measured in meters; period is measured in seconds.

The Greek letter lambda λ is used to represent wavelength. Wavelength is equal to velocity multiplied by the time of one wave. This can be stated thus:

$$\lambda = vt$$

The value of velocity can be substituted in the formula:

$$\lambda = 300 \times 10^6 \times t \quad \text{or} \quad \lambda = 3 \times 10^8 \times t$$

where

λ = wavelength in meters
t = time in seconds

This formula can be used to solve for the distance electromagnetic energy travels during the time of one wave.

Time and frequency are inversely proportional. As frequency increases, the period required for one wave decreases. This can be stated:

$$f = \frac{1}{t} \qquad \begin{array}{l} f = \text{frequency (in hertz)} \\ t = \text{time in seconds} \end{array}$$

Transposing the equation,

$$t = \frac{1}{f}$$

If time is equal to 1 divided by frequency, we can substitute $1/f$ for time in the wavelength formula. After substitution, the wavelength formula becomes:

$$\lambda = 3 \times 10^8 \times \frac{1}{f}$$

This reduces to

$$\lambda = \frac{3 \times 10^8}{f}$$

Now we use the formula to solve a problem. Suppose we want to know the wavelength of a 60-Hz sine wave. We use the formula:

$$\lambda = \frac{3 \times 10^8}{60 \text{ Hz}}$$

$$\lambda = 5 \times 10^6 \text{ m, or } 5{,}000{,}000 \text{ m}$$

At a frequency of 60 Hz, an electron moves 5 million meters (3125 miles) during the period of one sine wave.

Transposing the formula again, we have:

$$f = \frac{3 \times 10^8}{\lambda}$$

To determine the frequency of an AC wave that has a wavelength of 5 m, we use the formula:

$$f = \frac{3 \times 10^8}{\lambda}$$

$$f = \frac{3 \times 10^8}{5 \text{ m}}$$

$$f = 6 \times 10^6 \text{ Hz, or } 60 \text{ MHz}$$

If the wavelength is 0.5 m, the result is:

$$f = \frac{3 \times 10^8}{0.5 \text{ m}}$$

$$f = 600 \text{ MHz}$$

As Figure 14-8 shows, there is a definite relationship between wavelength, frequency, and time. If two of these characteristics are known, the third can be found. Figure 14-8 represents a frequency of 5 Hz. The period for one cycle is one-fifth second. The relationship of wavelength to distance traveled in 1 s is also one-fifth second.

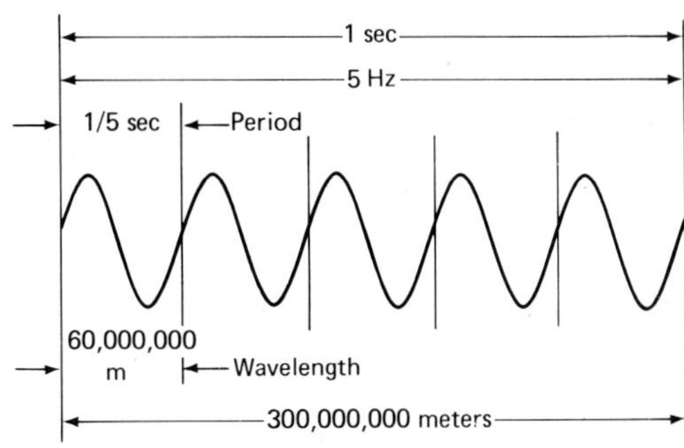

FIGURE 14-8

Let's examine the relationship between frequency and time more closely. This relationship can be stated by:

$$t = \frac{1}{f}$$

Transposing this formula results in:

$$f = \frac{1}{t}$$

It is evident that frequency and time are reciprocals; they are inversely proportional to each other. If we know that a frequency is 20 MHz, and we wish to determine the time of a cycle, we use:

$$t = \frac{1}{f}$$

$$t = \frac{1}{20 \text{ MHz}}$$

$$t = 0.00000005 \text{ s}$$

This is an awkward way to state time. To simplify the expression, it is stated either as 0.05 μs or as 50 nanoseconds (ns). To avoid having to convert units to prefixes, these values can be solved directly:

$$f \text{ (in MHz)} = \frac{1}{t} \text{ (in } \mu\text{s)}$$

$$f \text{ (in kHz)} = \frac{1}{t} \text{ (in ms)}$$

$$t \text{ (in } \mu\text{s)} = \frac{1}{f} \text{ (in MHz)}$$

$$t \text{ (in ms)} = \frac{1}{f} \text{ (in kHz)}$$

Self-Check

Answer each item by indicating whether the statement is true or false.

21. ___ The velocity of an electromagnetic wave is 3×10^8 meters per second.

22. ___ Time and frequency are reciprocals of each other.

23. ___ Time and period are inversely proportional.

24. ___ Wavelength is the distance energy travels during the period of one wave.

25. ___ In discussing high frequencies, time is often stated in prefix values.

Summary

A comparison of DC and AC reveals that AC has a continually changing amplitude and periodic reversal of current flow. In the case of DC, current flows in the same direction at all times. Thus we say that DC is unidirectional and that AC is bidirectional.

The transmission of household current and its characteristics were discussed. The frequency of transmission of this current is 60 Hz. The AC values of peak, peak-to-peak, and effective voltages were explored; conversion between different values was explained and formulas were provided. The heating effects of AC and DC were compared.

The waveshape used to represent AC is the sine wave. The horizontal axis of a wave represents the time required for one cycle of the wave. The duration of the wave, and its period, were found to be inversely proportional to the frequency of the wave. Frequency was defined as the number of cycles per second, and the cycle as one complete rotation of a radius through 360 degrees.

Frequencies were identified by band. The audio frequency range is 20 Hz to 20 kHz. Frequencies with characteristics similar to those of DC were identified as those below 20 Hz. Radio frequencies were identified as those in the range above 20 kHz and below 300 GHz. Three Power frequencies received special attention because of their use in the commercial transmission of heavy electrical current.

The wavelength of an AC wave was defined as the distance energy travels during one cycle. Further discussion revealed the relationship of sine wave, frequency, period, time, duration, and wavelength.

Review Questions and/or Problems

1. The two main characteristics of alternating current are:
   (a) periodic changes in amplitude and continuous current reversal.
   (b) a constant amplitude and periodic reversal of current.
   (c) constantly changing amplitude and periodic change of direction.
   (d) periodic flow and continually changing amplitude.

2. The difference between the maximum positive voltage and the maximum negative voltage of a sine wave is the
   (a) effective value.
   (b) peak-to-peak value.
   (c) average value.
   (d) peak value.

3. Which of the following has the same heating capacity as an equal amount of DC?
   (a) effective
   (b) peak
   (c) average
   (d) peak-to-peak

4. Another term used to identify E_{eff} is
   (a) peak.
   (b) peak-to-peak.
   (c) average.
   (d) RMS.

5. What term is used to indicate the number of alternating current cycles per second?
   (a) frequency
   (b) amplitude
   (c) RMS value
   (d) reversal

6. Current becomes negative at the same time voltage becomes positive. Therefore, current
   (a) lags voltage by 90 degrees.
   (b) leads voltage by 90 degrees.
   (c) is 180 degrees out of phase with voltage.
   (d) is in phase with voltage.

7. Two broad-frequency bands make up the frequency spectrum; they are
   (a) DC and AC.
   (b) audio and radio.
   (c) power and RF.
   (d) radio and television.

8. The frequency that separates AF from RF is
   (a) 20 Hz.
   (b) 20 kHz.
   (c) 20 MHz.
   (d) 20 mHz.

9. All power frequencies fall within the AF band.
   (a) true
   (b) false

10. Calculate the wavelength for a 6-MHz sine wave.
    (a) 5 m
    (b) 50 m
    (c) 500 m
    (d) 5000 m

11. Calculate the frequency of an 0.05-m wavelength.
    (a) 6 kHz
    (b) 6 MHz
    (c) 6 GHz
    (d) 600 Hz

12. Solve for the time of a 2500-Hz wave.
    (a) 400 ms
    (b) 400 s
    (c) 400 μs
    (d) 400 ns

13. Calculate the frequency of a wave whose period is 10 μs.
    (a) 1 kHz
    (b) 10 kHz
    (c) 100 kHz
    (d) 100 MHz

14. Wavelength is normally measured in
    (a) distance.
    (b) velocity and time.
    (c) meters.
    (d) miles.

15. The wavelength of a wave is determined by
    (a) velocity and time.
    (b) velocity per second.
    (c) frequency and time.
    (d) period and duration.

16. Which formula is used to calculate wavelength?
    (a) $\lambda = \frac{V}{t}$
    (b) $\lambda = V \times f$
    (c) $\lambda = \frac{f}{V}$
    (d) $\lambda = \frac{V}{f}$

17. Convert 170-V peak to effective voltage.
    (a) 120 V
    (b) 60 V
    (c) 240 V
    (d) 480 V

18. Convert 340-V peak-to-peak to effective voltage.
    (a) 120 V
    (b) 60 V
    (c) 240 V
    (d) 480 V

19. Convert 100-V effective to peak-to-peak voltage.
    (a) 141.4 V
    (b) 282.8 V
    (c) 70.7 V
    (d) 35.35 V

20. Convert 200-V effective to peak voltage.
    (a) 141.4 V
    (b) 282.8 V
    (c) 70.7 V
    (d) 35.35 V

15

The Oscilloscope

Introduction and Applications

In Chapter 14 we discussed the sine wave and its waveshape. It is convenient to be able to view the waveshape and measure its operating frequency. Often, it is necessary to view the peaked wave used to time the operation of a complex radar or a computer—procedures that are possible with an oscilloscope such as that shown in Figure 15-1.

The oscilloscope can perform several functions. It can measure the peak-to-peak value of voltage. It can be used to determine the frequency of operation for circuits, or it can plot the voltage characteristics of a signal against time to display a sine wave or some other waveshape. Many oscilloscopes can display simultaneously two, three, or several waveshapes. This allows comparison of the waves to ensure that they are identical. The ability to view and examine waveshapes cannot be stressed too much. The proper use of an oscilloscope is an important part of your work as an electronics technician.

There are many different manufacturers of oscilloscopes, or "scopes," as they are often called. Some scopes are basic, while others are complex; but all have certain characteristics in common. It is these characteristics that we concentrate on in this chapter.

Objectives

Each student is required to:

1. When provided with an oscilloscope, test leads, and a signal source,
   (a) Set signal generators to predetermined frequencies.
   (b) Properly connect leads to the scope and signal sources.
   (c) Set the controls on the oscilloscope to obtain a signal suitable for analysis.
   (d) Confirm the frequency of the input signal.
2. When given a sine-wave input signal, use an oscilloscope to
   (a) Measure peak-to-peak voltage.
   (b) Determine the duration of a sine wave.
   (c) Use duration to calculate a sine-wave frequency.
   (d) Determine sweep time.
   (e) Convert sweep time to sweep frequency.

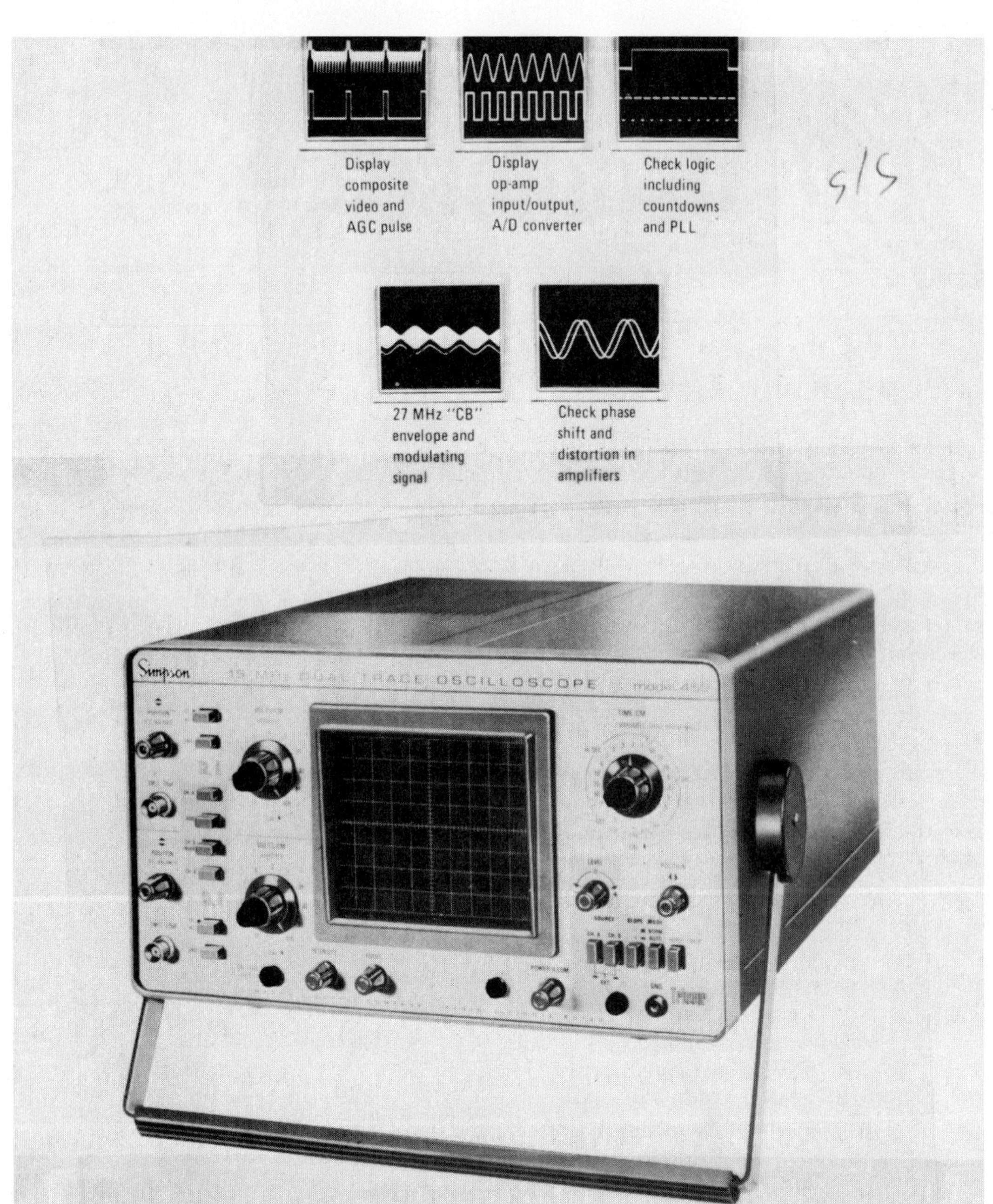

FIGURE 15-1

Modern Oscilloscope with Representative Waveshapes (Photo Courtesy of Simpson Electric Company)

3. Use a triggered dual-trace oscilloscope to measure the difference between phase angles of two sine waves.
4. Use data taken from the scope to convert time to frequency and frequency to time.

The Basic Oscilloscope

All oscilloscopes have a cathode ray tube (CRT), which presents a wave shape for observation. The CRT is similar to the picture tube of a TV set. The main differences are:

1. The scope's CRT has a phosphorus coating which retains the image portrayed on it for longer periods than does the CRT used in a TV set.
2. The image displayed on the scope CRT results from the electrostatic action of *horizontal* and *vertical deflection plates.*
3. Internally, the CRT produces a beam of electrons that are fired at the face of the CRT at high speed. As the electrons move through the tube they are shaped into a tiny stream; as they strike the face of the scope, they form a small dot. The electron beam is the key to operation of the CRT. In the discussion that follows, we are concerned with this beam and the way it is used to display the signals desired.

To portray signals of different frequency and amplitude, the scope has an array of controls. The more sophisticated the scope, the more controls it has for selecting and presenting the waveshape. Figure 15-2 is an oscilloscope of the type used in many electronic shops. The controls used may differ in location and type, but their functions are similar to those discussed.

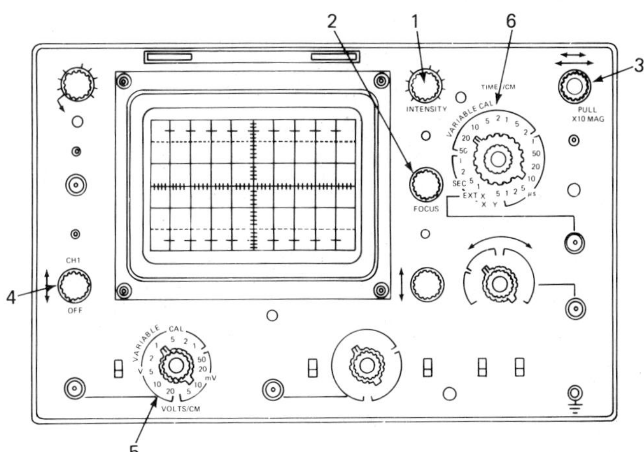

FIGURE 15-2

Controls common to all oscilloscopes are:

1. *intensity* Changes the brightness of the signal displayed. Varying the control can cause the signal to become dim or bright. It should be set so that the trace is a thin, well-defined line of medium brightness. Each operator may view this trace differently.

2. *focus* Works in conjunction with the intensity control. Varying the control helps us obtain a well-defined trace for presenting a waveshape. The focus control is used to adjust the sharpness of the dot. Some scopes also have an *astigmatism* control, which ensures that all parts of the sweep are in focus at the same time.
3. *horizontal position* (↔) Used to move the trace and waveshape horizontally (left or right) across the face of the CRT. It allows positioning a waveshape so a specific portion can be examined closely.
4. *vertical position* (↕) Used to vary the trace and waveshape vertically (up or down) on the face of the CRT and to establish a reference point from which amplitude can be read correctly.
5. *volts/cm* Adjusts the amplitude of the signal applied to the vertical-deflection plates. Variation of this control inserts resistance (attenuation) in the path of the input signal at the vertical input. The volts/cm control has a variable (*variable calibration*) control immediately behind it, which is operated in conjunction with the volts/cm switch. With the CAL control in the calibrate position, selection of a position on the volts/cm control sets the vertical deflection on the CRT to the setting selected. For example, selection of 1 V/cm on the volts/cm switch sets the graticule (the grid pattern shown on the face of the CRT) to measure 1 V/cm of height. On scopes with more than one sweep, there will be a separate volts/cm control for each sweep.
6. *time/cm* Used with a calibration control to select a calibrated time per centimeter along the horizontal graticule. Setting the control to 1 ms/cm sets each centimeter of length to a calibrated time of 1 ms. This allows the horizontal sweep to be used for accurate measurement of time.

On most scopes, the lines on a graticule are separated by 1 cm, both horizontally and vertically. All time and voltage settings made using the volts/cm and time/cm controls are calibrated to supply accurate voltage and time readings when the calibrated control for each is set to the calibrated position. Each line of the graticule is not only divided into centimeters, but each centimeter is further divided to increase the accuracy of voltage and time readings. A graticule usually consists of a pattern 10 cm wide and 8 cm high.

A scope is equipped with test probes for connecting the circuit to be tested to the scope's *vertical-input* jack. Some probes have *direct* and 10X capabilities. In the direct position, a signal is coupled to a scope without being attenuated, or reduced. When the 10X attenuator is used, the amplitude of the input signal is reduced by 10; in other words, if a sine wave is 100 V peak-to-peak, connecting it to the scope with a 10X probe reduces its amplitude to 1/10, or 10 V peak-to-peak.

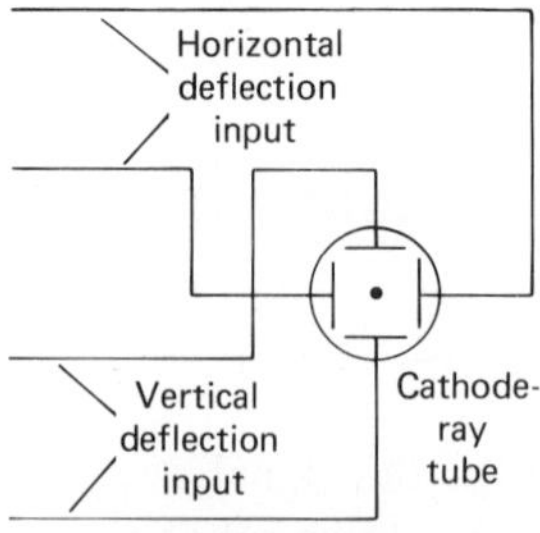

FIGURE 15-3

Deflection plates have already been mentioned. Figure 15-3 is a diagram of the plates' relative positions. Notice that two plates are labeled *horizontal* and that two others are labeled *vertical*; they are placed in the same position relative to each other that they would be in if you were looking at them through the face of the CRT.

We consider the vertical plates first. With the electron midway between them, the bottom plate grounded, and a positive voltage on the top plate, the electron is attracted upward; with a negative voltage on the top plate, the electron is repelled downward. The action on the vertical plates is a result of the signal being applied to the vertical input.

Now we consider the effect of horizontal deflection plates. Remember that an electron has a negative charge. An electron placed midway between the two plates (Figure 15-4) can be made to move right or left by electrostatic attraction or repulsion. Placing ground on a plate and the positive voltage on the other plate causes the electron to be attracted to the positive voltage. If the voltage is negative, the electron is repelled toward ground.

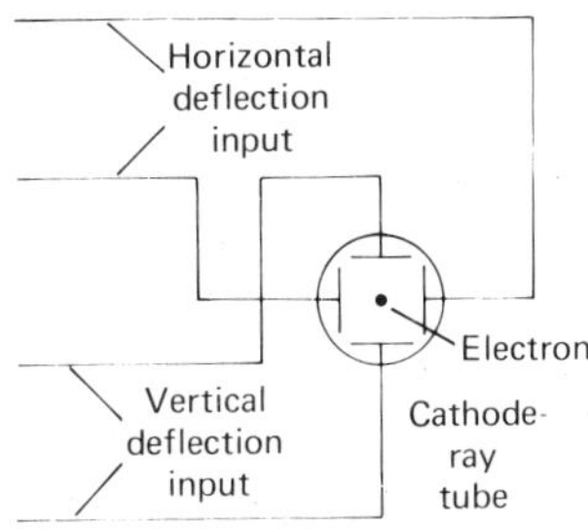

FIGURE 15-4

To analyze waveshapes, we cause the dot formed by the electron beam to move across the face of the CRT, thus forming a continuous line. For this purpose, a scope has a *sweep circuit* that causes the spot to sweep across the screen. Actually, the dot is first positioned on the left of the CRT, using the *horizontal position control*. Deflection plates are then used, along with sweep circuits, to sweep the spot. At higher frequencies the movement is fast enough that the line appears to be continuous. The phosphorus coating on the CRT screen glows long enough for the sweep to appear as a line. The ability of a CRT to glow after the electron beam has moved on is known as *persistence*. It is the horizontal sweep that is calibrated in time/cm. The horizontal sweep is also calibrated with the input signal, which causes the input signal to appear stationary on the screen.

Most scopes have other controls as well, which are explained in your training program or your operator's manual.

Self-Check

Answer each item by indicating whether the statement is true or false.

1. ___ The waveshape to be viewed is connected to the vertical-input jack.

2. ___ The focus control can be used to vary the brightness of the sweep.

3. ___ A 10X probe is used to divide the input signal by 10.

4. ___ Horizontal- and vertical-position controls change the reference voltage applied to the deflection plates, which, in turn, change the sweep position.

5. ___ Calibrated time readings result from setting the volts/cm control.

Scope Measurements and Calculations

Measuring Time

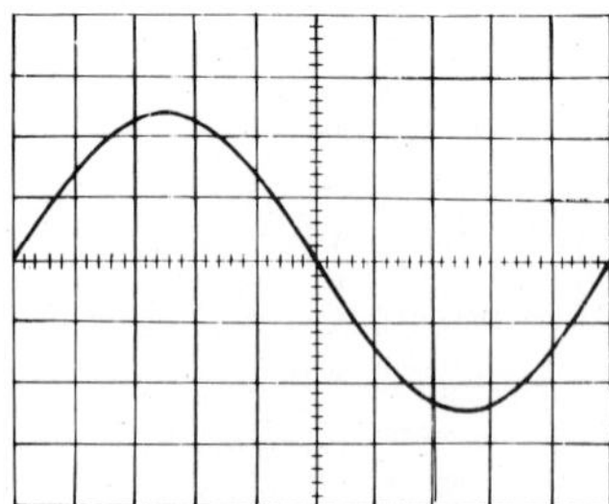

FIGURE 15-5

Time is measured by using the *horizontal scale* of the graticule. Figure 15-5 shows a graticule 10 cm long, with one cycle of a sine wave displayed. If you wish to measure the time of one alternation, you

1. Count the number of centimeters along the horizontal scale used to display one alternation.
2. In this case, one alternation uses 5 cm.
3. Check the setting of the time/cm control. Assume that it is set to 2 ms. *Note:* Make sure the calibration control is set to the CAL position.
4. Multiply the centimeters from step 2 by the 2 ms in step 3 to determine time for one alternation.

   $5 \text{ cm} \times 2 \text{ ms} = 10 \text{ ms}$

5. Convert the results to the time of 1 cycle by counting the number of centimeters and multiplying by 2.

   $10 \text{ cm} \times 2 \text{ ms} = 20 \text{ ms}$

   This could also have been found by multiplying the time of one alternation by 2, or by counting the centimeters for 1 sine wave.
6. *Note:* Time is calculated by multiplying the setting of the time/cm control by the number of centimeters of a sine wave. Here, the time/cm control sets the time it takes the electron beam to move 1 cm.

Calculating Frequency

The frequency of the sine wave under discussion can be calculated thus:

$$f = \frac{1}{t}$$

where f equals frequency (in hertz) and t equals time (in seconds). The mathematical solution is:

$$f = \frac{1}{20 \text{ ms}}$$

$$f = 50 \text{ Hz}$$

Action	Entry	Display
ENTER	20, EE, +/–,3	20 – 03
PRESS	1/X	5 01, or 50 Hz

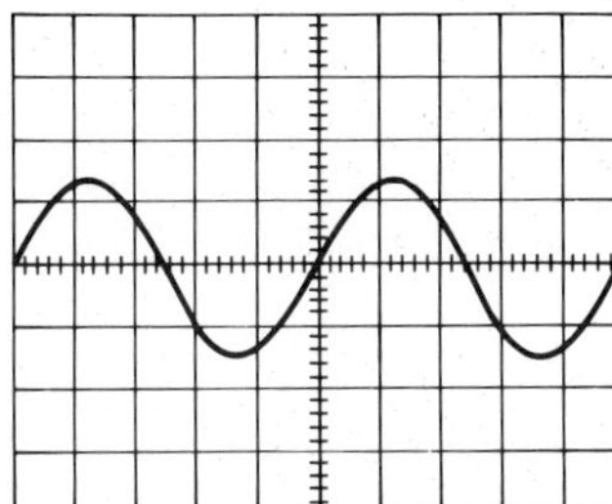

FIGURE 15-6

To obtain the most accurate measurement of time, we set the time/cm control so that one sine wave is as near 10 cm as possible. Make sure the variable time/cm control is in the CAL position; if it isn't, the reading will be wrong. (See Figure 15-6.) Now let us measure the time of one cycle and the frequency of the sine wave when the time/cm control is set to 5 μs/cm.

1. The sine wave is 5 cm long.
2. The time of 1 sine wave is 5 cm $\times$ 5 μs = 25 μs.
3. Frequency = 1/25 μs = 40 kHz.

Calculating Sweep Frequency

Figure 15-7 presents a graticule with three sine waves displayed. We have just determined the time of one sine wave by counting the number of centimeters and multiplying that number by the setting of the time/cm control. This is an important step in determining the sweep frequency of the oscilloscope. First, we compare the signal frequency and the *sweep frequency* displayed.

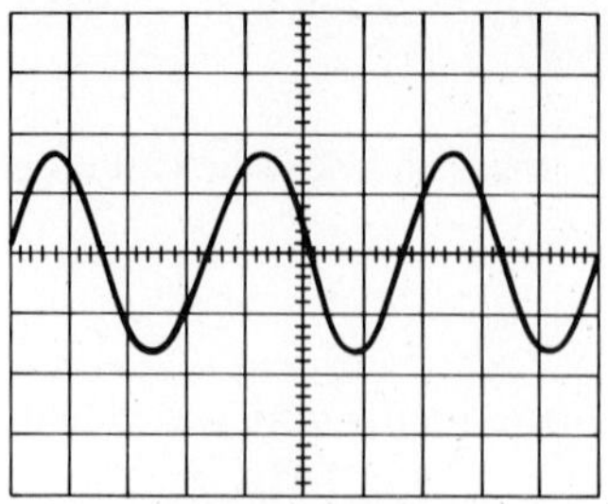

FIGURE 15-7

Picture the face of a CRT. Time is required for the beam to move from the left side to the right side. Sweep time is established by the calibrated setting of a time control. Assume that the time/cm control is set to 25 ms/cm. For the beam to sweep across the face of the scope, 10 cm × 25 ms = 250 ms is required. Notice the similarity between this number and the time of the sine wave just calculated; the process of converting the time to a frequency is identical.

$$\text{frequency} = \frac{1}{\text{sweep time}} = \frac{1}{250 \text{ ms}} = 4 \text{ Hz}$$

With the time required for the beam to sweep across the screen at 250 ms, the sweep frequency is 4 Hz.

Now we use Figure 15-7 to determine both the frequency of the sine wave and the sweep frequency of the scope. We assume that the setting on the time/cm control is not calibrated, making the measurement of time undependable; thus we need to determine sweep frequency. We do, however, have a calibrated signal generator, which is used to supply the 1-kHz input shown on the scope.

1. Convert the signal frequency to time using $t = 1/f = 1$ ms.
2. Count the number of sine waves displayed: 3.
3. Find sweep time as follows: Sweep time equals

$$\text{time for 1 sine wave (1 ms)} \times \text{number of sinewaves (3)} = 3 \text{ ms}$$

Note: With three sine waves displayed, sweep time must be three times as long as the time for one sine wave.

4. Convert sweep time to frequency using $f = 1/t = 333.3$ Hz.
5. Anytime that more than one sine wave is displayed, sweep frequency is less than signal frequency.

Following is a simple method to use in calculating sweep frequency when signal frequency is known:

1. Count the number of waveshapes on the face of the scope.
2. Divide the signal frequency by the number of sine waves present.

Measuring Amplitude

When a scope is used to measure voltages, remember that it measures *peak-to-peak* value. To convert this reading to *effective* voltage, use the formula:

$$E_{eff} = 0.3535 \times E_{pk\text{-}pk}$$

Example: An indication on the scope of 200 V peak to peak can be converted to RMS or *effective* as follows:

$$E_{eff} = 0.3535 \times 200 \text{ V}$$

$$E_{eff} = 70.7 \text{ V}$$

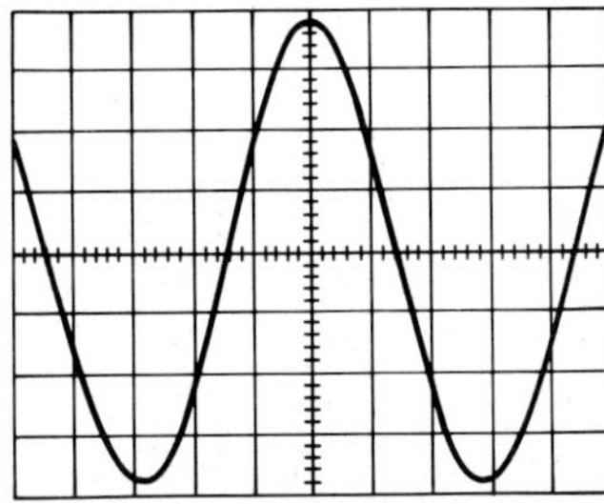

FIGURE 15-8

The graticule is used to measure voltage amplitude. Examine Figure 15-8, where a signal is displayed. To obtain this type of display, adjust the time/cm and volts/cm controls so that the signal has at least one negative peak and one positive peak displayed. Amplitude should be set to the highest volts/cm without causing the peaks to disappear.

To make calculation of amplitude easier, adjust the vertical position so the negative peak of the sine wave is along one line near the bottom of the graticule. Adjust the horizontal position until the positive peak is along the calibrated centerline of the graticule. With the signal thus positioned, proceed as follows:

1. Count the number of centimeters of height occupied by the signal. In this example, amplitude = 7.6 cm.
2. Multiply the number of centimeters by volts/cm. Here, we assume that the control has been calibrated and set to 10 V/cm. Therefore,

$$E_{pk\text{-}pk} = 7.6 \text{ cm} \times 10 \text{ V/cm} = 76 \text{ V}$$

3. Convert 76 V peak to peak to E_{eff}:

$$E_{eff} = 0.3535 \times E_{pk\text{-}pk}$$

$$E_{eff} = 26.866 \text{ V}$$

4. Remember the highest setting on the volts/cm control. With the number of centimeters (vertical) limited to 8, we are limited to measuring the voltage less than the number of centimeters times volts/cm. If the setting is 20, we are limited to 160 V peak to peak. Most modern oscilloscopes, however, come with an attenuator probe that divides a signal by 10 when it is in use. If a 10X probe is used, the maximum voltage measured is extended to 1600 V. For more accurate readings at low voltages, we can switch the attenuator on or off as needed.

Measuring DC Voltage

Different procedures are required in measuring DC voltage. When a DC signal is applied to the scope, a straight line is displayed. Unless precautions are taken, it is difficult to tell whether we are looking at the normal sweep or a DC voltage sweep. Figure 15-9 is used to explain measurement of DC.

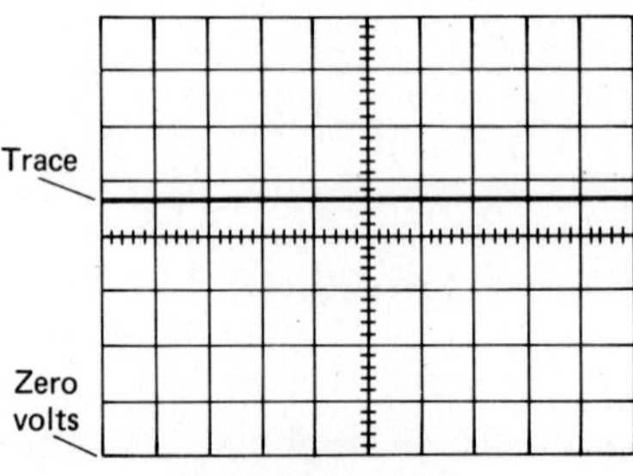

FIGURE 15-9

1. Before connecting the input signal, position the trace to its zero-volts DC point. Follow this procedure:
   (a) Some scopes have a DC-AC-GND switch. If the scope you are using has such a switch, and you wish to measure positive DC voltages, place the switch at GND and adjust the sweep to coincide with the bottom line on the graticule. The bottom line is now set to zero volts. Now move the switch from GND to DC.
   (b) On scopes that lack the switch, physically ground the input probe to the scope chassis and adjust the sweep to the bottom (zero-volts) line of the graticule.
   (c) To measure negative DC voltages, the zero reference should be adjusted to the top line of the graticule.
2. Connect the DC source to the vertical input. Observe that the sweep deflects upward for positive DC and downward for negative DC.
3. Adjust the volts/cm control for maximum deflection without having the sweep disappear. Assume that this control is set to 10 V/cm.
4. Count the centimeters of deflection. In Figure 15-9, the deflection is 4.6 cm.
5. Multiply the volts/cm control setting by the number of centimeters of deflection:

$$DCV = 10\ V/cm \times 4.6\ cm = 46\ V$$

6. Again, the maximum voltage that can be measured is limited by the 8 cm of the graticule. Without external attenuation, the scope is limited to 160 V:

$$8 \text{ cm} \times 20 \text{ V/cm} = 160 \text{ V}$$

Using the 10X probe, maximum voltage measured can be 1600 V.

Phase Measurements

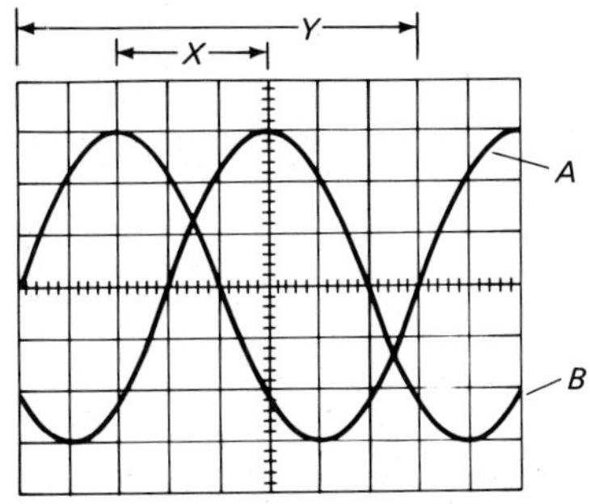

FIGURE 15-10

Scopes with *dual-trace* capability allow an extra measurement—*phase*. It is often useful to know the phase difference between two sine waves. Figure 15-10 presents a graticule with two sine waves displayed. Each sine wave is displayed by a separate sweep. To measure the phase difference:

1. Adjust the horizontal position control so both sweeps begin on the left vertical line of the graticule.
2. Adjust the vertical position control so that sweep A is superimposed on sweep B.
3. Adjust the volts/cm controls so the sine waves have the same amplitude.
4. Adjust the time/cm control so sine wave A equals 8 cm along the horizontal axis (Figure 15-10).
5. Compare the two sine waves.
   (a) Count the number of centimeters required for a complete sine wave (360 degrees):

   $$\text{Distance } Y = 8 \text{ cm}$$

   (b) Divide:

   $$\frac{360 \text{ deg}}{8 \text{ cm}} = 45 \text{ deg}$$

   (c) Count the number of centimeters separating the positive peaks of the two sine waves:

   $$\text{Distance } X = 3 \text{ cm}$$

   (d) Multiply:

   $$45 \text{ deg} \times 3 \text{ cm} = 135 \text{ deg}$$

Here, sine wave A *leads* sine wave B by 135 degrees.

In Figure 15-11, we solve another example. Repeat steps 1 through 4 in the list just given and compare the two sine waves.

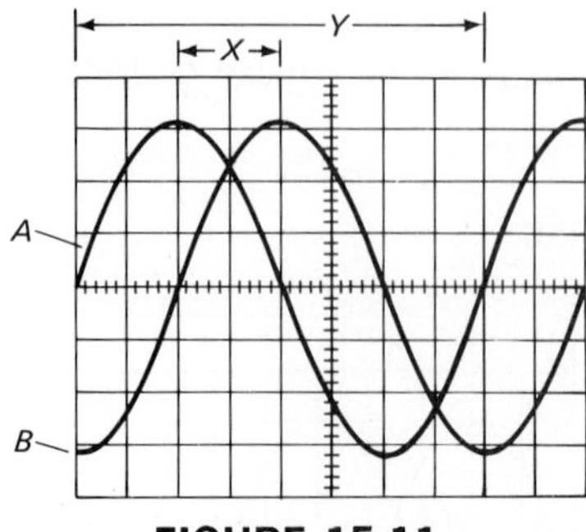

FIGURE 15-11

1. Count the number of centimeters for one sine wave (360 degrees):

   number of cm = 8

2. Divide:

   $$\frac{360 \text{ deg}}{8 \text{ cm}} = 45 \text{ deg}$$

3. Count the number of centimeters separating positive peaks.

   Distance X = 2 cm

4. In step 2 we determined that 1 cm = 45 deg. When we multiply,

   2 cm × 45 deg = 90 deg

In this example, sine wave A leads sine wave B by 90 degrees. We could also say that B *lags* A by 90 degrees.

Self-Check

Answer each item by indicating whether the statement is true or false.

6. ___ DC voltage is presented on the scope as a sine wave.
7. ___ Sweep frequency is always larger than signal frequency.
8. ___ RMS values of AC voltage are presented on the scope.
9. ___ It is possible to calculate phase difference using a dual-trace scope.
10. ___ Signal frequency can be calculated by taking the reciprocal of the time for 1 cycle.

11. ____ Negative DC causes the sweep to deflect upward.

12. ____ The time/cm control can be used to adjust vertical amplitude.

13. ____ To measure the time of a sine wave accurately, we calibrate the volts/cm control.

14. ____ Phase relation is a time relationship.

15. ____ Before measuring AC voltage, zero volts must be located on the graticule.

Summary

The basic oscilloscope was introduced and its operation was explained. The positioning of input signals and the controls required to change amplitude and time calibration were explained.

We also discussed the procedures used with an oscilloscope in determining different waveshape characteristics. Specific procedures for determining the following were covered; they are:

1. Measuring the time of a sine wave.
2. Determining the frequency of an input signal.
3. Determining the sweep frequency of a scope.
4. Measure AC voltage amplitude.
5. Measure DC voltage.
6. Approximate the phase difference between signals.

Review Questions and/or Problems

1. The correct procedure for determining effective voltage when taking measurements with an oscilloscope is
   (a) peak-to-peak multiplied by 0.707.
   (b) peak multiplied by 0.636.
   (c) peak multiplied by 1.414.
   (d) peak-to-peak multiplied by 0.3535.

2. The correct procedure for calculating frequency from a scope measurement is
   (a) sine-wave length, in centimeters, multiplied by the setting of the time/cm control.
   (b) time for 1 sine wave divided by 1.
   (c) 1 divided by the time for 1 sine wave.
   (d) 1 multiplied by the time for 1 sine wave.

3. The intensity control should be set so the
   (a) trace on the CRT is very bright.
   (b) trace is no brighter than necessary.
   (c) brightness suits the operator's desires.
   (d) intensity control is placed in its center position.

4. In Figure 15-12, the time/cm control is set to 10 ms/cm and the volts/cm control to 10 V/cm. What is the effective voltage of the input?
   (a) 21.2 V
   (b) 42.42 V
   (c) 10.6 V
   (d) 82.82 V

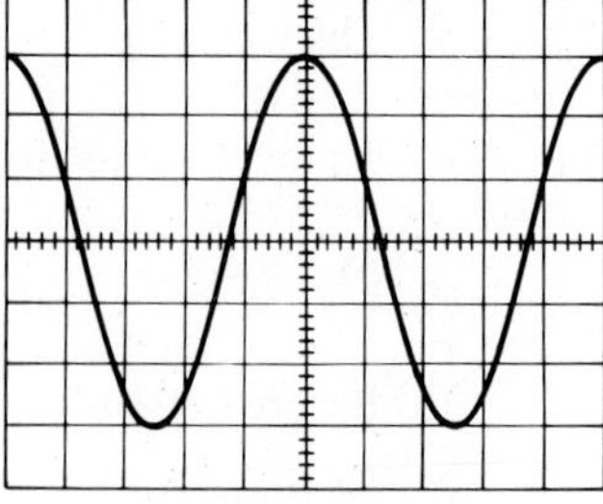

FIGURE 15-12

5. Use Figure 15-12. Time/cm is set to 10 ms/cm and volts/cm is set to 5 V/cm. What is the frequency of the sine wave?
   (a) 20 Hz
   (b) 10 Hz
   (c) 200 Hz
   (d) 5 Hz

6. In Figure 15-12, the time/cm control is set to 10 ms/cm. What is the scope's sweep frequency?
   (a) 2 Hz
   (b) 10 Hz
   (c) 200 Hz
   (d) 5 Hz

7. Check Figure 15-13. This scope is set to measure –DC voltage. How much voltage is applied to the vertical input if the time/cm is set to 10 ms/cm and volts/cm to 25 V/cm?
   (a) 125 V
   (b) 50 V
   (c) 75 V
   (d) 30 V

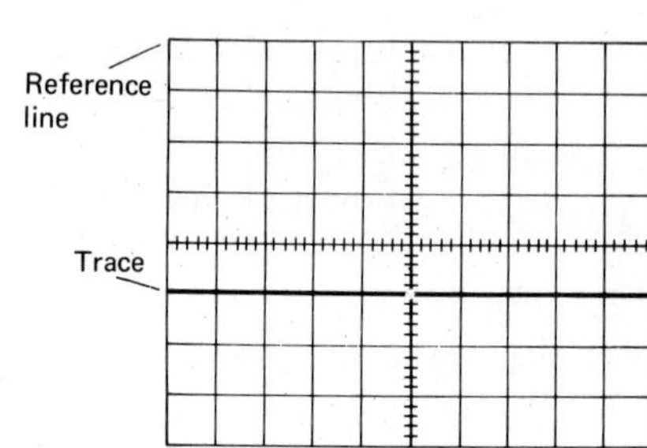

FIGURE 15-13

8. The variable time/cm control increases or decreases the number of sine waves that appear on the scope.
   (a) true
   (b) false

9. The focus control can be used to adjust the sharpness of the trace on an oscilloscope.
   (a) true
   (b) false

10. The vertical position control increases or decreases the amplitude of the signal displayed on the screen.
    (a) true
    (b) false

16

Inductors, Inductance, and Inductive Reactance

Introduction

Electricity and magnetism are fundamental to electronics. As we saw in Chapter 12, Oersted discovered the relationship of electricity and magnetism. It then became possible to develop large power-distribution systems, communications networks, and many other systems common today.

Oersted discovered that a current-carrying conductor is surrounded by a magnetic field, by demonstrating that a compass near the current-carrying conductor is affected by the magnetic field. Stopping the current causes the magnetic field to disappear. When the current is started again, the field reappears. When the direction of current flow is reversed, the magnetic field's polarity is reversed.

In this chapter we look at some of the ways magnetism and current affect electronics circuitry. The first way to do this is by *inductance.*

Objectives

Each student is required to:

1. Define the following terms and write their symbols:
   (a) Inductance.
   (b) Counter-EMF.

(c) Self-induction.
(d) Mutual inductance.
(e) henry.
(f) Inductive reactance.

2. List four factors that affect the inductance of a coil and explain the effect of each on inductance.
3. Compute the total inductance of a circuit, given a schematic diagram.
4. Identify the schematic symbol and letter used to identify a coil in a schematic.
5. Identify the schematic symbols for, and define the characteristics of
   (a) Audio-frequency coils.
   (b) Coils used at power frequencies.
   (c) Radio-frequency coils.
6. Describe the phase relationship of voltage and current in an inductor.
7. Explain the effect on coil construction of
   (a) Copper losses.
   (b) Hysteresis losses.
   (c) Eddy-current losses.
8. Explain the method used to compensate for each loss in number 7.
9. Explain how changes in frequency and coil inductance affect inductive reactance.
10. Calculate inductive reactance, given the necessary schematic diagrams or other data.

Inductors

As we saw in Chapter 12, when a current-carrying conductor is wound into a coil of several turns, an electromagnet is created. An electromagnet is a type of inductor. The terms *inductor, choke,* and *coil* are interchangeable. The term *coil* defines construction, and the term *inductance* defines electrical effect. Most people are familiar with *coil* because of its use in automobiles. A choke is involved in power-supply circuitry.

An inductor (coil) is a device that opposes any change in current flow.

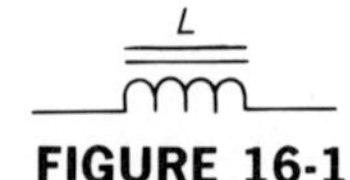

FIGURE 16-1

Note: An inductor does not oppose current flow itself, only a change in the amount of current flow. Because AC current changes continually, a coil continually opposes the change. The schematic symbol is shown in Figure 16-1, where the letter L indicates inductance in a circuit. Inductance is measured in *hcnries* (named for the American physicist Joseph Henry) and is represented by the letter H; 1 henry of inductance is present when:

1 volt of EMF is induced by current that is changing at the rate of 1 ampere per second.

In 1960 a decision was made to standardize electronic abbreviations. At that time the abbreviation for inductance was determined to be the uppercase letter H. This abbreviation is not to be confused with the Greek letter Eta(H).
For induction to occur, magnetic lines of flux must cut a conductor.

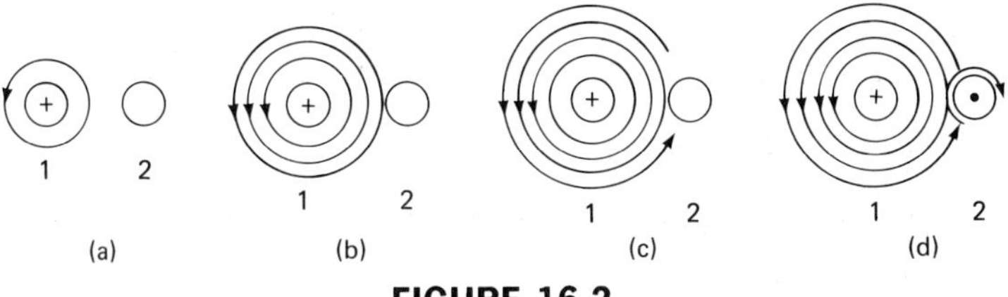

FIGURE 16-2

Two parallel conductors are illustrated in Figure 16-2. Conductor 1 has current flow into the page; conductor 2 has no flow. As current increases in conductor 1, a magnetic field expands around the conductor (Figure 16-2b). If the current increases enough, the field will expand to the point where it cuts conductor 2 (Figure 16-2c). When magnetic lines cut conductor 2, a voltage is induced in conductor 2, and a magnetic field builds around the conductor. The magnetic field around conductor 2 opposes the field around conductor 1 (Figure 16-2d). Using the left-hand rule, we find that current in conductor 2 flows out of the page; this current is opposite that in conductor 1. When a conductor induces current and voltage in an adjacent but unconnected conductor, we have the process known as *mutual induction.* Voltage is induced by *magnetic linkage.* This term gives the designator for inductance, which is *L* for linkage. The magnetic field that builds around conductor 2 opposes the field that caused it to be induced. The voltage induced in conductor 2 opposes the original voltage.

A coil is constructed by winding turns of a conductor so that an electromagnet is formed. All coils are wound around some type of frame. The center of this frame is called the core. In some coils the core will have soft iron inserted, in others the core is air. See Figure 16-3 for an example of an air-core coil. Explained in the figure are three conditions required for induction.

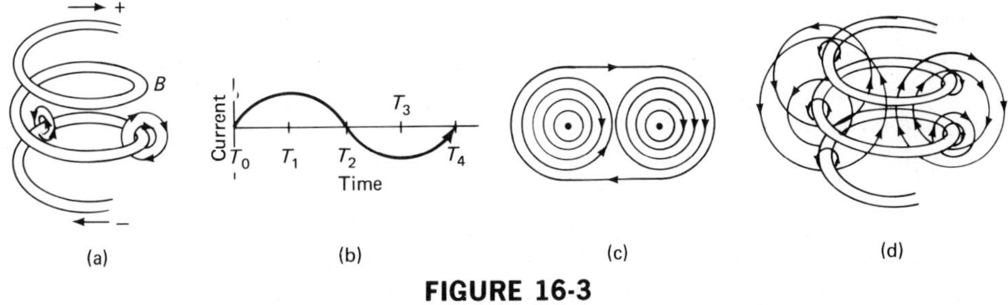

FIGURE 16-3

Figure 16-3a shows a coil with two turns. If AC is applied to the coil, the current wave resembles the wave in Figure 16-3b. Between T_0 and T_1, current increases; between T_1 and T_2, current decreases, returning to zero. At T_2, current reverses and flows in the opposite direction. From that point to T_3, current again increases; from T_3 to T_4, it decreases to zero. For each cycle of input current, these steps are repeated.

In a coil where many turns of the same conductor are involved, other factors must be considered. Figure 16-3a contains two loops. If we use the same analysis used to explain mutual induction, the explanation is as follows. Figure 16-3c is a cutaway view of the two loops at points A and B. A small current starting to flow is equal in either conductor. In each conductor, current flows out of the page, and a magnetic field builds around the conductor. The size and direction of the fields are identical, but as current increases and the fields expand, at the point where the two fields approach each other, we find that the lines move in opposite directions. From our study of magnetism, we already know that magnetic lines moving in opposite directions aid each other.

In a coil, when two opposite voltages oppose each other, the second is called a *counter-EMF* or CEMF. To proceed, we must understand the term:

> ***Counter-EMF is the force caused by induction that is in direct opposition to the initial current.***

For induction to occur, a magnetic field, a conductor, and relative motion must be present. When they are present, we have fields around the coil illustrated by Figure 16-3d and in larger coils, such as that in Figure 16-4.

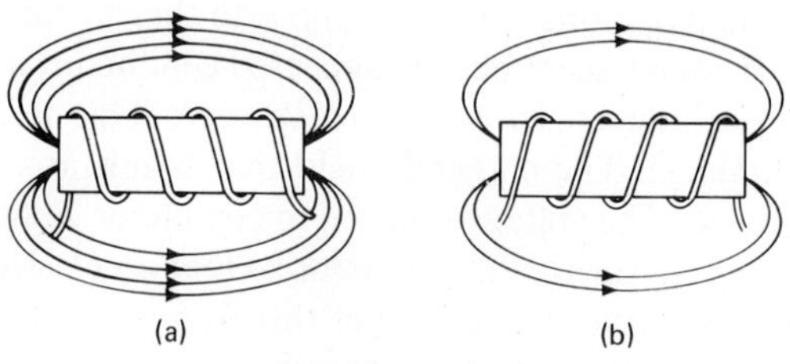

FIGURE 16-4

As alternating current flows through a coil, the magnetic field expands and collapses with each alternation of current. This meets two requirements for induction, magnetic field and relative motion. When the field is large enough to cut the adjacent loop, the third factor—a conductor—is added. In this case, induction occurs in the same conductor that supports the original current flow. The induced current is 180 degrees out of phase with the original current and will oppose its flow. Voltage induced in this manner is the CEMF discussed earlier. With each change in current; increases, decreases, and/or reversals, the coil's magnetic field also changes. The CEMF that results from these changes is the property that causes a coil to oppose any change in current flow.

When current is increasing, the coil uses energy from the power source and stores it in its electromagnetic field. When the current reaches its maximum point, the amount of energy stored is also at its maximum. As current decreases, the magnetic field collapses and returns the stored energy to the power source. Coils do not dissipate appreciable amounts of power.

The act of a conductor inducing voltage and current into itself is called *self-induction;* its symbol is the same as for inductance—*L.*

Self-Check

Complete each item by inserting the words and/or numbers needed to make a true and complete statement.

1. Two other names of an inductor are ______________ and ______________.
2. Cores used for inductors are usually either air or ______________.
3. For voltage to be induced in a conductor, it must be moved through a ______________.
4. The process of an inductor inducing a voltage in an adjacent conductor is called ______________.
5. A voltage that opposes the voltage that caused it to be induced is ______________.

Inductance

Four physical conditions are involved in all coils, which affect the device's inductance:

1. The number of turns of the coil.
2. The type of material used in the core.
3. The diameter-to-length ratio.
4. The method of winding.

Number of Turns

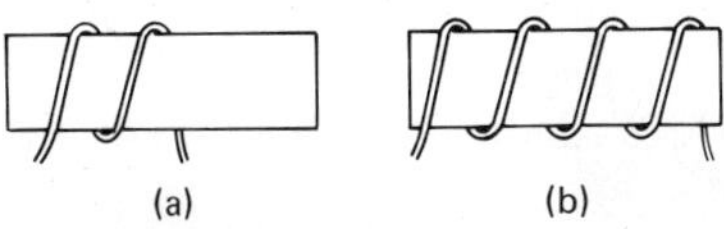

FIGURE 16-5

The coil shown in Figure 16-5a has only two turns. The field around either loop can only induce voltage (CEMF) in one other loop. Look at Figure 16-5b. It has four turns. The field of a loop can cut three other loops, thus inducing a CEMF in each. The CEMF opposes any change in current flow, with the result that

Inductance of a coil is directly proportional to the number of turns of the coil.

Type of Core Material

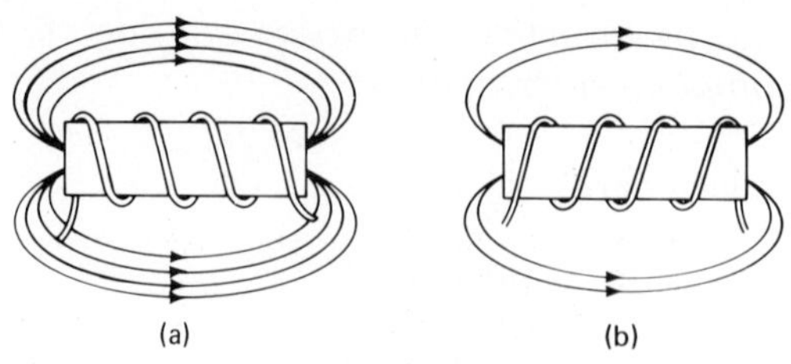

FIGURE 16-6

The diagram shown in Figure 16-6 represents two coils with different types of core. Coil a has a soft-iron core; coil b has an air core. Because soft iron is much more permeable than air, it conducts far more magnetic lines of flux. Such an increase results in higher inductance. Therefore,

> ***The inductance of a coil is directly proportional to the permeability of the material used as the core.***

Diameter-to-Length Ratio

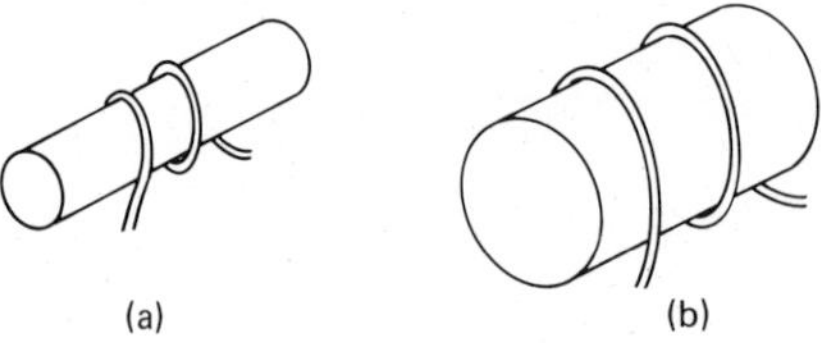

FIGURE 16-7

In Figure 16-7, both coils are of the same length; but the diameter of coil b is twice that of coil a. Thus the core of coil b has a much larger cross-sectional area, which results in less reluctance. The lower reluctance allows more lines of force to conduct through the core, and that increases the inductance of coil b so it is much larger than coil a.

> ***The inductance of a coil is inversely proportional to the cross-sectional area of its core.***

The opposite would be true if the core of the two coils were the same diameter and if the length varied. Making the core longer without adding windings *decreases* the coil's inductance.

> ***The inductance of a coil is inversely proportional to the length of its core.***

When the diameter-to-length ratio is considered, the following rule applies:

The inductance of a coil is inversely proportional to the diameter-to-length ratio of the core.

Method of Winding

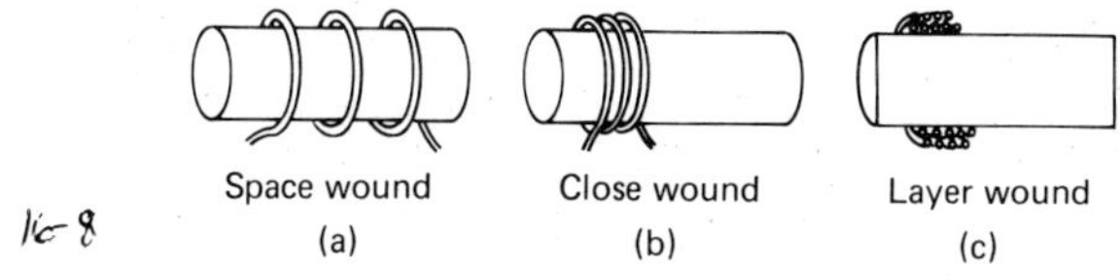

FIGURE 16-8

Figure 16-8 shows three methods commonly used to wind coils. Coil a has three turns spaced quite far apart. The lack of linkage results in a small amount of counter-EMF. When the loops are moved closer to each other, as they are in coil b, inductance increases because coil length is decreasing, allowing flux linkage to increase. In coil c, the windings are both closely spaced and layered. This type of winding concentrates the windings, and maximum linkage is achieved.

In all cases, counter-EMF is directly proportional to inductance. The more inductance a coil has, the more CEMF it induces. Thus, the larger the inductance, the greater the opposition to changes in current.

The inductance of a coil is directly proportional to the amount of voltage it induces; it is inversely proportional to the relationship of change in current and time. This can be stated by the formula,

$$L = \frac{V_L}{\left(\frac{\Delta i}{\Delta t}\right)}$$

where

L = inductance (in henries)
V_L = induced voltage (in volts)
Δi = change in current (in amps)
Δt = change in time (in seconds)

Note: The delta symbol (Δ) means "the change in."

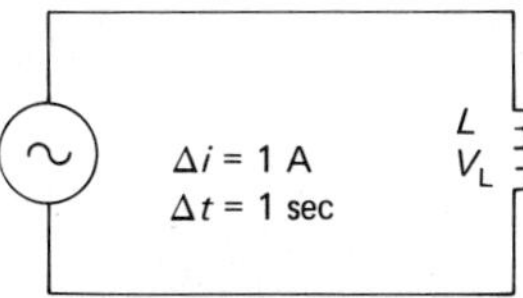

FIGURE 16-9

This fact is illustrated in Figure 16-9. In the circuit, current changes at the rate of 1 A per second induce 1 V across L. What, then, is the inductance of L?

Formula:

$$L = \frac{V_L}{\left(\frac{\Delta i}{\Delta t}\right)}$$

$$L = \frac{1}{\left(\frac{1A}{1\text{ s}}\right)}$$

$$L = \frac{1}{1}$$

$$L = 1\text{ H}$$

Suppose the same circuit has 100 V induced, with a 20-mA change of current over 50 μs. What does the coil's inductance equal?

Formula:

$$L = \frac{V_L}{\frac{\Delta i}{\Delta t}} = \frac{100\text{ V} \times \Delta t}{\Delta i}$$

$$L = \frac{100\text{ V} \times 50\ \mu\text{s}}{20\text{ mA}}$$

$$L = 250\text{ mH}$$

This formula can be transposed into a form suitable for solving induced voltage:

$$V_L = L\frac{\Delta i}{\Delta t}$$

If, in a circuit, $L = 20$ mH and current changes 1 A in 5 ms, what is the induced voltage?

Formula:

$$V_L = L\frac{\Delta i}{\Delta t}$$

$$V_L = 2 \times 10^{-2} \times \frac{1}{5 \times 10^{-3}}$$

$$V_L = 4\text{ V}$$

When current change is very fast, as when a switch is opened or closed, induced voltage can become very high. This can sometimes be noticed when a switch is opened on an inductive circuit; an *arc* occurs across the switch (Figure 16-10).

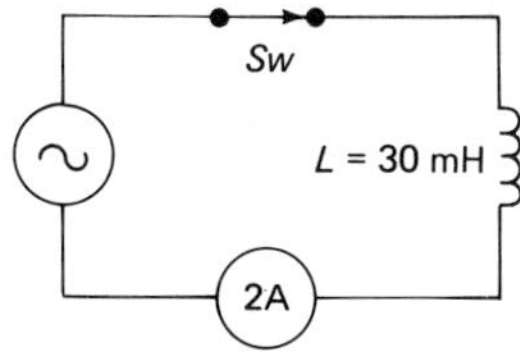

FIGURE 16-10

When a switch is opened, current immediately attempts to go to zero. Actually, because of the coil's opposition, time is required for this to happen. Let us say it takes 5 μs for current to reach zero. That means the change in current is 2 A in 5 μs across the 30-mH coil. What does the induced voltage equal?

Formula:

$$V_L = L\frac{\Delta i}{\Delta t}$$

$$V_L = 3 \times 10^{-2} \times \frac{2}{5 \times 10^{-6}}$$

$$V_L = 1.2 \times 10^4 \text{ or } 12{,}000 \text{ V}$$

The induced voltage of 12,000 V is applied across the switch. Thus it is easy to see how such high voltage can cause an arc across the switch. Though not a normal occurrence, it is of sufficient importance that induced voltage must be considered when we design inductive circuits.

Total Inductance in a Series Circuit

Many circuits contain more than one coil. Inductors may be connected either in series or in parallel; in some complex circuits, they may be connected in series-parallel. The rules for calculating total inductance are the same as those used to calculate total resistance.

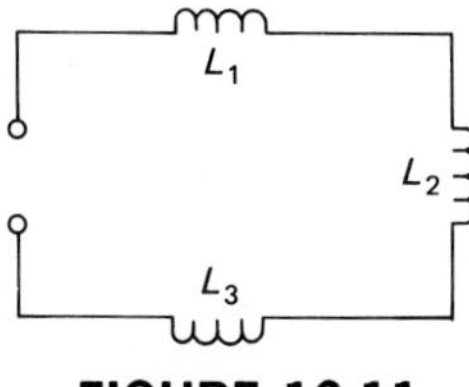

FIGURE 16-11

Figure 16-11 contains three coils connected in series in an AC circuit. Notice that the current has only one path; that means the circuit operates under the rules for series circuits. Therefore, the opposition of one coil is added to that of all other coils to provide total opposition. The formula for calculating total inductance, L_t, is:

$$L_t = L_1 + L_2 + L_3$$

Total Inductance in Parallel Circuits

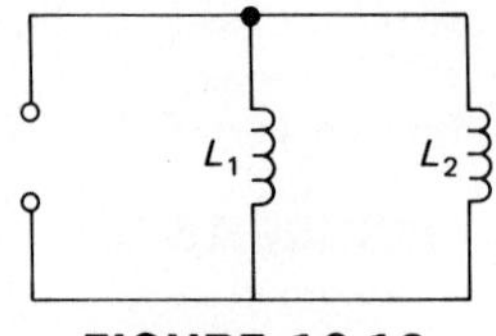

FIGURE 16-12

The circuit in Figure 16-12 contains two coils in parallel. There are two paths for current flow. To find inductance of this circuit, we use one of the following formulas:

$$\frac{1}{L_t} = \frac{1}{L_1} + \frac{1}{L_2}$$

$$L_t = \frac{L_1 \times L_2}{L_1 + L_2}$$

$$L_t = \frac{L}{N}$$

The L/N formula is only for branches containing inductors of equal size.

The solution of total-inductance problems is identical to that used for resistance problems. The only difference is that coils are often measured in milli and micro units. Take care that you notice the prefix.

Self-Check

Complete each item by inserting the words and/or numbers needed to make a true and complete statement.

6. What four factors affect the inductance of a coil?
   (a)
   (b)
   (c)
   (d)

7. A coil's inductance is ______________ proportional to the permeability of its core.

8. Of the three methods of winding coils, ______________ provides the highest inductance, when other factors are equal.

9. A circuit contains four series-connected coils; therefore, its total inductance equals the ______________.

10. The inductance of a coil is ______________ proportional to its diameter-to-length ratio.

Inductive Reactance

For operation at different frequencies, coils of suitable amounts of inductance must be made. To explain these operating characteristics, we first review the three categories of frequencies:

Audio frequencies (AF)	20 Hz to 20 kHz
Radio frequencies (RF)	20 kHz to 300 GHz
Power frequencies	50 Hz, 60 Hz, 400 Hz

Figure 16-13 illustrates the schematic symbol for each type of coil. Coil a is audio, coil b is power, and coil c is radio frequency. Note that coil c differs from the others, in that it has no iron core.

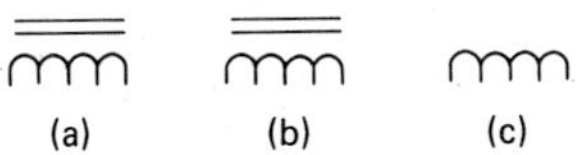

FIGURE 16-13

Note: Parallel lines alongside the coil symbol represent a core of soft iron. Its air core allows it to be designed for smaller amounts of inductance.

Because 50 Hz, 60 Hz, and 400 Hz are all in the audio range, the power and audio coil symbols are identical. The construction of a power coil requires much heavier conductors to handle the heavier current. This makes the power coil heavier and larger than the audio coil.

To gain a better understanding of the three types, consider the following:

Type	*Characteristic*
Power-frequency coil	Laminated iron core Large-diameter conductors Large Numerous turns
Audio-frequency coil	Laminated-iron core Small-diameter conductors Small Numerous turns
Radio-frequency coil	Air core Conductors that are very small Very small Very few turns

Power and audio coils differ only in the size of their conductors and in their physical size. RF coils differ greatly.

Coils and Direct Current

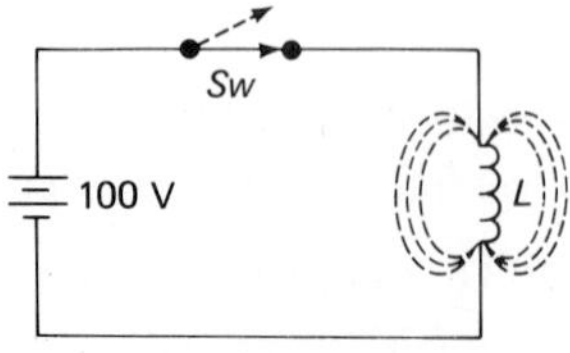

FIGURE 16-14

As has been stated, a coil opposes any change in current flow. When a coil is connected to a DC source (Figure 16-14), and the current is switched on and off, current changes immediately. Let us follow the sequence of events. When the switch is turned on, current begins to flow and a field builds around each turn. As current increases, these fields expand and induct CEMF into each other. The induced voltage and current oppose the original (battery) current. Some delay occurs before the current reaches its maximum. During the delay, the magnetic field expands and stores energy in its electromagnetic field. When current stops increasing, the magnetic field stops expanding. At that time, there is no magnetic line movement; so no CEMF is induced, with the result that, after DC current reaches its maximum, the coil offers no reactance to the circuit. The only opposition from the coil is the small amount of its resistance, the result of using copper-alloy conductors as windings. When the switch is opened; the coil's magnetic field collapses and the power it stored is returned to the power source.

Coils and Alternating Current

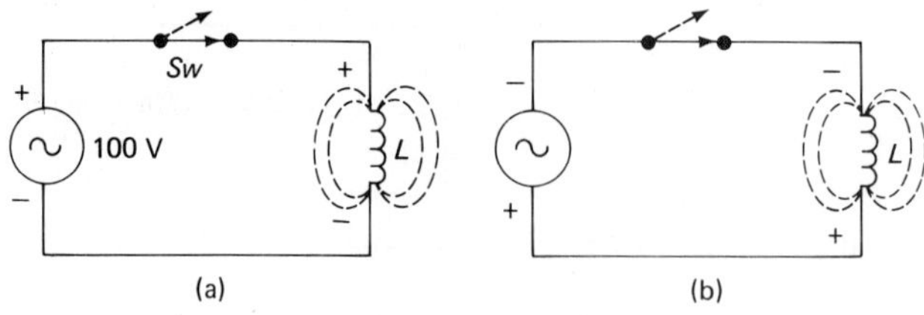

FIGURE 16-15

Replacing the DC source with an AC source results in a circuit similar to that shown in Figure 16-15. With AC applied, the coil reacts differently than it does with DC applied. When the voltage polarity is that shown in Figure 16-15a, the current and the magnetic field increase as they did in the DC circuit. The coil takes energy from the source and stores it in its electromagnetic field. The difference occurs when the current reaches maximum. At that point, current doesn't stop but immediately decreases. When current decreases, the magnetic field collapses, thus cutting the loops of the coil in the opposite direction. The resulting current, in an attempt to prevent current flow from decreasing, aids the current already present and returns energy to the circuit. When the voltage reaches zero, it begins to go negative, which results in a circuit with the polarities shown in Figure 16-15b. At this point, current reaches maximum and starts to decrease. CEMF is always of such polarity that it opposes the tendency of current to increase or decrease. This action is continual; as long as the coil operates with AC voltage applied, its magnetic field will expand and collapse at the same frequency as the AC frequency.

The opposition to change in current flow that results from CEMF is called *inductive reactance* (X_L).

Inductive reactance is the opposition that an inductor (coil) presents to a change in current.

Inductive reactance in a circuit is affected by

1. Amount of inductance (L).
2. Frequency of operation (f).

When we know the frequency and inductance of a circuit, we can calculate the inductive reactance present, using

$$X_L = 2\pi fL$$

To demonstrate the effect of changes in inductance on inductive reactance, we consider two coils. Both operate at a frequency of 1 kHz. The first is shown in Figure 16-16. To calculate X_L, we proceed as follows:

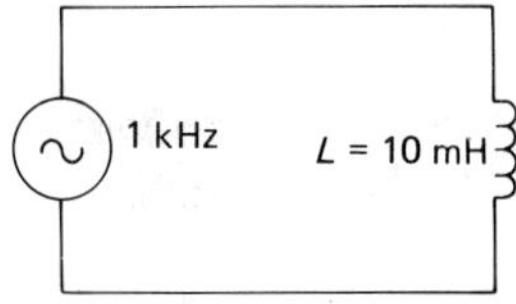

FIGURE 16-16

Formula:

$$X_L = 2\pi fL$$

$$X_L = 2 \times 3.14 \times 1 \times 10^3 \times 1 \times 10^{-2}$$

$$X_L = 6.28 \times 10^1 \text{ or } 62.8\ \Omega$$

Action	Entry	Display
ENTER	2	2
PRESS	×	2
PRESS	π	3.1416
PRESS	×	6.2832
ENTER	1, EE, 3	1 03
PRESS	×	6.2832 03
ENTER	10, EE, +/−, 3	10 − 03
PRESS	=	6.2832 01

6.2832 01 converts to a prefix of 62.8 Ω.

The same circuit is presented in Figure 16-17, except that the coil has 10 H of inductance:

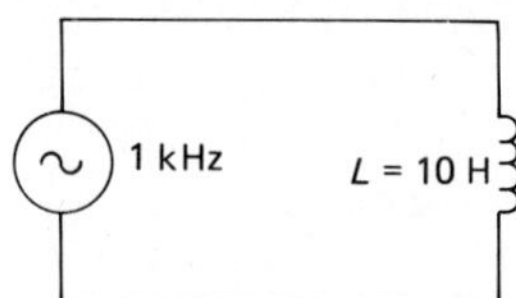

FIGURE 16-17

Formula:

$$X_L = 2\pi f L$$

$$X_L = 2 \times 3.14 \times 1 \times 10^3 \times 1 \times 10^1$$

$$X_L = 6.28 \times 10^4 \text{ or } 62.8 \text{ k}\Omega$$

Action	Entry	Display
ENTER	2	2
PRESS	×	2
PRESS	π	3.1416
PRESS	×	6.2832
ENTER	1, EE, 3	1 03
PRESS	×	6.2832 03
ENTER	10	10
PRESS	=	6.2832 04

6.2832 04 converts to a prefix of 62.83 kΩ.

Notice that as inductance increases, so does inductive reactance. Therefore,

The inductive reactance of a circuit is directly proportional to the inductance of the coil.

To illustrate the effect of frequency on inductive reactance in a circuit, we solve two problems. Both circuits contain 10-mH coils; the first operates at 1 kHz, and the second at 100 kHz. The first circuit is identical to the one in Figure 16-16 where:

$$X_L \text{ at } 1 \text{ kHz} = 62.8\ \Omega$$

Figure 16-18 presents this circuit at the operating frequency of 100 kHz. In this circuit,

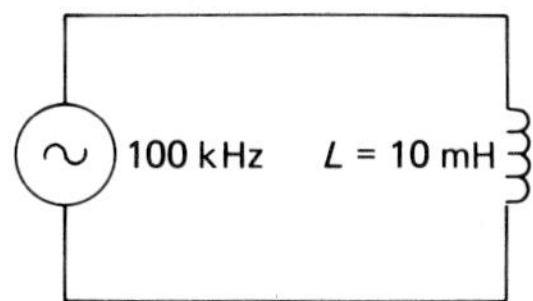

FIGURE 16-18

Formula:

$$X_L = 2\pi fL$$

$$X_L = 2 \times 3.14 \times 1 \times 10^5 \times 1 \times 10^{-2}$$

$$X_L = 6.28 \times 10^3 \text{ or } 6.28\ \text{k}\Omega$$

Action	*Entry*	*Display*
ENTER	2	2
PRESS	×	2
PRESS	π	3.1416
PRESS	×	6.2832
ENTER	100, EE, 3	100 03
PRESS	×	6.2832 05
ENTER	10, EE, +/−, 3	10 − 03
PRESS	=	6.2832 03

6.2832 03 converts to a prefix of 6.28 kΩ at 100 kHz.

Notice that when the frequency applied to the circuit increases, the inductive reactance also increases. Therefore,

The inductive reactance of a coil is directly proportional to the circuit's operating frequency.

The formula X_L can be converted into one of two other forms that can be used to solve for either frequency f or inductance L when the remaining quantities are given:

$$L = \frac{X_L}{2\pi f}$$

$$f = \frac{X_L}{2\pi L}$$

The first formula is illustrated, using Figure 16-19, where $f = 500$ Hz and $X_L = 10$ kΩ.

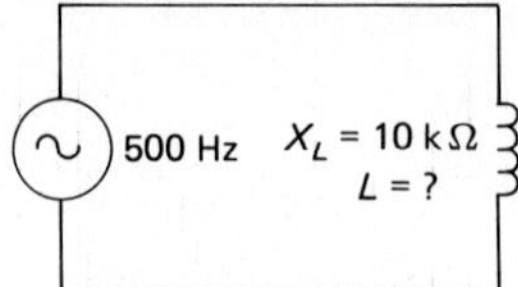

FIGURE 16-19

Formula:

$$L = \frac{X_L}{2\pi f}$$

$$L = \frac{1 \times 10^4}{2 \times 3.14 \times 5 \times 10^2}$$

$$L = 3.183 \text{ H}$$

Action	Entry	Display
ENTER	10, EE, 3	10 03
PRESS	÷	1 04
PRESS	(	1 04
ENTER	2	2
PRESS	×	2 00
PRESS	π	3.1416 00
PRESS	×	6.2832 00
ENTER	500	500
PRESS	)	3.1416 03
PRESS	=	3.1830 00

3.1830 00 converts to a prefix of 3.18 H.

To illustrate the calculation of frequency, we use Figure 16-20. Note that $X_L = 10$ kΩ and $L = 10$ mH.

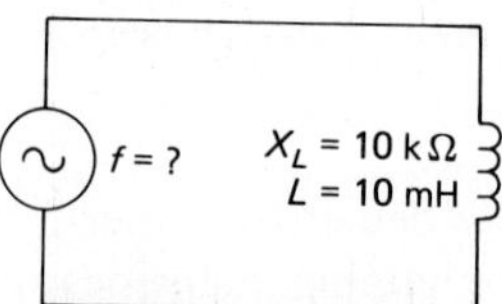

FIGURE 16-20

Formula:

$$f = \frac{X_L}{2\pi L}$$

$$f = \frac{1 \times 10^4}{2 \times 3.14 \times 1 \times 10^{-2}}$$

$$f = 1.5915 \times 10^5 \text{ or } 159 \text{ kHz}$$

Action	*Entry*	*Display*
ENTER	10, EE, 3	10 03
PRESS	÷	1 04
PRESS	(	1 04
ENTER	2	2
PRESS	×	2 00
PRESS	π	3.1416 00
PRESS	×	6.2832 00
ENTER	10, EE, +/−, 3	10 − 03
PRESS	)	6.2832 − 02
PRESS	=	1.5915 05

1.5915   05 converts to a prefix of 159 kHz.

We have already discussed the use of the three forms of the formula that state the relationship of inductance, frequency, and inductive reactance. With these formulas, we can solve for inductance, frequency, or inductive reactance, when other values are known.

Ohm's Law, Kirchhoff's Laws, and Inductive Circuits

Ohm's law and Kirchhoff's laws apply to inductive circuits in the same way that they apply to resistive circuits. The Ohm's-law formulas for the circuits are:

$$E_t = I_t \times X_{Lt}$$

$$I_t = \frac{E_t}{X_{Lt}}$$

$$X_{Lt} = \frac{E_t}{I_t}$$

Note: Calculator solutions provide answers that vary slightly from the values shown here. This is due to the fact that rounding off was done during the solutions.

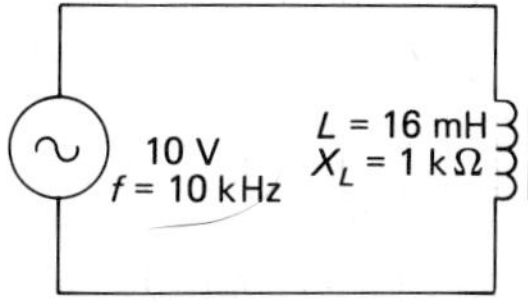

FIGURE 16-21

The circuit in Figure 16-21 can be used to solve for the total current in this circuit. To calculate total current, proceed as follows:

$$I_t = \frac{E_t}{X_L}$$

$$I_t = \frac{10\ \text{V}}{1\ \text{k}\Omega}$$

$$I_t = 10\ \text{mA}$$

From Kirchhoff's current law, we know that current in a series circuit is the same at all points in the circuit. Thus current flow at all points in this circuit equals 10 mA.

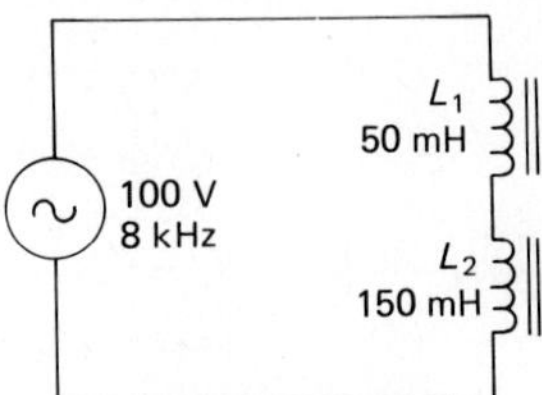

FIGURE 16-22

Examine Figure 16-22. Note that an AC power source, two coils with their inductance, and frequency values are given. Note, too, that this circuit is a series-inductive voltage divider which operates similarly to a resistive-voltage divider. We use these values to solve for total current and the individual voltage drops of each coil. To do this, we

Solve for X_{L1}:

$$X_{L1} = 2\pi f L_1$$

$$X_{L1} = 2 \times 3.14 \times 8\ \text{kHz} \times 5 \times 10^{-2}$$

$$X_{L1} = 2.5\ \text{k}\Omega$$

Solve for X_{L2}:

$$X_{L2} = 2\pi f L_2$$

$$X_{L2} = 2 \times 3.14 \times 8\ \text{kHz} \times 150 \times 10^{-3}$$

$$X_{L2} = 7.5\ \text{k}\Omega$$

Calculate L_t:

$$L_t = L_1 + L_2$$

$$L_t = 50\ \text{mH} + 150\ \text{mH}$$

$$L_t = 200\ \text{mH}$$

Solve for X_{Lt}:

$$X_{Lt} = 2\pi f L_t$$

$$X_{Lt} = 2 \times 3.14 \times 18\ \text{kHz} \times 200\ \text{mH}$$

$$X_{Lt} = 10\ \text{k}\Omega$$

or

$$X_{Lt} = X_{L1} + X_{L2}$$

$$X_{Lt} = 2.5\ \text{k}\Omega + 7.5\ \text{k}\Omega$$

$$X_{Lt} = 10\ \text{k}\Omega$$

Calculate I_t:

$$I_t = \frac{E_t}{X_{Lt}}$$

$$I_t = \frac{100\ \text{V}}{10\ \text{k}\Omega}$$

$$I_t = 10\ \text{mA}$$

We can use these calculations to solve for the voltage drop across each coil. We know that $I_t = 10$ mA, that $X_{L1} = 2.5\ \text{k}\Omega$, and that $X_{L2} = 7.5\ \text{k}\Omega$.

Calculate E_{L1}:

$$E_{L1} = I_t \times X_{L1}$$

$$E_{L1} = 10\ \text{mA} \times 2.5\ \text{k}\Omega$$

$$E_{L1} = 25\ \text{V}$$

Calculate E_{L2}:

$$E_{L2} = I_t \times X_{L2}$$

$$E_{L2} = 10\ \text{mA} \times 7.5\ \text{k}\Omega$$

$$E_{L2} = 75\ \text{V}$$

Examination of these two voltage drops reveals that the largest coil drops the largest part of the applied voltage. This too is similar to the resistive divider, where the largest resistor dropped the largest voltage. Kirchhoff's voltage law can be proved:

$$E_t = E_{L1} + E_{L2}$$

$$E_t = 25\ \text{V} + 75\ \text{V}$$

$$E_t = 100\ \text{V}$$

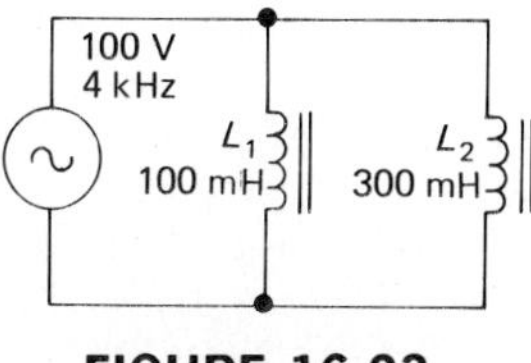

FIGURE 16-23

Now we turn our attention to Figure 16-23. In this circuit, the inductance of two parallel coils, a voltage, and a frequency are given. These values are used to demonstrate the analysis of this circuit for total inductance, inductive reactances, and total current:

Solve for L_t:

$$\frac{1}{L_t} = \frac{1}{L_1} + \frac{1}{L_2}$$

$$\frac{1}{L_t} = \frac{1}{100 \text{ mH}} + \frac{1}{300 \text{ mH}}$$

$$L_t = 75 \text{ mH}$$

Solve for X_{L1}:

$$X_{L1} = 2\pi f L_1$$

$$X_{L1} = 2 \times 3.14 \times 4 \text{ kHz} \times 100 \text{ mH}$$

$$X_{L1} = 2.5 \text{ k}\Omega$$

Solve for X_{L2}:

$$X_{L2} = 2\pi f L_2$$

$$X_{L2} = 2 \times 3.14 \times 4 \text{ kHz} \times 3 \times 10^{-1}$$

$$X_{L2} = 7.5 \text{ k}\Omega$$

Solve for X_{Lt}:

$$\frac{1}{X_{Lt}} = \frac{1}{X_{L1}} + \frac{1}{X_{L2}}$$

$$\frac{1}{X_{Lt}} = \frac{1}{2.5 \text{ k}\Omega} + \frac{1}{7.5 \text{ k}\Omega}$$

$$X_{Lt} = 1875 \ \Omega$$

From earlier circuits, we know that all parallel branches drop the voltage applied to that part of the circuit. Therefore,

$$E_{L1} = E_t = 100 \text{ V}$$

$$E_{L2} = E_t = 100 \text{ V}$$

With these values, we can solve for:

I_{L1}:

$$I_{L1} = \frac{E_t}{X_{L1}}$$

$$I_{L1} = \frac{100 \text{ V}}{2.5 \text{ k}\Omega}$$

$$I_{L1} = 40 \text{ mA}$$

I_{L2}:

$$I_{L2} = \frac{E_t}{X_{L2}}$$

$$I_{L2} = \frac{100 \text{ V}}{7.5 \text{ k}\Omega}$$

$$I_{L2} = 13.3333 \text{ mA}$$

I_t:

$$I_t = I_{L1} + I_{L2}$$

$$I_t = 40 \text{ mA} + 13.3333 \text{ mA}$$

$$I_t = 53.3333 \text{ mA}$$

Then we can prove:

$$X_{Lt} = \frac{E_t}{I_t}$$

$$X_{Lt} = \frac{100 \text{ V}}{53.3333 \text{ mA}}$$

$$X_{Lt} = 1875 \ \Omega$$

When we examine these solutions and compare them to those used in analyzing resistive circuits, we find that they are identical except for *resistance* being replaced by *reactance.*

Phase Relationships

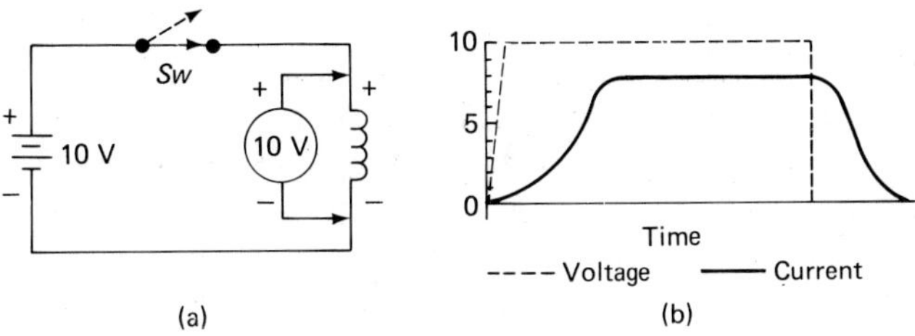

FIGURE 16-24

Refer to Figure 16-24a. When a coil is connected to a DC source, and the switch is closed, applied voltage is dropped across the coil. The voltmeter drops the full 10 V, as indicated by the dashed line in Figure 16-24b. The CEMF opposes the change in current flow, causing some delay between dropping E_a and reaching maximum current, represented by the solid line in Figure 16-24b. When the current has reached maximum, it no longer changes. Therefore, there is no reactance in the coil. The next change in current occurs when the switch is opened. Applied voltage drops to zero and current tries to decrease; but as soon as current tries to decrease, the magnetic field begins to collapse. Collapsing lines cut the coil's conductors and induce a voltage that opposes the decrease of current. The coil returns energy to the circuit in an attempt to continue current flow in the original direction. As the magnetic field decays, current flow gradually decreases and eventually reaches zero. Both voltage and current action are shown in Figure 16-24b.

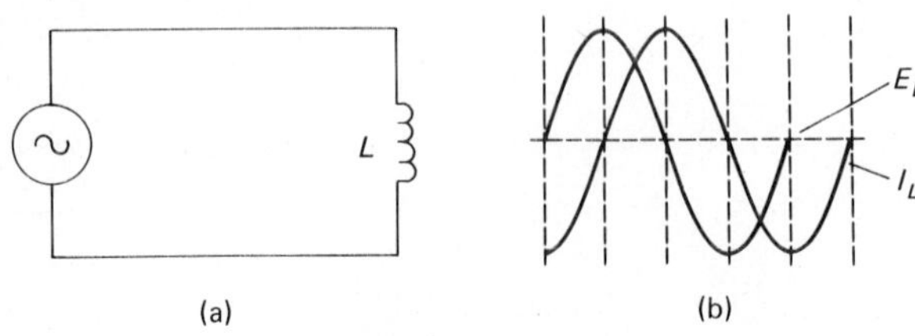

FIGURE 16-25

When AC is applied to a coil, as it is in Figure 16-25a, the coil attempts to react as it did to DC. The difference is that AC is continually changing and current cannot catch up with the voltage change. In fact, when voltage reaches its maximum, current is still increasing, and it will continue to increase even though voltage is beginning to decrease. In an inductor, current always lags voltage by 90 degrees. Figure 16-25b illustrates a comparison of phase for voltage and current in the AC circuit shown in Figure 16-25a. The fact that current lags voltage by 90 degrees can also be stated:

Voltage leads current by 90 degrees in an inductive circuit.

This relationship is easier to remember as ELI. In this abbreviation, *L* represents inductance; *E* in ELI occurs before *I*. The positions *E* and *I* can be read: voltage (*E*) leads current (*I*) by 90 degrees in an inductor (*L*).

Self-Check

Answer each item by indicating whether the statement is true or false.

11. ___ Power-frequency coils are similar to audio-frequency coils.

12. ___ In an ideal AC circuit containing a coil, voltage leads current by 90 degrees.

13. ___ An audio-frequency coil has an air core.

14. ___ An audio-frequency coil has larger conductors than a power-frequency coil.

15. ___ Iron cores are used in audio-frequency coils.

16. ___ Inductors are devices that oppose changes in voltage.

17. ____ the inductive reactance of a coil is directly proportional to the coil's inductance.

18. ____ Frequency and X_L are inversely proportional.

19. ____ As the frequency applied to an inductive circuit is increased, current flow in the circuit increases.

20. ____ Coil inductance and number of turns are directly proportional.

21. ____ Total inductance in a series-inductive circuit can be calculated by adding the separate inductance values.

22. ____ Circuits containing inductors are analyzed, using the same formulas as were used for resistive circuits.

23. ____ Total current in an inductive circuit can be solved by using the formula $I_t = E_t/X_{Lt}$.

24. ____ The total inductance of a parallel circuit can be determined by finding the sum of all the inductances and taking the reciprocal of the sum.

25. ____ In a series-inductive voltage divider, the smallest coil drops the largest voltage.

Losses in Coils

The reactive component does not dissipate power as a resistor does. Because of the way coils are constructed, however, some loss occurs. These losses are of three types:

1. Power (also called copper losses).
2. Hysteresis.
3. Eddy current.

Because of their construction, power- and audio-frequency coils are more subject to losses than are RF frequency coils. The first loss discussed is power loss. The conductor used to wind the coil has some resistance. This resistance opposes both DC and AC. It is called *copper loss* because the conductors are usually made of copper. The formula for power (current squared times resistance) can be applied to the coil to determine the loss from using copper conductors. Copper losses can be minimized by using conductors of larger wire, which have less resistance.

The second type of loss is that of hysteresis. In Figure 16-26, note that the coil's core is composed of sheets of soft iron laminated with shellac. Because material used in the core is magnetic, the core is magnetized when the coil is conducting. When AC is applied, the field constantly reverses. This causes the molecules in the core to reverse in order to maintain their alignment with the field. The friction created by reversing molecules causes the core to dissipate heat. The power, or heat, thus dissipated is the *hysteresis* loss. To reduce hysteresis loss, we must use a core material of higher permeability. The core will then pass more lines with less friction.

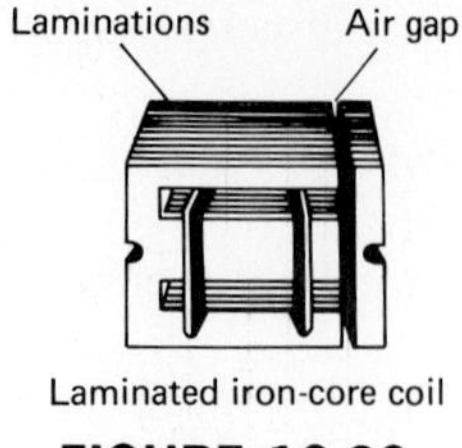

Laminated iron-core coil

FIGURE 16-26

The third type of loss is that of *eddy current*. This loss is also associated with the core. Lines of force cutting the core induce small currents. As these currents flow, or eddy, from one lamination to another, power is lost. To reduce eddy-current losses, the metal plates used in the core are laminated which retards the flow of current between core sections (Figure 16-26).

Problems with Coils

Coils are subject to the same problems as other components. Opens and shorts predominate. A coil can be checked by using an ohmmeter to measure the resistance of windings. An open coil reads infinity, a shorted coil zero Ω.

A totally shorted coil is rarer than a partially shorted one. In a partially shorted coil, the insulation has broken down between the windings. This condition can be verified when the ohmmeter reads a resistance less than that listed in the specifications for a good coil of the same type.

As is true of all electronic components, care must be taken when handling coils. Breaking the insulation's covering can lead to problems in the future. Subjecting a coil to currents greater than those for which it was designed can cause its insulation to deteriorate.

Self-Check

Answer each item by indicating whether the statement is true or false.

26. ___ Copper losses result from improper lamination.

27. ___ An open coil reads zero ohms on an ohmmeter.

28. ___ Eddy currents flow between plates in the core of a coil.

29. ___ Hysteresis losses occur in a coil's core.

30. ___ Copper losses can be reduced by using larger wire to wind the coil.

Summary

Inductance is the property of a coil that opposes a change in current flow. The symbol for inductance is L. Inductance is measured in henries (H).

Self-induction and mutual inductance were discussed, and the CEMF resulting from one conductor inducing a voltage in an adjacent conductor was explained. The three factors— magnetic field, conductor, and relative motion were discussed, and the part each plays in induction was explained.

The effect of adding inductors in series and parallel was discussed. When coils are in series, total inductance is found by adding the inductance of all coils. In the case of parallel coils, total inductance is calculated by using the reciprocal formula, as was done with resistors.

Four factors that affect the inductance of a coil were introduced and discussed: core material, length-to-diameter ratio, number of turns, and the method of winding the coil.

The reactance of a coil was explained. The term *inductive* reactance and the symbol X_L indicate this reactance. The effect of frequency and inductance on inductive reactance was explained. We found that as frequency increases, X_L increases, and that when inductance increases, inductive reactance increases.

Three losses that occur in coils were discussed. The reason for each loss was explained, as were the methods used to reduce the effect of each type.

Review Questions and/or Problems

1. _______________ occurs when a coil induces a counter-EMF in its own loops?
   (a) Mutual induction
   (b) Counter-EMF
   (c) Self-induction
   (d) Flux linkage

2. The henry is a unit of measure for:
   (a) inductance.
   (b) counter-EMF.
   (c) flux linkage.
   (d) current.

3. Which of the following has *no* effect on the inductance of a coil?
   (a) number of turns
   (b) diameter-to-length ratio
   (c) method of winding
   (d) conductor material

4. Which coil usually has an iron core?
   (a) audio frequency
   (b) power frequency
   (c) radio frequency
   (d) both (a) and (b)

5. Which of the following has the heaviest wire windings?
   (a) RF coils
   (b) AF coils
   (c) power coils
   (d) none of these

6. Voltage leads current by 90 degrees in a purely inductive circuit.
   (a) true
   (b) false

7. Power, copper, and hysteresis losses are the three different types of loss that occur in coils.
   (a) true
   (b) false

8. Which of the following losses can be reduced by using a core with high permeability?
   (a) hysteresis
   (b) copper
   (c) power
   (d) eddy current

9. As the frequency applied to a coil is increased, X_L
   (a) increases.
   (b) decreases.
   (c) remains the same.

10. If the inductance of a coil is increased, the coil's X_L
    (a) increases.
    (b) decreases.
    (c) remains the same.

11. Refer to the current flow in Figure 16-27. Considering the direction of (a), in what direction does the current of (b) flow?
    (a) into the page
    (b) out of the page

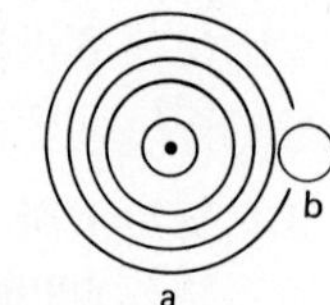

FIGURE 16-27

12. Refer to Figure 16-28, and calculate the total inductance of the circuit.
    (a) 3 mH
    (b) 22 mH
    (c) 12 mH
    (d) 20 mH

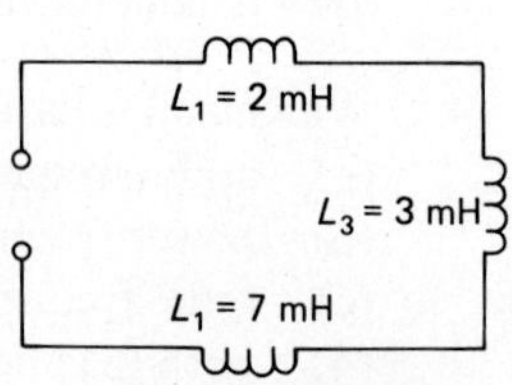

FIGURE 16-28

13. Refer to Figure 16-29, and calculate the total inductive reactance for the circuit.
    (a) 47 Ω
    (b) 470 Ω
    (c) 4.7 kΩ
    (d) 47 kΩ

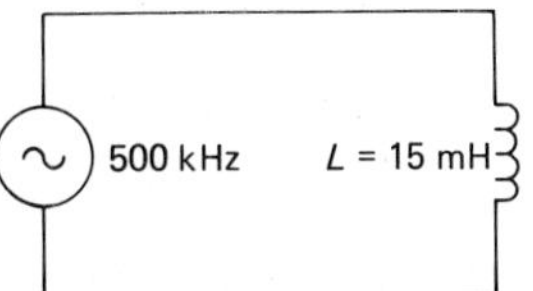

FIGURE 16-29

14. In Figure 16-30, calculate the applied frequency.
    (a) 159 Hz
    (b) 1590 Hz
    (c) 15.9 kHz
    (d) 159 kHz

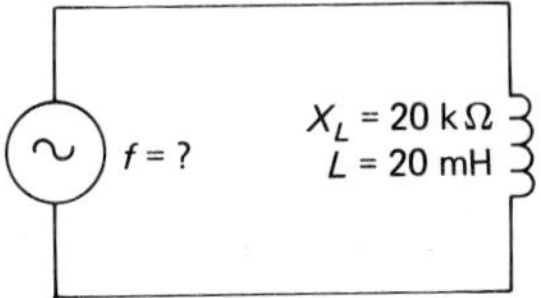

FIGURE 16-30

15. Refer to Figure 16-31; solve for total inductance.
    (a) 9 mH
    (b) 2 mH
    (c) 18 mH
    (d) 3 mH

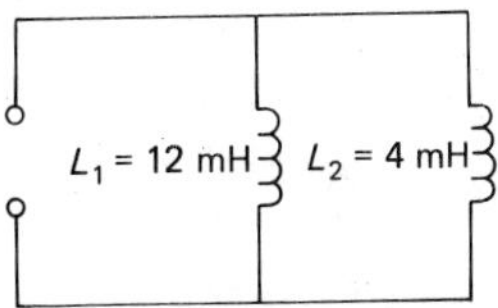

FIGURE 16-31

16. Refer to Figure 16-32, and solve for the total inductive reactance of the circuit.
    (a) 9 kΩ
    (b) 2 kΩ
    (c) 18 kΩ
    (d) 3 kΩ

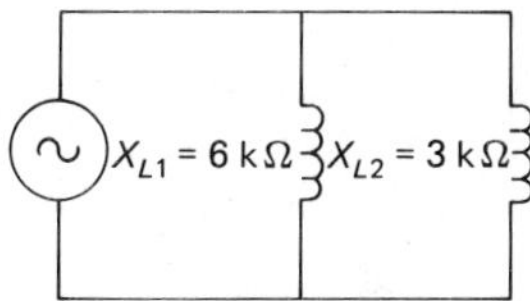

FIGURE 16-32

17. How much current flows in the circuit shown in Figure 16-33?
    (a) 4.0 mA
    (b) 0.4 mA
    (c) 4.0 A
    (d) 0.4 A

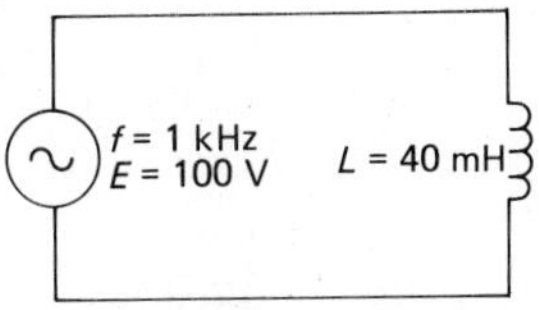

FIGURE 16-33

18. Refer to Figure 16-34. If the applied frequency decreases, what happens to the total current?
    (a) Increases
    (b) Decreases
    (c) Remains the same

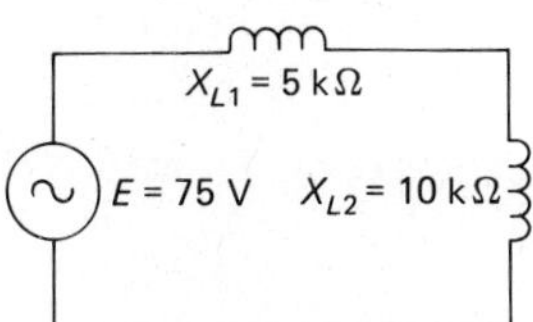

FIGURE 16-34

19. Refer again to Figure 16-34. What is the voltage drop on L_2?
    (a) 75 V
    (b) 25 V
    (c) 50 V
    (d) 37.5 V

20. In Figure 16-35, solve for total current of the circuit.
    (a) 10 A
    (b) 6.36 A
    (c) 10 mA
    (d) 6.36 mA

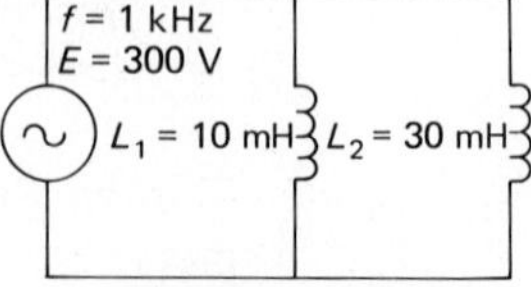

FIGURE 16-35

21. Solve for X_L of the coil in Figure 16-36.
    (a) 301.6 Ω
    (b) 3016 Ω
    (c) 30.16 Ω
    (d) 301.6 kΩ

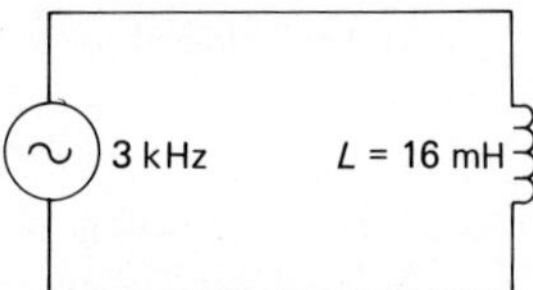

FIGURE 16-36

22. Solve for X_{Lt} in Figure 16-37.
    (a) 3.33 kΩ
    (b) 333 Ω
    (c) 15 kΩ
    (d) 1.5 kΩ

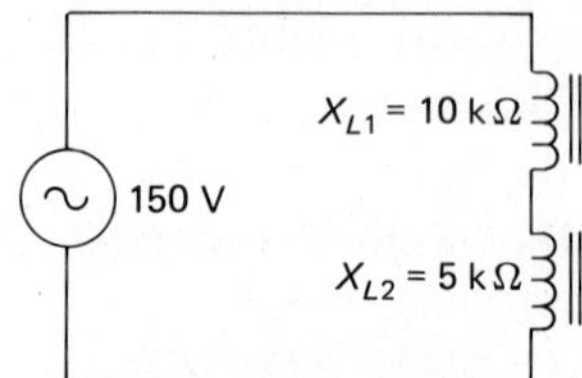

FIGURE 16-37

23. Calculate total current for the circuit in Figure 16-37.
    (a) 10 mA
    (b) 1 mA
    (c) 4.5 mA
    (d) 10 A

24. Use Figure 16-37, and solve for E_{L1}.
    (a) 50 V
    (b) 100 V
    (c) 150 V
    (d) 0 V

25. In Figure 16-38, solve for total inductance, L_t.
    (a) 149 mH
    (b) 1490 mH
    (c) 14.9 mH
    (d) 1.49 mH

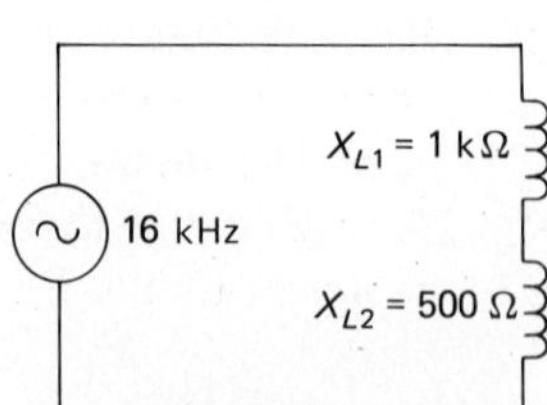

FIGURE 16-38

17

Transformers

Introduction

In other chapters we discussed inductance, electromagnetism, counter-EMF, magnetic fields, coils, and inductors. In this chapter we continue our study. The chapter is devoted to the effect produced when a coil driven by an AC source is used to induce a voltage, current, and a magnetic field in another coil. Induction between coils is possible because of an action called *mutual induction.* Mutual induction was mentioned in Chapter 16; it is explained in this chapter.

By manufacturing coils with specific characteristics, it is possible to obtain highly accurate amounts of induction from one coil to another. By such manufacture and application, we can cause the voltage of a driven coil to be increased or decreased in another coil. By using these devices, we can use one AC source with a transformer to provide AC voltage and current to one or several different circuits.

Transformers are produced in many shapes and sizes. Figure 17-1 shows several types. This assortment contains transformers capable of working at power frequencies as low as 50 Hz, at audio frequencies, and at radio frequencies as high as 100 kHz. Some can conduct heavy current loads, others small loads.

Objectives

Each student is required to:

1. Explain the operation of transformers.
2. When given voltage ratios or turns ratios, calculate step-up and step-down ratios for
   (a) Voltage.
   (b) Current.
   (c) Impedance.
3. Draw and identify the schematic symbols for four types of transformers.
4. Explain the use of an ohmmeter in checking transformers for opens, shorts, and partial shorts.
5. Demonstrate the use of equipment to test transformer windings for correct current, voltage, and resistance.

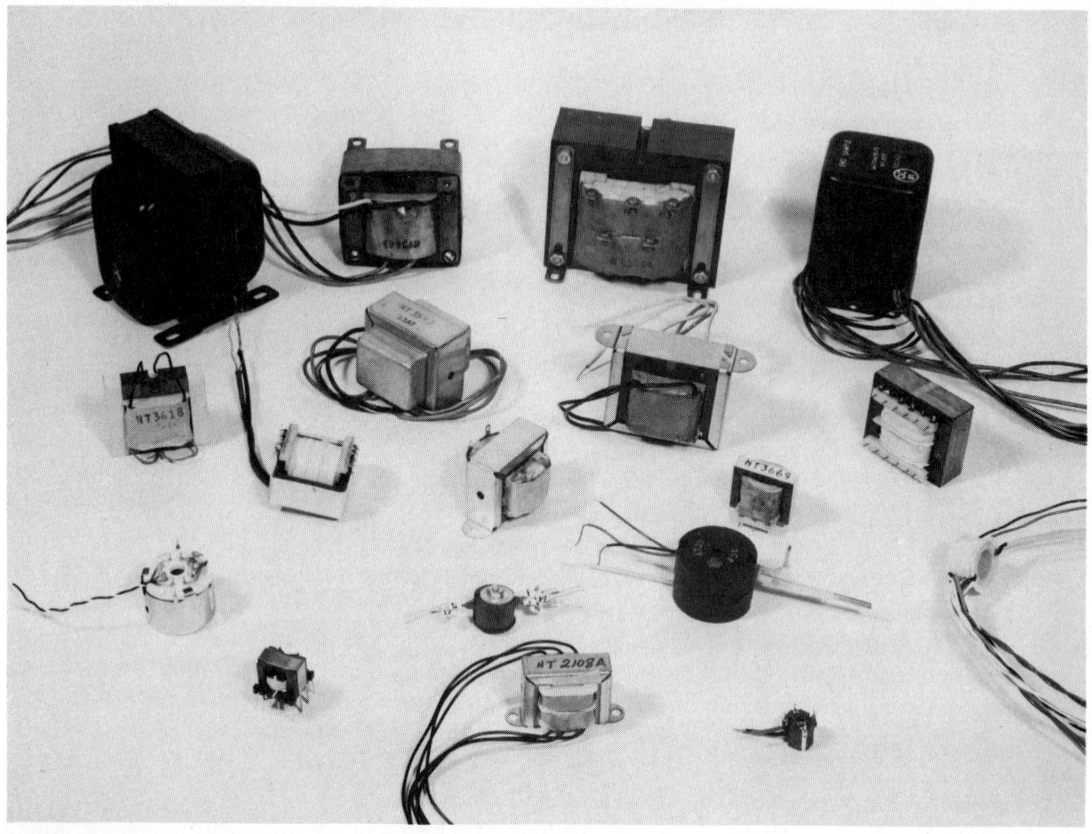

FIGURE 17-1
Assorted Transformers Designed for Operation at Frequencies from 50 Hz to 100 kHz (Photo Courtesy National Transformers Corporation)

6. Define
   (a) Turns ratio.
   (b) Coefficient of coupling.
   (c) Transformer action.
   (d) Mutual induction.
   (e) Step-up ratio and step-down ratio.
   (f) Reflected impedance.

Theory of Operation

A transformer is nothing more than two coils positioned so that a coil supplies energy to another coil by magnetic coupling. To better understand this action, let us review some principles.

Figure 17-2a illustrates two conductors placed side by side. One conductor has current flow indicated by a dot. From rules already established, we know that this represents current flowing "out of the page." By using the left-hand rule, we find the direction of the magnetic field. Point your left thumb in the direction of current flow, and your fingers will circle the conductor in the direction of rotation of the magnetic field. Therefore, conductor a has a field which rotates clockwise around it. As the current in conductor a increases, the field expands until it strikes conductor b. The lines that cut b in Figure 17-2b cause a magnetic field to be induced which opposes the original field and which revolves counterclockwise. With the magnetic field known, we point the fingers of the left hand in the direction of the field; the thumb then points in the direction that current flows. The induced current in b flows *into* the page.

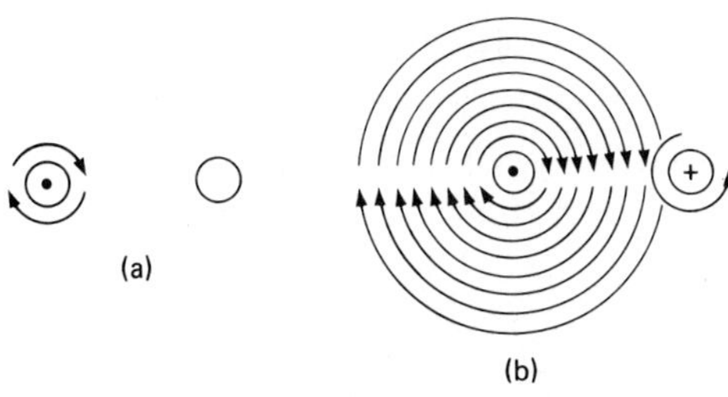

FIGURE 17-2

If we apply this rule in Figure 17-3, we find that current flowing through turn 1 induces current in turn 2 which opposes the original current supplied by the power source.

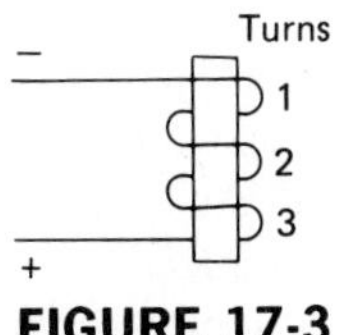

FIGURE 17-3

Induced current opposes the current that caused it to be induced. This is why an inductor is defined as a device that opposes any change in the flow of current. In other chapters this opposition is called inductive reactance. Because a coil is wound wire, it has *resistance*. The combination of reactance and resistance results in *impedance*.

Impedance is the total opposition offered by a circuit to the flow of alternating current at a given frequency.

Earlier you learned how to locate the north pole of an electromagnet. When we know the direction of current in a coil, we can locate its north pole by

Pointing the fingers of the left hand in the direction that current flows across the face of the coil; the thumb then points toward the north pole.

Figure 17-4 illustrates a coil whose north and south poles have been located. Notice that the battery current enters the top of the coil, flows across the face (front) of the coil, then disappears behind the coil. Pointing the left fingers in a left-to-right position on the page enables you to see that the thumb points toward the bottom of the page; the north pole is at the bottom of the coil.

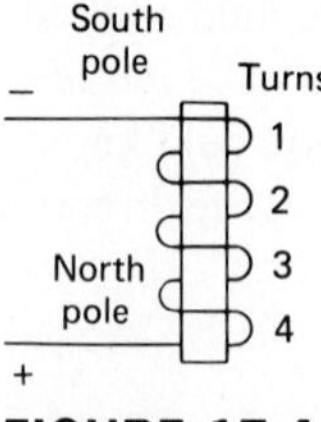

FIGURE 17-4

This process can be reversed. If we know where the north pole is, we point the left thumb toward it. Then the fingers point in the direction that current flows across the face of the coil.

Lenz's Law

Lenz's law states that

The EMF induced in an inductor always opposes the current change that caused it to be induced.

This means the magnetic field of an induced voltage opposes the magnetic field of the source that caused it to be produced. Figure 17-5 demonstrates Lenz's law. With current entering L_1 at point A, the north pole of L_1 is at the bottom of the coil. As current in L_1 increases, its magnetic field expands outward. When the field cuts L_2, a voltage is induced in L_2. Lenz's law states that the magnetic field around L_2 opposes the field around L_1. For these two fields to be opposed, the north pole of each coil must be located at the bottom of the drawing—because

Like poles repel.

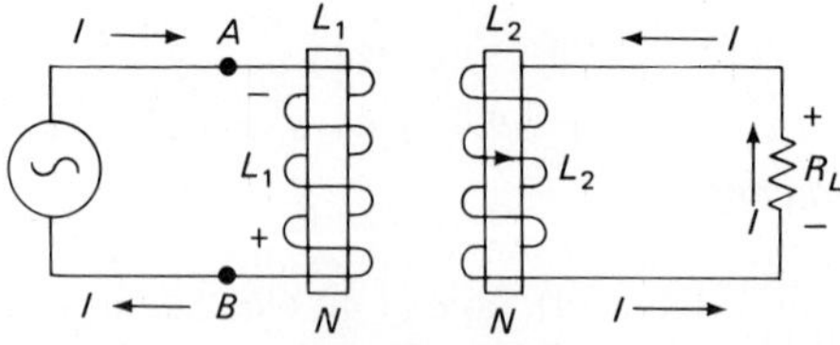

FIGURE 17-5

To verify this statement, grasp the coil with your left thumb pointing to the bottom; your fingers will indicate the direction of current flow across the face of the coil. This test reveals that current in both coils flows from left to right across the faces of the coils. By following the current path in L_2, we see that current flows out of the coil at the top and back into it at the bottom. Current in R_L is from top to bottom.

Observing Figure 17-5, we see that L_1 is the load device for the AC power source. During the alternation when current enters the top of L_1, current leaves the top of L_2. This results in both coils having negative (−) polarity at the top and positive (+) polarity at the bottom. L_1 accepts current from the AC source, but L_2 acts as an AC source, supplying current to R_L.

In Figure 17-6, we see that the winding of L_2 has been changed. The winding goes behind the core at the top of L_2 and over the face at its bottom. By performing the analysis just discussed, we see that current flows away from the bottom of L_2 and returns at the top. Thus the current in R_L is opposite that in Figure 17-5, with the flow of current from bottom to top. These two diagrams illustrate that current through R_L is a result of the relationship of L_1 and L_2 windings.

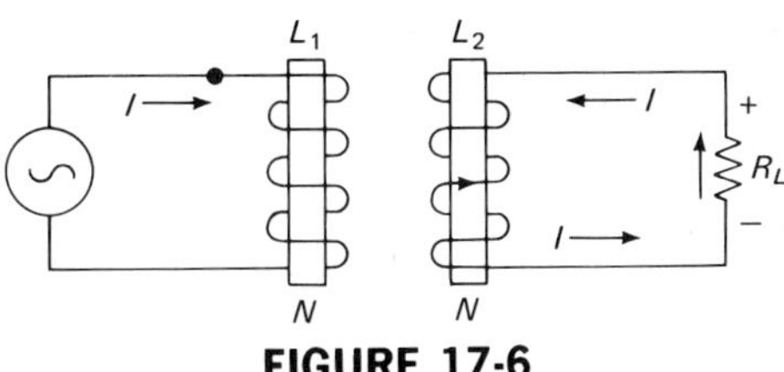

FIGURE 17-6

In Figure 17-7, coils L_1 and L_2 are combined, forming a *transformer*. Coil L_1 transfers electrical energy into coil L_2 by means of the mutual induction that exists between the coils.

A transformer is a device that transfers electrical energy from one circuit to another by magnetic coupling (mutual induction).

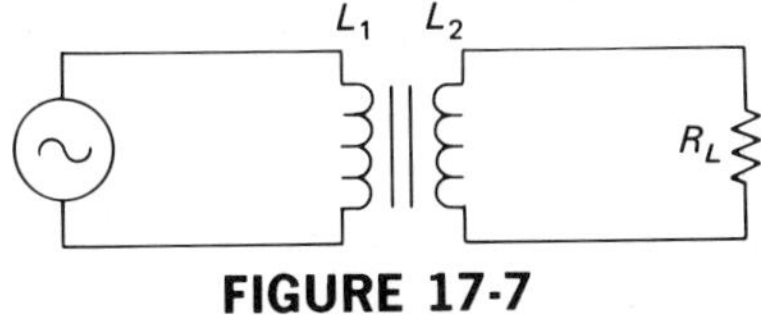

FIGURE 17-7

The part of the transformer, L_1, that draws power from the AC source is the *primary winding*. The coil that supplies energy to R_L is the *secondary winding*. The *turns ratio* determines the amount of voltage transferred. This term refers to the number of turns, or loops, wound in the primary, compared to the number of turns in the secondary. The primary is the "sending" coil, and the secondary, the "receiving" coil.

Transformers can be used to *step up* or *step down* both voltage and current. Transformers whose secondary voltage is higher than the primary voltage are called *step-up transformers*. Transformers whose secondary voltage is lower than their primary voltage are called *step-down transformers*.

Step-up and step-down transformers are shown in Figure 17-8. Diagram a is step-up and diagram b is step-down; they are a simple approach to turns ratio that serves to illustrate these transformers. Diagram a has four turns on its secondary winding for each turn on the primary winding. This ratio is referred to as a 1:4 ratio step-up transformer. The transformer in diagram b is opposite, with four turns on its primary for one on its secondary; it is a 4:1 ratio step-down transformer. The schematic symbol for each type of transformer is shown; the symbol indicates two coils separated by a core of soft iron. Figure 17-8 shows both windings wound around the same core, the most common method of transformer construction. Some transformer windings contain thousands of turns. Because it is impossible to represent all the turns for each winding in one symbol, a standard representative symbol is used. The step-up/step-down ability of the transformer is shown by a ratio. In buying a transformer, the technician selects one capable of delivering a specific voltage and current from its secondary winding. To provide the desired voltage and current output, the turns ratio and all other characteristics must be correct.

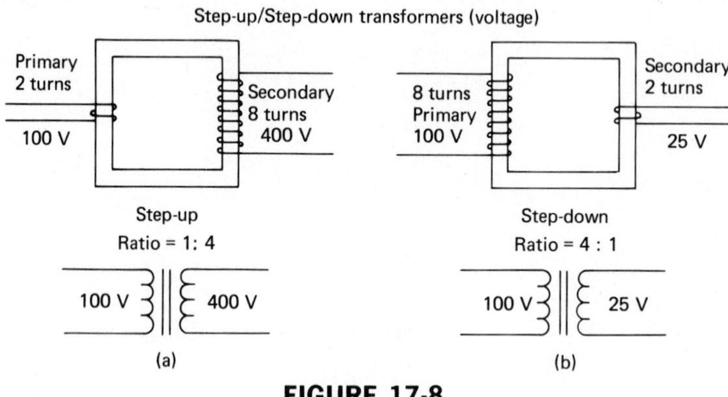

FIGURE 17-8

An ideal transformer has 100% efficiency of coupling between primary and secondary. The voltage and turns ratios of a transformer are closely related. For example, with 100 V AC input to the primary of each transformer in Figure 17-8, the secondary voltages vary greatly. Secondary voltage in diagram a is 400 V; in b, it is 25 V. Regardless of the voltage applied to the primary of diagram a, the output voltage is four times that of the primary. The secondary voltage of b is always one-fourth the voltage applied to the primary.

The voltage across each winding of a transformer is proportional to the number of turns in the winding. These characteristics can be stated:

$$\frac{N_p}{N_s} = \frac{E_p}{E_s}$$

where

E_p = voltage applied to primary

E_s = voltage present across secondary

N_p = number of turns in primary

N_s = number of turns in secondary

By substituting values from Figure 17-8b in this formula, we can illustrate application of the ratio.

Mathematical Solution	Action	Entry	Display
$\frac{N_p}{N_s} = \frac{E_p}{E_s}$	ENTER	2	2
$\frac{8}{2} = \frac{100}{E_s}$	PRESS	×	2
$8E_s = 200$	ENTER	100	100
$E_s = 25$	PRESS	=	200
	PRESS	÷	200
	ENTER	8	8
	PRESS	=	25

Coefficient of Coupling

Not all magnetic lines of force cut the secondary winding; therefore, transformers are not 100% efficient. The number of lines that do cut the secondary winding, compared to the total number of lines around the primary, is the *coefficient of coupling.* Figure 17-9 shows a transformer whose primary emits 10,000 magnetic lines of force. Of these, 9000 cut the secondary, while 1000 do not. Dividing total lines emitted by the primary into the number of lines that cut the secondary results in:

$$\text{Coefficient of coupling} = \frac{N_s}{N_p}$$

$$\text{Coefficient of coupling} = \frac{9000}{10{,}000}$$

$$\text{Coefficient of coupling} = 0.9 \text{ or } 90\%$$

Thus, 90% of the primary's energy is coupled to the secondary winding.

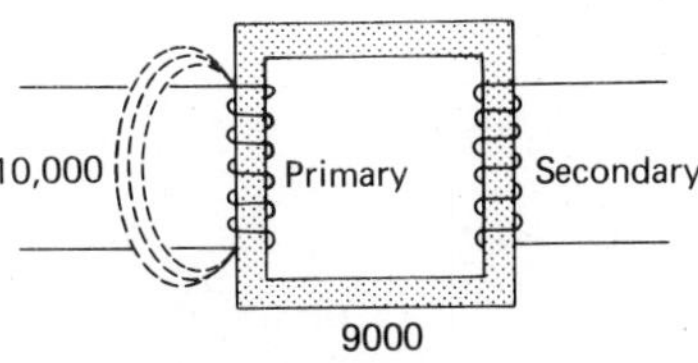

FIGURE 17-9

In designing transformers, it is fairly easy to calculate the coefficient of coupling and compensate for it by adding turns to the primary or the secondary until the desired output voltage becomes available. The result is a transformer that operates as though it were 100% efficient. That is why technicians do not concern themselves with the coefficient; compensation was made during manufacture. A transformer purchased, which can provide 24 V at a maximum current of 200 mA, has already been compensated to provide the output. Our only concern is identifying the voltage, current, and impedance characteristics of the circuit and buying the appropriate transformer.

Impedance in Transformers

Impedance, a new term, is associated with alternating current. From earlier discussions, we know that coils react to and oppose the flow of AC. We have also established that resistance opposes AC and DC equally well, but that X_L is not a factor in DC circuitry. In AC circuitry, two oppositions are present—resistance and reactance. The term *impedance* (Z) is used to state that both resistance and reactance are present in a circuit. Impedance is discussed more fully later in this book. We introduce the concept at this point to illustrate the role of impedance in the windings of a transformer.

> ***Impedance (Z) is the total opposition that a circuit offers to the flow of alternating current at a given frequency.***

We can apply Ohm's law to transformers to determine impedance. This is done by replacing resistance (R) with impedance (Z) in the Ohm's-law formula

$$Z = \frac{E}{I}$$

When using this equation to calculate primary impedance its form would be:

$$Z_p = \frac{E_p}{I_p}$$

and secondary impedance by use of:

$$Z_s = \frac{E_s}{I_s}$$

Refer to Figure 17-10. This circuit illustrates how these formulas can be used to find the impedance for each winding of a transformer.

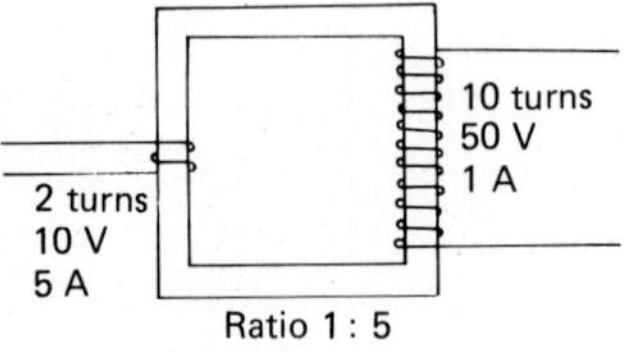

FIGURE 17-10

Ohm's law can be applied as follows:

$$Z_p = \frac{E_p}{I_p} = \frac{10\text{ V}}{5\text{ A}} = 2\ \Omega$$

$$Z_s = \frac{E_s}{I_s} = \frac{50\text{ V}}{1\text{ A}} = 50\ \Omega$$

In some cases, however, we may need to calculate impedance without our knowing either voltage or current. The impedance of a transformer is directly proportional to the square of the turns ratio. In other words, if a transformer has a 1:2 ratio, its impedance ratio would be 1:4. Stated in formula form this is

$$\frac{Z_p}{Z_s} = \frac{N_p^2}{N_s^2}$$

This formula can be used when calculating the impedance or turns of either winding, primary or secondary. For example, if a transformer has a turns ratio of 2:3 and a primary impedance of 10 kΩ, we can calculate secondary impedance as follows:

$$\frac{Z_p}{Z_s} = \frac{N_p^2}{N_s^2}$$

$$Z_s = \frac{Z_p \times N_s^2}{N_p^2}$$

$$Z_s = \frac{(10\text{ k}\Omega) \times 9}{4} = \frac{90{,}000}{4} = 22.5\text{ k}\Omega$$

Note that the impedance ratio is 4:9, which is the square of the turns ratio 2:3.

We know that voltage is directly proportional to the turns ratio; therefore, they have the same ratio. Realizing this we can see that the voltage ratio could be used in place of the turns ratio to calculate impedances. This formula would be

$$\frac{Z_p}{Z_s} = \frac{E_p^2}{E_s^2}$$

Assume that we have a transformer whose turns ratio is 2:3; its impedance ratio would be 4:9.

The rule is:

Impedance is directly proportional to the square of either the turns ratio or the voltage ratio.

There are times when it is necessary to match the impedance of a transformer's winding to other circuits. To do this the correct turns ratio must be selected. Assume that we want to connect to circuits using a transformer to match Z. The output of one circuit is 16 Ω and the input to the other circuit is 100 Ω. Remember:

The impedance ratio is directly proportional to the square of the turns ratio of the transformer.

Knowing that this is true we can state:

The turns ratio is directly proportional to the square root of the impedance ratio.

By extracting the square root of each impedance of the circuits to be matched, we can find the turns ratio:

Primary turns $= \sqrt{16} = 4$

Secondary turns $= \sqrt{100} = 10$

This tells us that the primary must have four turns for each ten turns in the secondary. When stating this ratio it can be reduced further and stated as a 2:5 ratio. Knowing this we could purchase a transformer having a 2:5 ratio and use it in matching the impedance of these two circuits where the input to the primary is 16 Ω and the output to the next circuit must be 100 Ω.

Current in Transformers

Earlier in this chapter, we discussed the effect of transformers on voltage. Although the effect on current is similar, it differs sufficiently that it is discussed separately.

Clearly, current in the primary winding can control current in the secondary winding. For instance, if current in the primary increases, the magnetic field around the primary becomes stronger, thus inducing more current in the secondary.

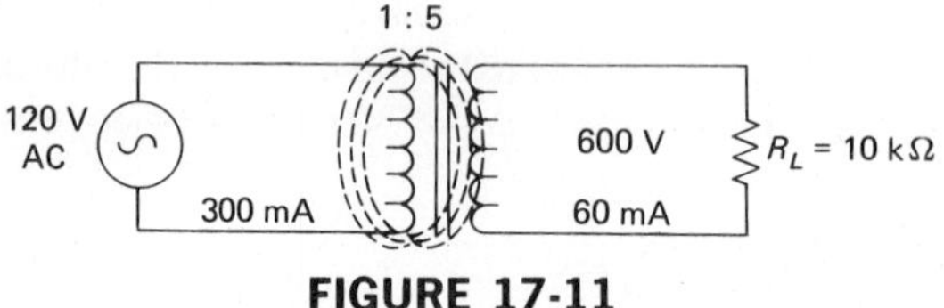

FIGURE 17-11

Secondary current also controls primary current, though that may not be seen as easily. Figure 17-11 is used to illustrate the control a secondary winding has over a primary winding. Restating some rules already stated should make this action more understandable:

1. When current flows in a secondary coil, a magnetic field is created around that coil.
2. The field around a secondary coil builds outward and cuts the turns of the primary coil.
3. Lines cutting the primary coil induce a voltage in the primary coil.
4. The magnetic field caused by the induction of this voltage opposes the field around the secondary coil (an application of Lenz's law).
5. Thus, the north pole of the newly induced field must be the same as the pole that already exists around the primary.
6. The result is that the newly induced current is added to the current already flowing.
7. Therefore, any increase in secondary current causes a corresponding increase in primary current, and any decrease in secondary current causes a decrease in primary current.

Due to the control of each winding over the other, it is possible to protect both windings of the transformer with one fuse. The primary winding is in the applied voltage circuit; thus, we place the fuse there. If current stops in the primary, no current can be induced in the secondary. The control of current exercised by the windings ensures that the current *ratio* existing between the two windings remains constant even with variations in load. The ability of the secondary winding to influence the primary is known as *reflected impedance.*

In discussing impedance earlier, we stated the rule that impedance is directly proportional to the square of either the turns ratio or the voltage ratio. With this ratio affecting the opposition (impedance) present in each winding, we can see the effect on current of the ratio. Remember that current and impedance (opposition) are inversely proportional; that therefore, greater impedance in a winding means the current in that winding is less. On the basis of that statement, we can say:

Current and impedance are inversely proportional.

Let us take this statement a step further. If the secondary winding has more windings than the primary, the secondary will have more impedance and thus less current. If the secondary has fewer windings, its impedance will be less and its current will be greater. The relationship of primary current to secondary current is the opposite of voltage relationships. The relationships are:

A transformer that steps up voltage will step down current.

A transformer that steps down voltage will step up current.

The turns ratio remains the key; it governs the extent to which current is stepped up or down. Current ratio is the reciprocal of voltage ratio. The diagrams in Figure 17-11 represent this. The ratios that govern current are:

$$\frac{N_p}{N_s} = \frac{I_s}{I_p}$$

$$\frac{E_p}{E_s} = \frac{I_s}{I_p}$$

where

N_p = number of primary turns

N_s = number of secondary turns

E_p = voltage on primary

E_s = voltage on secondary

I_s = current in secondary

I_p = current in primary

By substituting the values in Figure 17-11, we can see that the statements are correct.

Electrical Power in Transformers

Secondary current can control primary current. As this current increases, the power dissipated by the load must increase. The power used by the load is supplied by the primary; therefore, any increase in secondary power produces a corresponding increase of primary power.

We use the values in Figure 17-11 and the power formula, $P = I^2 \times R$, to determine the primary and secondary power of the ideal transformer circuit.

Power primary $I_p \times E_p = 300 \text{ mA} \times 120 \text{ V} = 3.6 \text{ W}$

Power secondary $I_s \times E_s = 60 \text{ mA} \times 600 \text{ V} = 3.6 \text{ W}$

From this, we see that primary power equals secondary power. Because the power used in the secondary is supplied by the primary, the following rule is true for our ideal transformer:

Primary power equals secondary power.

In equation form:

$$P_p = P_s$$

In this solution, we see that in the ideal transformer, primary power does indeed equal secondary power.

Reflected Impedance

We have discussed the control secondary current has over primary current. Let us see how this fact affects the operation of a transformer. Suppose the load is shorted on the secondary. If it is, the following will happen:

1. Secondary current increases rapidly.
2. Secondary magnetic field expands rapidly.
3. Many lines cut the primary.
4. Current in the primary increases rapidly.
5. Primary current exceeds the fuse's rating and it burns in two.
6. Current stops in both windings.
7. A short in the secondary is reflected back to the primary as a short, causing the fuse to blow.

Suppose the load connected to the secondary opens. Such an action causes:

1. Secondary current to drop to zero.
2. The secondary's field to drop to zero.
3. No magnetic lines to cut the primary.
4. Primary current to drop.

This demonstrates that any trouble occurring in the secondary will be reflected to the primary and will affect the primary current.

Self-Check

Complete each item by inserting the words and/or numbers needed to make a true and complete statement.

1. ______________ is defined as the process of transferring electrical energy from one circuit to another by using a magnetic field.

2. A transformer has four turns in its secondary for each turn in its primary. It is called a step ______________ transformer.

3. The percentage of magnetic coupling that occurs between a primary winding and a secondary winding is referred to as ______________.

4. A 4:1 step-down transformer has 2 A of current in its primary. Secondary current equals ______________ A.

5. A 1:4 step-up transformer has 1 kΩ of primary impedance. Secondary impedance equals ______________ Ω.

Types of Transformers

Figure 17-12 illustrates the schematic symbols for four basic types of transformers:

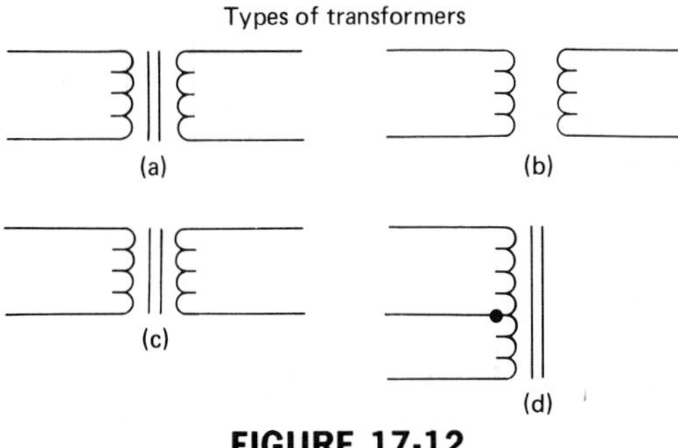

FIGURE 17-12

a. *Audio transformer.*
Designed to operate across the audible range of 20 Hz to 20 kHz. An audio-frequency transformer is shown in Figure 17-13.

b. *RF transformer.*
Designed to operate at radio frequencies above 20 kHz. An RF transformer is shown in Figure 17-14.

FIGURE 17-13
Assorted Audio Transformers

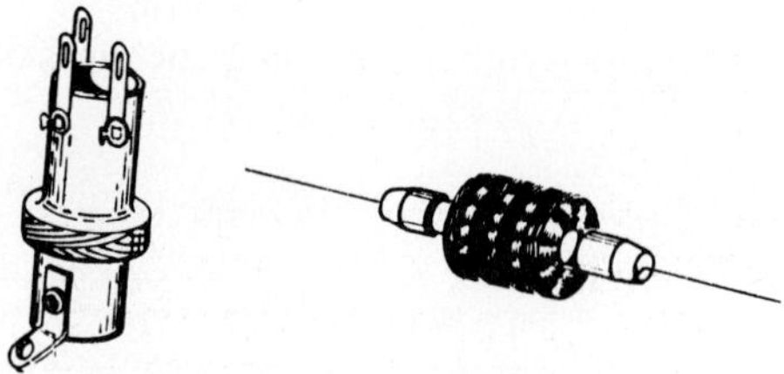

FIGURE 17-14
Assorted Radio Frequency Devices

c. *Power transformer.*
Designed to operate at specific frequencies of 50 Hz, 60 Hz, and 400 Hz. A power transformer is shown in Figure 17-15.

d. *Autotransformer.*
Identifies a method used to wind this type of transformer. It is distinguishable from the others only because of its winding and schematic symbol.

Figure 17-12a shows the schematic symbol for an audio-frequency (AF) transformer. This type of transformer is designed to operate at frequencies within the audible range of the human ear, when those frequencies are converted to mechanical ones. The audible range is considered 20 Hz to 20 kHz. Most AF transformers can handle a wider band of frequencies, but such transformers are classed as *audio* because these frequencies are included and because they can be passed with little distortion. Audio transformers have iron cores and relatively high inductance.

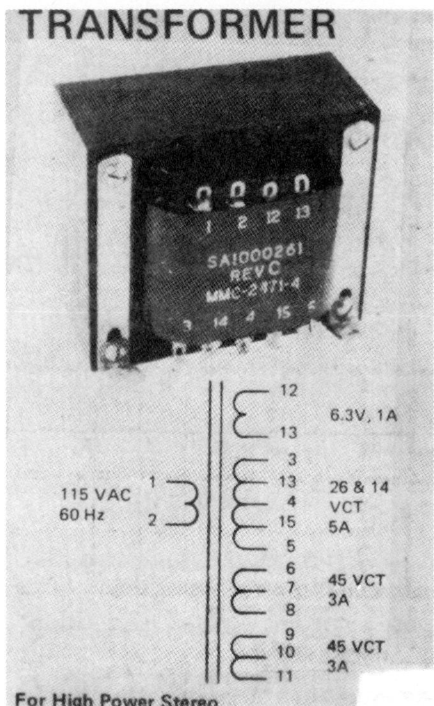

FIGURE 17-15
Multiple-Secondary Power Transformer with Schematic

Figure 17-12b represents a radio-frequency (RF) transformer. The RF transformer is used with the high frequencies used in radio communications. Because of the direct relationship between frequency and X_L, an audio transformer has such high opposition at these frequencies that it blocks the signal. Many precautions have been taken to reduce the inductance of RF transformers. The soft-iron core has been removed, the number of turns reduced, and the diameter-to-length ratio altered—with the result that RF transformers have low inductance.

Figure 17-12c shows a power transformer. Power transformers are designed specifically for use with systems operating at the commercial power frequencies of 50, 60, and 400 Hz. Household current in the United States operates at 60 Hz; therefore, all power transformers operate at this frequency. The 50-Hz system is common in Europe; 400 Hz is used primarily in aircraft and submarines. These transformers operate at low frequencies that are part of the audio band. Because their inductance must be high, an iron core is inserted. The main difference between the operation of an audio transformer and a power transformer is in the amount of current each must handle. The high-power requirements of power transformers means that their primary and secondary must have conductors of large diameter.

Figure 17-12d represents an autotransformer. The autotransformer differs from the other transformers in that it has only one winding. At some point in the winding, one turn is connected to another conductor, which serves as the third connection for the transformer. This results in a transformer that can be used with the entire coil acting as the primary, with a small part used as the secondary for voltage step-down purposes. The transformer can also be used with the small winding as the primary, and the entire winding as the secondary for voltage step-up purposes. The autotransformer is suitable for operation over wide frequency ranges, as long as its design provides for this range of frequencies.

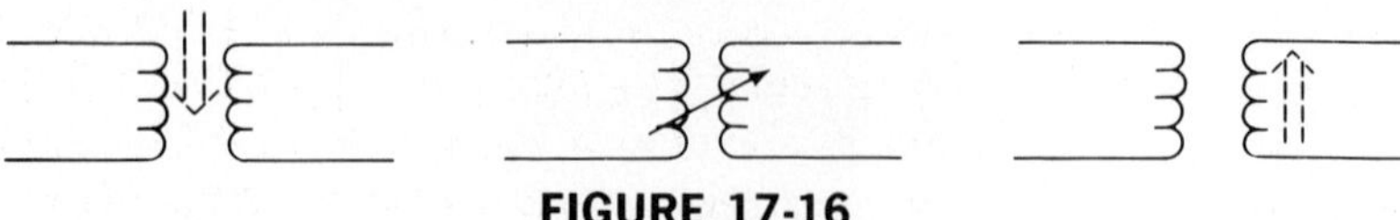

FIGURE 17-16

Some transformers in RF and autotransformer classifications are variable. Figure 17-16 illustrates the symbols used most often to indicate variable transformers (often called *tuned transformers*). Tuning is accomplished by mounting a small piece of soft iron on a threaded device. By screwing the threads in or out, the iron core can be inserted in or extracted from the air core of the transformer. Changing the position of the iron core changes the coil's inductance. As inductance changes, so does the frequency of operation.

Self-Check

Complete each item by inserting the words and/or numbers needed to make a true and complete statement.

6. Name four types of transformers.
   (a)
   (b)
   (c)
   (d)

7. Audio transformers have ____________ cores and RF transformers have ____________ cores.

8. The main difference between an audio transformer and a power transformer is the amount of ____________ they must handle.

9. Transformers are tuned by the movement of ____________ core into and out of their centers.

10. A power transformer designed for use with household current in the United States is designed to operate at ____________ Hz.

11. The main three frequencies used for commercial power distribution are:
    (a) ____________ Hz
    (b) ____________ Hz
    (c) ____________ Hz

Multiple Secondary Transformers

In some applications, a transformer has more than one secondary. This type of transformer is most often a power transformer but some RF transformers also have more than one secondary. Figure 17-17 shows the schematic diagram for a two-secondary-power transformer. It is not uncommon for power transformers to have two, three, four, or more secondaries. Each secondary is designed to supply voltage and current to a different load. A single primary must supply enough power for all the secondaries to use. All windings are wrapped around a common core, and they share the same magnetic field. The circuit in Figure 17-17 acts as two separate transformers, providing two separate power sources. The primary must be wound with large-diameter conductors capable of supplying the power required by a multiple-secondary power transformer.

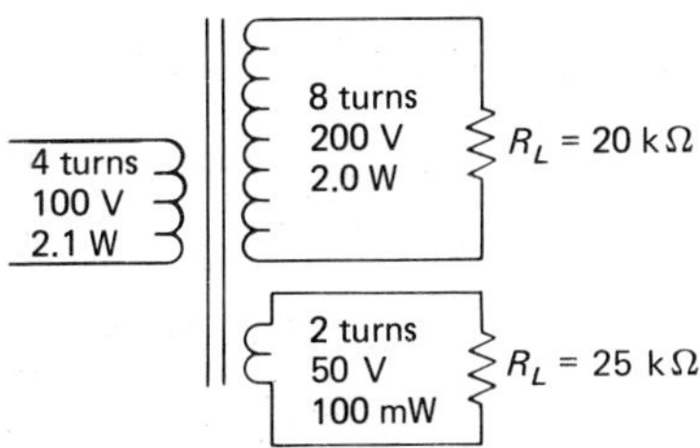

FIGURE 17-17

In some applications, the third winding of a transformer is called a *tertiary winding* (Figure 17-18). This winding is often used to couple the output of one circuit to another circuit. Tertiary windings are used in such circuits as oscillators.

FIGURE 17-18

Troubleshooting Transformers

Frequently, it is necessary to troubleshoot transformers, using an ohmmeter. Figure 17-19 represents a transformer with an open secondary.

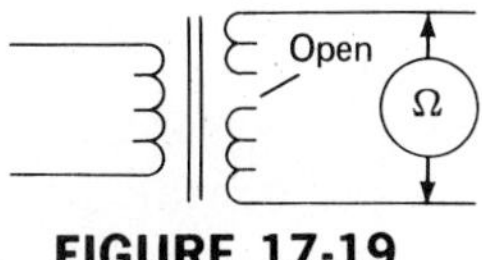

FIGURE 17-19

An open winding can be located by connecting the ohmmeter as shown in the diagram. When the winding is open, the ohmmeter indicates infinity. If the winding is not open, the ohmmeter will indicate a low resistance. This resistance represents the wire resistance of the conductors used in the winding. The primary can be checked by following the same procedure. *Caution:* Remove all power from the secondary, and disconnect one side of the winding to be checked.

The voltmeter can be used to check for an open transformer winding. Make sure the primary has the correct applied voltage. In checking the secondary voltage, if the meter indicates 0 V, either the primary or the secondary is open. Otherwise, induction occurs, and voltage is present in the secondary.

Figure 17-20 is a winding with several turns shorted, a common occurrence in transformers. Heat buildup in the winding can cause the insulating material to break down, which allows current to flow between windings, thus causing a short. A short, in turn, causes a decrease in induction and in output voltage. The winding contains numerous turns of wires, and several turns must be shorted before the change is detected by an ohmmeter. For this reason, an ohmmeter is not very accurate for making this check, though it may be of some value. Turn off the power and isolate the winding. Connect the ohmmeter and compare its indication to the ohmic value of a good winding, as stated in the specifications for the transformer.

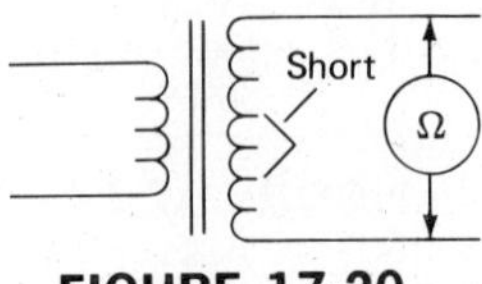

FIGURE 17-20

A more accurate method of checking transformer quality is to connect the primary to an accurate AC voltage and measure secondary voltage. Shorted turns in either winding cause secondary voltage to be low.

Determining Step-Up and Step-Down

To determine whether a transformer is a step-up or a step-down transformer, use an ohmmeter. The meter indications are:

Step-down.
Primary resistance is *high* and secondary resistance is *low.*

Step-up.
Primary resistance is *low* and secondary resistance is *high.*

A voltmeter can be used to make these checks. To do this,

1. Connect a known voltage to the primary.
2. Measure the output (secondary) voltage.
   (a) If the secondary voltage is higher than the primary voltage, the transformer is of the *step-up* type.

(b) If the secondary voltage is lower than the primary voltage, the transformer is of the step-down type.

Note: Make sure the voltage connected to the primary does not exceed its rating and that it is capable of operating at that frequency.

Self-Check

Answer each item by indicating whether the statement is true or false.

12. ____ An ohmmeter can be used to determine whether a transformer is step-up or step-down.

13. ____ An ohmmeter that reads 0 Ω across a transformer winding indicates that the winding is open.

14. ____ It is difficult to detect a shorted transformer winding, using an ohmmeter, because a good winding has low DC resistance.

15. ____ A multisecondary transformer has one secondary winding with many turns of wire.

16. ____ An audio transformer can be identified by its tertiary winding.

Summary

In this chapter, we have discussed the following terms:

1. *Transformer action.*
The process of transferring electrical energy from one circuit to another by magnetic coupling.

2. *Step-up transformer.*
A transformer with a higher voltage on its secondary than is applied to its primary.

3. *Step-down transformer.*
A transformer with lower voltage on its secondary than is applied to its primary.

4. *Turns ratio.*
The ratio of turns in the primary to the number of turns in the secondary.

5. *Coefficient of coupling.*
The percentage of magnetic-line coupling between the primary and the secondary.

6. *Transformers.*
The types are audio, radio, power, and autotransformer.

7. *Reflected impedance.*
The apparent impedance across the primary of a transformer when current flows in the secondary.

8. *Autotransformer.*
A transformer with a distinctive winding.

9. Transformers that step up voltage and step down current.
10. Transformers that step down voltage and step up current.
11. The procedures for using a voltmeter and an ohmmeter to check transformers for
    (a) Open windings.
    (b) Shorted turns.
    (c) Step-up characteristics.
    (d) Step-down characteristics.

Review Questions and/or Problems

1. In a step-up transformer, the
    (a) secondary voltage is less than primary voltage.
    (b) primary voltage and secondary voltage are equal.
    (c) primary voltage is less than secondary voltage.
    (d) current is equal in each winding.

2. A transformer has a 7:1 ratio. In a transformer that is 100% efficient, primary voltage is 210 V. What is the secondary voltage?
    (a) 7 V
    (b) 30 V
    (c) 300 V
    (d) 1470 V

3. The symbol in Figure 17-21 represents
    (a) an audio transformer.
    (b) an autotransformer.
    (c) a power transformer.
    (d) an RF transformer.

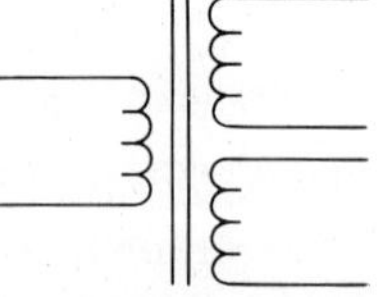

FIGURE 17-21

4. A transformer has 110 V and 2 A in its primary, and 880 V on its secondary. What is its secondary current?
    (a) 25 mA
    (b) 250 mA
    (c) 2 A
    (d) 16 A

5. Which of the following illustrates a step-down transformer?
    (a) $E_p = 120$ V; $E_s = 120$ V
    (b) $E_p = 120$ V: $E_s = 500$ V
    (c) $E_p = 200$ V; $E_s = 120$ V

6. Figure 17-22 represents a good transformer and its specified resistances. Which of the following ohmmeter indications indicates a partially shorted secondary?
   (a) 70 Ω between points 1 and 2
   (b) 0 Ω between points 3 and 4
   (c) 0 Ω between points 1 and 2
   (d) 110 Ω between points 3 and 4

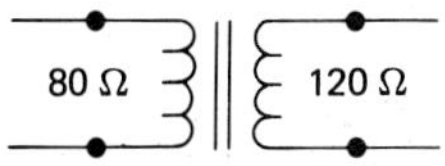

80 Ω 120 Ω

FIGURE 17-22

7. If the transformer in Figure 17-23 has 100% efficiency, what is its primary power?
   (a) 50 W
   (b) 100 W
   (c) 150 W
   (d) 200 W

8. In Figure 17-23, solve for the primary impedance of the circuit.
   (a) 100 Ω
   (b) 50 Ω
   (c) 200 Ω
   (d) 12.5 Ω

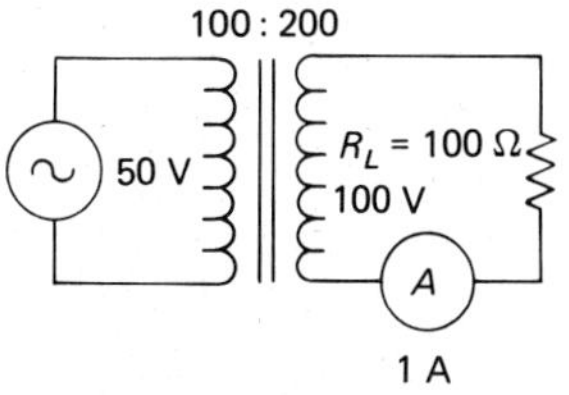

100 : 200
50 V
R_L = 100 Ω
100 V
A
1 A

FIGURE 17-23

9. Many transformers are designed and used to match the impedance between two separate circuits.
   (a) true
   (b) false

10. In a step-down transformer, secondary impedance is ____________ primary impedance.
   (a) higher than
   (b) lower than
   (c) equal to

18

Capacitors, Capacitance, and Capacitance Reactance

Introduction

We have studied inductive components and their reaction to alternating current. Inductive components are used in AC circuits because they possess a property that allows them to oppose any change in current. Because alternating current changes continually, the components continually react. Inductors store energy in an electromagnetic field.

At this point, another component is introduced. The *capacitor*: It has a property that enables the capacitor to oppose any change in voltage. In AC circuits voltage is continually changing and the capacitor reacts continually to the change. A capacitor stores electrical energy in an electrostatic field.

Capacitance in a circuit depends on several factors. The effect of capacitance is opposite that of inductors, and the two effects are 180 degrees out of phase. Capacitive reactance is present and can be predicted. Its effect is a valuable factor in the design of electronic circuits. Without capacitive and inductive reactances, it would be impossible to build the communication circuits used today.

Circuits have three properties—resistance, inductance, and capacitance—and a thorough understanding of each is necessary. Resistance and inductance have already been covered; we start our study of capacitance here.

Objectives

Each student is required to:

1. Define
   (a) Capacitance.
   (b) Capacitance reactance.
   (c) Farad.
   (d) Dielectric.
   (e) Capacitor DC voltage rating.
2. Identify the schematic symbols for:
   (a) A capacitor.
   (b) An electrolytic capacitor.
   (c) An adjustable capacitor.
   (d) A variable capacitor.
3. Identify the symbols for
   (a) Capacitance.
   (b) Total capacitance.
   (c) Capacitance reactance.
4. Solve for total capacitance in series strings and parallel networks.
5. Solve for total capacitive reactance when capacitance and frequency values are given.
6. Draw simple schematics using capacitors as part of the circuits.
7. Convert capacitance values to microfarads and picofarads.
8. Name the factors that affect the capacitance of a capacitor.
9. Explain the effect of voltage rating and state what determines the voltage rating of a capacitor.

Capacitance

A *capacitor* is an electronic device made of two conductors separated by an insulator, or dielectric. Capacitors have a property that allows them to oppose a change in voltage by storing a charge that opposes the applied voltage. *Capacitance* is the property of a component that accepts and stores a charge in an electrostatic field. All conductors in operating circuits form capacitors with other nearby conductors.

A capacitor is formed when two conductors are separated by a dielectric. Any time two conductors operate near each other, they form a capacitor, and capacitance exists between them. Therefore, capacitors can store electrical energy in the electrostatic field.

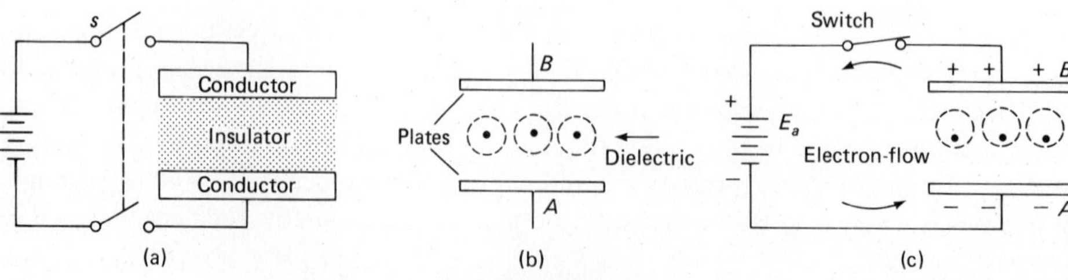

FIGURE 18-1

The diagram of a simple capacitor is shown in Figure 18-1a. To understand what happens in a capacitor, we must understand and apply the principles of electrostatics. In their uncharged condition, the atoms that make up the dielectric of the capacitor are in a neutral state; the electrons orbit the nucleus, as is normal (Figure 18-1b).

Three atoms can illustrate what happens in a capacitor. When a voltage is applied to the capacitor as shown in Figure 18-1c, a polarity is set up. The potential applied to the capacitor's plates attracts electrons to the positive potential and repels them from the negative potential. This distorts the atoms' orbits. Called *orbital stress,* this distortion is proportional to the strength of the applied voltage. Electrons forced nearer plate B, due to distortion, release electrons from plate B, causing them to flow to the battery's positive pole. The resulting positive ions cause plate B to have a positive charge, and the extra electrons at plate A give it a negative charge. The change in voltage felt at plate A is passed to plate B through its effect on the atoms in the dielectric. Electrons from a voltage source *do not* flow through the dielectric; the force exerted by the source passes from one plate to another by *electrostatic* effect. (A small number of electrons may flow in the dielectric; this leakage is not desirable.) Energy is stored in a dielectric in the form of *orbital stress.* Millions of atoms are under stress, and the total energy they can store is considerable. Energy charged in the dielectric is referred to as *capacitor voltage* (E_C); it is sometimes equal to the applied voltage (E_a). When $E_C = E_a$, the capacitor is said to be fully *charged.* At this point, E_C and E_a act as opposing power sources. The result is an infinite resistance to DC, which causes DC current to stop. When that happens, the capacitor is said to "block DC."

Should the switch be opened, the electrons that have accumulated at plate A will be trapped and will remain there until an escape route is provided. This condition also causes the electrostatic field to continue under stress and to repel electrons from plate B, which gives it a positive charge. The charge stored on a capacitor can be retained indefinitely. The charge on the capacitor can be removed only by providing a path for the electrons on plate A to move to plate B, where they neutralize the positive ions concentrated there.

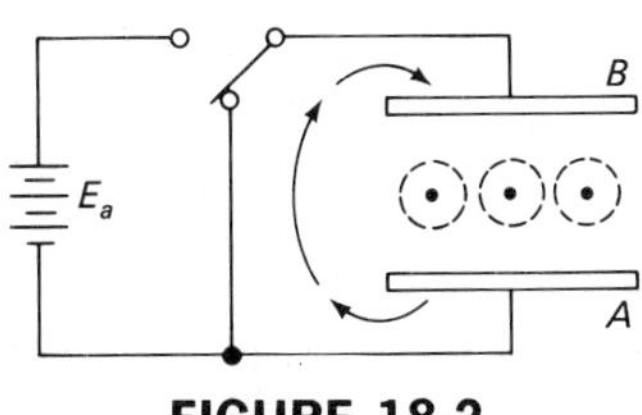

FIGURE 18-2

Placing the switch in the position shown in Figure 18-2 provides a path for the capacitor to *discharge.* Electrons move through the short and neutralize positive ions at plate B. When the capacitor is fully discharged, the two plates are returned to the neutral state, and no difference in potential exists. At this time, all atoms are relieved of their orbital stress and return to normal.

The amount of charge stored by a capacitor is directly proportional both to the applied voltage and to the *capacity* of the capacitor. For a given capacitor, the ratio of charge (Q) to voltage applied, E, is the measure of capacitor action called *capacitance* (C). This can be stated by the equation:

$$Q = C \times E$$

where

Q = amount of charge
E = voltage applied

The unit of measure for capacitance is the *Farad* (f), named for the English chemist and physicist Michael Faraday (1791–1867). One farad equals the capacity of a capacitor to store 1 coulomb (6.28×10^{18} electrons) when 1 V is applied. At present, it is not practical to use capacitors of 1 farad; in recent years, however, the lower voltages used in operating circuits have made larger capacitances more practical. Today, it is not uncommon to see capacitors with a capacitance of 0.02 farad. In most applications, capacitance values in microfarads and picofarads are common.

The capacitance (C) of a capacitor is determined by distance between conductive plates, plate area, and type of dielectric. The equation can be stated as

$$C = \frac{kA}{d}$$

where

A = plate area (in square centimeters)
d = distance between plates (in centimeters)
k = dielectric constant

See Table 18-1.

TABLE 18-1

Dielectric Material	*Dielectric Constant (k)*	*Dielectric Strength (volts per 0.001 in.)*
Air (Vacuum)	1.0	80
Fiber	6.5	50
Glass	6.0	500
Mica	4.2	200
Castor oil	6.0	2000
Paper	4.7	380
(1) Beeswaxed	3.1	1800
(2) Paraffined	2.2	1200
Ceramics	80–1200*	335–2000

**Specific values depend on the composition of a material.*

Plate area determines the number of electrons available for use in the current action of a capacitor. A plate of greater area accepts and stores a greater charge; therefore, plate area is directly related to the capacity (Q) of a capacitor.

Distance between plates is inversely proportional to the amount of energy the capacitor can store. Increasing the distance between plates reduces the amount of energy the capacitor can store. This conforms to Coulomb's law, which established that as the distance between charged bodies increases, the force between them decreases.

The dielectric used has a direct effect on the amount of energy a capacitor can store. Each dielectric can survive application of a specific voltage before it breaks down and allows current to flow through it. Each material has a different atomic structure which controls the number of atoms available for orbital stress. All dielectric materials are classified according to their *dielectric constant* (k), a constant arrived at by comparing dielectric materials under vacuum conditions to determine their ability to accept and retain a charge. Each dielectric is compared to air, which has a constant of 1, and numbers are assigned for purposes of comparison. The higher the value of this constant, the greater the capacitor's ability to accept and store a charge.

The dielectric constants for some common dielectrics are given in Table 18-1.

Voltage Rating

In addition to capacitance, every capacitor has a *voltage rating.* This rating indicates the amount of voltage that can be applied across the capacitor without the dielectric breaking down and allowing current to flow. The amount of voltage that can be applied depends on the dielectric material used. The voltage at which the dielectric no longer insulates the two plates from each other is called *breakdown* voltage. Table 18-1 also gives the breakdown voltages for materials that are one-thousandth inch thick.

A capacitor's voltage rating is the maximum DC voltage that can be applied continuously to the capacitor. A capacitor marked 600 V DC should operate at 600 V DC on a continuous basis without damage to the capacitor.

The capacitance and voltage rating are usually printed on a capacitor itself; some capacitors, however, are marked with a color code similar to that used for resistors (see Appendix H).

Capacitors

In manufacturing capacitors, three factors are considered: (1) type and thickness of dielectric; (2) spacing between plates; and (3) the area of the plates. These factors also determine the size and shape of capacitors. Capacitors are classified as either *fixed* or *variable.*

A *fixed capacitor* provides a fixed amount of capacitance for use at a desired voltage rating. Manufactured to fit most circuits, such capacitors are usually classified according to the type of dielectric—paper, oil, mica, ceramic, tantalum, or electrolytic (not a type of dielectric).

Variable capacitors are used in circuits where the amount of capacitance must be adjustable. Variable capacitors are usually named according to their function, for example, tuning, trimmer, padder, or neutralizing.

Paper

Capacitors with paper dielectric are inexpensive and are in wide use. The low cost and size of paper capacitors make them ideal for many applications. The paper used as their dielectric is usually waxed and porous. Thus paper capacitors are seldom used in circuits operating above 600 V DC. Plates for this type of capacitor are long strips of aluminum foil separated by a dielectric of waxed paper. The two plates and the dielectrics are rolled together (Figure 18-3). Once rolled, they are sealed with wax or some other material to keep out moisture.

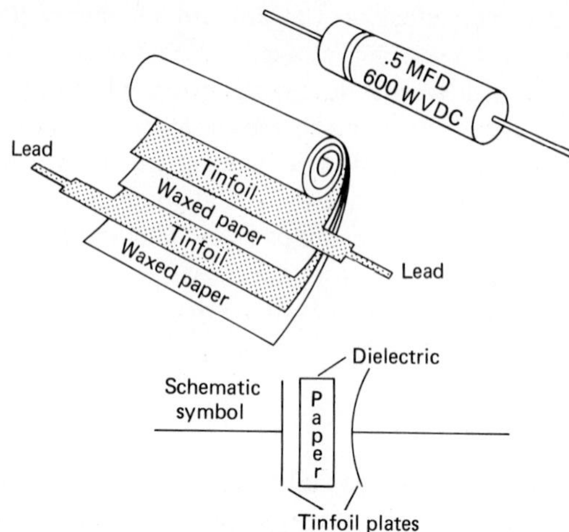

FIGURE 18-3

Oil

Oil capacitors are constructed similarly to paper capacitors. Oil has a much higher breakdown voltage than paper. This allows oil capacitors to operate at medium-high voltages. Oil capacitors can function in circuits that handle large amounts of power, such as a transmitter.

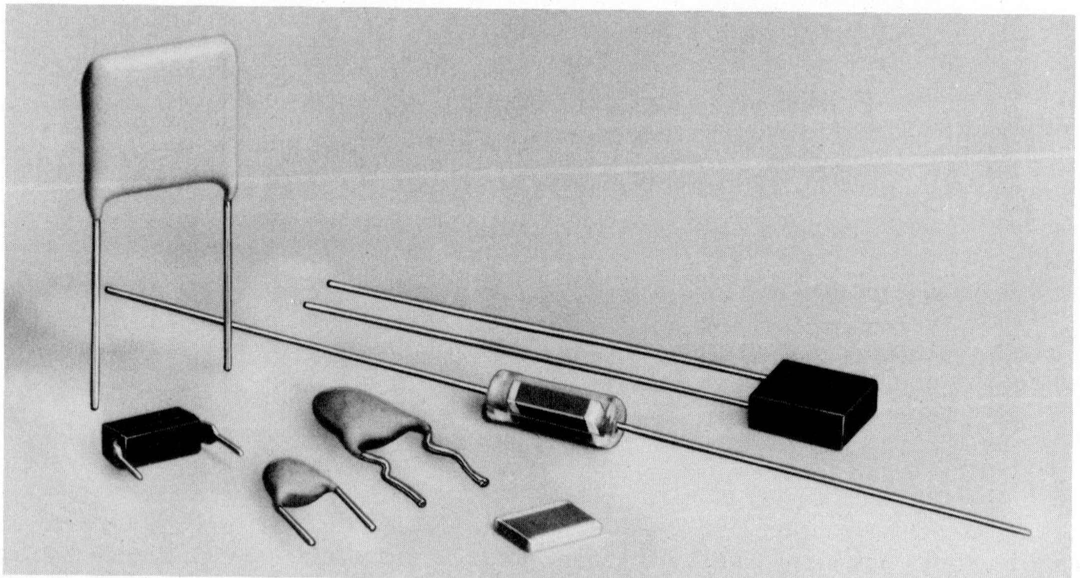

FIGURE 18-4
An Assortment of Ceramic Capacitors (Picture Courtesy of Centralab, Inc., A North American Phillips Company)

Mica

For very small capacitances—those in the range 5 to 50,000 picofarads (pf)—it is common to use mica as the dielectric. Mica capacitors can operate at voltage ratings up to 7500 V DC and with high-frequency circuits. This combination of capacitance and breakdown voltage allows small mica capacitors to be used in place of bulky paper ones. Their construction involves laminating foil and mica to obtain the desired capacitance and working voltage. The finished unit is enclosed in a bakelite or plastic case, which makes it compact and durable.

Ceramic

With the advent of high-frequency communication circuits, the need arose for small capacitors with high-voltage ratings and relatively small capacitance values. *Ceramic capacitors* were designed to fill this need. Ranging from about 0.5 pf to 0.01 μf, these capacitors can be used in low-power high-voltage circuits at up to 30,000 V DC. They are used extensively in television and two-way communication systems. Their construction involves coating a hollow ceramic cylinder inside and out with silver paint. The paint becomes the capacitor's plates; leads are then connected to the plates at each end of the tube. The ceramic separates the two silver plates and acts as a dielectric. When completed, these capacitors, which are small, resemble resistors. Figure 18-4 shows some ceramic capacitors.

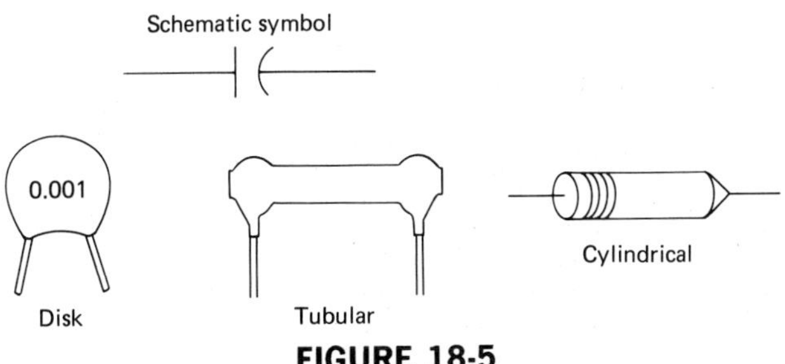

FIGURE 18-5

Electrolytic

The basic construction of an *electrolytic capacitor* is shown in Figure 18-6. The metal container serves as the negative connection to the capacitor and the electrolyte as the negative plate. The positive (plate) connection of this capacitor is an aluminum rod inserted in the center of the capacitor. The rod is surrounded by a film of oxide, which serves as the dielectric. Electrolytic capacitors have definite polarities that must be observed when connecting them in a circuit. If the current is reversed in this type capacitor, the oxide coating on the positive plate will break down, allowing current to flow through the capacitor. Heat buildup in the capacitor can cause it to explode.

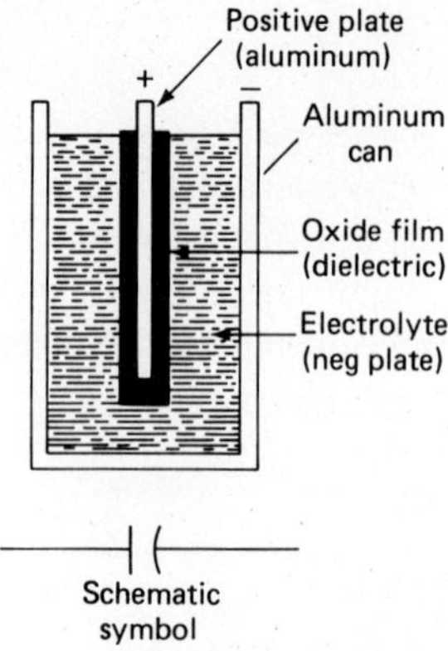

FIGURE 18-6

Electrolytic capacitors for AC use are constructed so that two units are placed back to back, and they share a negative plate. An oxide coating is placed on each positive plate. The AC electrolytic capacitor is designed so that it presents a dielectric to each polarity of the sine wave. This is used to accomplish a phase shift for starting AC motors. Unless otherwise noted, all references to electrolytic capacitors are to those with DC polarity.

The dielectric (oxide film) is extremely thin, usually about 0.1 inch. This thickness allows greater capacitance than for the other types discussed. Another advantage is that this very large capacitance is housed in a smaller space. Many computer-grade capacitors today have 20,000 μf of capacitance.

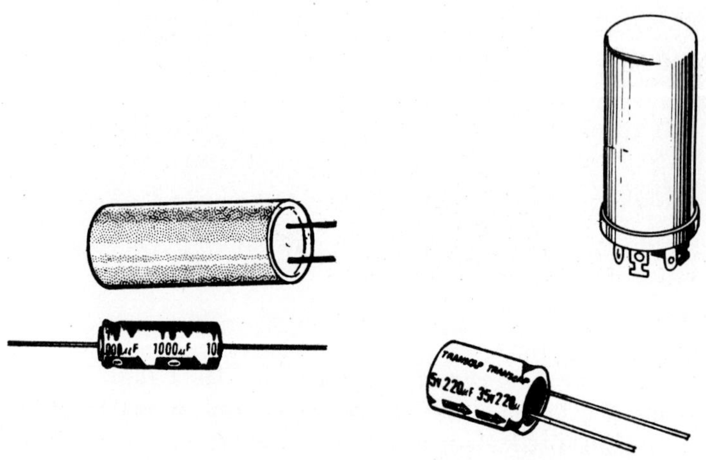

FIGURE 18-7
An Assortment of Electrolytic Capacitors

Electrolytic capacitors used with high-power high-frequency transmitters have an oil electrolyte. Called "wet electrolytes," they must be mounted vertically to avoid spilling oil through the breather. The dry electrolyte was developed to overcome the disadvantages of wet-electrolyte capacitors. In this type of capacitor, two flexible sheets of aluminum are separated by a gauze or paper that has been impregnated with a jellylike electrolyte. Construction is similar to that of paper capacitors. Single or multiple capacitors can be placed in a single "can." In some dry electrolytic capacitors, plates are etched to increase the area of the plate that comes in contact with the electrolyte, resulting in greater capacitance.

Electrolytic capacitors are used extensively in low-voltage power supplies. Capacitance, voltage rating, and polarity are stamped on the unit. On cans that contain more than one capacitor, a coding system is used to identify the various capacitors. When checked with an ohmmeter, this type of capacitor has a resistance of 1 MΩ to 2 MΩ. This compares to the 200 MΩ to 500 MΩ characteristic of a paper capacitor. Figure 18-7 shows electrolytic capacitors.

Variable

In circuits that operate over a band of frequencies it is necessary to have a way to vary capacitance. *Variable capacitors* such as the one shown in Figure 18-8 were designed. As frequencies become higher, the amount of capacitance used is much smaller. Variable capacitors of 500 pf can be used in a low-frequency circuit, whereas a 10-pf value can be used at very high frequencies.

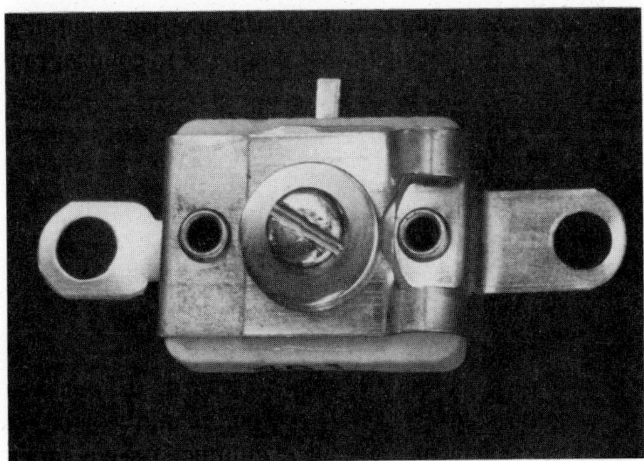

FIGURE 18-8
A Variable Capacitor

In some variable capacitors, air is the dielectric. Power and voltage determine the size of the capacitor used. Air has a low breakdown voltage, which means the plates must be spaced farther apart as the voltage is increased. The thin plates are made of aluminum, thus reducing size and weight. Strong and corrosion-resistant aluminum is ideal. One set of plates (the stator) is stable, and the other set (the rotor) is movable. The rotor is connected to a shaft, which allows it to be moved. When rotated, the rotor's plates pass between the stator's plates without touching them. This increases or decreases the plate area and varies the capacitance. Figure 18-9 illustrates an air-dielectric variable capacitor of the type used to tune an AM/FM radio. Note the schematic symbol.

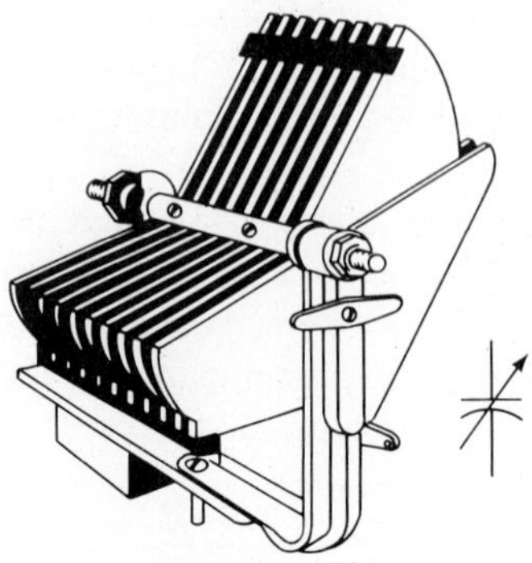

FIGURE 18-9

Another type variable capacitor is the *adjustable capacitor*. This type of variable capacitor is used to set the amount of capacitance during maintenance and to leave it in that position indefinitely. Adjustable capacitors are made of mica or ceramic dielectrics. Mica capacitors consist of a spring-metal plate separated from another plate by a thin slice of mica. A screwdriver-type adjust is connected to the plate, which can be varied, using a screwdriver. (See Figure 18-10.)

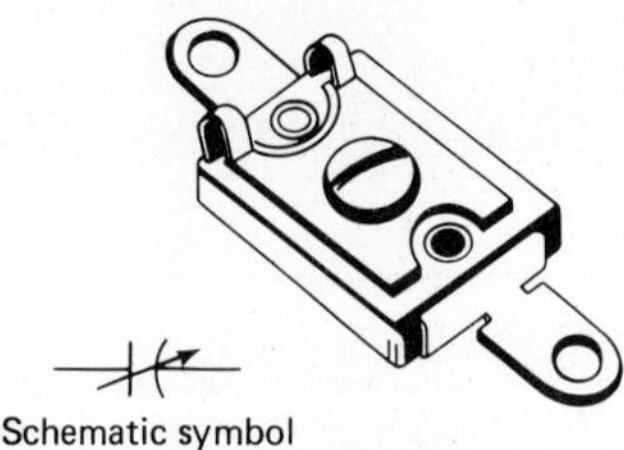

FIGURE 18-10

Adjustable capacitors are classified in two groups, padders and trimmers; either can be used to fine tune a circuit. Ceramic capacitors are small and can be adjusted with a screwdriver. Their plates are set in ceramic crystal and are rated up to 500 V DC (Figure 18-11).

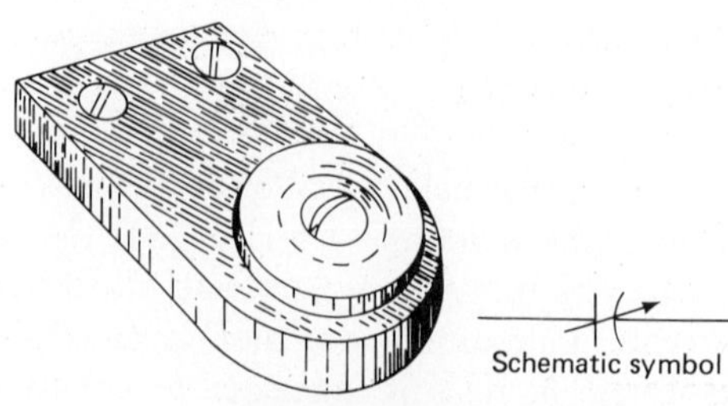

FIGURE 18-11

Self-Check

Answer each item by indicating whether the statement is true or false.

1. ___ A capacitor consists of two conductors separated by a dielectric.
2. ___ Capacitors that have an air dielectric have high capacitance.
3. ___ Energy stored in a capacitor results from orbital stress.
4. ___ Capacitors block DC and pass AC.
5. ___ In a fully charged capacitor, $E_C = E_a$.
6. ___ E_C and E_a act as series-aiding power sources.
7. ___ All electrolytic capacitors use oil dielectrics.
8. ___ A trimmer capacitor is connected in series with a larger variable capacitor.
9. ___ Farad is the unit of measure for capacitance.
10. ___ When space is a problem, paper capacitors can be used in place of ceramic capacitors.

Phase Relationships

When connected to an AC voltage source, a capacitor reacts in predictable ways. Like a coil, a capacitor reacts in certain ways. Its phase relations are predictable, as are its opposition to a specific frequency.

Remember that a capacitor blocks DC, but AC can be passed, because of the capacitor's ability to charge and discharge to any change in voltage. For AC to be passed, it is not necessary for current to flow through the dielectric. Because of the applied voltage, electrons accumulate on the negative plate. Electrons in the dielectric are repelled by electrostatic repulsion, which pushes them nearer the positive plate. The nearness of these electrons to the positive plate causes other electrons to be dislodged from the positive plate and to flow to the positive pole of the power source, thus completing the current loop. In effect, AC has been passed, although current did not flow in the capacitor.

A fully charged capacitor has an electrostatic field whose voltage is equal to the applied voltage. These voltages—E_C and E_a—act as opposing power sources (Figure 18-12). In diagram a, an ideal capacitor is connected to a variable DC power source, whereas diagram b represents current and voltage relationships in the capacitor. A dashed line represents the flow of current from and to the battery during charge and discharge. A solid line indicates voltage across the capacitor. Elapsed time is indicated by T_0, T_1, and so on.

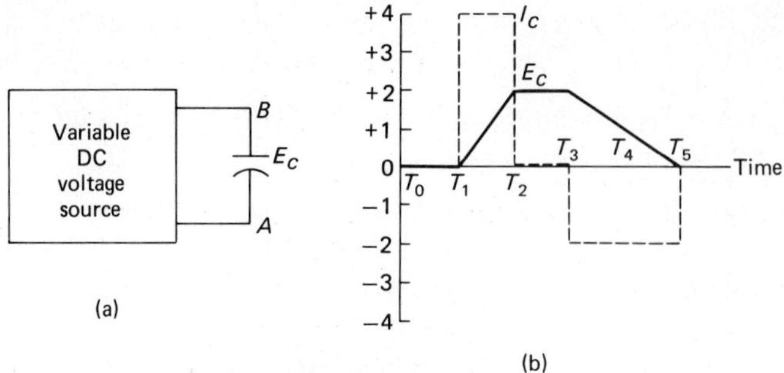

FIGURE 18-12

With the power source off, voltage and current are both zero, as shown between points T_0 and T_1. At the time voltage is applied, T_1, the charge on the capacitor is zero. With zero charge, $E_C = 0$ V, and no opposition exists between the power source and the capacitor. When applied voltage is increased (T_1), maximum current is allowed to flow, which begins to charge the capacitor. As applied voltage increases (at a linear rate), electrons accumulate on plate A and leave plate B at the same linear rate. This results in E_C increasing at a linear rate and opposing E_a. The steady increase in E_a causes a steady rate of current flow, which continues as long as the applied voltage increases (T_1 to T_2).

When the increase in E_a stops, and the voltage becomes constant, current flow stops. From T_2 to T_3, applied voltage is a constant value; current flow is zero. During this time, $E_C = E_a$; this remains true as long as the applied voltage does not change. When E_C and E_a are equal in voltage, but opposite in polarity, their opposition prevents the flow of current.

At T_3, the applied voltage begins to decrease at a linear, but slower rate. As soon as E_a changes, the capacitor acts to couple the change and current flows in the circuit. As the voltage gradually decreases between T_3 and T_5, the capacitor continually couples the change and is discharging. The discharge current is steady, as is the decrease in E_a. Because the discharge time, T_3 to T_5, is twice as long as the charge time, T_1 to T_2, current flow is half as much. Note that as the applied voltage rose, the capacitor charged at a rate of $+4$; and that as the applied voltage decreased, the discharge current rate was -2.

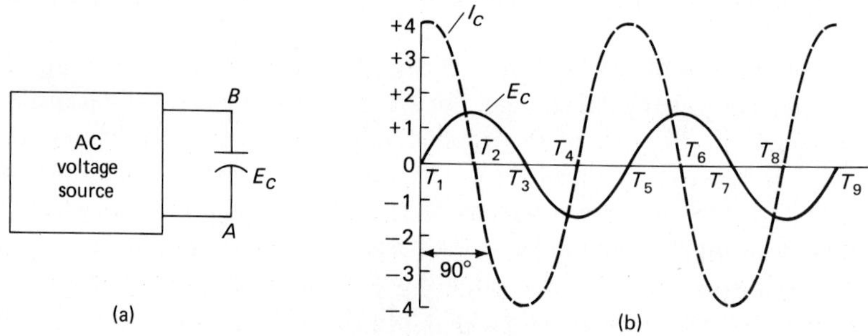

FIGURE 18-13

So much for the effect of DC on a capacitor. Now let us see what effect AC has on the same circuit. Observe Figure 18-13a. At T_1, the charge current ($+4$) rate is the same as it was in the DC circuit. Because E_a and E_C are parallel, their voltages must be equal. As E_a increases to a positive peak, E_C increases to a positive peak; but at the same time, I_C is decreasing to zero. When E_a and E_C reverse and decrease to zero, I_C increases to a negative peak. At T_3, I_C is maximum negative; E_a is zero and E_C is zero. Between T_3 and T_4, E_a and E_C reverse polarity and start increasing toward a negative peak; meanwhile, I_C starts decreasing to zero. At T_4, E_a and E_C are maximum negative, and I_C is zero. From T_4 to T_5, E_a and E_C decrease to zero; but I_C increases in a positive direction again. At T_5, E_a and E_C are zero, and I_C is maximum positive. This completes one sine wave. When we compare the second sine wave to the first, we see that the effect on the capacitor is identical. Note that during the period T_4 to T_6, current flow in the capacitor is the reverse of what it is during the period T_2 to T_4.

Notice, too, that when E_C is maximum (positive or negative), I_C is zero (T_2, T_4, T_6, and T_8). At these points, the applied voltage is maximum, which causes the capacitor to be charged to maximum. Since E_C is opposite polarity to the applied voltage, it acts as a series-opposing power source blocking any current flow. When E_a and E_C are zero, I_C is maximum, resulting in the fastest change in capacitor charge. Capacitor current is directly proportional to the rate of change of the voltage across the capacitor.

During the first 90 degrees of the sine wave, T_1 to T_2, electrons accumulate at plate A. These electrons produce orbital stress in the atoms of the dielectric, which forces electrons at plate B to be dislodged and flow to the source.

Capacitors store energy during all periods of voltage increase, then discharge energy during all periods of voltage decrease.

In Figure 18-13b, we see that between periods T_1 to T_2, T_3 to T_4, T_5 to T_6, and T_7 to T_8, the voltage across the capacitor is rising to either a positive or a negative peak. In all cases, however, capacitor current leads capacitor voltage by 90°. Current does not flow through the capacitor; but, by distorting atoms in the dielectric, energy can be stored in and released from the capacitor. The capacitor's polarity of charge may be either positive or negative. As E_a is increased, the number of distorted atoms increases. As E_a decreases, the number of these atoms decreases. When E_a is zero, the number of distorted atoms is zero. During the period in which the number of distorted atoms decreases, energy stored as orbital stress is returned to the power source.

In our study of coils, we established that

Voltage leads current by 90° in a coil.

and that

A coil opposes any change in current flow.

A similar condition exists in the case of a capacitor. In discussing Figure 18-12a and Figure 18-12b, we established that as soon as DC voltage is applied, current is maximum and voltage is minimum across the capacitor. Over a period of time, current drops to zero and E_C equals E_a. When the switch is opened, if a path for current is present, energy stored in the capacitor's electrostatic field is returned to the source. The effect is similar when the voltage applied to a capacitive circuit is supplied by an AC power source. If voltage starts at zero, as shown in Figure 18-14, current immediately flows at the maximum rate. As voltage increases, current decreases. After 90° of the sine wave, E_C is maximum and equals E_a; but I_C has decreased to zero. From this, we can state that in a capacitor,

Current leads voltage by 90°.

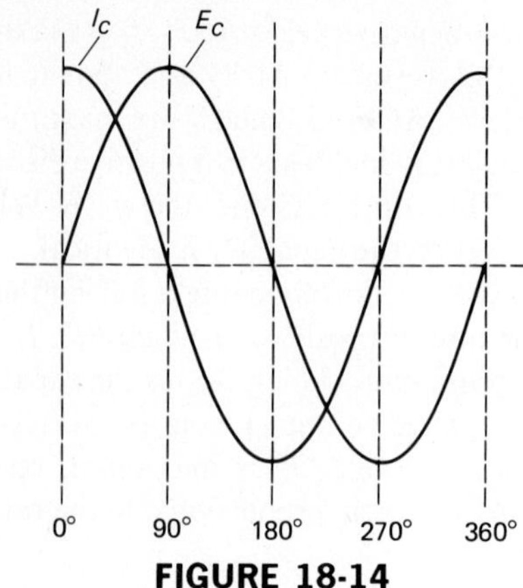

FIGURE 18-14

We have already established that

Capacitors oppose any change in voltage.

In inductive circuits, we introduced the term *ELI* to help us remember that voltage leads current. Now we take this a step further. The term can now be

ELI the ICE man.

where

E = voltage
L = inductance**
I = current

and

I = current
C = capacitance**
E = voltage

The double asterisks represent the devices, and the position of E and I in each word represents the lead and lag portions of the analyses. Because, in *ICE*, I leads E across C, we can state,

Current leads voltage across a capacitor by 90°.

Capacitive Reactance

The opposition of a capacitor to current flow is called *capacitive reactance* and is indicated by the symbol X_C. What factors determine a capacitor's reactance? In Figure 18-15, we see a capacitor in series with an AC source and an AC ammeter. When the capacitor (C) is charged (Q) to the voltage

(E), the capacitor stores an amount of energy equal to its capacity times voltage: $Q = CE$. The same amount of energy is stored in each alternation of the AC sine wave (Figure 18-15b). The amount of time allowed for charging depends on the frequency, where $f = 1/t$. If we double the frequency without changing the capacitance, C, or the voltage, E, the same amount of energy is stored in half the time (Figure 18-15c). In this case, the same amount of energy is stored during each alternation, but it takes only half the time. To do this, the capacitor current must be twice the current discussed. Cutting the frequency in half has the opposite effect (Figure 18-15d). The peak voltage charged is the same, but the time required for charging is doubled, which means that half as much current flows.

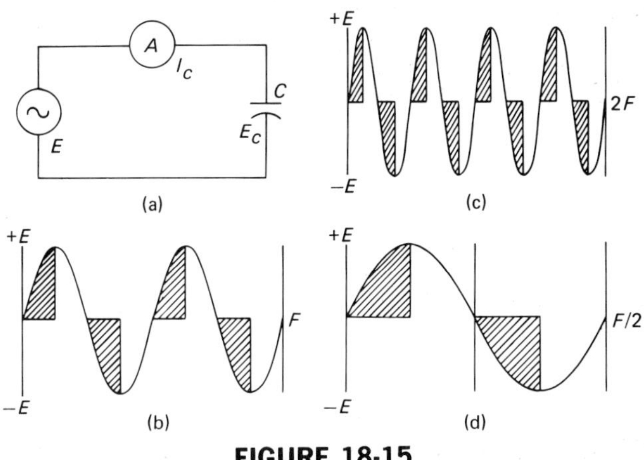

FIGURE 18-15

Frequency is a factor that affects capacitive reactance. As frequency increases, current increases and capacitive reactance decreases.

Capacitive reactance is also affected by the capacity of a capacitor. Capacitors with large capacities allow more current to flow when charging to the same applied voltage. If more current is allowed to flow, the opposition (X_C) to current flow must decrease. As capacitance is increased, capacitive reactance is decreased.

Capacitive reactance, X_C, can be stated by the following formula:

$$X_C = \frac{1}{2\pi fC}$$

where

X_C = capacitive reactance (in ohms). *Note:* This opposition does not dissipate power.
2π = 6.2831853
f = frequency (in hertz)
C = capacitance (in farads)

Both equation and explanation point to the fact that two variables affect capacitive reactance: frequency and capacitance. The X_C formula can also be used to find frequency or capacitance if the other quantities are known.

Capacitance:

$$C = \frac{1}{2\pi f X_C}$$

Frequency:

$$f = \frac{1}{2\pi C X_C}$$

These are used in mathematical and calculator solutions of capacitance, capacitance reactance, and frequency problems. Before starting, though, you should be aware of the method used with your calculator. The procedures for solving X_C are:

Step 1: Multiply the quantities $2\pi fC$.
Step 2: Take the reciprocal of the number.

In solving these problems, solve the denominator first; then take the reciprocal of that value.

EXAMPLE 1

Solve for X_C when $f = 6$ kHz and $C = 0.05$ μf.

Mathematical solution:

$$X_C = \frac{1}{2\,\pi f C}$$

$$X_C = \frac{1}{2\,(3.14) \times 6 \times 10^3 \times 5 \times 10^{-8}}$$

$$X_C = \frac{1}{1.8849556 \times 10^{-3}}$$

$$X_C = 530.5\ \Omega$$

Calculator solution:

Action	*Entry*	*Display*
ENTER	2	2
PRESS	×	2
PRESS	π	3.14159
PRESS	×	6.28318
ENTER	6,EE,3	6 03
PRESS	×	3.76991 04
ENTER	.05,EE,+/−,6	0.05 − 06
PRESS	=	1.88496 − 03
PRESS	1/X	5.30516 02

When converted to a prefix, 5.30516 02 equals 530.5 Ω.

EXAMPLE 2

Solve for X_C when $C = 0.05\ \mu f$ and $f = 30$ kHz.

Mathematical solution:

$$X_C = \frac{1}{2\pi fC}$$

$$X_C = \frac{1}{2\,(3.14) \times 3 \times 10^4 \times 5 \times 10^{-8}}$$

$$X_C = \frac{1}{9.42 \times 10^{-3}}$$

$$X_C = 106\ \Omega$$

Calculator solution:

Action	*Entry*	*Display*
ENTER	2	2
PRESS	×	2
PRESS	π	3.14159
PRESS	×	6.28318
ENTER	30,EE,3	30 03
PRESS	×	1.88495 05
ENTER	.05,EE,+/−,6	0.05 − 06
PRESS	=	9.42478 − 03
PRESS	1/X	1.06103 02

When converted to a prefix, 1.06103 02 equals 106 Ω.

In the two examples presented 0.05-μf capacitors were operated at different frequencies. Notice that at the high frequency, X_C is lower than it is at the low frequency. This illustrates the rule that,

In an alternating current circuit, when capacitance is constant and frequency increases, capacitive reactance decreases; when frequency decreases, capacitive reactance increases.

The next two examples use a constant frequency and a varying capacitance. They illustrate the following rule:

In an alternating current circuit, when frequency is constant and capacitance increases, capacitive reactance decreases; when capacitance decreases, capacitive reactance increases.

EXAMPLE 3

Solve for X_C when $f = 1$ kHz and $C = 0.5$ μf.

Mathematical solution:

$$X_C = \frac{1}{2\,\pi f C}$$

$$X_C = \frac{1}{2 \times 3.14 \times 1 \times 10^3 \times 5 \times 10^{-7}}$$

$$X_C = \frac{1}{3.14 \times 10^{-3}}$$

$$X_C = 318\ \Omega$$

Calculator solution:

Action	*Entry*	*Display*
ENTER	2	2
PRESS	×	2
PRESS	π	3.14159
PRESS	×	6.28318
ENTER	1,EE,3	1 03
PRESS	×	6.28318 03
ENTER	.5,EE,+/−,6	0.5 − 06
PRESS	=	3.14159 − 03
PRESS	1/X	3.18309 02

When converted to a prefix, 3.18309 02 equals 318.3 Ω.

EXAMPLE 4

Solve for X_C when $f = 1$ kHz and $C = 0.005$ μf.

Mathematical solution:

$$X_C = \frac{1}{2\,\pi f C}$$

$$X_C = \frac{1}{2 \times 3.14 \times 1 \times 10^3 \times 5 \times 10^{-9}}$$

$$X_C = \frac{1}{3.14 \times 10^{-5}}$$

$$X_C = 31{,}847\ \Omega$$

Calculator solution:

Action	Entry	Display
ENTER	2	2
PRESS	×	2
PRESS	π	3.14159
PRESS	×	6.28318
ENTER	1,EE,3	1 03
PRESS	×	6.28318 03
ENTER	.005,EE,+/−,6	0.005 − 06
PRESS	=	3.14159 − 05
PRESS	1/X	3.18310 04

When converted to a prefix, 3.18310 04 equals 31,831 Ω.

It is possible to use other forms of this equation to find frequency or capacitance. Let us look at an example of each.

EXAMPLE 5

Solve for frequency when $C = 0.01\ \mu$f and $X_C = 100$ kΩ.

Mathematical solution:

$$f = \frac{1}{2\pi C X_C}$$

$$f = \frac{1}{2 \times 3.14 \times 1 \times 10^{-8} \times 1 \times 10^{5}}$$

$$f = \frac{1}{6.28 \times 10^{-3}}$$

$$f = 159 \text{ Hz}$$

Calculator solution:

Action	Entry	Display
ENTER	2	2
PRESS	×	2
PRESS	π	3.14159
PRESS	×	6.28318
ENTER	.01,EE,+/−,6	0.01 − 06
PRESS	×	6.28318 − 08
ENTER	100,EE,3	100 03
PRESS	=	6.28318 − 03
PRESS	1/X	1.59154 02

When converted to a prefix, 1.59154 02 equals $f = 159$ Hz.

EXAMPLE 6

Calculate the capacitance for a circuit in which $f = 10$ kHz and $X_C = 100$ kΩ.

Mathematical solution:

$$C = \frac{1}{2\pi f X_C}$$

$$C = \frac{1}{2 \times 3.14 \times 1 \times 10^4 \times 1 \times 10^5}$$

$$C = \frac{1}{6.28 \times 10^9}$$

$$C = 159 \text{ pf}$$

Calculator solution:

Action	*Entry*	*Display*
ENTER	2	2
PRESS	×	2
PRESS	π	3.14159
PRESS	×	6.28318
ENTER	10,EE,3	10 03
PRESS	×	6.28318 04
ENTER	100,EE,3	100 03
PRESS	=	6.28318 09
PRESS	1/X	1.59154 – 10

When converted to a prefix, 1.59154 – 10 equals 159 pf.

As for X_C, if frequency changes, the amount of capacitance needed for a desired opposition must change. When X_C is held constant, and frequency is changed, the capacitor's value must be varied.

Self-Check

Answer each item by indicating whether the statement is true or false.

11. ___ Capacitance reactance is measured in ohms.

12. ___ As frequency increases, capacitive reactance decreases.

13. ___ Capacitor size and capacitive reactance are directly proportional.

14. ___ When a DC voltage is applied to a capacitor, current flow is maximum at the first instant but decreases to zero after some time has elapsed.

15. ___ Increasing a capacitor's size causes capacitive reactance to decrease.

Total Capacitance

Adding a capacitor to a circuit affects both the total capacitance and the capacitive reactance of the circuit. Capacitors can be added in series or parallel. The effect is different for each connection. Both types of connections are discussed.

Parallel-Connected Capacitors

Trimmer capacitors are represented in Figure 18-16. This arrangement is used extensively in communications circuitry. A small capacitor is placed in parallel with a larger capacitor to allow fine tuning of the circuit. When capacitors are connected in parallel, total capacitance equals the sum of all capacitor values. In operating circuits, conductors form capacitors with other conductors in the same circuit. The conductors become an additional plate area for capacitance, in parallel with the capacitance already available in the circuit design. At some frequency this *stray capacitance* becomes a considerable problem and must be taken into consideration. A padder capacitor is shown in Figure 18-18. Adding capacitors in parallel has the same effect as using a capacitor with a larger plate area. To determine total capacitance of parallel capacitors, use the following equation:

$$C_t = C_1 + C_2 + C_3 \ldots C_n$$

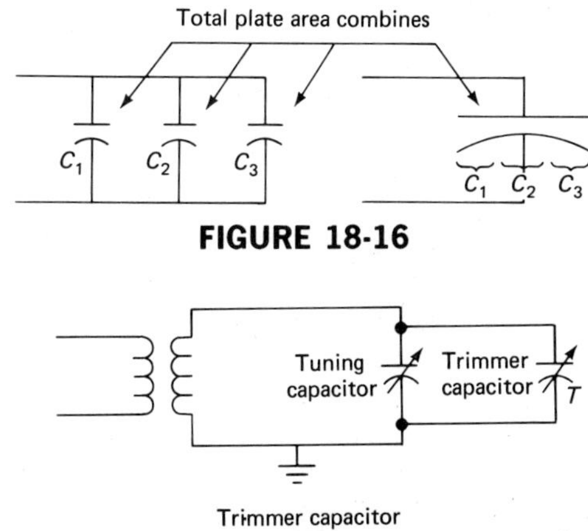

FIGURE 18-16

FIGURE 18-17

This equation is the same as that used for series resistors and coils. Connecting capacitors in parallel causes total capacitance, C_t, to increase. As C_t increases, capacitive reactance, X_C, decreases, resulting in less opposition to changes in voltage and allowing more current to flow.

Series-Connected Capacitors

When calculating total capacitance, we must treat series-connected capacitors like parallel-connected coils or resistors (Figure 18-18). For circuits containing any number of series capacitors, we can use:

$$\frac{1}{C_t} = \frac{1}{C_1} + \frac{1}{C_2} + \frac{1}{C_3} \cdots \frac{1}{C_n}$$

For two series capacitors, we can use the product-over-sum formula:

$$C_t = \frac{C_1 \times C_2}{C_1 + C_2}$$

For capacitors of equal value, we use the *C/N* formula:

$$C_t = \frac{C}{N}$$

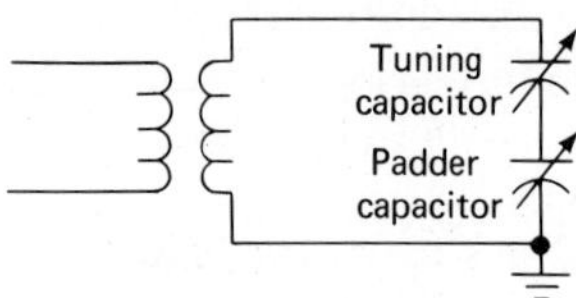

FIGURE 18-18

Let's take a look at each formula as it applies to a circuit. In each example, we use capacitors with a capacitance of 0.06 μf.

When three 0.06-μf capacitors are connected in parallel, the total capacitance equals:

Mathematical solution:

$$C_t = C_1 + C_2 + C_3$$

$$C_t = 6 \times 10^{-8} + 6 \times 10^{-8} + 6 \times 10^{-8}$$

$$C_t = 18 \times 10^{-8}$$

$$C_t = 0.18\ \mu\text{f}$$

Calculator solution:

Action	*Entry*	*Display*
ENTER	.06,EE,+/−,6	0.06 − 06
PRESS	+	6 − 08
ENTER	.06,EE,+/−,6	0.06 − 06
PRESS	+	1.2 − 07
ENTER	.06,EE,+/−,6	0.06 − 06
PRESS	=	1.8 − 07

When converted to a prefix, 1.8 − 07 equals 0.18 μf.

To solve a capacitive circuit with three 0.06 μf capacitors in series, we use the reciprocal formula:

Mathematical solution:

$$\frac{1}{C_t} = \frac{1}{C_1} + \frac{1}{C_2} + \frac{1}{C_3}$$

$$\frac{1}{C_t} = \frac{1}{6 \times 10^{-8}} + \frac{1}{6 \times 10^{-8}} + \frac{1}{6 \times 10^{-8}}$$

$$\frac{1}{C_t} = 5 \times 10^{7}$$

$$C_t = 2 \times 10^{-8}$$

$$C_t = 0.02\ \mu f$$

Calculator solution:

Action	*Entry*	*Display*
ENTER	.06,EE,+/−,6	0.06 − 06
PRESS	1/X	1.66667 07
PRESS	+	1.66667 07
ENTER	.06,EE,+/−,6	0.06 − 06
PRESS	1/X	1.66667 07
PRESS	+	3.33333 07
ENTER	.06,EE,+/−,6	0.06 − 06
PRESS	1/X	1.66667 07
PRESS	=	5 07
PRESS	1/X	2 − 08

When converted to a prefix, 2 − 08 equals 0.02 μf.

When two 0.06-μf capacitors are connected in series, and we wish to calculate total capacitance, we use:

Mathematical solution:

$$C_t = \frac{C_1 \times C_2}{C_1' + C_2}$$

$$C_t = \frac{6 \times 10^{-8} \times 6 \times 10^{-8}}{6 \times 10^{-8} + 6 \times 10^{-8}}$$

$$C_t = \frac{3.6 \times 10^{-15}}{1.2 \times 10^{-7}}$$

$$C_t = 3 \times 10^{-8}$$

$$C_t = 0.03\ \mu f$$

Calculator solution:

Action	Entry	Display
ENTER	.06,EE,+/−,6	0.06 − 06
PRESS	×	6 − 08
ENTER	.06,EE,+/−,6	0.06 − 06
PRESS	=	3.6 − 15
PRESS	÷	3.6 − 15
PRESS	(	3.6 − 15
ENTER	.06,EE,+/−,6	0.06 − 06
PRESS	+	6 − 08
ENTER	.06,EE,+/−,6	0.06 − 06
PRESS	)	1.2 − 07
PRESS	=	3 − 08

When converted to a prefix, 3 − 08 equals 0.03 μf.

The capacitors in the last example are of equal value (0.06 μf); thus we can use the formula:

Mathematical solution:

$$C_t = \frac{C}{N}$$

$$C_t = \frac{6 \times 10^{-8}}{2}$$

$$C_t = 3 \times 10^{-8}$$

$$C_t = 0.03\ \mu\text{f}$$

Calculator solution:

Action	Entry	Display
ENTER	.06,EE,+/−,6	0.06 − 06
PRESS	÷	6 − 08
ENTER	2	2
PRESS	=	3 − 08

When converted to a prefix, 3 − 08 equals 0.03 μf.

Remember, we have been working with capacitors, not capacitive reactance. We have already noted that capacitance and X_C are inversely proportional. As capacitance increases, capacitive reactance decreases; as capacitance decreases, capacitive reactance increases. In cases where capacitive reactance is known, we can calculate total capacitive reactance, using formulas and methods identical to those used in solving total-resistance problems. For series X_C values, total X_C

equals the sum of all separate capacitive reactances:

$$X_{Ct} = X_{C1} + X_{C2} + X_{C3}$$

For any number of parallel capacitors whose X_C is known, we can use the reciprocal formula:

$$\frac{1}{X_{Ct}} = \frac{1}{X_{C1}} + \frac{1}{X_{C2}} + \frac{1}{X_{C3}}$$

For two parallel capacitors of different reactances, we use the product-over-sum formula:

$$X_{Ct} = \frac{X_{C1} \times X_{C2}}{X_{C1} + X_{C2}}$$

Finally, for capacitors with the same X_C, we use the C/N formula:

$$X_{Ct} = \frac{X_{C1}}{N}$$

Capacitive Circuits, Ohm's Law, and Kirchhoff's Laws

In solving voltage, current, and reactance problems, Ohm's law and Kirchhoff's laws remain valid. Ohm's-law formulas are:

$$E_t = I_t \times X_{Ct}$$

$$I_t = \frac{E_t}{X_{Ct}}$$

$$X_{Ct} = \frac{E_t}{I_t}$$

Remember: Capacitors do not dissipate power. Any power stored in the electrostatic field during charge is returned to the power source during discharge.

We now consider capacitive circuits from the standpoint of Ohm's and Kirchhoff's laws. The circuit in Figure 18-19 is used to calculate the circuit's total current:

$$I_t = \frac{E_t}{X_C}$$

$$I_t = \frac{100 \text{ V}}{318.3 \text{ k}\Omega}$$

$$I_t = 314\ \mu\text{A}$$

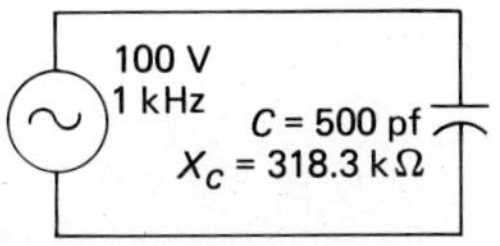

FIGURE 18-19

Note: Your calculator solutions will vary slightly from the results listed here, because of the rounding off of values done in the solutions.

We know that current is the same at all points in a series circuit. With that in mind, we can state that, in Figure 18-19, 314 μA of current is flowing at every point.

In Figure 18-20, note that the circuit contains an AC power source and two capacitors. The values provided are voltage, capacitance, and frequency. To solve for total current and voltage drops, we first solve for X_{C1}, X_{C2}, and X_{Ct}:

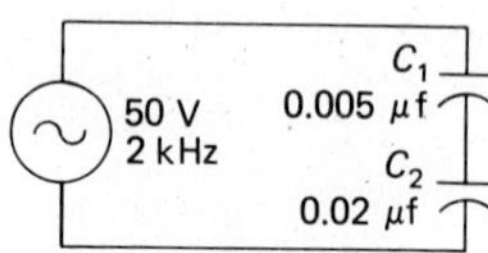

FIGURE 18-20

Solve for X_{C1}:

$$X_{C1} = \frac{1}{2\pi f C_1}$$

$$X_{C1} = \frac{1}{2 \times 3.14 \times 2\text{ kHz} \times .005\ \mu\text{f}}$$

$$X_{C1} = 15.9\text{ k}\Omega$$

Solve for X_{C2}:

$$X_{C2} = \frac{1}{2\pi f C_2}$$

$$X_{C2} = \frac{1}{2 \times 3.14 \times 2\text{ kHz} \times .02\ \mu\text{f}}$$

$$X_{C2} = 4\text{ k}\Omega$$

Solve for X_{Ct}:

$$X_{Ct} = X_{C1} + X_{C2}$$

$$X_{Ct} = 15.9\text{ k}\Omega + 4\text{ k}\Omega$$

$$X_{Ct} = 19.9\text{ k}\Omega$$

Then calculate I_t:

$$I_t = \frac{E_t}{X_{Ct}}$$

$$I_t = \frac{50\text{ V}}{19.9\text{ k}\Omega}$$

$$I_t = 2.512\text{ mA}$$

Now that all the values are available, we can solve for the voltage drop on C_1 and C_2. Note that this circuit is a voltage divider that uses capacitors for voltage division.

Solve for E_{C1}:

$$E_{C1} = I_t \times X_{C1}$$

$$E_{C1} = 2.512 \text{ mA} \times 15.9 \text{ k}\Omega$$

$$E_{C1} = 40 \text{ V}$$

Solve for E_{C2}:

$$E_{C2} = I_t \times X_{C2}$$

$$E_{C2} = 2.512 \text{ mA} \times 4 \text{ k}\Omega$$

$$E_{C2} = 10 \text{ V}$$

These values can be used to prove our solutions. We apply Kirchhoff's voltage law:

Solve for E_t:

$$E_t = E_{C1} + E_{C2}$$

$$E_t = 40 \text{ V} + 10 \text{ V}$$

$$E_t = 50 \text{ V}$$

Note that in a capacitive voltage divider, the smallest capacitor has the largest capacitive reactance and drops the largest voltage.

Now we turn to a parallel capacitive circuit. The circuit in Figure 18-21 has two parallel capacitances, total voltage, and frequency as its givens. These values are used to solve for currents and capacitance reactances of the circuit.

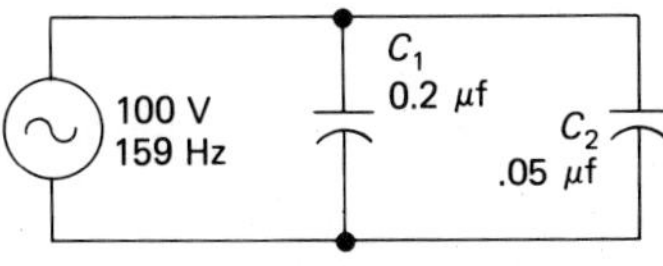

FIGURE 18-21

Remember: In a parallel circuit, each branch drops the voltage applied to that part of the circuit. In this case, each capacitor drops 100 V.

With these knowns, we analyze the circuit:

Solve for X_{C1}:

$$X_{C1} = \frac{1}{2\,\pi f C_1}$$

$$X_{C1} = \frac{1}{2 \times 3.14 \times 159 \text{ Hz} \times .2\ \mu\text{f}}$$

$$X_{C1} = 5 \text{ k}\Omega$$

Solve for X_{C2}:

$$X_{C2} = \frac{1}{2\,\pi f C_2}$$

$$X_{C2} = \frac{1}{2 \times 3.14 \times 159 \text{ Hz} \times .05\ \mu\text{f}}$$

$$X_{C2} = 20 \text{ k}\Omega$$

Solve for X_{Ct}:

$$\frac{1}{X_{Ct}} = \frac{1}{X_{C1}} + \frac{1}{X_{C2}}$$

$$\frac{1}{X_{Ct}} = \frac{1}{5 \text{ k}\Omega} + \frac{1}{20 \text{ k}\Omega}$$

$$\frac{1}{X_{Ct}} = \frac{1}{0.00025}$$

$$X_{Ct} = 4 \text{ k}\Omega$$

With these results, we can calculate currents in the circuit. Because each capacitor is in a separate branch, it has its own current.

Solve for I_{C1}:

$$I_{C1} = \frac{E_t}{X_{C1}}$$

$$I_{C1} = \frac{100 \text{ V}}{5 \text{ k}\Omega}$$

$$I_{C1} = 20 \text{ mA}$$

Solve for I_{C2}:

$$I_{C2} = \frac{E_t}{X_{C2}}$$

$$I_{C2} = \frac{100 \text{ V}}{20 \text{ k}\Omega}$$

$$I_{C2} = 5 \text{ mA}$$

Solve for $I_t = I_{C1} + I_{C2}$:

$$I_t = 20 \text{ mA} + 5 \text{ mA}$$

$$I_t = 25 \text{ mA}$$

or

$$I_t = \frac{E_t}{X_{Ct}}$$

$$I_t = \frac{100 \text{ V}}{4 \text{ k}\Omega}$$

$$I_t = 25 \text{ mA}$$

Solve for C_t:

$$C_t = C_1 + C_2$$

$$C_t = 0.2\ \mu\text{f} + 0.05\ \mu\text{f}$$

$$C_t = 0.25\ \mu\text{f}$$

The solutions discussed in this section should help you understand capacitive circuits.

Self-Check

Answer each item by indicating whether the statement is true or false.

16. ___ In a circuit with five series-connected capacitors, total capacitance is less than the capacitance of the smallest capacitor.

17. ___ Capacitors connected in parallel are treated the same way as resistors connected in series.

18. ___ Use the reciprocals of Ohm's-law equations to solve capacitive circuits.

19. ___ When solving for total capacitance, we treat series capacitors the same as we do series resistors.

20. ___ When calculating total capacitive reactance, we treat series-capacitive reactances the same as we do series resistors.

21. ___ Ohm's law and Kirchhoff's laws apply to a capacitive circuit in the same way they apply to resistive circuits.

22. ___ To solve for current in a capacitive circuit, we divide voltage by total capacitance.

23. ___ Each branch of a parallel-capacitive circuit drops applied voltage.

24. ___ In a series-capacitive voltage divider, the capacitor with the smallest capacitance drops the most voltage.

25. ___ To calculate total capacitance reactance of a series-capacitive circuit, we add individual capacitive reactances.

Troubleshooting Capacitors

Periodically, problems develop that require some knowledge of capacitor-troubleshooting procedures. We can use an ohmmeter to make checks of capacitors to determine whether a capacitor is good, bad, open, shorted, or leaky.

Set the ohmmeter to its highest *ohms* range. The capacitor must be isolated from the circuit and it must be discharged. Then we proceed as we did in checking resistance. An ohmmeter connected across a capacitor acts as a power supply; its internal batteries charge the capacitor. Although this may happen quickly, in most cases the meter's pointer will deflect sufficiently for observation of the current's action. If the pointer deflects over part of the scale and then returns to infinity, the capacitor is good. If the ohmmeter check is not conclusive, the best procedure is to replace the suspect capacitor with one known to be good. In the case of large-value capacitors, no deflection indicates that the capacitor is open. If the ohmmeter indicates a resistance other than infinity after being connected for some time, the capacitor is shorted, or "leaky." In either case, it is bad and must be replaced.

A capacitor can become defective through age even when in a spare-parts inventory. In some cases, capacitors stored for a long time are bad when purchased.

A capacitor can be damaged by subjecting it to a voltage higher than the voltage it is rated for, resulting in a high resistance path (leakage) through the capacitor's dielectric where current flows. The electrolyte in an electrolytic capacitor can dry out because of heat and age. A dry electrolyte can act as a high resistance instead of a dielectric, and allow current to leak. Because some electrons flow in the dielectric, a capacitor with high resistance to DC is said to be "leaky."

Exposing capacitors to voltages that are too high can cause their dielectric to break down because of an arc through the dielectric. This causes the dielectric to heat and burn, resulting in a carbon path between the plates. The carbon acts as a conductor, which shorts the plates. The result is a shorted capacitor.

Moisture and oxidation can cause a capacitor to be leaky or shorted. Around the terminals of some electrolytic capacitors, a corrosive material develops. This is usually caused by the dielectric operating under heavy load and thus swelling, which separates the seal on the "can." Some capacitors arc, or short, only when high voltage is applied, and some power-supply capacitors "hum" after extended use; the latter causes the distortion heard in the speakers of some stereo systems. The capacitors can no longer remove all the voltage fluctuations from the power supply. Any number of symptoms—a sound or a smell, or something that just doesn't look right—may alert you to a capacitor that is going bad. In other words, troubleshooting capacitive circuits improves with experience.

Self-Check

Answer each item by indicating whether the statement is true or false.

26. ___ An ohmmeter connected across a good capacitor deflects, then returns to zero.

27. ____ It is not necessary to test a newly purchased capacitor.

28. ____ An arc through a dielectric can cause a capacitor to open.

29. ____ A leaky capacitor loses its electrolyte by separation in the can.

30. ____ An ohmmeter connected across a shorted capacitor indicates a low resistance.

Summary

Capacitors consist of two conductors separated by a dielectric. Once connected in a DC circuit, a capacitor charges to a voltage equal to the DC voltage source. When fully charged, the capacitor acts as a series-opposing power source to the DC source; thus no DC current flows. Should change occur in the voltage from the DC source, a capacitor either charges or discharges an amount equal to the change. Once the DC source is removed, the capacitor, if provided a path for current flow, returns its stored energy to the source.

A capacitor stores electrical energy in its electrostatic fields by a process called orbital stress. The larger the number of atoms under stress, the more energy stored by the capacitor. Capacitors do not use power, but return the energy to the circuit.

Several types of capacitors are available, each with a different application. Low frequencies use large-capacitor values, whereas higher frequencies use lower values. Capacitors are used in communication circuits for tuning purposes; capacitors are of both fixed and variable types.

Capacitive reactance, a factor in all AC circuits, is the opposition of the capacitor to a change in voltage. Capacitive reactance is affected by capacitor size and frequency and is inversely proportional to changes in either.

Connecting capacitors in parallel causes total capacitance to increase. Connecting capacitors in series causes total capacitance to decrease.

Troubleshooting capacitors is similar to troubleshooting resistors.

Review Questions and/or Problems

1. The electrolyte in a capacitor
   (a) must be heated before current can flow.
   (b) is the dielectric.
   (c) is half the thickness of the conductor.
   (d) is the conductor.

2. Capacitance can be increased by
   (a) increasing the dielectric constant.
   (b) using a different conductor.
   (c) moving the plates farther apart.
   (d) none of the above

3. Two major classifications of capacitors are
   (a) paper and electrolytic.
   (b) mica and ceramic.
   (c) fixed and variable.
   (d) paper and air core.

4. A capacitor that shows the polarity of its connection is a ceramic capacitor.
   (a) true
   (b) false

5. Voltage rating of a capacitor is determined by the type of dielectric used in it.
   (a) true
   (b) false

6. Capacitors are rated according to their
   (a) frequency and polarity.
   (b) operating voltage and current.
   (c) operating voltage and frequency.
   (d) operating voltage and capacitance.

7. As operating frequency is increased,
   (a) capacitance decreases.
   (b) capacitance increases.
   (c) capacitive reactance increases.
   (d) capacitive reactance decreases.

8. A circuit contains a 0.01-μf capacitor and is operating at 400 Hz. What is its capacitive reactance?
   (a) 3978.8 Ω
   (b) 796 Ω
   (c) 397.8 Ω
   (d) 39788 Ω

9. Capacitors connected in series cause total capacitance to
   (a) increase.
   (b) decrease.
   (c) remain the same.

10. When capacitors are connected in series, capacitive reactance
    (a) increases.
    (b) decreases.
    (c) remains the same.

11. A circuit operating at 100 Hz contains two series capacitors, 10 μf and 5 μf. What does X_{Ct} equal?
(a) 4.77 Ω
(b) 47.7 Ω
(c) 477 Ω
(d) 4774 Ω

12. Five capacitors, all rated at 10 μf, are connected in series. What is their total capacitance?
(a) 50 μf
(b) 10 μf
(c) 2 μf
(d) 0.1 μf

13. As the frequency applied to a capacitive circuit is increased, total current
(a) increases.
(b) decreases.
(c) remains the same.

14. A 12-μf capacitor can be replaced by
(a) two 6 μf capacitors in series.
(b) two 24 μf capacitors in parallel.
(c) one 30 μf and one 20 μf capacitors in series.
(d) any of the above.

15. A shorted capacitor has a dielectric with a path for current flow.
(a) true
(b) false

16. Solve for I_t in Figure 18-22.
(a) 20 A
(b) 20 mA
(c) 200 A
(d) 200 mA

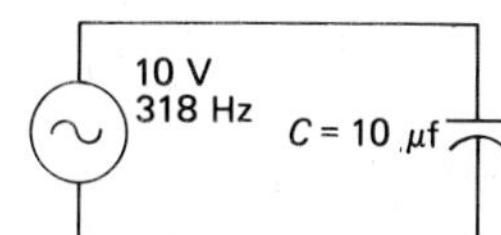

FIGURE 18-22

17. In Figure 18-23, calculate X_{Ct}.
(a) 12 kΩ
(b) 50 kΩ
(c) 20 kΩ
(d) 30 kΩ

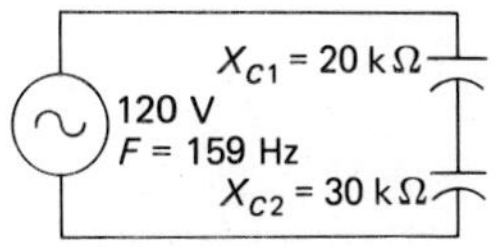

FIGURE 18-23

18. The total current of the circuit in Figure 18-23 is
    (a) 10 mA.
    (b) 4 mA.
    (c) 2.4 mA.
    (d) 6 mA.

19. In the circuit in Figure 18-23, what does E_{C1} equal?
    (a) 48 V
    (b) 72 V
    (c) 200 V
    (d) 300 V

20. Refer to Figure 18-24, and solve for total capacitance in the circuit.
    (a) 50 μf
    (b) 20 μf
    (c) 30 μf
    (d) 12 μf

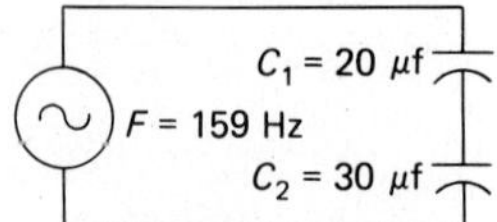

FIGURE 18-24

19

Mathematics for AC Circuits

Introduction

Throughout this book, we have taken a limited mathematical approach to the explanation of circuits. All circuits were explained from the viewpoint of mathematics, with the approach being as simple as possible. With the added complexity of AC circuits, the math becomes more complex. Many teachers approach the study of AC circuits from one point of view, others from another quite different viewpoint. This chapter presents the approaches available.

In presenting these approaches, the aim is to provide students with mathematical tools that help him or her learn AC circuitry. Advanced AC circuitry can be analyzed by any of several methods. Although it is possible to simplify the process considerably by learning and using suitable math skills, it remains our intention to present only methods that are easy to learn and simple to use.

Methods are presented that are suitable for solving a problem in more than one way. Some students will find one method preferable to another, just as teachers do. Each method is explained and its use is demonstrated.

Objectives

Each student is required to:

1. Identify the following quantities:
   (a) Scalar.
   (b) Polar notation.
   (c) Rectangular notation.
2. Use the Pythagorean theorem to solve right triangles.
3. Describe the rotation of vectors as affected by
   (a) Multiplication by -1.
   (b) Multiplication by $\sqrt{-1}$.
   (c) Multiplication by $-\sqrt{-1}$.

4. Explain the use of $-j$ and $+j$ as *j operators*.
5. State vectorial quantities in
   (a) Rectangular notation.
   (b) Polar notation.
6. Solve vector- and/or right-triangle problems, using
   (a) Graphical methods.
   (b) The Pythagorean theorem.
   (c) Rectangular notations.
   (d) Polar notations.
7. Convert rectangular notations to polar form.
8. Convert polar notations to rectangular form.

Scalar Quantities

A quantity that can be stated by a number and that does not require a statement of direction is called a *scalar* quantity. This is the type of quantity we have been studying so far. Scalar quantities are in magnitude only; they can be added or subtracted algebraically. No consideration of direction is necessary. In the case of DC circuits, force is always exerted in the same direction; therefore, we can disregard direction. Following are some examples of scalar quantities and their mathematical manipulation.

50 miles + 20 miles = 70 miles

75 cents + 15 cents = 90 cents

40 kΩ + 20 kΩ = 60 kΩ

20 mA + 30 mA = 50 mA

12 pf + 15 pf = 27 pf

Vectors and Vector Analysis

In AC circuits, voltage, current, and other quantities have both magnitude and direction. It is convenient to represent quantities of this type by a vector.

A vector is a graphical quantity that has both magnitude and direction.

Graphically, a vector is drawn at an angle from a reference axis or reference vector. The point where the vector and axis (or vectors) intersect is the point of origin. The vector (line) drawn has two components, length (magnitude in units) and direction (in degrees). A typical vector diagram is illustrated in Figure 19-1.

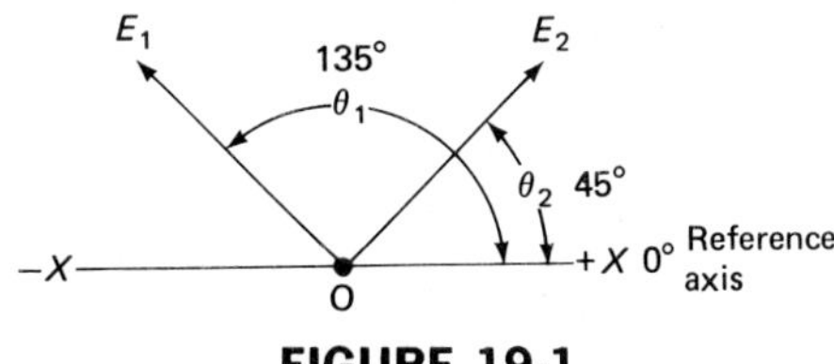

FIGURE 19-1

These vectors have a common origin (point O), but they have different angles. Notice that the reference axis is at zero degrees; vector direction is stated with respect to this reference. The angle between the axis and each vector is labeled as a separate angle theta, $\angle\theta_1$ and $\angle\theta_2$. A counterclockwise (CCW) vector rotation results in a positive (+) angle, whereas clockwise (CW) rotation results in a negative (−) angle. E_1 has rotated +135° CCCW and E_2 has rotated +45° CCW. We can say that E_1 leads E_2 and the reference by the number of degrees separating them. For example,

E_1 leads the reference axis by 135.
E_1 leads E_2 by 90.
E_2 leads the reference axis by 45°.

Vector position can also be stated by *lagging* statements:

E_2 lags E_1 by 90°.
Reference axis lags E_1 by 135°.
Reference axis lags E_2 by 45°.

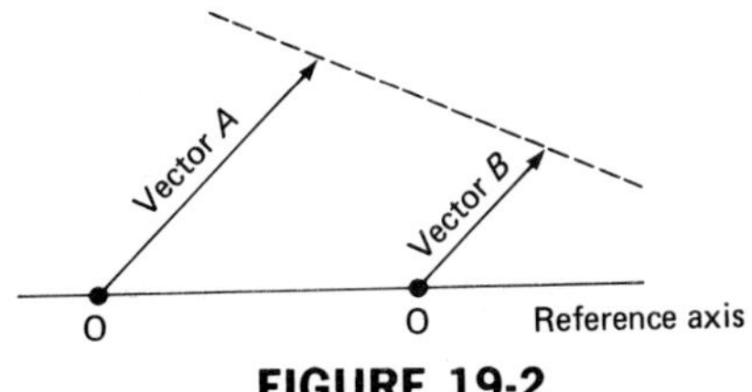

FIGURE 19-2

Figure 19-2 illustrates two vectors said to have the *same direction. Note:* The dashed line that connects their ends does not cross the reference axis. If both vectors have their origin on the reference axis, and the line connecting their ends does not cross the reference, the vectors have the same direction. When vectors are drawn as shown in Figure 19-3 they are said to have *opposite direction. Note:* The line connecting their ends crosses the reference axis.

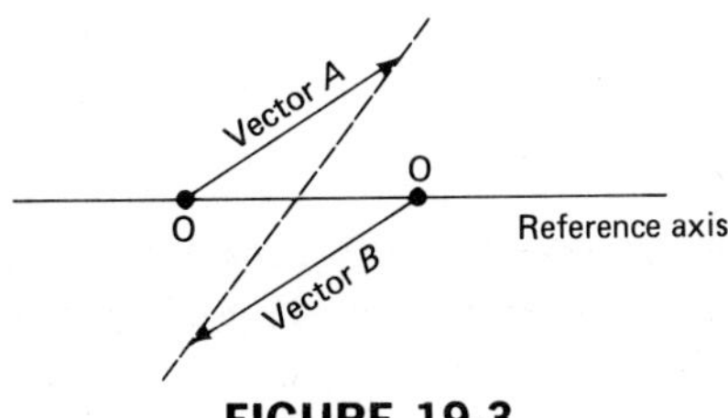

FIGURE 19-3

Vectors are said to be equal when they have the same magnitude and the same direction. Figure 19-4 illustrates *equal vectors*. Any given vector can be replaced by an equal vector. Vectors are often combined by addition or subtraction, to get a single vector called a *resultant vector*. The resultant can be obtained either graphically or mathematically.

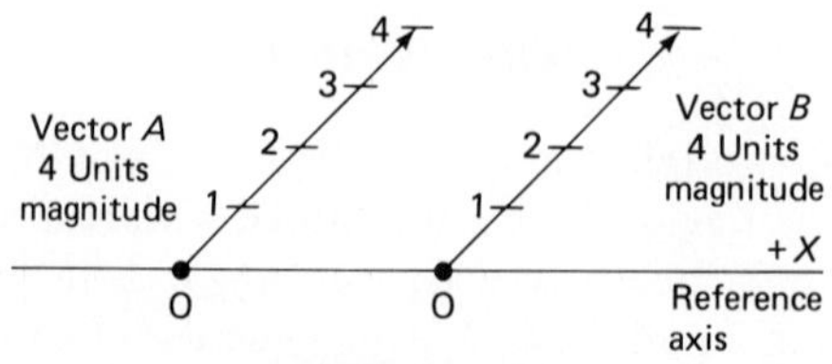

FIGURE 19-4

Addition of Same-Direction Vectors

In Figure 19-5a, two vectors are shown that have the same direction. Vector 1 has four units magnitude and vector 2 has three. These vectors can be added by either graphical or mathematical methods. To add vectors graphically,

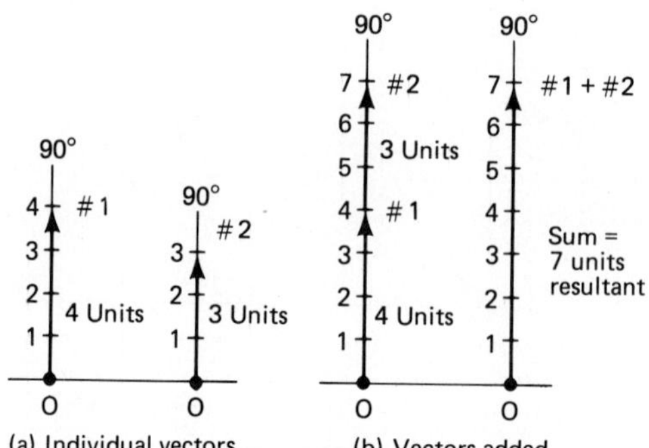

FIGURE 19-5

1. Take either vector and move its origin to the end of the other vector.
2. Redraw the vector, starting at this point, as follows:
   (a) If vectors are in the same direction, extend the length of the original vector.
   (b) If vectors are in the opposite direction, draw the new vector in the opposite direction to the original vector.
3. Add the two magnitudes (units) algebraically; the result equals the length of the resultant vector.

When vectors 1 and 2 in Figure 19-5b are added, the resultant vector is seven units long. The two vectors are represented at the left, and the single resultant is at the right. The resultant is a vector seven units long at $+90°$.

Addition of Opposite-Direction Vectors

Figure 19-6a shows two vectors of opposite directions. Notice that the lengths of the vectors are still 3 and 4. Vector 1, however, is at +90° and vector 2 is at −90°. Moving vector 2 to the end of vector 1 and superimposing it on vector 1 cancels three units of the latter. The resulting vector is one unit at +90°, as shown in Figure 19-6b.

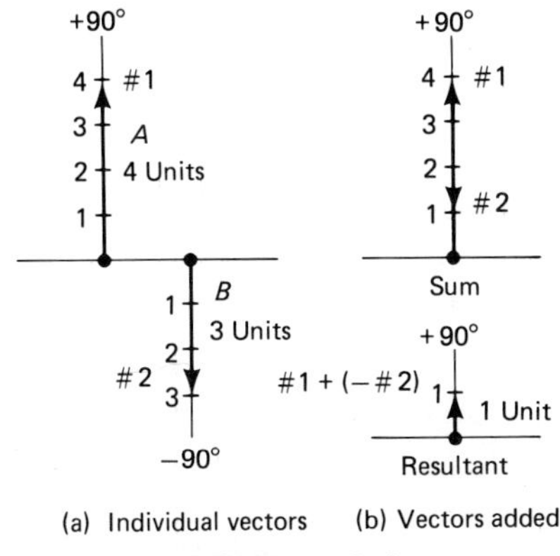

(a) Individual vectors (b) Vectors added

FIGURE 19-6

Addition of Vectors 90° Apart

The drawing in Figure 19-7a illustrates two vectors whose directions differ by 90°. To add these vectors, they must be rearranged so they share the same point of origin and form a right angle (Figure 19-7b). Note that side A of the triangle is the four-unit +90° vector and that side B is the three-unit 0° vector.

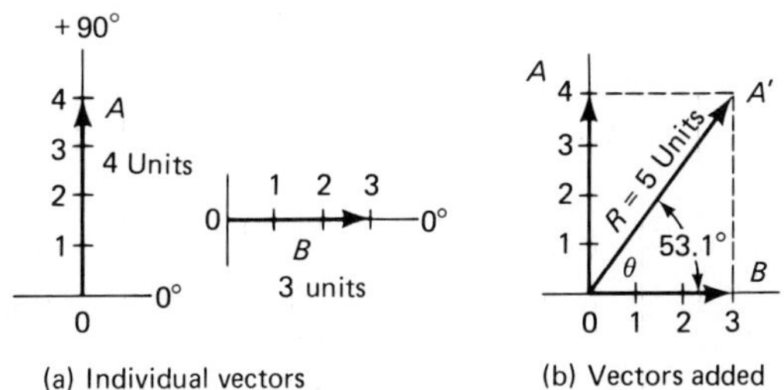

(a) Individual vectors (b) Vectors added

FIGURE 19-7

To find the result, a vector (four units at +90°) is drawn that intersects the end of the three-unit vector, illustrated by the dashed line. Drawing a diagonal connecting the origin to the tip of the dashed line results in a vector the physical length of the diagonal. Measurement of the physical length of vector R results in a length of five units. With a protractor, we measure angle θ as 53.1°. The result is 5 units at 53.1°.

Mathematically, vectors can be added by using the Pythagorean theorem.

In a right triangle, the square of the hypotenuse is equal to the sum of the squares of the other two sides of the triangle.

or

$$c^2 = a^2 + b^2$$

$$c = \sqrt{a^2 + b^2}$$

where

c = hypotenuse of right triangle

a = altitude of right triangle

b = base of right triangle

In using the Pythagorean theorem, we must make sure all items are stated in the same units—volts, ohms, and so on. In Figure 19-7b, vector 1 (side a) is four units at $+90°$, and vector 2 (side b) is three units at 0°. To solve for the hypotenuse (side c), we use mathematical and calculator solutions.

Mathematical solution:

$$c = \sqrt{a^2 + b^2}$$

$$c = \sqrt{4^2 + 3^2}$$

$$c = \sqrt{25}$$

$$c = 5 \text{ units}$$

Calculator solution:

Action	*Entry*	*Display*
ENTER	4	4
PRESS	X^2	16
PRESS	+	16
ENTER	3	3
PRESS	X^2	9
PRESS	=	25
PRESS	$\sqrt{X}$	5

This gives us the unit length of the resultant vector but not its angle of rotation. The method used to solve for the angle is discussed later in this chapter.

Solution of a right triangle can also be done by comparing the ratio between different sides as viewed with respect to either acute angle (an angle less than 90°). With respect to θ in Figure 19-8, we can state three ratios, called trigonometric functions of angle θ. For our purposes, the most important

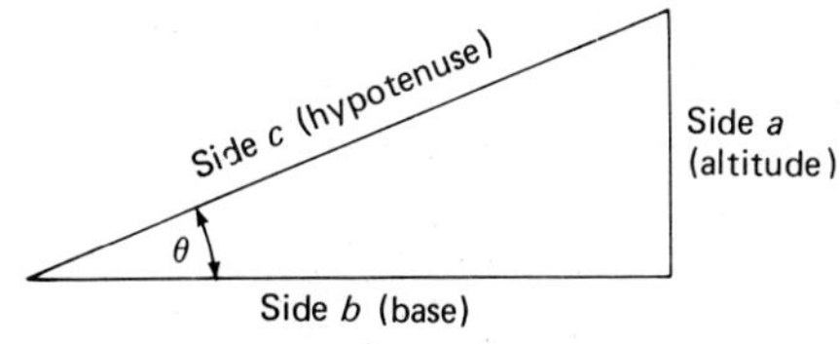

FIGURE 19-8

trigonometric functions are sine, cosine, and tangent (abbreviated sin, cos, and tan, respectively). The trigonometric functions for θ are given in Table 19-1.

TABLE 19-1

$\sin\theta =$ side a/hypotenuse or a/c
$\cos\theta =$ side b/hypotenuse or b/c
$\tan\theta =$ side a/side b or a/b

To solve for the angle of rotation (Figure 19-7b), we use side $a = 4$ and side $b = 3$. To find the tangent of angle θ, we proceed:

Mathematical solution:

$$\tan\theta = \frac{a}{b}$$

$$\tan\theta = \frac{4}{3}$$

$$\tan\theta = 1.3333$$

Action	*Entry*	*Display*
ENTER	4	4
PRESS	÷	4
ENTER	3	3
PRESS	=	1.3333

Both solutions reduce the ratio to a numerical figure, 1.3333. This tangent function of the angle can be converted to the angle either by checking a math table or by using a calculator. *Note:* Some calculators use INV-tan, whereas others use ARC-tan for the operation. ARC (INV) and tan keys are separate keys.

Action	*Entry*	*Display*
ENTER	1.3333	1.3333
PRESS	INV(ARC)	1.3333
PRESS	tan	53.1232

This results in the value measured earlier, 53.1°.

If we are given an angle in degrees and wish to convert it to a trigonometric function, we use the calculator:

Action	Entry	Display	Function
ENTER	53.1	53.1	
PRESS	sin	0.79968	sine
ENTER	53.1	53.1	
PRESS	cos	0.6004	cosine
ENTER	53.1	53.1	
PRESS	tan	1.3318	tangent

Sin, cos, and tan functions equal the values indicated.

Using the sine function (0.79968) and side a, we calculate the resultant vector. First, we transpose the sine function (Table 19-1), isolating the hypotenuse (side c). Transposing the sine function results in:

$$\text{Hypotenuse} = \frac{\text{side } a}{\sin \theta}$$

Mathematical solution:

$$\text{Hypotenuse} = \frac{\text{side } a}{\sin \theta}$$

$$\text{Hypotenuse} = \frac{4}{0.79968}$$

$$\text{Hypotenuse} = 5.002$$

Action	Entry	Display
ENTER	4	4
PRESS	÷	4
ENTER	.79968	0.79968
PRESS	=	5.002

The 0.002 error results from rounding off the sine function.

We could just have easily used the cosine function (0.6004) to solve for the vector. To do this, we transpose the cosine formula, to get:

$$\text{Hypotenuse} = \frac{\text{side } b}{\cos \theta}$$

Mathematical solution:

$$\text{Hypotenuse} = \frac{\text{side } b}{\cos\theta}$$

$$\text{Hypotenuse} = \frac{3}{0.6004}$$

$$\text{Hypotenuse} = 4.997$$

Action	*Entry*	*Display*
ENTER	3	3
PRESS	÷	3
ENTER	.6004	0.6004
PRESS	=	4.99669

The 0.00031 error results from rounding off the cosine function.

This has been a mathematical look at vector algebra. In subsequent sections of this chapter, we refer math of this type to actual AC-circuit solutions.

Self-Check

Answer each item by indicating whether the statement is true or false.

1. ____ Scalar quantities are numerical in nature, but direction is included as part of their statement.

2. ____ Vectors can be added graphically.

3. ____ The Pythagorean theorem can be used to add vectors mathematically that are 90° apart.

4. ____ When vectors are subtracted graphically, the −90° vector is superimposed on the +90° vector.

5. ____ Vectors in the same direction are positioned so a line connecting their ends crosses the reference axis.

j Operators

The sign of any number can be changed without affecting the magnitude of the statement. To convert a positive quantity to a negative one, all we have to do is multiply the quantity by −1. For example, +20 A can be converted by multiplying: −1 × 20 A = −20 A. To convert −30 A, multiply −30 A × −1, which results in +30 A. Remember that multiplying unlike signs results in a minus and that multiplying like signs results in a positive. In either case, the sign of operation is changed, but the magnitude is not affected.

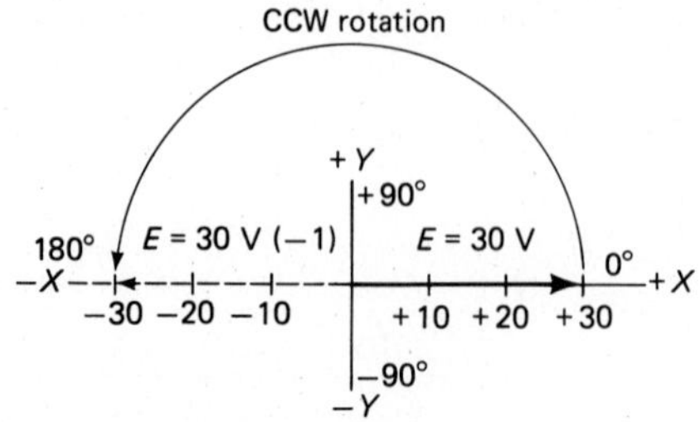

FIGURE 19-9

This can be illustrated by application to the vector diagram shown in Figure 19-9. When the 30-V vector placed on reference axis X is multiplied by −1, the vector rotates 180° and takes the position shown by the dashed line. In other words, multiplication of a vector quantity by −1 reverses its direction. When the −30-V vector is multiplied by −1, it rotates 180° and becomes a +30-V vector. Many problems in AC circuits involve rotation of vectors through 90° only. Two examples are:

1. ***Voltage leads current by 90° in an inductive circuit.***
2. ***Current leads voltage by 90° in a capacitive circuit.***

Mathematically, it is possible to multiply a quantity by the square root of −1, $\sqrt{-1}$, which results in 90° CCW vector rotation. Again, the conventional direction of rotation is CCW, and the resultant vector moves to a position that leads its original position by 90°.

Examine Figure 19-10. The E (voltage) vector that lies on the X (reference) axis has a magnitude of 30 V. When this vector is multiplied by $\sqrt{-1}$, it is rotated to the +90° position represented by the vector illustrated on the positive Y (+Y) axis. Multiplication of this vector by $-\sqrt{-1}$ causes it to rotate −90° and take the position shown on the −Y axis.

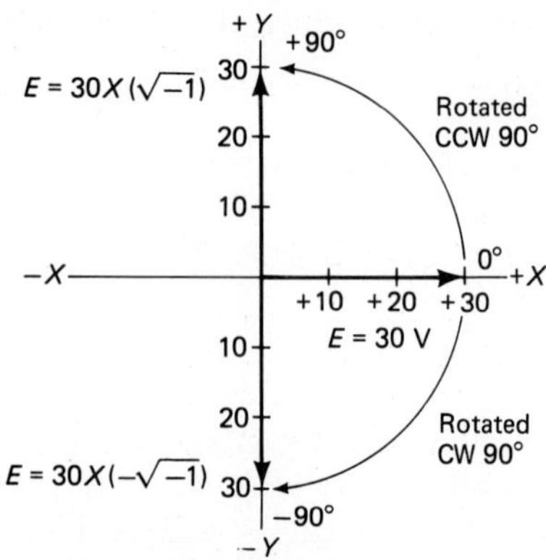

FIGURE 19-10

In electronics, $\sqrt{-1}$ and $-\sqrt{-1}$ denote direction only. Because it is awkward to use these symbols in calculations, we use j and $-j$ instead, with j representing $\sqrt{-1}$ and $-j$ representing $-\sqrt{-1}$. These symbols are called *operators*. The j operator represents a vector lying along the +Y (+90°) axis; the $-j$ operator represents a vector lying along the −Y (−90°) axis. The rules covering j operators are:

j operators are used to indicate the direction of a vector as it compares to the reference (+X) axis (0°).

Multiplication of a vector by $\sqrt{-1}$ (or +j) causes the vector to rotate 90° CCW.

Multiplication of a vector by $-\sqrt{-1}$ (or −j) causes the vector to rotate 90° CW.

To indicate that a vector has been rotated +90°, the vector is plotted on the +Y axis. In fact, this vector has been multiplied by $\sqrt{-1}$, a rotation signified by the $+j$ operator. In Figure 19-11, the 30-V vector has been rotated to the $+j$ (+90°) position and is located on the +Y axis. The same vector has also been rotated to the $-j$ (−90°) position and is plotted on the −Y axis. In using the j operator, we identify the three vectors represented in Figure 19-11 as:

Standard position (+X) axis stated as +30 V
+90° [+Y] axis stated as $+j30$ V
−90° [−Y] axis stated as $-j30$ V

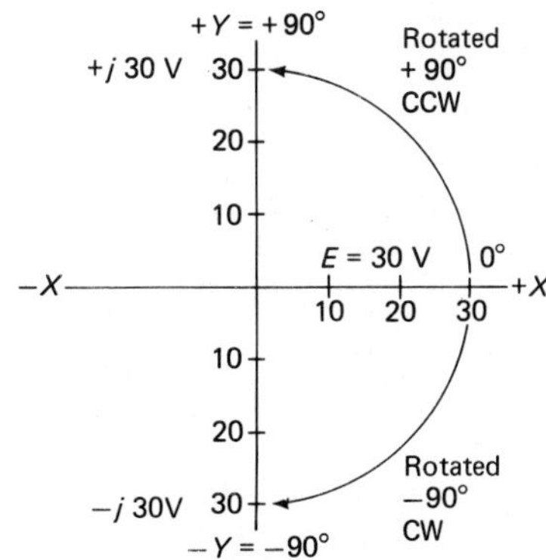

FIGURE 19-11

Self-Check

Answer each item by indicating whether the statement is true or false.

6. ____ To rotate a vector 180° CW, we multiply it by $-\sqrt{-1}$.

7. ____ The sign of a negative quantity can be changed to a positive sign by multiplying it by $-\sqrt{-1}$.

8. ____ Quantities multiplied by $\sqrt{-1}$ are assigned $+j$ operators.

9. ____ A j operator indicates the direction of a vector when it is compared to the reference (+X) axis.

10. ____ To change the quantity +30 V to −30 V, we multiply it by $\sqrt{-1}$.

Rectangular Notation

When we work with quantities that have no direction or that have the same direction, the math procedures used are the same as those we have been using for scalar notations. When we wanted to add two resistance values (5 Ω and 10 Ω), we wrote them:

$$R_t = R_1 + R_2$$

$$R_t = 5\ \Omega + 10\ \Omega$$

$$R_t = 15\ \Omega$$

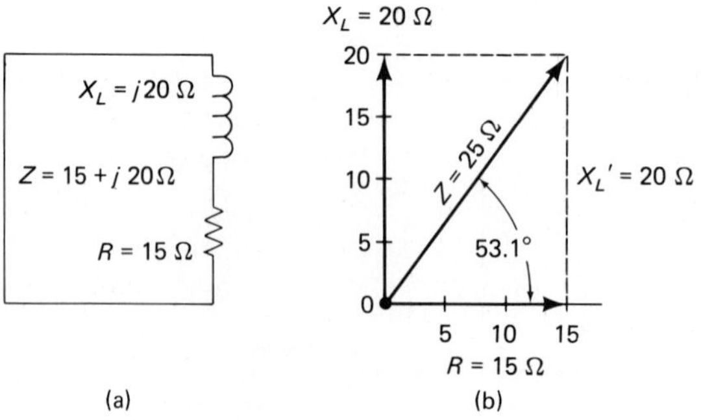

FIGURE 19-12

The method used to get this solution is simple algebraic addition. When we wish to add the reactances in an AC circuit, the two quantities do not have the same phase, or direction. Algebraic addition is not sufficient for this problem; the addition must be performed by use of vectorial addition. Refer to Figure 19-12a, noting that the circuit contains a resistor and a coil. Resistance and reactance are given. When the vectors representing the two values are plotted, the vector diagram is similar to that in Figure 19-12b. Note that the resistance is plotted in the "standard position" (reference axis) and has a direction of 0°. X_L, plotted on the +Y vector, has a leading direction of +90°. X_L is written as $j20\ \Omega$, which indicates the +90° CCW rotation of the vector. With two quantities at right angles, their addition can be stated as follows:

$Z = R + jX_L$ for inductive circuits (equation 19-1)

$Z = R - jX_C$ for capacitive circuits (equation 19-1)

(*Note:* Either formula may be referred to as equation 19-1 for solutions later in this chapter. The form you use is determined by whether the circuit is inductive or capacitive.)

Equation 19-1 consists of:

Z = impedance (in ohms)

R = resistance (in ohms)

X_L = inductive reactance (in ohms)

X_C = capacitive reactance (in ohms)

Note: X_L is stated as $+j$, and X_C is stated as $-j$.

Let us use the values given in Figure 19-12 to state the impedance of this circuit in j-operator form:

$$Z = 15 + j20\ \Omega$$

The impedance of this circuit can be read: "Impedance equals 15 Ω of resistance added vectorially to 20 Ω of inductive reactance." Note that the right side of the equation is stated in the same form as the Phythagorean theorem. It can be inserted in the equation:

$$Z = \sqrt{R^2 + X_L^2} \qquad \text{(equation 19-2)}$$

In electronics, any complex number is a statement that includes both reactive, X_C or X_L, and nonreactive (resistive) components. Equation 19-1, a complex number, is necessary because two quantities that are 90° out of phase cannot be added in normal algebraic terms. Only their sum can be indicated. The plus sign in equation 19-1 merely indicates direction; it is not a mathematical sign of operation. When an equation is stated in this form, it is said to be stated in *rectangular notation,* or rectangular form. Figure 19-13 shows two series circuits, the components in each circuit, and the rectangular notation that states their impedance.

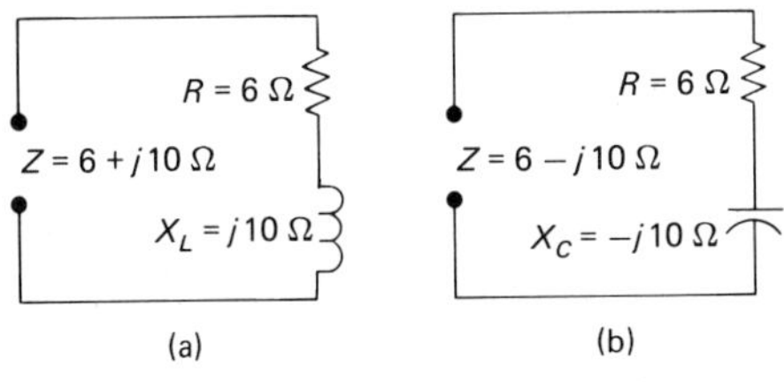

FIGURE 19-13

We can state the following rules, which pertain to current and voltage distribution in RC and RL circuits. Remember,

E_L leads current by 90° in an inductor.
E_C lags current by 90° in a capacitor.

Figure 19-14 presents two parallel circuits, the amount of current flow in each branch, and rectangular notation, which states the total current of the circuit.

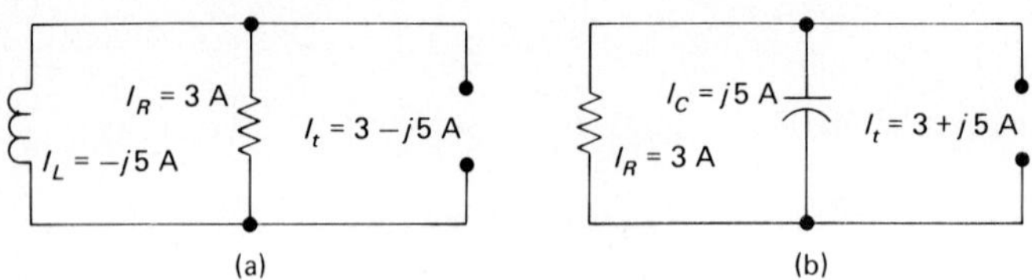

FIGURE 19-14

A reminder:

I_L lags voltage by 90° in an inductor.
I_C leads voltage by 90° in a capacitor.

Self-Check

Answer each item by indicating whether the statement is true or false.

11. ____ Two inductive reactances can be added, using simple algebraic addition.

12. ____ Rectangular notation is necessary to state the vectorial addition of quantities 90° out of phase.

13. ____ X_C is stated as a $-j$ operator.

14. ____ The complex number $20 + j15$ tells us to add the two numbers.

15. ____ Complex numbers may be used to express impedance and total current in a series RL circuit.

Mathematics with Complex Numbers

Complex numbers can be combined algebraically when they are in rectangular form. For example, impedances written in complex-number form can be added. Figure 19-15 illustrates two impedances connected in series, $R_1 - jX_C$ and $R_2 + jX_L$. Total impedance for this circuit can be found by addition:

$$Z_t = Z_1 + Z_2$$

$$Z_t = (R_1 - jX_C) + (R_2 + jX_L)$$

$$Z_t = (R_1 + R_2) + j(X_L - X_C)$$

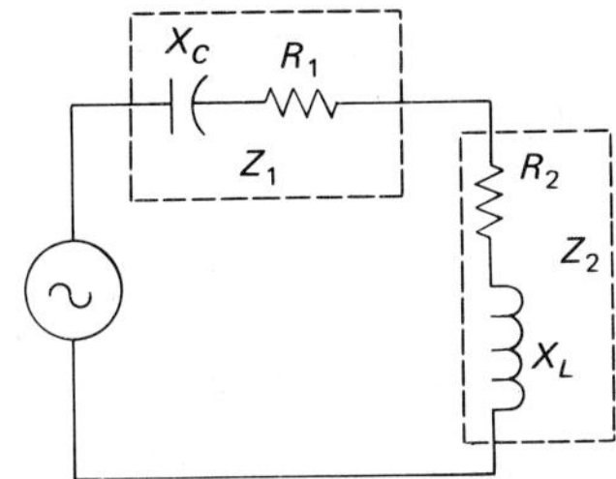

FIGURE 19-15

To find the total impedance, Z_t, in Figure 19-16, we write the separate impedances in rectangular notation. Remember: Resistance is always plotted at 0° and is positive; X_L results in $+j$; and X_C results in $-j$. Thus the three impedances for this circuit are:

$$Z_1 = 20 + j70\ \Omega$$

$$Z_2 = 10 + j10\ \Omega$$

$$Z_3 = 50 - j20\ \Omega$$

$$Z_t = 80 + j60\ \Omega$$

We used addition to calculate the rectangular form of Z_t. Note that all resistances are positive and that, when we use arithmetic addition, they add to 80 Ω. Two reactances are positive and one is negative; when the three are combined by algebraic addition, the result is 60 Ω. They are illustrated:

Resistance:

$$20\ \Omega + 10\ \Omega + 50\ \Omega = 80\ \Omega$$

Reactance:

$$(+j70\ \Omega) + (+j10\ \Omega) + (-j20\ \Omega) = +j60\ \Omega$$

The result is one impedance, Z_t, which can be stated as 80 Ω of resistance plus 60 Ω of inductive reactance, which can be stated in rectangular form:

$$Z_t = 80 + j60\ \Omega$$

To determine the magnitude of this vector, we must use the Pythagorean theorem. The rectangularly notated quantity can be substituted in the theorem as follows:

$$Z_t = \sqrt{R_2 + X_L^2}$$

$$Z_t = \sqrt{80^2 + 60^2}$$

$$Z_t = \sqrt{10{,}000}$$

$$Z_t = 100\ \Omega$$

In rectangular form, the vector Z_t is described in terms of two sides of a right triangle. A rectangularly notated impedance states the amount of impedance that results from resistance and the amount that results from reactance. It does not, however, state the magnitude of the impedance vector. To solve for magnitude, the Pythagorean theorem must be used.

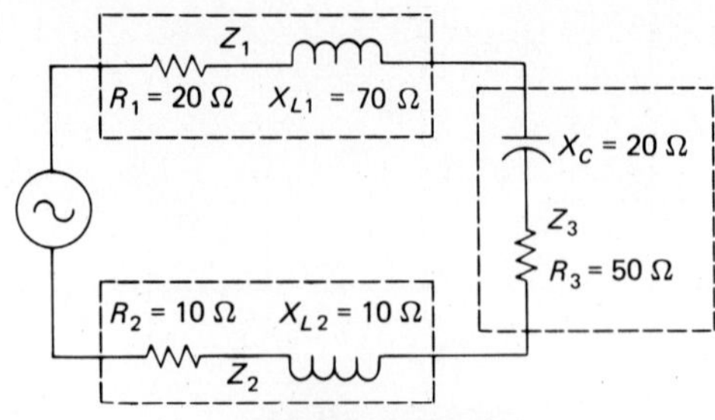

FIGURE 19-16

Self-Check

Answer each item by indicating whether the statement is true or false.

16. ____ Impedance is the total opposition contained in an AC circuit.

17. ____ Rectangularly notated complex numbers can be combined by addition.

18. ____ Capacitance reactance is represented by $-j$ in a rectangular notation.

19. ____ All resistance values are in phase when solving for total impedance of a circuit.

20. ____ Total X_C must be subtracted from total X_L when calculating the total reactance of a circuit.

Polar Notation

Up to now, we have been discussing a method for solving a resultant vector. Now we need a method suitable for expressing the quantities included in the resultant vector. In Figure 19-17, vector B is described in rectangular form by the statement $3 + j4$. This is the vector that results when vectors are added graphically. A vector equal to X_L is drawn as a dashed line, with its origin at the end of the R vector. Vector B results when point O is connected to the top of the dashed line, which is equal to X_L in direction and magnitude.

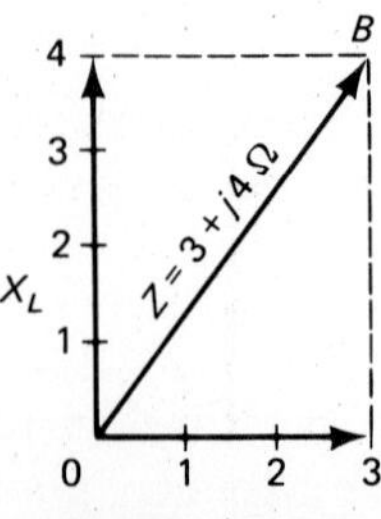

FIGURE 19-17

As we have seen, this vector can be described in combined form if its magnitude and angle of direction are included in the statement. Earlier, we found that the vector's magnitude was 5 units (Ω) long and that the angle θ was 53.1°. Figure 19-18 illustrates the same vector stated in *polar notation.* In this form it is stated as $5\underline{/+53.1^\circ}$. It is said to have a *polar* magnitude of 5 units and an angle of 53.1°.

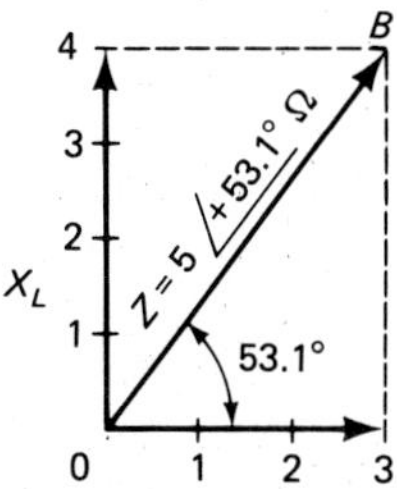

FIGURE 19-18

Vectors expressed in polar form can be graphed directly, without constructing a parallelogram. Examples are the graphing of vectors as shown in Figure 19-19:

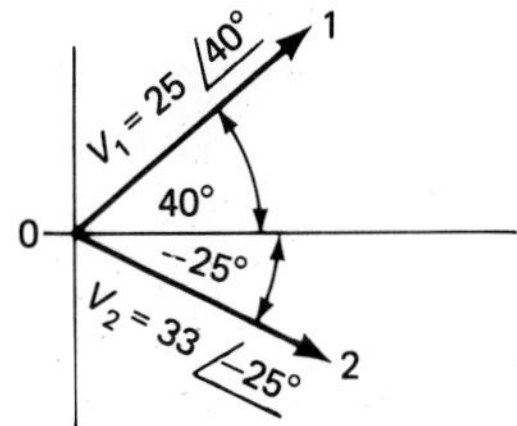

FIGURE 19-19

Vector 1:

$$25\underline{/+40^\circ}\ \text{V}$$

Vector 2:

$$35\underline{/-25^\circ}\ \text{V}$$

Notice that vector 1 is plotted at +40° and that vector 2 is plotted at −25°, with respect to the reference axis. This means that vector 1 has been rotated +40° and vector 2 −25°.

Note that polar and rectangular notations are a matter of convenience; they are used to describe circuit conditions from the point of view of mathematics and electricity. The vector in Figure 19-18 can be described in either form:

$$Z_t = 5\underline{/+53.1^\circ}\ \Omega$$

or

$$Z_t = 3 + j4\ \Omega$$

Self-Check

Answer each item by indicating whether the statement is true or false.

21. ____ Polar notation is another way of stating a vector's magnitude.

22. ____ Rectangular notation expresses both magnitude and direction.

23. ____ Polar notation is the method used to describe resultant vectors that have been solved by using a graph and a protractor.

24. ____ Rectangular notation states the length of two sides of a right triangle.

25. ____ To solve a rectangularly notated right triangle, it is necessary to construct a right triangle and use the Pythagorean theorem.

Vectorial Multiplication and Division

Vectors stated in polar form lend themselves to multiplication and division. To multiply polar-form vectors, we multiply the magnitude (scalar quantity) and add the angles, using algebraic addition. An example is multiplication of $V_1 \times V_2$:

where

$$V_1 = 25\underline{/+30^\circ}\text{ V}$$

$$V_2 = 20\underline{/+40^\circ}\text{ V}$$

then

$$V_3 = V_1 \times V_2$$

$$V_3 = 25\underline{/+30^\circ} \times 20\underline{/+40^\circ}$$

$$V_3 = 25 \times 20\underline{/30^\circ + 40^\circ}$$

$$V_3 = 500\underline{/+70^\circ}\text{ V}$$

Thus

$$V_3 = V_1 \times V_2$$

$$V_3 = 25\underline{/+30^\circ} \times 20\underline{/+40^\circ}$$

$$V_3 = 500\underline{/+70^\circ}\text{ V}$$

Other examples:

$$25\underline{/+15^\circ} \times 3\underline{/-40^\circ} = 75\underline{/-25^\circ}\text{ V}$$

$$10\underline{/-70^\circ} \times 2\underline{/+10^\circ} = 20\underline{/-60^\circ}\text{ V}$$

$$14\underline{/+25^\circ} \times 3\underline{/-5^\circ} = 42\underline{/+20^\circ}\text{ V}$$

$$20\underline{/+45^\circ} \times 2\underline{/+15^\circ} = 40\underline{/+60^\circ}\text{ V}$$

To perform division of polar notations, we

1. Divide the numerator's magnitude by the magnitude of the denominator.
2. Subtract the angles algebraically.

In one example, we divide V_1 by V_2:

$$V_1 = 30\underline{/+50^\circ}\text{ V}$$

$$V_2 = 10\underline{/+25^\circ}\text{ V}$$

Stated in problem form, these become:

$$V_3 = \frac{V_1}{V_2}$$

$$V_3 = \frac{30\underline{/+50^\circ}}{10\underline{/+25^\circ}}$$

$$V_3 = 3\underline{/+25^\circ}\text{ V}$$

which are determined as follows:

$$\text{Magnitude} = \frac{30}{10} = 3$$

$$\text{Angle } \theta = \underline{/+50^\circ} - \underline{/+25^\circ} = \underline{/+25^\circ}$$

Other examples:

$$\frac{40\underline{/+30^\circ}}{20\underline{/-10^\circ}} = 2\underline{/+40^\circ}$$

$$\frac{15\underline{/-20^\circ}}{3\underline{/+30^\circ}} = 5\underline{/-50^\circ}$$

Self-Check

Answer each item by indicating whether the statement is true or false.

26. ____ Vectors can be combined by multiplying their magnitudes and subtracting the angles algebraically.

27. ____ Vectors can be divided by dividing their magnitudes and subtracting the angles algebraically.

28. ____ Two vectors have magnitudes of 50 and 15; after multiplication, the resultant magnitude is 75.

29. ____ To subtract algebraically, the sign of the subtrahend must be changed and the quantities added.

30. ____ Rectangularly notated quantities are used to multiply and divide vectors.

Rectangular-to-Polar Conversions

It is possible to convert to polar form vectors stated in rectangular form. The procedure involves adding rectangular components and determining the angle (or phase angle) θ. The circuit shown in Figure 19-20a is a series RL circuit with 50 V AC applied. Its impedance can be stated by the rectangular notation,

$$Z_t = 60 + j80\ \Omega$$

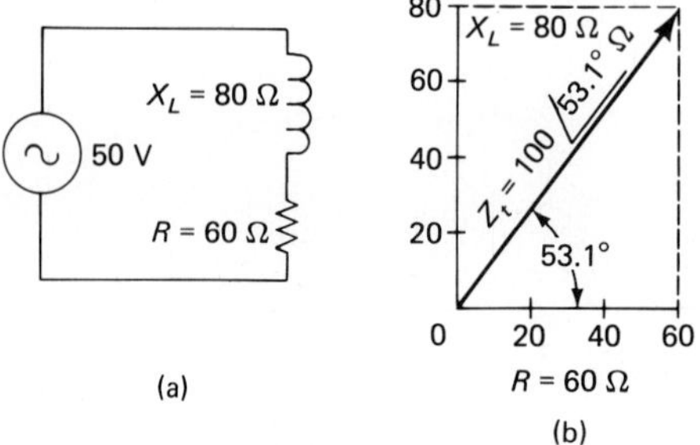

FIGURE 19-20

Figure 19-20b presents the vector diagram for X_L and R of the circuit. Earlier, we saw that the tangent of the phase angle which results from these vectors equals the ratio X_L divided by resistance. Use a calculator to find phase angle θ. *Note:* Some calculators use INV/tan, while others use ARC/tan. The phase angle θ equals $+53.1°$ for this inductive circuit.

Action	*Entry*	*Display*
ENTER	80	80
PRESS	÷	80
ENTER	60	60
PRESS	=	1.3333
PRESS	ARC (INV)	1.3333
PRESS	tan	53.1

To determine magnitude, use the Pythagorean theorem:

$$Z_t = \sqrt{R_2 + X_L^2}$$

Action	Entry	Display
ENTER	60	60
PRESS	X^2	3600
PRESS	+	3600
ENTER	80	80
PRESS	X^2	6400
PRESS	=	10000
PRESS	$\sqrt{X}$	100

The vector magnitude is 100. When stated in polar notation, the impedance vector becomes

$$Z_t = 100\angle{+53.1°}\ \Omega$$

The rectangularly notated vector $60 + j80\ \Omega$ has been converted to polar notation.

Figure 19-21a presents a series RC circuit whose resistance and reactance values are identical to those already given. Figure 19-21b presents the vectorial diagram for this circuit. The rectangular notation of the resultant vector is

$$Z_t = 60 - j80\ \Omega$$

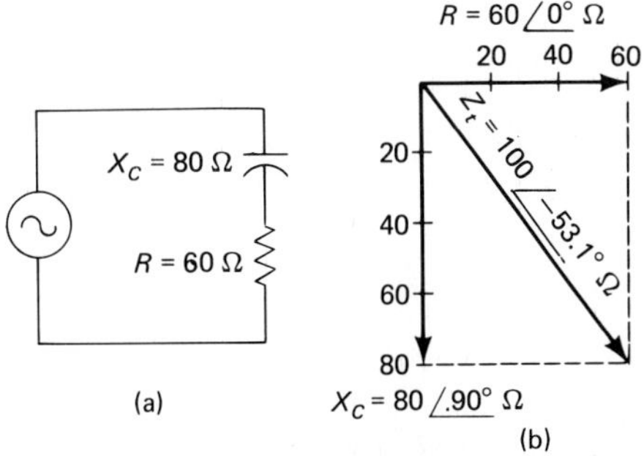

FIGURE 19-21

We solve this vector for Z_t and phase angle, using the procedures discussed.

Mathematical solution:

$$\tan\theta = \frac{X_C}{R}$$

$$\tan\theta = \frac{80}{60}$$

$$\tan\theta = 1.3333$$

To find angle $\theta = 53.1°$, you can use your calculator

Action	Entry	Display
ENTER	80	80
PRESS	÷	80
ENTER	60	60
PRESS	=	1.3333
PRESS	ARC (INV)	1.3333
PRESS	tan	53.1

Because the circuit is capacitive, angle $\theta = -53.1°$.

To solve for magnitude, use the Pythagorean theorem.

Mathematical solution:

$$Z_t = \sqrt{R^2 + X_C^2}$$

$$Z_t = \sqrt{60^2 + 80^2}$$

$$Z_t = \sqrt{3600 + 6400}$$

$$Z_t = \sqrt{10{,}000}$$

$$Z_t = 100\ \Omega$$

Action	Entry	Display
ENTER	60	60
PRESS	X^2	3600
PRESS	+	3600
ENTER	80	80
PRESS	X^2	6400
PRESS	=	10000
PRESS	$\sqrt{X}$	100

The polar notation that represents the impedance of this circuit is

$$Z_t = 100\underline{/-53.1°}\ \Omega$$

Self-Check

Answer each item by indicating whether the statement is true or false.

31. ____ To convert X_L and R to angle θ, use the sine function.

32. ____ The Pythagorean theorem is used to find the total impedance of a circuit.

33. ___ Vector diagrams for the circuits in Figures 19-20 and 19-21 are identical.

34. ___ In Figure 19-21, X_C lags R by 90°.

35. ___ X_C and R are 180° out of phase in an RC circuit.

Polar-to-Rectangular Conversions

The conversion of polar-notated quantities to rectangular notation involves the use of trigonometric functions similar to those already discussed. The functions are listed for review purposes:

$$\sin\theta = \frac{X_L}{Z_t} \quad \text{or} \quad \frac{X_C}{Z_t}$$

$$\cos\theta = \frac{R_t}{Z_t} \quad \text{or} \quad \frac{R_t}{Z_t}$$

$$\tan\theta = \frac{X_L}{R_t} \quad \text{or} \quad \frac{X_C}{R_t}$$

By transposition, we get other formulas which help us determine the rectangular components of a circuit. For an inductive circuit, they are:

$$X_L = Z_t \sin\theta \qquad \text{(equation 19-3)}$$

$$R_t = Z_t \cos\theta \qquad \text{(equation 19-4)}$$

Equation 19-1 has been introduced. For an RL circuit, the equation is

$$Z_t = R + jX_L$$

For an RC circuit, it is

$$Z_t = R - jX_C$$

Let us use equation 19-1 and substitute the X_L and R_t values of equations 19-3 and 19-4. Note that the impedances from equations 19-3 and 19-4 are identified as Z'. In this representation, Z represents vector magnitude and Z' represents impedance, in numerical value:

$$Z = R + jX$$

$$Z = Z'\cos\theta + Z'_j\sin\theta$$

Factor Z' from right side, and we have:

$$Z = Z'(\cos\theta + j\sin\theta) \qquad \text{(equation 19-5)}$$

This solution is illustrated in Figure 19-22. Note that, in this figure the only quantity given is the polar-notated vector for Z_t, which is stated as $25\underline{/+40°}\ \Omega$. The solution is:

$$Z = Z'(\cos\theta + j\sin\theta)$$

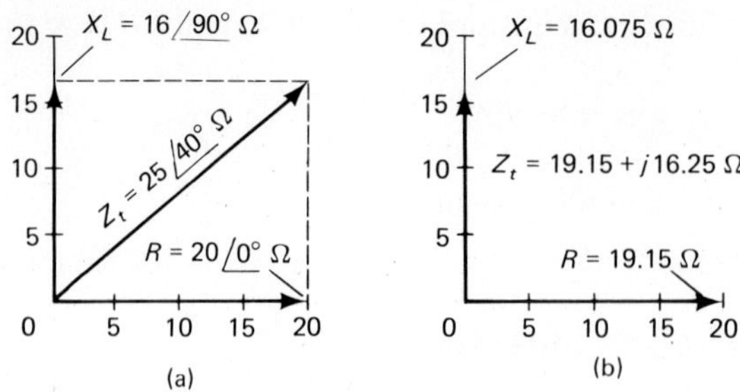

FIGURE 19-22

Mathematical solution:

$$Z = 25\,[(\cos\angle +40°) + j(\sin\angle +40°)]$$

$$Z = 25[(0.766) + j(0.643)]$$

$$Z = (25 \times 0.766) + j(25 \times 0.643)$$

$$Z = 19.15 + j16.075\ \Omega$$

First, we determine the trigonometric functions:

Action	*Entry*	*Display*
ENTER	40	40
PRESS	cos	0.766
ENTER	40	40
PRESS	sin	0.643

Now these functions can be used:

Action	*Entry*	*Display*
ENTER	25	25
PRESS	×	25
ENTER	.766	0.766
PRESS	=	19.15*
ENTER	25	25
PRESS	×	25
ENTER	.643	0.643
PRESS	=	16.075**

**resistance, in Ω*
***jX_L, in Ω*

Both solutions result in a rectangularly notated vector of $19.15 + j16.075\ \Omega$. Had the circuit been capacitive, the only difference would be the use of a $-j$ operator.

Self-Check

Answer each item by indicating whether the statement is true or false.

36. ____ The sine function can be used to calculate resistance, R, if impedance, Z, and angle θ are known.

37. ____ The result of polar-to-rectangular conversions can be checked by using the Pythagorean theorem.

38. ____ The cosine function of an angle can be used to find X_C, when resistance, R, and angle θ are known.

39. ____ The polar notation $45\underline{/-45°}$ converts to $31.82 - j31.82$.

40. ____ The cosine value of a polar notation always represents a *resistance* value at a phase angle of 0°.

Summary

In this chapter we discussed vectors and their classification. Vectors were defined as having both direction and magnitude. The graphical method and the Pythagorean theorem, for vectorial addition, were explained and examples presented.

Any quantity can be changed from positive to negative or negative to positive by multiplying it by -1. To indicate that a vector has been multiplied by -1, it is rotated 180° and plotted in the opposite direction. Vectors can be rotated 90° by multiplying either $\sqrt{-1}$ or $-\sqrt{-1}$. Multiplying by $\sqrt{-1}$ causes the vector to rotate $+90°$ and *lead* the reference axis. Multiplying by $-\sqrt{-1}$ causes the vector to rotate $-90°$ and *lag* the reference axis. Vectors that lead the reference vector by 90° are assigned the $+j$ operator to indicate this fact; those that lag by 90° are assigned the $-j$ operator to indicate their position. When a vector is described, using a j operator, we are using *rectangular notation.* Because they relate two sides of a right triangle, these notations state a vector's characteristics. To determine their direction and magnitude, it is necessary to solve mathematically.

It is also possible to use *polar notation* for vectors. These notations indicate magnitude (in units) and direction (in degrees). A polar notation need not be stated in relation to a triangle. This notation is the result of vectorial solutions that employ either graph or trigonometric methods.

Mathematical manipulation of vectors was discussed, and addition, subtraction, multiplication, and division of vectors were covered.

The method used to convert rectangular notations to polar form was explained, as was a method of converting polar to rectangular.

Review Questions and/or Problems

1. A series RC circuit has 15 Ω of resistance and 10 Ω of X_C. Which statement describes its impedance?
   (a) $-15 + j10\ \Omega$
   (b) $15 - j10\ \Omega$
   (c) $15 + j10\ \Omega$
   (d) $-15 - j10\ \Omega$

2. A series RL circuit contains 20 Ω of resistance and 12 Ω of X_L. What is Z_t?
   (a) $20 - j12\ \Omega$
   (b) $-20 + j12\ \Omega$
   (c) $-20 - j12\ \Omega$
   (d) $20 + j12\ \Omega$

3. A series AC circuit has Z_t of $6 + j8\ \Omega$. State this in polar form.
   (a) $10\angle{+53^\circ}\ \Omega$
   (b) $10\angle{+37^\circ}\ \Omega$
   (c) $10\angle{-53^\circ}\ \Omega$
   (d) $10\angle{-37^\circ}\ \Omega$

4. Convert the statement $5 - j5\ \Omega$ to polar form.
   (a) $-7\angle{-45^\circ}\ \Omega$
   (b) $-7\angle{+45^\circ}\ \Omega$
   (c) $7\angle{-45^\circ}\ \Omega$
   (d) $7\angle{+45^\circ}\ \Omega$

5. Convert the polar statement $15\angle{+53^\circ}$ to rectangular form.
   (a) $9 + j12$
   (b) $9 - j12$
   (c) $-9 - j12$
   (d) $-9 + j12$

6. Convert $20\angle{-37^\circ}$ to rectangular form.
   (a) $-16 - j12$
   (b) $-16 + j12$
   (c) $16 - j12$
   (d) $16 + j12$

7. Add: $(4 - j3) + (8 + j9)$. What do you get?
   (a) $4 - j12$
   (b) $4 + j12$
   (c) $12 + j12$
   (d) $12 + j6$

8. Add: $(6 - j4) + (2 + j6)$. What do you get?
   (a) $4 - j2$
   (b) $8 + j2$
   (c) $8 - j10$
   (d) $8 + j10$

9. Multiply the following complex numbers: $12\angle +20^\circ \times 3\angle +5^\circ$.
   (a) $15\angle +100^\circ$
   (b) $36\angle -100^\circ$
   (c) $36\angle +25^\circ$
   (d) $15\angle +25^\circ$

10. Multiply: $15\angle -20^\circ \times 2\angle +5^\circ$.
   (a) $30\angle -15^\circ$
   (b) $30\angle +15^\circ$
   (c) $17\angle -100^\circ$
   (d) $17\angle +100^\circ$

11. Divide: $\dfrac{14\angle +30^\circ}{2\angle +15^\circ}$.
   (a) $7\angle -15^\circ$
   (b) $7\angle +45^\circ$
   (c) $7\angle -45^\circ$
   (d) $7\angle +15^\circ$

12. Divide: $\dfrac{6\angle -20^\circ}{3\angle +10^\circ}$.
   (a) $2\angle -10^\circ$
   (b) $2\angle -30^\circ$
   (c) $2\angle +30^\circ$
   (d) $2\angle +10^\circ$

13. A series RCL circuit contains $R_1 = 2\ \Omega$, $R_2 = 8\ \Omega$, $X_C = 40\ \Omega$, and $X_L = 60\ \Omega$. Calculate Z_t and state your answer in rectangular form.
   (a) $10 - j20\ \Omega$
   (b) $10 + j10\ \Omega$
   (c) $10 - j10\ \Omega$
   (d) $10 + j20\ \Omega$

14. Use the values in question 13 to state Z_t in polar form.
   (a) $22.36\angle +63.4^\circ\ \Omega$
   (b) $14.1\angle +45^\circ\ \Omega$
   (c) $14.1\angle -45^\circ\ \Omega$
   (d) $22.36\angle -63.4^\circ\ \Omega$

20

Series RC, RL, and RCL Circuits

Introduction

Connecting resistance and reactive components in the same AC circuit has a much different effect than connecting resistors only. Both components oppose the change in current flow that continually occurs. The sum of the two oppositions, however, is much different. The term *impedance* refers to the total opposition provided by resistive and reactive components.

Circuits that contain capacitance and resistance have characteristics considerably different from those of resistance and inductance circuits. Circuits with resistance and capacitance are called *RC circuits,* those with inductance and resistance, *RL circuits.* When all three components, inductance, capacitance, and resistance are present, the circuit becomes an *RCL circuit.* When they are present in a series AC circuit, total impedance has still other characteristics.

Solving these circuits involves understanding the effect of reactances and the phase relations within an AC circuit.

Objectives

Each student is required to:

1. Draw vector diagrams of series RC, RL, and RCL circuits.
2. Solve series RC, RL, and RCL circuits for impedance.
3. Apply Ohm's law in analyzing series RC, RL, and RCL circuits.
4. Use trigonometry in solving:
    - (a) Voltage magnitude and direction for
        - (1) Total voltage.
        - (2) Capacitor voltage drop.
        - (3) Resistor voltage drop.
        - (4) Inductor voltage drop.
    - (b) Current magnitude and direction.
    - (c) Impedance magnitude and direction.

5. Analyze series RC, RL, RCL circuits for
   (a) Apparent power.
   (b) True power.
   (c) Power factor.

Impedance

Impedance (Z) is the total opposition of combined amounts of resistance and reactance in an AC circuit. It is the total opposition to the flow of alternating current. Impedance may include any combination of resistance, inductive reactance, and/or capacitive reactance.

Impedance is the total opposition to current flow in an AC circuit.

The series circuit in Figure 20-1a contains an AC power source and a resistor. Voltage and current in a purely resistive circuit are in phase. Thus, as voltage increases, current increases. This rule was presented in our analysis of DC circuits. The analysis is identical when AC circuits contain only resistance. With AC applied, though, the current flow is not equal at all times, and periodically changes direction. Figure 20-1b represents the relationship of E_R and I_R in the circuit. The two sine waves pass through zero, and reach their peaks at precisely the same time; they are in phase. In any circuit, current through a resistor is in phase with the voltage dropped by the resistor.

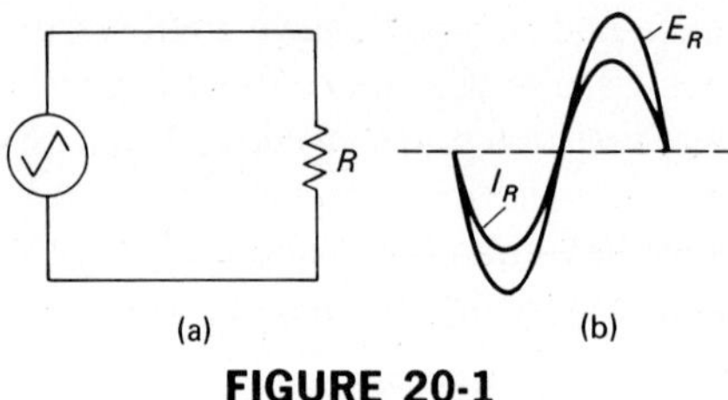

FIGURE 20-1

Figure 20-2a shows a circuit with an AC power source and a capacitor. Figure 20-2b shows the phase relation across the capacitor. Note that I_C leads E_C by 90°. In any circuit, current through a capacitor leads the voltage dropped by that capacitor by 90°. This can also be stated: Voltage on a capacitor lags the current through the capacitor by 90°. In other words, current in a capacitor reaches maximum at a time equal to one-fourth the duration of the sine wave earlier than capacitor voltage reaches maximum.

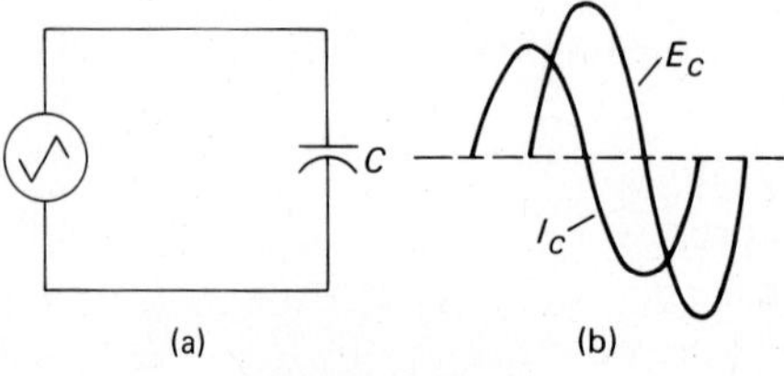

FIGURE 20-2

Figure 20-3a is a series circuit with an AC power source, resistor, and capacitor. Current has one path; the resistor and capacitor have the same current, but phase relationships discussed earlier still apply. To plot the phase relationships for this type of circuit, current is used as the reference. Resistor and capacitor voltage sine waves are then plotted, using I as their reference. Figure 20-3b shows the phase relationships. E_R is in phase with I, but E_C is 90° out of phase with both E_R and I. From this, we know that E_C lags both I and E_R by 90°.

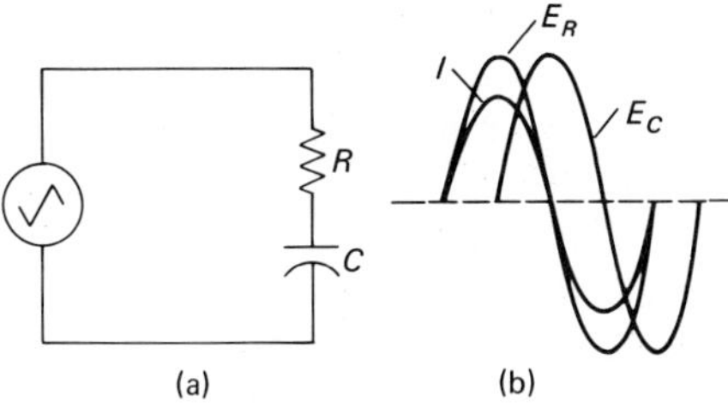

FIGURE 20-3

In Chapter 19 we discussed vectors and their manipulation. Vectors can be used to illustrate phase relationships in a circuit of this type. Let us analyze some circuits, comparing their sine waves and vector analysis.

A series RC circuit is a series circuit with an AC power source, resistance, and capacitance. A simple RC circuit is presented in Figure 20-4. Resistance, R, in this circuit is 8 Ω and capacitive reactance, X_C, is 6 Ω. The total impedance of the circuit results from a combination of the opposition caused by the capacitor's reactance, X_C, and the resistance, R. When AC is applied to a capacitor, voltage and current are 90° out of phase. Some characteristics of the circuit in Figure 20-4a are:

1. Current must be equal at all points of a series circuit. Therefore, $I_t = I_C = I_R$.
2. Current and voltage in a resistor are in phase. As current increases, the voltage drop on the resistor increases.
3. Current and voltage are out of phase across a capacitor.

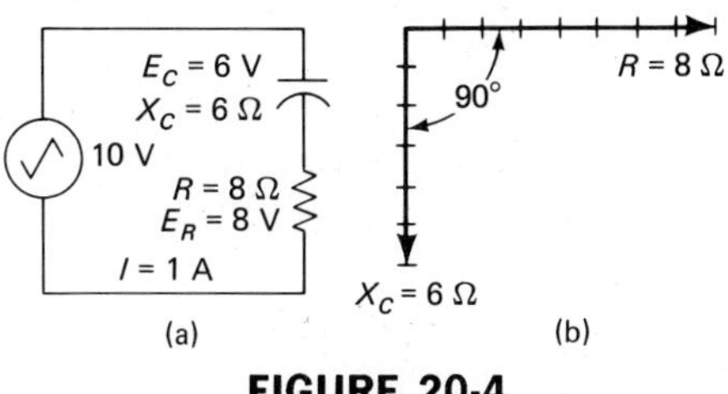

FIGURE 20-4

In a capacitor the phase angle, θ, is 90°, with current leading voltage. In an RC circuit, capacitor current, I_C, leads capacitor voltage, E_C, by an angle of 90°. Insertion of R makes it impossible to calculate impedance the same way it is calculated for resistive circuits. Mathematically, we use *vectorial addition*. First, we take a closer look at the circuit we have been discussing.

It is common practice to use vectors that represent the quantities present in an AC circuit. A vector is defined as

A line used to denote magnitude and direction.

Figure 20-4b shows the vector diagram for resistance, R, and capacitive reactance, X_C, where capacitance reactance lags resistance by 90°. Notice that resistance is plotted on the +X axis. To represent X_C in a −90° position, we must plot it on the −Y axis.

In Figure 20-5a, note that current is plotted on the +X axis. The vector length represents the current (in amps) flowing in the circuit. Because I_R and E_R are always in phase, they can serve as a reference. Figure 20-5b is the resistance vector for this circuit; its length represents ohms of resistance in the circuit. Because I_C and E_C are 90° out of phase across the capacitor, with I_C leading, E_C is plotted on the −Y vector. The length and direction of the vector conform to the effect and amount of E_C dropped by the capacitor. Figure 20-5c contains vectors that represent I, E_R, and E_C. Figure 20-5d is a vector diagram derived by completing the rectangle outlined by the dashed lines. Once the rectangle is complete, a vector drawn from the origin to the opposite corner of the rectangle is equal, in angular direction and magnitude, to the voltage applied to the circuit. In this case, E_t equals 10 V at −36.9°.

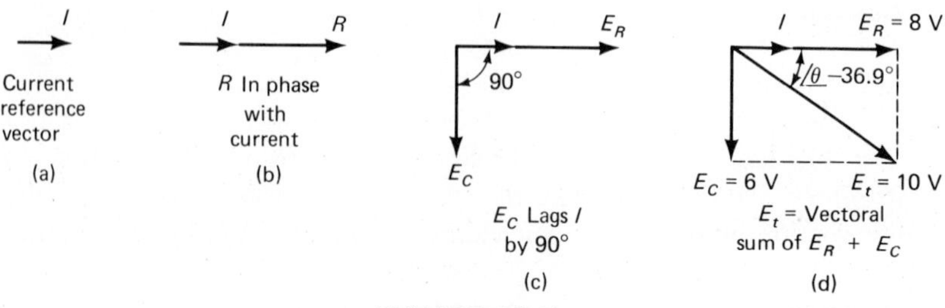

FIGURE 20-5

Note that E_t is out of phase with both E_R and E_C. The extent of the phase difference (phase angle) depends on the relationship of E_R to E_C. When they are equal, the phase difference is 45°. When E_C is larger, the phase angle is more than 45°; when E_R is larger, the angle is less than 45°.

The relationship of current to voltage is represented by the sine waves (Figure 20-6). Note that one sine wave represents both I_R and I_C. The sine wave representing E_R is in phase with I_C and I_R. A sine wave represents E_C, which is 90° out of phase with the other two sine waves.

Referring to Figures 20-4a and 20-4b, we continue our analysis of impedance. Figure 20-7 is used to represent the vectors involved with resistance, X_C, and impedance. When vectors representing resistance and X_C have been plotted, we can determine Z by constructing the rectangle and drawing the Z vector. The impedance and its phase angle can be measured physically. To determine the magnitude of Z, we use the same scale per unit of ohms that we used for R and X_C to measure the length of the diagonal vector. Its length equals impedance, Z. To measure the phase angle, we use a protractor and measure the number of degrees between the X axis and the Z vector. Note that when X_C and R are connected in the circuit, the phase angle is negative. When X_C is larger than R, the phase angle is larger than 45°. If X_C and R are equal, the phase angle is 45°. If R is larger, the angle is less than 45°.

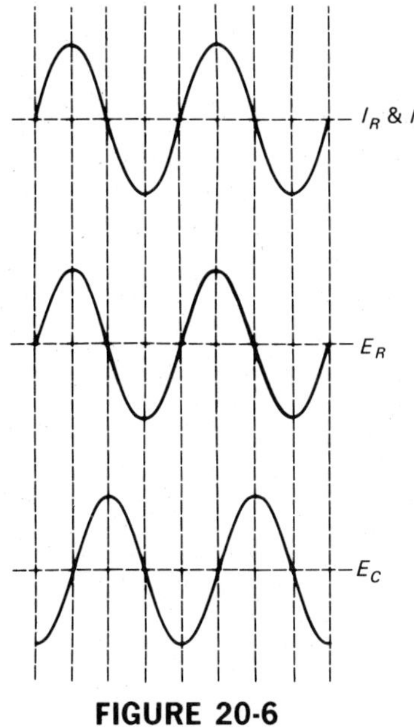

FIGURE 20-6

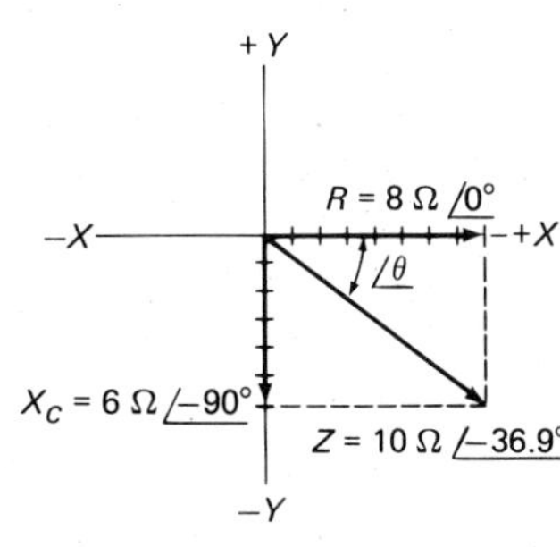

FIGURE 20-7

Figure 20-8 illustrates an RL circuit, where $X_L = 6\ \Omega$, $E_L = 60$ V, $E_R = 80$ V, and $R = 8\ \Omega$. The total voltage, impedance, and phase angle of this circuit can be determined the same way they were for the RC circuit. Inductor voltage leads current by 90°. Current is equal in both the resistor and the inductor; therefore, I_L equals I_R and the two are in phase. Inductor voltage, E_L, is plotted at 90° leading, which is on the +Y axis. This gives us the vectors shown in Figure 20-9a, which represent E_R, E_L, and E_t.

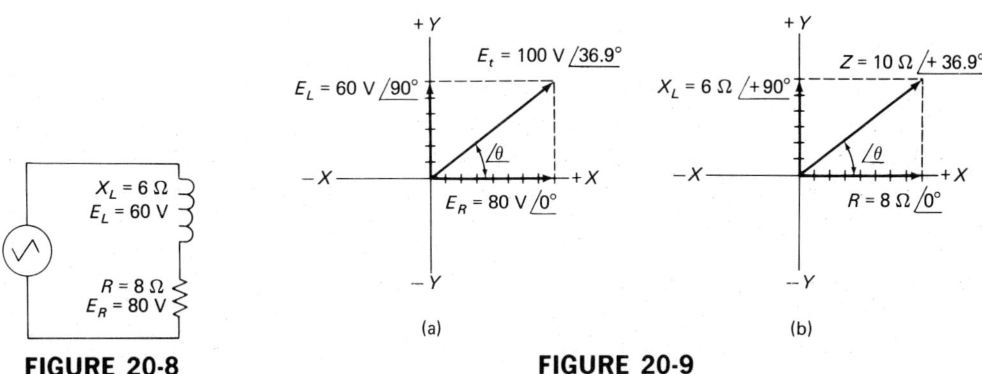

FIGURE 20-8

FIGURE 20-9

To determine Z and the phase angle, construct the vector diagram shown in Figure 20-9b so it represents X_L and R. Using these vectors, we constructed the rectangle (dashed lines) and drew the Z vector. Because R_t is larger than X_L, the phase angle is less than 45°. Had they been reversed, the phase angle would have been larger than 45°. When $R_t = X_L$, the phase angle is 45°.

Figure 20-10 shows three sine waves that represent the circuit-phase relationships. Sine wave a represents I_L and I_R. Sine wave b represents E_R, which is in phase with the current. Sine wave c represents E_L, which leads current by 90°. In both cases covered, E_C and E_L are out of phase with E_R to the same extent that they are out of phase with the circuit current. This is because I_R and E_R are in phase. In both cases, the *current* vector has been used as the reference vector.

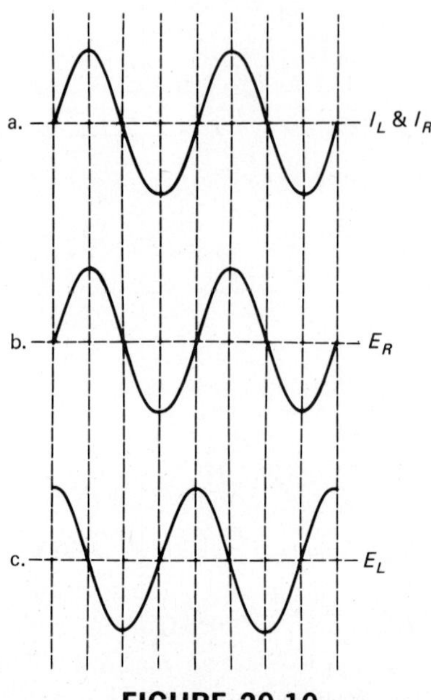

FIGURE 20-10

When R, X_C, and X_L are contained in the same circuit as that shown in Figure 20-11, the results differ considerably. When we consider these solutions, we find that X_C and X_L are 180° out of phase. When both devices are connected in the same circuit, they act as opposing power sources and cancel each other. Therefore, total impedance results from the combination of $R_t^2 + (X_C - X_L)^2$, added vectorially. $Z = \sqrt{R^2 + (X_L - X_C)^2}$

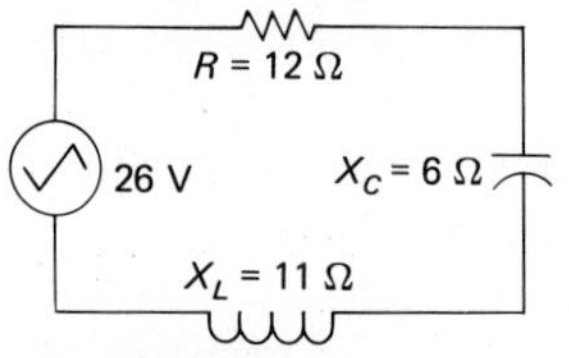

$R = 12\ \Omega$

26 V $\quad X_C = 6\ \Omega$

$X_L = 11\ \Omega$

FIGURE 20-11

When the vectors are plotted (Figure 20-12), the smaller of X_C or X_L must be subtracted from the larger. Once that has been done, construction of the rectangle, measurement of the Z vector, and measurement of the phase angle will be the same as that discussed. In this circuit, the Z vector is 13 Ω at a +22° phase angle; X_L is larger than X_C, resulting in a circuit with a positive phase angle. The circuit thus acts inductively even though capacitance is present. The opposition of the coil is strong enough to cancel X_C and operate as though no X_C were present. Had X_C been larger than X_L, the circuit would have operated capacitively with a negative phase angle.

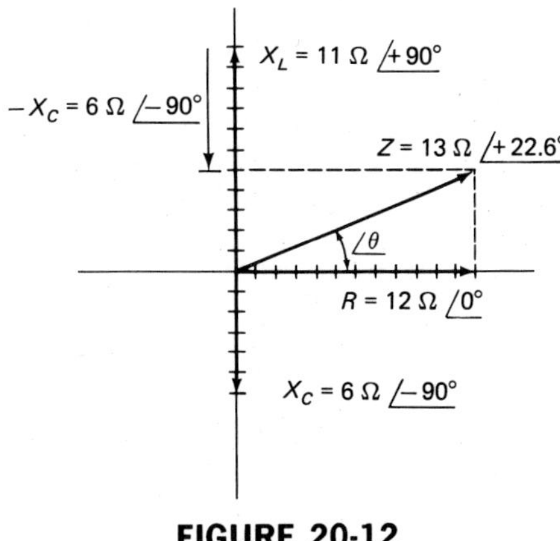

FIGURE 20-12

When capacitance, inductance, and frequencies are known in a circuit, we can solve for X_C or X_L for use in solutions. To simplify the explanation at this point, reactances are stated in ohms. Now that we know how to solve for Z in this circuit, let us look at the effect on Ohm's law.

The only difference in using Ohm's law to reach solutions is that impedance, Z, is used in place of R in the formulas. Thus the three forms of Ohm's law are:

$$I = \frac{E}{Z}$$

$$Z = \frac{E}{I}$$

$$E = I \times Z$$

Using these formulas to solve for values in Figure 20-11, we find:

$$I = \frac{E}{Z} = \frac{26\text{ V}}{13\ \Omega} = 2\text{ A}$$

$$E_R = I \times R = 2\text{ A} \times 12\ \Omega = 24\text{ V}$$

$$E_L = I \times X_L = 2\text{ A} \times 11\ \Omega = 22\text{ V}$$

$$E_C = I \times X_C = 2\text{ A} \times 6\ \Omega = 12\text{ V}$$

Self-Check

Select the correct option in each item below.

1. Which is the formula for Z?
   (a) $Z = X_C + R$
   (b) $Z = \frac{E_t}{I_t}$
   (c) $Z = \sqrt{X_C + R}$
   (d) $Z = E_t \times I_t$

2. Impedance is the sum of resistance plus reactance.
   (a) True
   (b) False

3. Current and voltage drop on the capacitor are in phase.
   (a) True
   (b) False

4. A vector has length and magnitude.
   (a) True
   (b) False

5. RC circuits have negative phase angles.
   (a) True
   (b) False

Trigonometric Solutions

You should now be familiar with the terms *impedance, phase,* and *reactance*. In this section, we discuss the methods of analyzing AC circuits mathematically. This involves simple trigonometry. You should be familiar with the terms sine, cosine, and tangent. Each term refers to a specific part of a right triangle (Figure 20-13). In comparing this triangle to an AC circuit, we see that the following applies:

Hypotenuse = impedance = labeled Z

Base = resistance = labeled R

Altitude = X_C or X_L = labeled X

Phase angle = angle theta = labeled θ

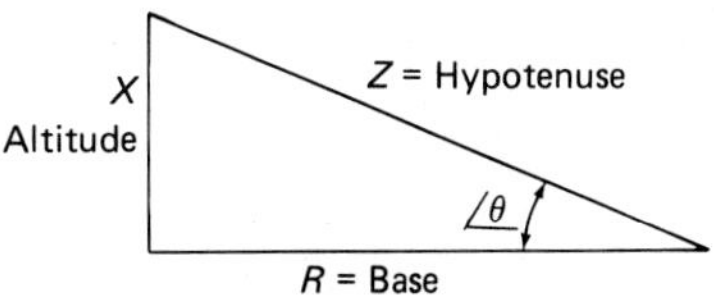

FIGURE 20-13

To illustrate the use of these factors with the trigonometric functions already mentioned, we consider that the sine of a right triangle is the ratio of the side of the triangle opposite angle θ divided by the hypotenuse:

$$\sin\theta = \frac{X}{Z}$$

The equations for other trigonometric functions are:

$$\cos\theta = \frac{R}{Z}$$

$$\tan\theta = \frac{X}{R}$$

These result in three equations we can use to analyze the circuit:

$$\sin\theta = \frac{X_C}{Z_t} \quad \text{or} \quad \frac{X_L}{Z_t}$$

$$\cos\theta = \frac{R_t}{Z_t}$$

$$\tan\theta = \frac{X_C}{R_t} \quad \text{or} \quad \frac{X_L}{R_t}$$

Refer to Figure 20-14. Figure 20-14a presents the same circuit as that shown in Figure 20-4; Figure 20-14b presents the vectors plotted in Figure 20-4. The problem given includes values for resistance and X_C. We can use the tangent function to determine the size of angle θ, because it requires use of the triangle's opposite and adjacent sides.

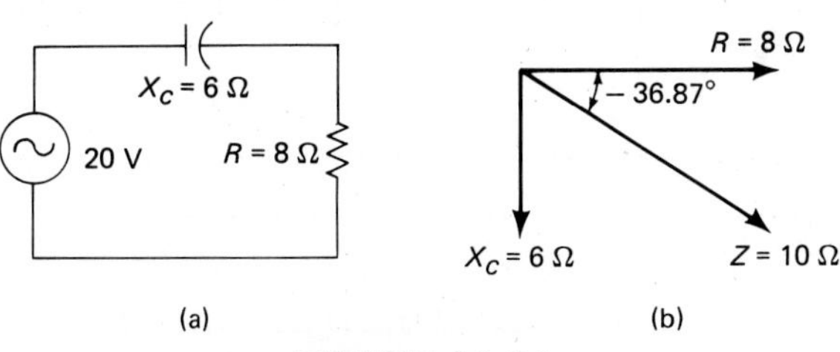

FIGURE 20-14

Mathematical solution:

$$\tan \theta = \frac{X_C}{R_t}$$

$$\tan \theta = \frac{6\ \Omega}{8\ \Omega}$$

$$\tan \theta = 0.75$$

We can use a table of trigonometric functions to convert this ratio to an angle measured in degrees. The problem becomes much easier and more convenient, however, when it is performed on the calculator.

Calculator solution:

Action	*Entry*	*Display*
ENTER	6	6
PRESS	÷	6
ENTER	8	8
PRESS	=	0.75

We now have the same value obtained in the mathematical solution. To use the calculator to convert the solution to its angle, we proceed:

Action	*Entry*	*Display*
ENTER	.75	0.75
PRESS	INV (ARC)*	0.75
PRESS	tan	36.8699

**Some calculator keys are marked "INV," others "ARC."*

This tells us that phase angle θ is $-36.8699°$.

The size of angle θ tells us the direction of the impedance vector. Before we solve for impedance, though, let us consider taking the angle 36.8699° and converting it to the tangent function, using a calculator:

Action	*Entry*	*Display*
ENTER	36.8699	36.8699
PRESS	tan	0.75

With the angle 36.8699° known, we can solve for sine and cosine functions:

Action	Entry	Display
ENTER	36.8699	36.8699
PRESS	sin	0.6
ENTER	36.8699	36.8699
PRESS	cos	0.8

Now that we know the size of the angle and its ratio, we can substitute in either of the other two formulas and solve for Z_t:

Mathematical solution:

$$\sin\theta = \frac{X_C}{Z_t}$$

Transposed:

$$Z_t = \frac{X_C}{\sin\theta}$$

Then:

$$Z_t = \frac{6\ \Omega}{0.60}$$

Result:

$$Z_t = 10\ \Omega$$

Note: Sine θ can also be found in trig tables.

Calculator solution:

Action	Entry	Display
ENTER	6	6
PRESS	÷	6
PRESS	(	6
ENTER	36.8699	36.8699
PRESS	sin	0.60000003
PRESS	)	0.60000003
PRESS	=	9.999994 Ω

The impedance of the circuit is 10 Ω at −36.87°.

We can use the cosine to check the calculator solution. The cosine, 36.8699° equals 0.8 is used in the solution.

Formula:

$$Z_t = \frac{R_t}{\cos\theta}$$

Action	Entry	Display
ENTER	8	8
PRESS	÷	8
ENTER	0.8	0.8
PRESS	=	10

Impedance equals 10 Ω at −36.87°. The two solutions check.

Now that we know $X_C = 6\ \Omega$, $R_t = 8\ \Omega$, and $Z_t = 10\ \Omega$, we can use Ohm's law to solve for other values, for instance,

$$I_t = \frac{E_t}{Z_t}$$

$$I_t = \frac{20\text{ V}}{10\ \Omega}$$

$$I_t = 2\text{ A}$$

The solution of series RL circuits is the same. The circuit shown in Figure 20-15 contains a 10-H coil and a 5-kΩ resistor. The operating frequency of the circuit is 40 Hz. Our first step in analyzing this circuit is to solve for X_L. To do this, we use

$$X_L = 2\pi f L$$

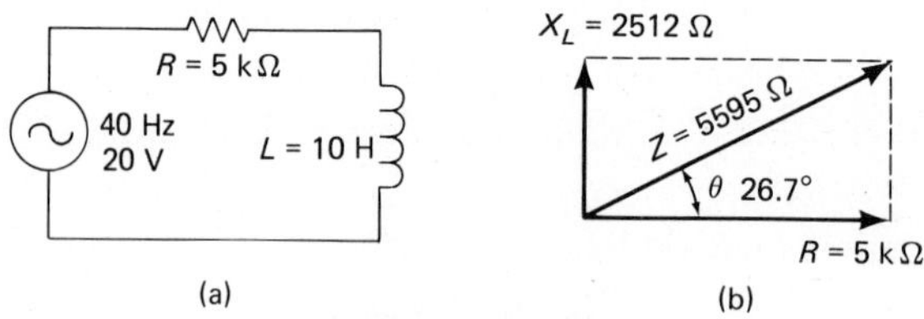

FIGURE 20-15

Mathematical solution:

$$X_L = 2 \times 3.14 \times 40\text{ Hz} \times 10\text{ H}$$

$$X_L = 2512\ \Omega$$

With X_L known, we can plot the resistance and inductive reactance vectors (Figure 20-15b). Angle θ is also shown. To find the size of the angle, we again use the tangent function of the angle. Done on the calculator, this appears:

Action	Entry	Display
ENTER	2512	2512
PRESS	÷	2512
ENTER	5, EE, 3	5 03
PRESS	=	5.024 − 01
PRESS	INV (ARC)	5.024 − 01
PRESS	tan	2.6675 01

Converting 2.6675 01 yields 26.675°.

Angle θ equals 26.6749°. Using the calculator to find the sine and the cosine:

Action	*Entry*	*Display*
ENTER	26.6749	26.675
PRESS	sin	0.4489
ENTER	26.6749	26.675
PRESS	cos	0.8936

These values can be used with other trig functions to solve for impedance. The phase angle is 26.7649°. Now we find Z_t:

Mathematical solution:

$$Z_t = \frac{R_t}{\cos \theta}$$

$$Z_t = \frac{5 \text{ k}\Omega}{0.8936}$$

$$Z_t = 5595 \ \Omega$$

Calculator solution:

Action	*Entry*	*Display*
ENTER	5, EE, 3	5 03
PRESS	÷	5 03
ENTER	.8936	0.8936
PRESS	=	5.5953 03

Converting 5.5953 03 yields 5595 Ω. Thus impedance equals 5595 Ω at 26.6749°.
To solve for I_t of the circuit, we use Ohm's law:

$$I_t = \frac{E_t}{Z_t}$$

$$I_t = \frac{20 \text{ V}}{5595 \ \Omega}$$

$$I_t = 0.00357 \text{ A or } 3.57 \text{ mA}$$

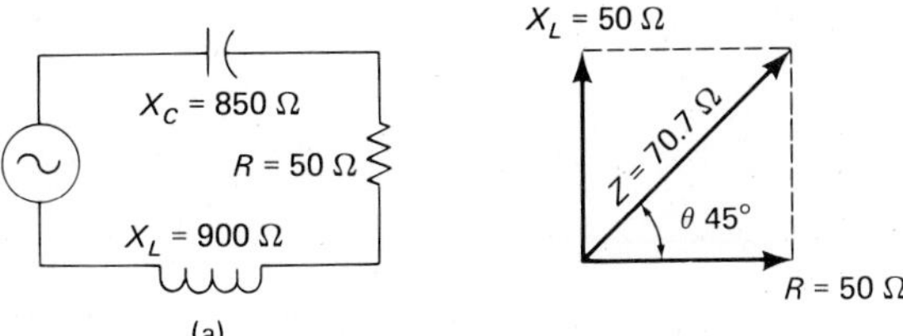

FIGURE 20-16

In solving the RCL circuit illustrated in Figure 20-16a, we must remember that the total impedance of this circuit is equal to the vectorial sum of all reactances and resistors. The first step in this solution requires subtraction of the smaller reactance from the larger. In this case, $X_L - X_C = 50\ \Omega$. When the resulting R_t and X_L vectors are plotted, we have Figure 20-16b. To determine the size of angle θ, we can use the tangent function:

$$\tan\theta = \frac{X_L}{R_t}$$

$$\tan\theta = \frac{50\ \Omega}{50\ \Omega}$$

$$\tan\theta = 1$$

Action	*Entry*	*Display*
ENTER	50	50
PRESS	÷	50
ENTER	50	50
PRESS	=	1
PRESS	INV (ARC)	1
PRESS	tan	45

Phase angle θ equals 45°. To find its sine and cosine, use the calculator as follows:

Action	*Entry*	*Display*
ENTER	45	45
PRESS	sin	0.7071
ENTER	45	45
PRESS	cos	0.7071

To solve Z_t, substitute in either of the other functions.

$$Z_t = \frac{R_t}{\cos\theta}$$

Action	*Entry*	*Display*
ENTER	50	50
PRESS	÷	50
ENTER	0.7071	0.7071
PRESS	=	70.7

Z_t equals 70.7 Ω at 45°.

Solve for I_t:

$$I_t = \frac{E_t}{Z_t}$$

$$I_t = \frac{100 \text{ V}}{70.7 \ \Omega}$$

$$I_t = 1.4 \text{ A}$$

Self-Check

Answer each item by indicating whether the statement is true or false.

6. ___ The sine of an angle equals the opposite side divided by the adjacent side.
7. ___ When we are given X_L and R_t of a circuit, we can use the tangent function to solve for impedance.
8. ___ Total current is solved, using Kirchhoff's law.
9. ___ A series RCL circuit has X_C larger than X_L; its phase angle will be a negative angle.
10. ___ To convert an angle to its sine, press the INV/ARC key on the calculator before pressing the sin key.

Pythagorean Theorem Analysis

In any study that includes analysis of right triangles, the Pythagorean theorem is useful. This theorem, based on the geometry of the triangle, states the relationship of the three sides of a right triangle. In Figure 20-17, note the following:

Side C = hypotenuse
Side B = base
Side A = altitude

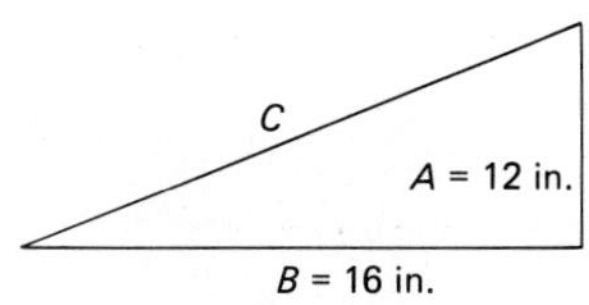

FIGURE 20-17

The Pythagorean theorem states:

> ***In any right triangle, the square of the hypotenuse is equal to the sum of the squares of the other two sides.***

Stated as an equation, the theorem is:

$$C^2 = A^2 + B^2$$

The Pythagorean theorem is a means of calculating either of three sides of a right triangle when the other two sides are known. By transposing and reducing the formula, we get the following three formulas, which are in the form most often used:

$$C = \sqrt{A^2 + B^2}$$

$$B = \sqrt{C^2 - A^2}$$

$$A = \sqrt{C^2 - B^2}$$

Refer to Figure 20-17. We use the data listed in the schematic to prove the accuracy of the three formulas. Note that side A = 12 in. and that side B = 16 in. To find side C, we proceed as follows:

$$C = \sqrt{A^2 + B^2}$$

Mathematical solution:

$$C = \sqrt{12^2 + 16^2}$$

$$C = \sqrt{144 + 256}$$

$$C = \sqrt{400}$$

$$C = 20 \text{ in.}$$

Action	*Entry*	*Display*
ENTER	12	12
PRESS	X^2	144
PRESS	+	144
ENTER	16	16
PRESS	X^2	256
PRESS	=	400
PRESS	$\sqrt{X}$	20

Side C of the triangle is 20 in. long. Now we use these values to prove the other two equations:

$$A = \sqrt{C^2 - B^2}$$

Action	Entry	Display
ENTER	20	20
PRESS	X^2	400
PRESS	−	400
ENTER	16	16
PRESS	X^2	256
PRESS	=	144
PRESS	$\sqrt{X}$	12

and:

$$B = \sqrt{C^2 - A^2}$$

Action	Entry	Display
ENTER	20	20
PRESS	X^2	400
PRESS	−	400
ENTER	12	12
PRESS	X^2	144
PRESS	=	256
PRESS	$\sqrt{X}$	16

We now apply the Pythagorean theorem to the calculation of impedance in an AC circuit (Figure 20-18). In this circuit, resistance equals 10 kΩ and X_C equals 13.3 kΩ. Thus, Z_t equals:

$$Z = \sqrt{R^2 + X_C^2}$$

Action	Entry	Display
ENTER	10, EE, 3	10 03
PRESS	X^2	1 08
PRESS	+	1 08
ENTER	12.5, EE, 3	12.5 03
PRESS	X^2	1.5625 08
PRESS	=	2.5625 08
PRESS	$\sqrt{X}$	1.60078 04

Converting 1.6007 04 to a prefix yields 16 kΩ. Therefore, Z_t equals 16 kΩ.

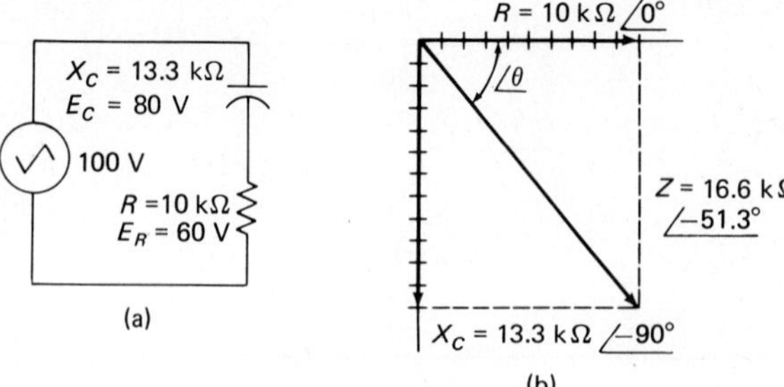

FIGURE 20-18

The Pythagorean theorem can be used with any reactive circuit to solve for E_t and Z_t. If we used the same circuit, the voltage solution would be:

$$E_t = \sqrt{E_R^2 + E_C^2}$$

Action	*Entry*	*Display*
ENTER	60	60
PRESS	X^2	3600
PRESS	+	3600
ENTER	80	80
PRESS	X^2	6400
PRESS	=	10000
PRESS	$\sqrt{X}$	100

Self-Check

Answer each item by indicating whether the statement is true or false.

11. ___ The Pythagorean theorem states a relationship between the sides of a triangle.

12. ___ The Pythagorean theorem can be used to solve for either side of a right triangle when the other two sides are known.

13. ___ The Pythagorean theorem is a convenient tool for calculating total voltage in an RC circuit.

14. ___ Voltage distribution in an AC circuit can be solved by adding individual voltage drops.

15. ___ If we use the Pythagorean theorem, we need not concern ourselves with the phase angle.

Power Distribution

Remember that in DC circuits and AC circuits that contain only resistors, voltage and current are always in phase. As voltage increases, current increases. Current is maximum when voltage is maximum. This makes calculation of power dissipation fairly simple. All current flow is used to dissipate power. *When we refer to "power," we are referring to the rate at which "practical work" is done.*

$$P = I^2 \times R$$

$$P = I \times E$$

$$P = \frac{E^2}{R}$$

AC reactive circuits present a different picture in the consideration of power. In capacitive and inductive circuits, current and voltage are *not* in phase. Current leads voltage in an RC circuit and lags voltage in an RL circuit. What is actually happening in a reactive circuit is that, on one alternation of the AC sine wave, the coil (or capacitor) takes energy from the source. On the next alternation, it discharges and returns energy to the circuit. If the circuit is purely reactive, the average power used is zero. We have no pure circuits, however; all circuits have some resistance, and that resistance dissipates power.

Now we have two different types of power in the circuit. In series RC, RL, and RCL circuits, capacitors and coils absorb power from the source, but they return the power to the same source during subsequent operations. Therefore, it is the resistor that dissipates the power. Application of the DC power formula results in *power apparent,* P_A. Power that is dissipated by the resistor is *true power,* P_T. The product of current and voltage across a reactive component is called VARs not watts. VAR stands for Volts-Amperes Reactive.

Let us look at these in a different way. True power is actual power dissipated by the resistor; it is consumption of power, just as those studied in resistive circuits are. Apparent power is the amount of power that *appears* to have been used. In fact, P_A is the vectorial sum of resistive-power dissipation and reactive power. Because reactive power is not dissipated, true power is never greater than apparent power. In AC resistive circuits, P_T and P_A are equal. In most reactive circuits, true power is less than power apparent. Resonant circuits are an exception.

Formulas used with power calculations for AC reactive circuits are:

Power true:

$$P_T = I_t \times E_R$$

$$P_T = I^2 \times R$$

$$P_T = \frac{E_R^2}{R}$$

$$P_T = I_t \times E_t \times \cos \theta$$

Power apparent:

$$P_A = I_t \times E_t$$

$$P_A = I^2 \times Z$$

$$P_A = \frac{E_t^2}{Z}$$

Both formulas can, of course, be transposed into other forms, as was done in the case of resistive circuits.

With this in mind, we proceed to another important part of AC reactive circuits: the *power factor,* or P_F.

The power factor is the ratio of true power to apparent power.

The ratio can be written:

$$P_F = \frac{P_T}{P_A} \qquad \text{(equation 1)}$$

We can transpose this equation:

$$P_A = \frac{P_T}{P_F} \qquad \text{(equation 2)}$$

or:

$$P_T = P_F \times P_A \qquad \text{(equation 3)}$$

Other useful equations are:

$$P_F = \cos \theta \qquad \text{(equation 4)}$$

$$P_F = \frac{R_t}{Z_t} \qquad \text{(equation 5)}$$

To emphasize that reactive power is *not* dissipated power, we do not measure its units in watts but rather in volt amperes. Remember that the formula for P_A is volts times amps, and you will get an idea of why the term is used. To apply the equation, we consider a circuit where P_A is 500 VA and P_F = 0.7071. To solve for P_T, we use equation 3:

$$P_T = P_F \times P_A$$

$$P_T = 0.7071 \times 500$$

$$P_T = 353.55 \text{ W}$$

The RL circuit in Figure 20-19 has $R = 40\ \Omega$ and $X_L = 30\ \Omega$, $I_t = 2$ A, and $E_t = 100$ V at a frequency of 1590 Hz. These values illustrate the use of equations 1–5.

$$P_A = I_t \times E_t$$

$$P_A = 2 \text{ A} \times 100 \text{ V}$$

$$P_A = 200 \text{ VA}$$

$$P_T = I_t^2 \times R$$

$$P_T = 2^2 \times 40$$

$$P_T = 4 \times 40$$

$$P_T = 160 \text{ W}$$

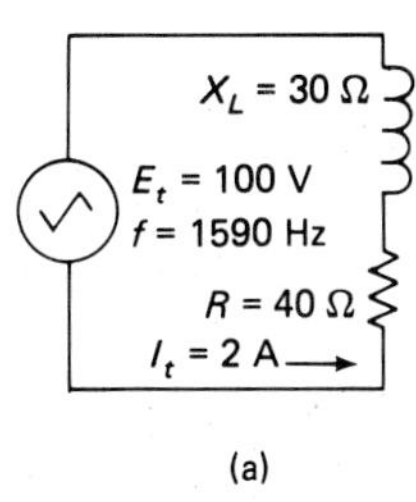

(a)

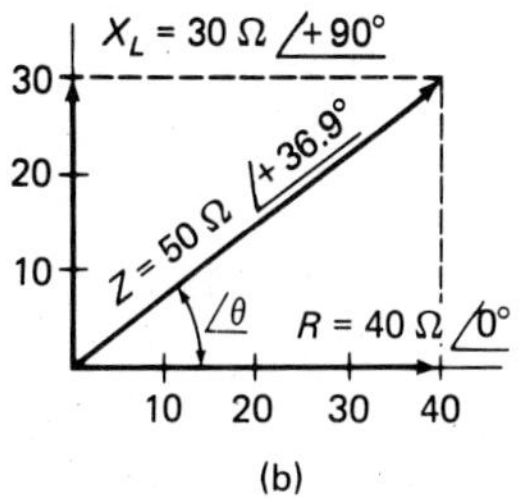

(b)

FIGURE 20-19

or:

$$P_F = \frac{P_T}{P_A}$$

$$P_F = \frac{160}{200}$$

$$P_F = 0.8$$

Given a circuit like this, we can also find the power factor, as follows:

$$P_F = \frac{R_t}{Z_t}$$

$$P_F = \frac{40}{50}$$

$$P_F = 0.8$$

and impedance:

$$Z_t = \frac{E_t}{I_t}$$

$$Z_t = \frac{100 \text{ V}}{2 \text{ A}}$$

$$Z_t = 50 \ \Omega$$

Remember that impedance has a phase angle. What angle would it be? What trigonometric function can be used to find the size of θ? Although we could use the sine, cosine, or tangent, we choose to use the sine:

$$\sin \theta = \frac{\text{opposite side}}{\text{hypotenuse}}$$

$$\sin \theta = \frac{X_L}{Z}$$

$$\sin \theta = \frac{30}{50}$$

$$\sin \theta = 0.6$$

Action	Entry	Display
ENTER	.6	0.6
PRESS	INV/ARC	0.6
PRESS	sin	36.9

$$Z_t = 50\underline{/36.9°}\ \Omega$$

Equation 4 can be used to find the power factor if we determine the cosine of θ. The ratio for the cos is:

$$\cos\theta = \frac{R}{Z}$$

$$\cos\theta - \frac{40}{50}$$

$$\cos\theta = 0.8$$

This also results in $P_F = 0.8$.

Solving a series RL circuit is identical to solving an RC circuit. The only difference is that a series RL circuit has a positive phase angle.

Only a slight difference occurs when we have a series RCL circuit like that shown in Figure 20-20. To find total reactance, we subtract X_C from X_L:

$$X_L - X_C = 800\ \Omega$$

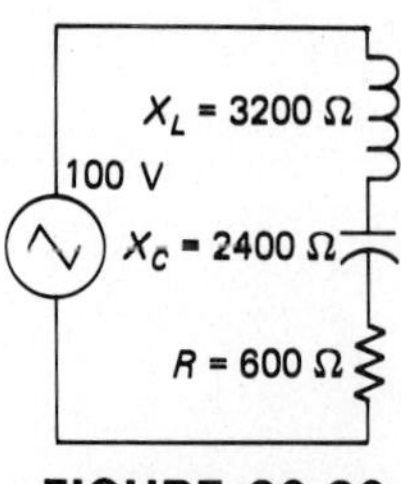

FIGURE 20-20

Continuing our solution, we use reactance and resistance to determine circuit impedance:

$$Z_t = \sqrt{R^2 + X^2}$$

$$Z_t = \sqrt{600^2 + 800^2}$$

$$Z_t = \sqrt{360{,}000 + 640{,}000}$$

$$Z_t = \sqrt{1{,}000{,}000}$$

$$Z_t = 1000\ \Omega$$

The power factor can be found:

$$P_F = \cos\theta$$

$$P_F = \frac{R_t}{Z_t}$$

$$P_F = \frac{600}{1000}$$

$$P_F = 0.6$$

We can use the calculator to convert 0.6 to θ:

Action	Entry	Display
ENTER	.6	0.6
PRESS	INV/ARC	0.6
PRESS	cos	53.1

$$Z_t = 1000\underline{/+53.1^\circ}\ \Omega$$

Continuing our solution, we find:

$$I_t = \frac{E_t}{Z_t}$$

$$I_t = \frac{100\ \text{V}}{1000}$$

$$I_t = 0.1\ \text{A or } 100\ \text{mA}$$

$$E_R = I_t + R_t$$

$$E_R = 0.1\ \text{A} \times 600\ \Omega$$

$$E_R = 60\ \text{V}$$

$$E_X = I_t \times (X_L - X_C)$$

$$E_X = 0.1\ \text{A} \times 800\ \Omega$$

$$E_X = 80\ \text{V}$$

$$P_A = I_t \times E_t$$

$$P_A = 0.1\ \text{A} \times 100\ \text{V}$$

$$P_A = 10\ \text{VA}$$

$$P_T = I_t \times E_R$$

$$P_T = 0.1 \text{ A} \times 60 \text{ V}$$

$$P_T = 6 \text{ W}$$

$$P_F = \frac{P_T}{P_A}$$

$$P_F = \frac{6}{10}$$

$$P_F = 0.6$$

Note that P_F is the same value already found. If a calculator is used to convert P_F to θ, the phase angle again results in angle θ equals 53.1°.

Self-Check

Answer each item by indicating whether the statement is true or false.

16. ____ The actual power dissipation of a series RC circuit is resistive power plus reactive power.

17. ____ The power factor is equal to the sine of θ.

18. ____ Power apparent is calculated by dividing power total by resistive power.

19. ____ The phase angle of a series RL circuit is negative.

20. ____ The impedance of the circuit in Figure 20-20 could have been solved by dividing total voltage by total current.

Summary

In this chapter we explored the conditions present when series RC, RL, and RCL circuits are used with an AC source. When we compared them to DC resistive circuits, we found that many aspects were unchanged. Because the capacitor and inductor have different reactances at different frequencies, however, we found that opposition to current flow resulted in a new formula. The total opposition in an RC, RL, or RCL circuit is expressed as impedance (symbol Z). The formula for calculating impedance involves vectorial addition of resistance and reactance.

With slight variations, the Pythagorean theorem is used to solve for impedance of the circuit and voltage distribution in series RC, RL, and RCL circuits. In either case, the total value is greater than either individual value but less than the sum of the two values.

Ohm's law can be used to solve the same circuit components as in DC circuitry, except that now we use impedance in place of resistance to solve total-circuit values. With AC applied to the circuit, a phase angle becomes part of our analysis because of the phase relations that now exist. Resistance and reactance values are used to solve for a trigonometric ratio, and the ratio is converted to the appropriate angle θ.

Power in series RC, RL, and RCL circuits was explored. We found that the power which appeared to be total power was not. The true-power dissipation in a reactive circuit occurs across the resistance. Reactive components do not dissipate power. This led to new terms:

Power true. ***The amount of power dissipated by the circuit's resistance. This is the total power dissipated as the reactive power is returned to the circuit by the field that caused it to be absorbed. A new power equation was introduced, which is*** $P_T = I_t \times E_t \times \cos\theta$.

Power apparent. ***The amount of power that appears to be consumed. Power apparent equals the product of total current and total voltage.***

Power factor. ***The ratio of true power to power apparent. The power factor is found by dividing resistive power by power apparent. We also found that the power factor equals*** $\cos\theta$.

Review Questions and/or Problems

1. An AC circuit contains a capacitor with 100 Ω X_C and has 50 V applied. What is I_t?
   (a) 0.5 A
   (b) 0.637 A
   (c) 0.707 A
   (d) 1.414 A

2. A series RC circuit has $R = 2$ kΩ and $X_C = 2$ kΩ. This makes the phase angle 45°. If resistance is doubled, what will angle θ do?
   (a) Increase
   (b) Decrease
   (c) Remain the same
   (d) Double

3. Which of the following is correct for impedance in an RC circuit?
   (a) Z_t is less than either X_C or R.
   (b) Z_t is more than $X_C + R$.
   (c) Z_t is less than $X_C - R$.
   (d) None of the above is correct.

4. The frequency applied to a series RC circuit suddenly doubles. What does the phase angle do?
   (a) Increase
   (b) Decrease
   (c) Remain the same

5. A coil has $X_L = 1\ \text{k}\Omega$. If its inductance and frequency are doubled, what happens to X_L?
   (a) It increases.
   (b) It decreases.
   (c) It remains the same.

6. A series RCL circuit has $R = 40\ \Omega$, $X_C = 90\ \Omega$, and $X_L = 60\ \Omega$. What is the circuit impedance?
   (a) $50\ \Omega$
   (b) $80\ \Omega$
   (c) $70\ \Omega$
   (d) $190\ \Omega$

7. A series RCL circuit has $R = 100\ \Omega$, $X_C = 200\ \Omega$, and $X_L = 300\ \Omega$. What is the phase angle θ?
   (a) 0°
   (b) 37°
   (c) 45°
   (d) 90°

8. Refer to Figure 20-21. Which of the following is true?
   (a) $Z = 256\ \Omega$
   (b) $Z = 265\ \Omega$
   (c) $Z = 165\ \Omega$
   (d) $Z = 156\ \Omega$

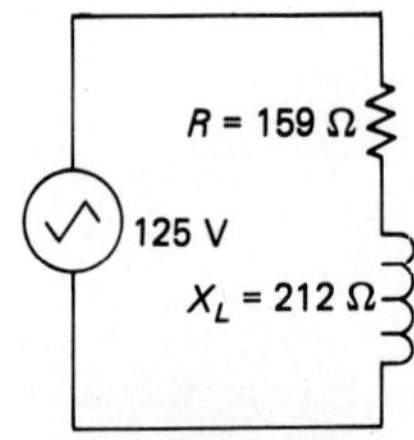

FIGURE 20-21

9. In Figure 20-21, what is angle θ?
   (a) 53.1°
   (b) 37.3°
   (c) 56.6°
   (d) 45°

10. In Figure 20-21, I_t equals
   (a) 47 mA
   (b) 0.47 A
   (c) 4.7 A
   (d) 4.7 mA

11. Refer to Figure 20-21. What does increasing the frequency cause?
   (a) The voltage across R remains the same.
   (b) I_t decreases.
   (c) Z_t decreases.
   (d) I_t increases.

12. Increasing the frequency applied to a series RCL circuit causes X_C to ______________ and X_L to ______________.
    (a) increase; increase
    (b) increase; decrease
    (c) decrease; decrease
    (d) decrease; increase

13. A series RCL circuit contains $R = 300\ \Omega$, $X_C = 300\ \Omega$, and $X_L = 300\ \Omega$. What is Z_t?
    (a) 300 Ω
    (b) 0 Ω
    (c) 900 Ω
    (d) 100 Ω

14. Which formula is correct for calculating the power factor?
    (a) $P_F = \dfrac{P_T}{\sin \theta}$
    (b) $P_F = \dfrac{P_T}{P_A}$
    (c) $P_F = \dfrac{P_T}{P_F}$
    (d) $P_F = P_T \times P_A$

15. In a series RC circuit,
    (a) the resistor dissipates true power.
    (b) the capacitor dissipates true power.
    (c) the power factor equals 1.
    (d) the power factor divides according to R_t and X_C.

16. In a series RC circuit, as X_C increases,
    (a) R_t decreases.
    (b) angle θ decreases.
    (c) the power factor decreases.
    (d) all of these are true.

17. A series RC circuit contains two capacitors; each capacitor has $X_C = 1000\ \Omega$. What is the circuit's total reactance?
    (a) 500 Ω
    (b) 1000 Ω
    (c) 2000 Ω
    (d) 1414 Ω

18. Apparent power is the
    (a) vector sum of R_t and $X_L + X_C$.
    (b) vector sum of X_L and X_C.
    (c) vector sum of $R_t + X_L$.
    (d) product of total current and total voltage.

19. The power factor is equal to
    (a) $\tan \theta$.
    (b) $\sin \theta$.
    (c) $\cos \theta$.
    (d) $\cot \theta$.

20. The power factor for the phase angle 57.9° is closest to
    (a) 0.053.
    (b) 0.53.
    (c) 0.513.
    (d) 0.0053.

21. As the phase angle is increased, the
    (a) P_F increases.
    (b) P_A remains the same.
    (c) P_F decreases.
    (d) P_T decreases.

22. What is the power factor for the circuit in Figure 20-22?
    (a) 0.9186
    (b) 0.3939
    (c) 0.3750
    (d) 0.5333

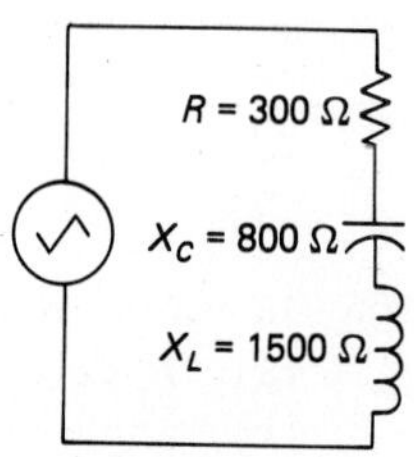

FIGURE 20-22

23. In a series RL circuit, inductance is doubled and frequency is cut in half. Which of the following is true?
    (a) X_L decreases.
    (b) X_L remains the same.
    (c) X_L increases.
    (d) Z_t decreases.

24. In Figure 20-23, which is correct?
    (a) $P_T = 48$ W
    (b) $P_T = 48$ VA
    (c) $P_T = 189$ W
    (d) $P_T = 189$ VA

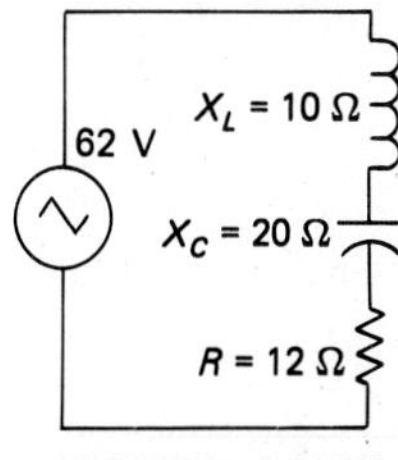

FIGURE 20-23

25. Refer to Figure 20-23. For this circuit, P_F is
    (a) 0.3888.
    (b) 0.768.
    (c) 0.8333.
    (d) 0.50.

21

Parallel RC, RL, and RCL Circuits

Introduction

In your study of series-reactive circuits, you have learned to solve circuits for impedance, voltage drops, current distribution, and phase angle. The vector analyses necessary to represent these have also been covered. In the series circuit, current is equal at all points in the circuit; this means that current is in phase at each component of the circuit. You have also learned that it was necessary to use the Pythagorean theorem to calculate voltage distribution and impedance. All vector diagrams were based on the fact that current is constant and is used as the zero reference. This happens because E_R and I_t are always in phase in a series-reactive circuit.

In this chapter we study the same components, but now arranged in parallel circuits. We already know that, in parallel circuits,

1. Total current equals the sum of all branch currents.
2. The voltage drop is equal on all parallel branches.
3. Total opposition is less than the opposition in the smallest branch.
4. The branch with the least opposition has the greatest current flow.
5. Voltage leads current by 90° in a coil.
6. Current leads voltage by 90° in a capacitor.
7. Current and voltage are in phase across a resistor.

Objectives

Given parallel RC, RL, and RCL circuits with component values included, the student is required to:

1. Calculate branch currents, using Ohm's law.
2. Calculate I_t by vector analysis.
3. Use Ohm's law to calculate impedance.

4. Draw vector diagrams that include
   (a) Resistor current.
   (b) Reactance current.
   (c) Total current.
   (d) Phase angle.
5. Use trigonometry to solve for
   (a) Total impedance.
   (b) Total current.
   (c) Phase angle.
   (d) Power factor.

Graphical Analysis

First, we use vectors to analyze circuits for current, voltage, and impedance. In the parallel circuit in Figure 21-1a, applied voltage is dropped across each branch. Because all voltage vectors are equal and in phase, we cannot use them as the basis for vector analysis; but branch currents have different phase relationships and can be plotted vectorially. What we can do is use the information given, as well as Ohm's law, to solve for branch currents.

$$I_R = \frac{E_t}{R} = \frac{60\ \text{V}}{3\ \Omega} = 20\ \text{A}$$

$$I_C = \frac{E_C}{X_C} = \frac{60\ \text{V}}{4\ \Omega} = 15\ \text{A}$$

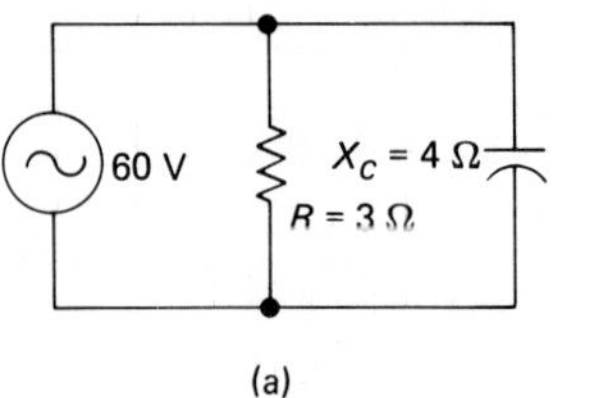

(a)

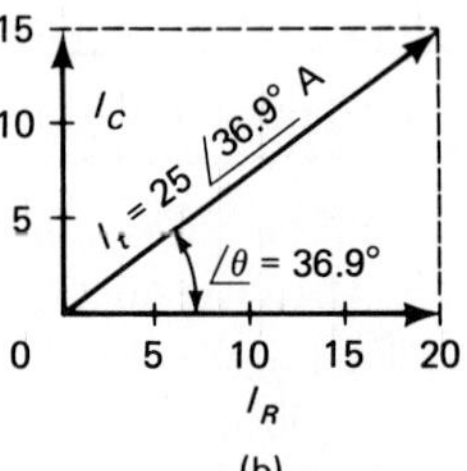

(b)

FIGURE 21-1

These values yield the current vectors,

$$I_R = 20\underline{/0°}\ \text{A} \qquad \text{(reference vector)}$$

$$I_C = 15\underline{/+90°}\ \text{A}$$

Remember: I_C leads E_C by 90°.

With E_C and E_R in phase, I_C leads I_R by 90° (Figure 21-1b). To solve for total current, we construct a rectangle like the one shown as dashed lines in the vector diagram. Draw the I_t vector to connect the origin of the current vectors with the opposite corner of the rectangle. Measurement of this vector's magnitude reveals that the total current is 25 A. To determine angle θ, we measure the angle with a protractor. We find that angle θ is $+36.9°$. The total current vector is:

$$I_t = 25\underline{/+36.9°}\text{ A}$$

By using total current, total voltage, and Ohm's law, we can calculate total impedance:

$$Z_t = \frac{E_t}{I_t}$$

$$Z_t = \frac{60\text{ V}}{25\text{ A}}$$

$$Z_t = 2.4\ \Omega$$

Did you notice:

1. The I_C vector leads I_R by 90°.
2. Impedance is less than the opposition in the smallest branch.
3. Total current is greater than current in either branch but less than the sum of the two branch currents.

The solution of RL circuits is identical to the analysis just completed, with the exception that I_L lags I_R by 90°. In Figure 21-2, diagram a presents a parallel RL circuit and diagram b shows the vector diagram. This circuit has $E_t = 40$ V, $R = 10\ \Omega$, and $X_L = 8\ \Omega$ as its givens. With this information, we can calculate the branch current, using Ohm's law:

$$I_R = \frac{E_t}{R} = \frac{40\text{ V}}{10\ \Omega} = 4\text{ A}$$

$$I_L = \frac{E_L}{X_L} = \frac{40\text{ V}}{8\ \Omega} = 5\text{ A}$$

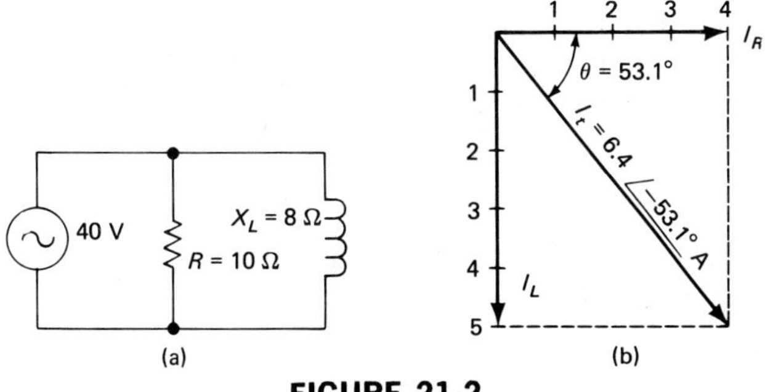

FIGURE 21-2

Vectors representing these values are stated:

$$I_R = 4\underline{/0°}\text{ A}$$

$$I_L = 5\underline{/-90°}\text{ A}$$

The vector diagram has been plotted in Figure 21-2b. To use this information to find total current, we construct the rectangle outlined in the vector diagram with dashed lines. When we have done that, we draw the diagonal from the origin to the opposite corner of the rectangle. The magnitude of this line is equal to the total current, and the angle that separates it from I_R is angle θ. Measurement of the physical size of these two values reveals that

$$I_t = 6.4\underline{/-53.1^\circ}\text{ A}$$

To calculate total impedance, we use these values and Ohm's law:

$$Z_t = \frac{E_t}{I_t}$$

$$Z_t = \frac{40\text{ V}}{6.4\text{ A}}$$

$$Z_t = 6.25\ \Omega$$

In this solution, note:

1. I_L vector lags I_R by 90°.
2. Impedance is less than the opposition in the smallest branch.
3. Total current is greater than the current in either branch but less than the sum of the two branch currents.

To understand parallel-reactive circuits you must understand circuits which contain resistance, inductive reactance, and capacitive reactance. A circuit of this type is shown in Figure 21-3. Each component occupies a separate branch. The circuit diagram has given values for E_t, R, X_C, and X_L. To solve the circuit, we first find each branch current:

$$I_R = \frac{E_t}{R} = \frac{160\text{ V}}{160\ \Omega} = 1\text{ A}$$

$$I_C = \frac{E_t}{X_C} = \frac{160\text{ V}}{80\ \Omega} = 2\text{ A}$$

$$I_L = \frac{E_t}{X_L} = \frac{160\text{ V}}{320\ \Omega} = 0.5\text{ A}$$

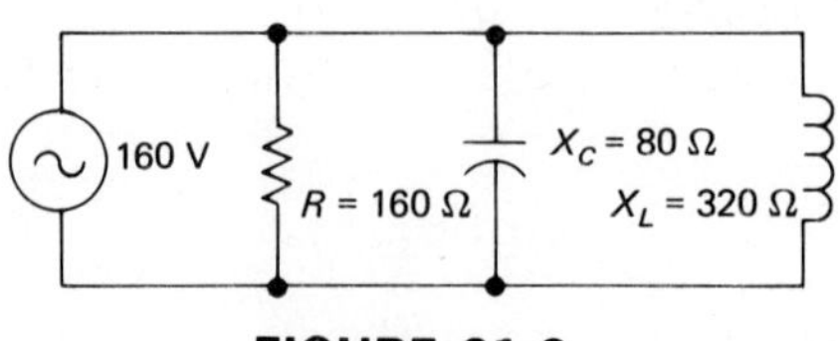

FIGURE 21-3

Recall that I_C and I_L are 180° out of phase, and will cancel. Therefore, the smaller must be subtracted from the larger (Figure 21-4a). Figure 21-4b represents the resultant vector diagram.

With the rectangle constructed and the I_t vector drawn, we measure the vector's magnitude and find a total current of 1.8 A. Measurement of phase angle θ gives $+56.3°$. The vector is then stated:

$$I_t = 1.8\angle{+56.3°}\text{ A}$$

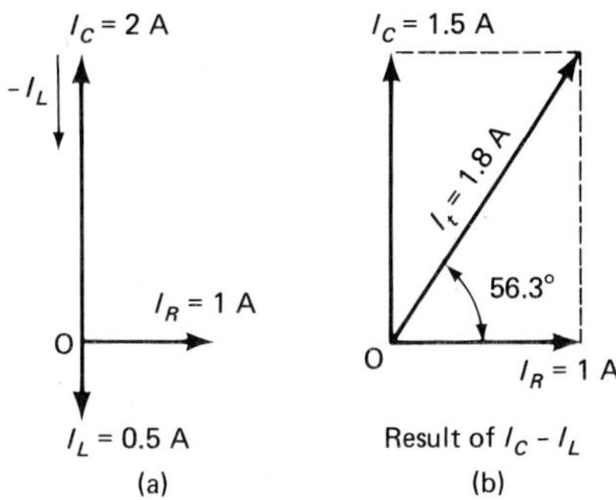

FIGURE 21-4

Total impedance is calculated:

$$Z_t = \frac{E_t}{I_t}$$

$$Z_t = \frac{160\text{ V}}{1.8\text{ A}}$$

$$Z_t = 88.8\ \Omega$$

Again, the results are:

1. I_C vector leads I_R by 90°.
2. Total current is greater than the current in either branch but less than the sum of the branch currents.

Self-Check

Answer each item by indicating whether the statement is true or false.

1. ____ Total reactive current equals the difference between I_C and I_L.

2. ____ In a parallel RC circuit, impedance can be plotted and solved vectorially.

3. ____ Voltage leads current by 90° across a capacitor.

4. ____ Total current in a parallel RC circuit is larger than the sum of the two branch currents.

5. ____ Total impedance in a parallel RL circuit is less than the opposition of any branch.

Trigonometric Analysis

Let us review the formulas introduced in Chapter 19. Here, though, they are stated in current terms, as illustrated by the right triangle in Figure 21-5. Note that I_X is used to denote either I_L or I_C.

$$I_t = \sqrt{I_R^2 + (I_L - I_C)^2}$$

$$I_t = \sqrt{I_R^2 + I_X^2}$$

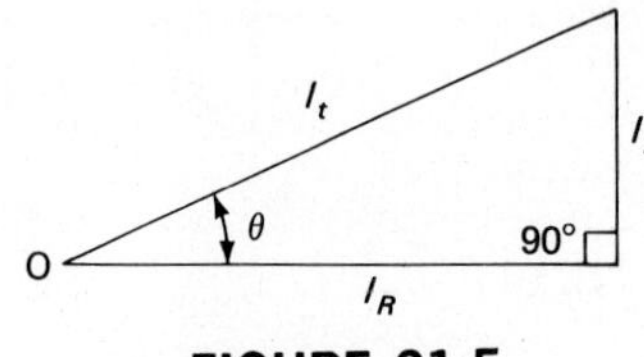

FIGURE 21-5

Using the three currents from the triangle in Figure 21-5, we can state each trigonometric function:

$$\sin \theta = \frac{I_X}{I_t}$$

$$\cos \theta = \frac{I_R}{I_t}$$

$$\tan \theta = \frac{I_X}{I_R}$$

These equations can be transposed to alternate forms for calculating I_t, I_X, or I_R:

$$I_t = \frac{I_X}{\sin \theta}$$

$$I_X = I_t \times \sin \theta$$

$$I_R = \frac{I_X}{\tan \theta}$$

In Figure 21-6a, the RC circuit has $I_R = 3$ A and $I_C = 2$ A. The vector diagram is the same as that shown in Figure 21-6b. This time, instead of taking physical measurements, we solve for I_t mathematically. We do this by:

$$I_t = \sqrt{I_R^2 + I_C^2}$$

$$I_t = \sqrt{3^2 + 2^2}$$

$$I_t = \sqrt{13}$$

$$I_t = 3.6 \text{ A}$$

or

$$\tan \theta = \frac{I_C}{I_R}$$

$$\tan \theta = \frac{2}{3}$$

$$\tan \theta = 0.6667$$

$$\text{angle } \theta = \text{INV tan} \quad 0.6667 = 33.7°$$

$$\sin \theta = \sin 33.7° = 0.5548$$

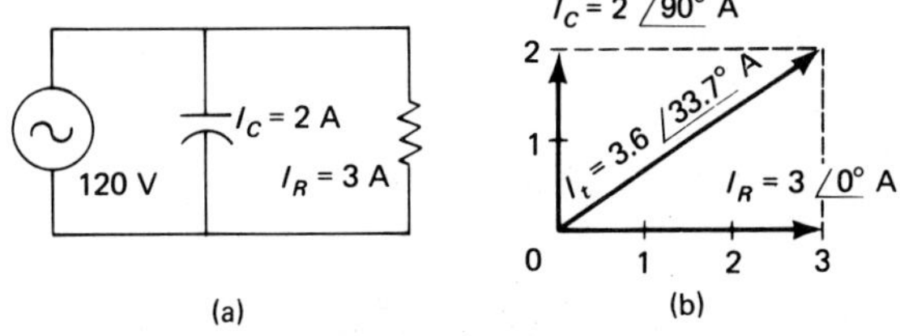

FIGURE 21-6

Transpose the equation:

$$\sin \theta = \frac{I_C}{I_t}$$

to get the equation:

$$I_t = \frac{I_C}{\sin \theta}$$

$$I_t = \frac{2 \text{ A}}{0.5548}$$

$$I_t = 3.6 \text{ A}$$

The vector that represents I_t can be stated:

$$I_t = 3.6\angle +33.7° \text{ A}$$

We now have enough data to use Ohm's law to solve for total impedance of the circuit:

$$Z_t = \frac{E_t}{I_t}$$

$$Z_t = \frac{120 \text{ V}}{3.6 \text{ A}}$$

$$Z_t = 33.3 \ \Omega$$

All the statements about the circuit characteristics found to be true in graphical analysis are true for this circuit.

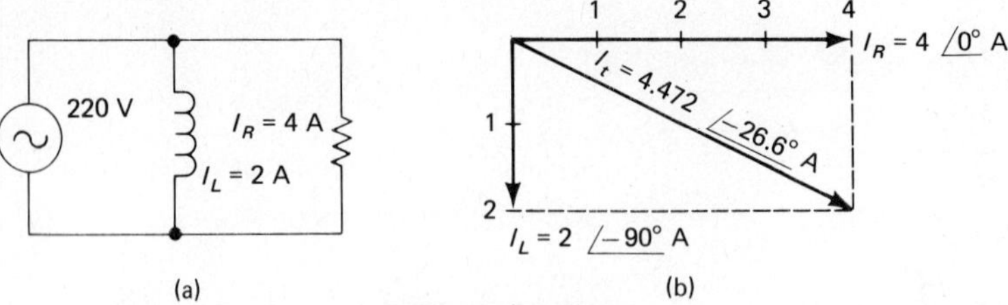

FIGURE 21-7

The solution of parallel RL circuits is identical, except that the phase angle is negative. We take some shortcuts in analyzing the circuit shown in Figure 21-7a.

Observe the vector diagram in Figure 21-7b. Notice that I_L lags I_R by 90°. To find the tangent of angle θ, we do the following:

$$\tan\theta = \frac{I_L}{I_R}$$

$$\tan\theta = \frac{2\text{ A}}{4\text{ A}}$$

$$\tan\theta = 0.5$$

$$\text{angle }\theta = \text{INV}\ \tan\quad 0.5$$

$$\text{angle }\theta = 26.56°$$

$$\sin\theta = \sin 26.56°$$

$$\sin\theta = 0.4472$$

$$\sin\theta = \frac{I_L}{I_t}$$

Therefore:

$$I_t = \frac{I_L}{\sin\theta}$$

$$I_t = \frac{2\text{ A}}{0.4472}$$

$$I_t = 4.472\text{ A}$$

I_t can also be calculated:

$$I_t = \sqrt{I_R^2 + I_L^2}$$

$$I_t = \sqrt{4^2 + 2^2}$$

$$I_t = \sqrt{20}$$

$$I_t = 4.472\text{ A}$$

$$Z_t = \frac{E_t}{I_t}$$

$$Z_t = \frac{220 \text{ V}}{4.472 \text{ A}}$$

$$Z_t = 49.2 \ \Omega$$

If we refer to the section on graphical analysis, we find that the values resulting from this analysis are identical to those obtained by the other method.

So much for RC and RL circuits. We now consider a circuit with branches that have all three components. Refer to Figure 21-8.

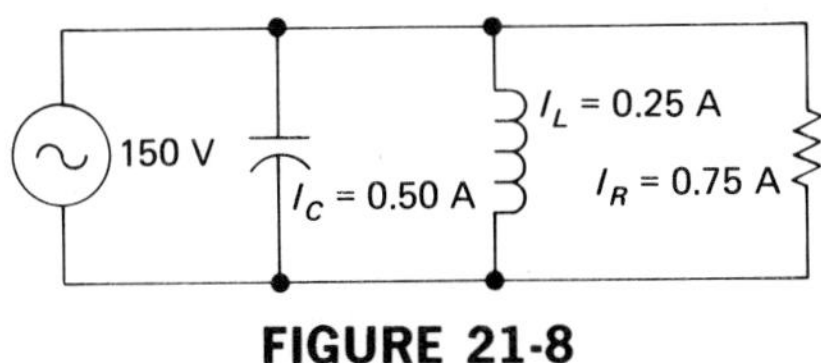

FIGURE 21-8

As we have seen, total reactive current results from the algebraic addition of the two reactive currents, because I_C and I_L are 180° out of phase and will cancel. Therefore,

$$I_X = I_C - I_L$$

$$I_X = 0.5 \text{ A} - 0.25 \text{ A}$$

$$I_X = 0.25 \text{ A}$$

The circuit acts capacitively because I_C is larger than I_L. This solution gives us the reactive vector; the resistive vector is given. The vector diagram for this circuit is shown in Figure 21-9.

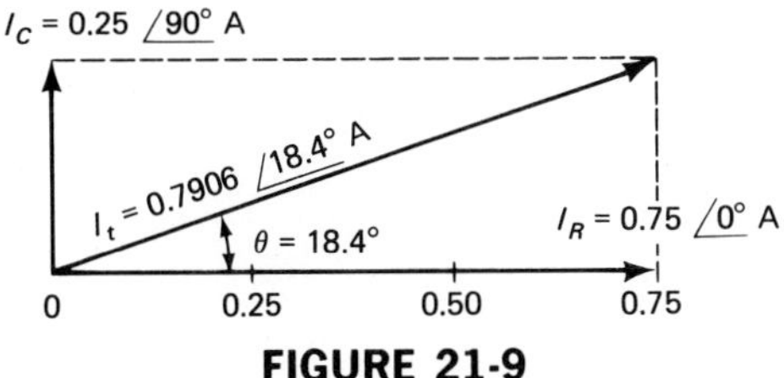

FIGURE 21-9

Continuing with the solution of this circuit, we use the vector diagram to find the tangent of θ:

$$\tan \theta = \frac{I_X}{I_R}$$

$$\tan \theta = \frac{0.25 \text{ A}}{0.75 \text{ A}}$$

$$\tan \theta = 0.3333$$

$$\text{angle } \theta = \text{INV tan } 0.3333$$

$$\text{angle } \theta = 18.435°$$

$$\sin \theta = \sin 18.435°$$

$$\sin \theta = 0.3162$$

$$I_t = \frac{I_X}{\sin \theta}$$

$$I_t = \frac{0.25 \text{ A}}{0.3162}$$

$$I_t = 0.7906 \text{ A}$$

I_t can also be calculated, using the Pythagorean theorem. The results are:

$$I_t = \sqrt{I_R^2 + I_x^2}$$

$$I_t = \sqrt{0.25^2 + 0.75^2}$$

$$I_t = \sqrt{0.0625 + 0.5625}$$

$$I_t = \sqrt{0.625}$$

$$I_t = 0.79056 \text{ A}$$

Note the closeness of the two results obtained for I_t.

$$Z_t = \frac{E_t}{I_t}$$

$$Z_t = \frac{150 \text{ V}}{0.79056 \text{ A}}$$

$$Z_t = 189.74 \, \Omega$$

Self-Check

Answer each item by indicating whether the statement is true or false.

6. ____ The Pythagorean theorem can be used to vectorially add current vectors in parallel RC circuits.

7. ____ When I_C and I_R are given in a parallel RC circuit, they can be used to determine the tangent function of angle θ.

8. ____ In a parallel RCL circuit, total current is the vectorial sum of the three branch currents.

9. ____ To solve for total impedance of any parallel RC, RL, or RCL circuit, use Ohm's law.

10. ____ Total current in a parallel RCL circuit is less than the largest branch current.

Power Dissipation in Parallel Reactive Circuits

Now we discuss power and the power factor for the circuits. Remember that reactive components *do not* dissipate power; resistance is the only component that dissipates power. The power dissipated by resistance is *true power, P_T*.

When reactive and resistive components are connected in the same circuit, the circuit appears to use more power than is actually dissipated. This is the *power apparent, P_A*, of the circuit. The ratio of power apparent to true power is the *power factor, P_F*.

The formulas used to calculate power in these circuits are:

$$P_T\ (\text{W}) = I_R \times E_R$$

$$P_A\ (\text{VA}) = I_t \times E_t$$

$$P_F\ (\text{in a decimal ratio}) = \frac{P_T}{P_A}$$

$$P_F = \cos\theta$$

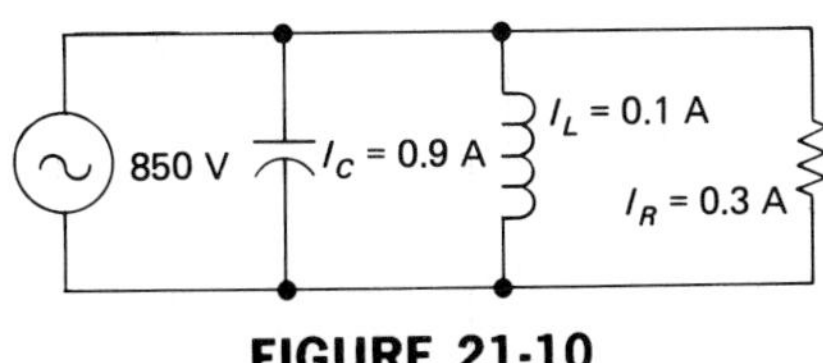

FIGURE 21-10

Use Figure 21-10 to calculate the values for true power, the power apparent, power factor, and angle θ.

$$I_t = \sqrt{I_R^2 + (I_L - I_C)^2}$$

$$I_t = \sqrt{0.3^2 + (0.1 - 0.9)^2}$$

$$I_t = \sqrt{0.3^2 + (-0.8)^2}$$

$$I_t = \sqrt{0.73}$$

$$I_t = 0.85 \text{ A}$$

$$Z_t = \frac{E_t}{I_t}$$

$$Z_t = \frac{850\text{ V}}{0.85\text{ A}}$$

$$Z_t = 1000\ \Omega$$

$$P_A = I_t \times E_t$$

$$P_A = 0.85\text{ A} \times 850\text{ V} = 722.5\text{ VA}$$

$$P_T = I_R \times E_R$$

$$P_T = 0.3\text{ A} \times 850\text{ V} = 255\text{ W}$$

$$P_F = \frac{P_T}{P_A}$$

$$P_F = \frac{255}{722.5}$$

$$P_F = 0.3529$$

$$P_F = \cos\theta$$

Therefore

$$\cos\theta = 0.3529$$

and

$$\text{angle } \theta = \text{INV} \ \cos \ 0.3529$$

$$\text{angle } \theta = 69.34°$$

Self-Check

Answer each item by indicating whether the statement is true or false.

11. ____ Reactive components are the only components that dissipate power.

12. ____ Dissipation of true power occurs the same way that power is dissipated in a DC circuit.

13. ____ With angle θ known, we can find the power factor by determining the sine of angle θ.

14. ____ The difference in power apparent and true power is the energy stored in the field of the reactive component, which is returned to the circuit.

15. ____ If I_R and I_L are known, true power can be calculated by using the Pythagorean theorem.

Summary

Parallel RC, RL, and RCL circuits are typical parallel circuits, in that many of the characteristics learned in DC are true here. The main difference lies in the need to use vectorial addition to calculate total current. Due to the existence of reactive components, current in different branches of a parallel circuit has different phases. In series-reactive circuits, voltages have different phases. The angle resulting from this phase difference is called the *phase angle*.

Some items to be remembered are:

$$\tan \theta = \frac{I_C - I_L}{I_R} \quad \text{or} \quad \frac{I_L - I_C}{I_R}$$

$$I_t = \frac{I_R}{\cos \theta}$$

$$I_t = \frac{I_X}{\sin \theta}$$

$$I_t = \sqrt{I_R^2 + (I_L - I_C)^2}$$

$$Z_t = \frac{E_t}{I_t}$$

$$P_T = I_R \times E_R$$

$$P_A = I_t \times E_t$$

$$P_F = \frac{P_T}{P_A}$$

Review Questions and/or Problems

1. Voltage drop across each branch of a parallel circuit is
   (a) equal to the sum of the voltage drops.
   (b) the same as total applied voltage.
   (c) the vector sum of the individual voltages.
   (d) dependent on frequency.

2. Total current in a parallel RC circuit can be found by using
   (a) $I_t = I_C + I_R$.
   (b). $I_t = I_C = I_R$.
   (c) the graphic method.
   (d) None of these

3. When plotting the vector diagram for a parallel RL circuit, which is the reference vector?
   (a) E_t
   (b) I_R
   (c) I_C
   (d) I_t

4. Z_t in a parallel RC circuit is:
   (a) equal to $R + X_C$.
   (b) less than X_C.
   (c) the vector sum of R and X_C.
   (d) less than R or X_C.

5. In a parallel RL circuit,
   (a) current and voltage are in phase in the resistor.
   (b) current leads voltage by 90° in the resistor.
   (c) current lags voltage by 90° in the resistor.

6. In a parallel RL circuit, increasing the frequency causes
   (a) I_t to decrease.
   (b) I_L to increase.
   (c) X_L to decrease.
   (d) I_R to increase.

7. In a parallel RL circuit, which can be used to calculate I_t?
   (a) $I_t = \dfrac{I_L}{\sin \theta}$
   (b) $I_t = \sqrt{I_R^2 + I_L^2}$
   (c) The graphic method
   (d) All of these

8. In a parallel RCL circuit, when constructing the vector diagram, we must
   (a) subtract the larger reactive current from the smaller.
   (b) subtract the smaller reactive current from the larger.
   (c) subtract: $X_L - X_C$.
   (d) subtract: $X_C - X_L$.

9. To calculate Z_t, we use
   (a) $Z_t = \dfrac{E_L}{I_L}$.
   (b) $Z_t = \dfrac{E_C}{I_C}$.
   (c) $Z_t = \dfrac{E_R}{I_R}$.
   (d) $Z_t = \dfrac{E_t}{I_t}$.

10. To calculate the power factor, we use
   (a) $P_F = I_t \times E_t$.
   (b) $P_F = I_R \times E_R$.
   (c) $P_F = \dfrac{P_A}{P_T}$.
   (d) $P_F = \dfrac{P_T}{P_A}$.

22

Transient Circuits

Introduction

Many circuits that a technician works with are used to control the timing of other circuitry. Some circuits are used to create special waveshapes for use in turning other circuits on and off. These are categorized as square, sawtooth, trapezoidal, rectangular, and peaked (trigger) waves. They are illustrated in Figure 22-1.

FIGURE 22-1

All waveshapes have one thing in common: their shape (duration and amplitude) is controlled by a transient circuit. The term *transient* refers to the time required to complete a change. The proper operation of waveshaping circuits depends on their response to transient voltages or currents. In this chapter we discuss basic transient circuits and their operation.

Objectives

Each student is required to:

1. When provided a schematic and period of time,
   (a) Calculate the percent of charge present on the capacitor in a series RC circuit.
   (b) Calculate the percent of discharge for a capacitor in a series RC circuit.
   (c) Determine the percent of current buildup in a series RL circuit.
   (d) Determine the percent of current decay during discharge of an RL circuit.
2. Compute time constants for
   (a) Series RC circuits.
   (b) Series RL circuits.
3. Determine current and voltage distribution in series RC and RL circuits at the end of a time period.
4. Calculate the time required for circuit current and voltages to reach predetermined levels.

5. Calculate component sizes required to
   (a) Produce waveshapes of specific amplitude.
   (b) Produce recurrent waves of a specific time base.
6. Given series RC and RL circuit schematics, match E_C, E_R, and E_L to
   (a) Waveshapes that result from long-time constants.
   (b) Waveshapes that result from short-time constants.
   (c) Waveshapes that result from medium-time constants.

Transient Voltage and Current

A *transient voltage,* or *current,* is the rapid change of voltage or current from one steady state to another steady state. *Transient interval* is the time required for the change in states. It is possible to observe the form of a waveshape when its transient voltage or current is graphed against its elapsed, or transient, time. It is also possible to observe the waveshape on an oscilloscope, which can plot and store transients.

Charging a Capacitor

Capacitance is the characteristic of a circuit or component which allows it to store an electrical charge in its electrostatic field. Charge is developed by the force of electrons building up on a negative plate, which act on the electrons in a dielectric, which then repel electrons from a second plate. The loss of electrons from the second plate creates a charge of positive ions (deficiency of electrons).

Electrons cannot move from place to place instantaneously; time must elapse. Therefore, all capacitors require time to charge. The amount of time required for a capacitor to charge depends on the amount of resistance in the path of the charging current and on the size of the capacitor.

Figure 22-2 illustrates a simple, series RC circuit connected to a DC power source. When the switch is closed, current begins to flow, and E_t is applied across the resistor and the capacitor. Because the capacitor, C, does not have a charge at the first instant, the amount of voltage dropped on the capacitor is limited by the resistor, R, as is the amount of current that can flow.

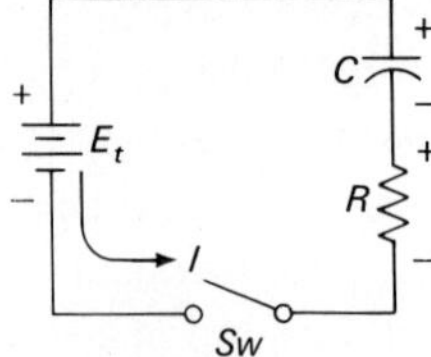

FIGURE 22-2

Remember: A capacitor opposes any change in voltage. With no charge, it has zero voltage and opposes the change caused by closing the switch. At the first instant, the resistor voltage equals E_t and the circuit current equals

$$I_t = \frac{E_t}{R_t}$$

At the first instant after the switch is closed, current flow is maximum; but as current begins to flow, a charge accumulates in the capacitor's electrostatic field. The accumulation of charge appears as a voltage drop on the capacitor. As the voltage drop on C increases, E_R decreases. As voltage decreases on the resistor, current must also decrease, since resistor current and voltage are always in phase. The resistor's size limits the charging current and determines the time required to charge the capacitor. When the capacitor is fully charged, current is zero. Two things affect capacitor charge time:

In a series RC circuit, the time required to charge the capacitor is directly proportional to the size of the capacitor and the resistor.

You may remember from another chapter that the charge on a capacitor can be calculated by using

$$Q = C \times E$$

According to this equation, when we hold voltage, E, constant and increase capacitance, C, the number of electrons required to charge the capacitor is increased. Let us pull this together:

1. More electrons are needed to charge large capacitors.
2. Charging current is controlled by E_t and series resistance.
3. Therefore, if current remains the same, more time is required to charge the capacitor.

The reverse is true when capacitance is decreased. In that case, with the same current, a smaller capacitor charges in less time.

Total voltage has no effect on the time required to charge a capacitor. When E_t changes, the charging current changes proportionally. This results in the capacitor charging to the new voltage in the same period of time as before. *Charging time* is not affected.

Time Constant

The resistance and capacitance contained in a circuit are the only factors that affect the time required for the capacitor to charge to a given percentage of E_t. The same is true for the discharge of a capacitor. Note, therefore, that time, resistance, and capacitance are related. This relationship can be stated:

$$1\ TC = R \times C$$

where

TC = time constant (time, in seconds)
R = total resistance (ohms) in the charge path
C = total capacitance (farads) in the charge path

Converted to a statement:

The product of resistance and capacitance equals one time constant.

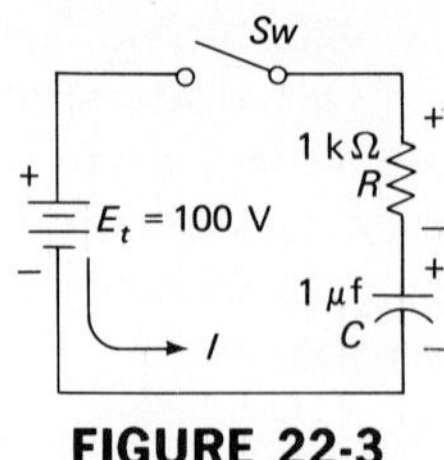

FIGURE 22-3

Figure 22-3 illustrates a simple RC circuit. The *time constant* of this circuit is the time it takes C to charge to E_t *if it continued to charge at its initial rate.*

As the capacitor accumulates a charge, however, the charging current decreases and the rate of charge decreases. The result of this decrease in rate of charge, means that it is impossible for the capacitor to charge to E_t in one time constant. In fact, the capacitor only charges to about 63% of E_t during the first time constant. Thus, we define the time constant as

The time required for a capacitor to charge to 63% of the available voltage.

During this same period, current decreases to 37% of the original current.

The information in Figure 22-3 is used to calculate the time constant for the circuit. Note that $C = 1\ \mu f$ and $R = 1\ k\Omega$. To calculate TC, proceed as follows:

$$TC = R \times C$$

$$TC = (1 \times 10^3)\ (1 \times 10^{-6})$$

$$TC = 1 \times 10^{-3}$$

$$TC = 1 \text{ ms}$$

The time required for the capacitor in this circuit to charge to 63% of the available voltage is 0.001 s, or 1 ms.

During the first time constant, C charges to 63% of E_t. In the circuit shown, E_C equals 63 V at the end of 1 TC. The difference between E_t and E_C is the voltage still available for charge. Therefore,

$$E_{\text{available}} = E_t - E_C$$

$$E_{\text{available}} = 100 \text{ V} - 63 \text{ V}$$

$$E_{\text{available}} = 37 \text{ V}$$

The opposition of 63 V (E_C) to 100 V (E_t) causes the charge current to *decrease by 63% from its original rate*. Therefore, the current for the second time constant is much lower than before. With these conditions existing, the capacitor charges to 63% of the available voltage during the second time constant and each subsequent time constant. At the end of two TC, E_C, and the available voltage are:

$$E_C = 63 \text{ V} + (0.63)(37 \text{ V})$$

$$E_C = 63 \text{ V} + 23.3 \text{ V}$$

$$E_C = 86.3 \text{ V}$$

$$E_{\text{available}} = E_t - E_C$$

$$E_{\text{available}} = 100 \text{ V} - 86.3 \text{ V}$$

$$E_{\text{available}} = 13.7 \text{ V}$$

During the third time constant, the capacitor charges to 63% of the available voltage. At the end of the third *TC*, the following conditions exist:

$$E_C = 86.3 \text{ V} + (0.63)(13.7 \text{ V})$$

$$E_C = 86.3 \text{ V} + 8.63 \text{ V}$$

$$E_C = 94.93 \text{ V}$$

$$E_{\text{available}} = E_t - E_C$$

$$E_{\text{available}} = 100 \text{ V} - 94.93 \text{ V}$$

$$E_{\text{available}} = 5.07 \text{ V}$$

During the fourth *TC*, the capacitor charges to 63% of 5.07 V. The following conditions exist after four time constants:

$$E_C = 94.93 \text{ V} + (0.63)(5.07 \text{ V})$$

$$E_C = 94.93 \text{ V} + 3.2 \text{ V}$$

$$E_C = 98.13 \text{ V}$$

$$E_{\text{available}} = E_t - E_C$$

$$E_{\text{available}} = 100 \text{ V} - 98.13 \text{ V}$$

$$E_{\text{available}} = 1.87 \text{ V}$$

During the fifth *TC*, the capacitor charges to 63% of 1.87 V. The following conditions exist at the end of five time constants:

$$E_C = 98.13 \text{ V} + (0.63)(1.87 \text{ V})$$

$$E_C = 98.13 \text{ V} + 1.18 \text{ V}$$

$$E_C = 99.31 \text{ V}$$

$$E_{\text{available}} = 0.69 \text{ V}$$

Thus, a capacitor using this charging process never becomes fully charged to E_t. With the tolerances allowable in electronics, however, we consider the capacitor fully charged after five time constants.

For the circuit in Figure 22-3, it takes 5 ms for the capacitor to fully charge. At this time, $E_C = E_t$ and $I_t = 0$.

Rate of Charge and Discharge

With capacitance and resistance remaining constant, the time for one time constant remains constant regardless of how much the applied voltage changes. The capacitor charges to E_t in five *TC*. Applied voltage and resistance, however, determine the *rate of charge*. It is this rate that allows the capacitor to fully charge in the same period of time for each applied voltage. As applied voltage is varied, the rate of charge varies in direct proportion to the change in voltage.

Up to now, we have been discussing what happens when a capacitor charges. Now we consider the capacitor's discharge. The same conditions exist during discharge. Consider the discharge of C in Figure 22-3.

First TC
E_C decreases by 63% of E_t, or 37 V

Second TC
E_C decreases 63% to 13.7 V

Third TC
E_C decreases 63% to 5.07 V

Fourth TC
E_C decreases 63% to 1.87 V

Fifth TC
E_C decreases 63% to 0.69 V

At the end of five time constants, the capacitor is fully discharged. Variations in the amount of voltage the capacitor is charged to affect the rate of discharge. The time required to discharge the capacitor completely can be changed only by variations in R or C.

The Universal Time Constant Chart

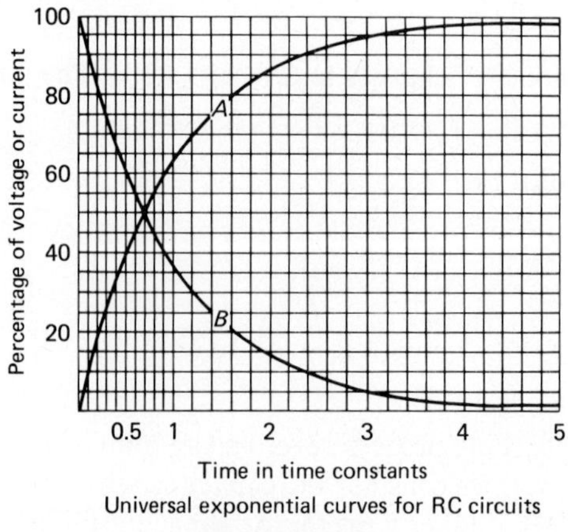

Universal exponential curves for RC circuits

FIGURE 22-4

Figure 22-4 shows a chart, the Universal Exponential Curves for RC and RL circuits, commonly called the *universal time constant chart.* For the RC circuit, this chart plots the percent of change in E_C on curve A and the percent of change in E_R and I_t on curve B. Note that the vertical side of the chart is labeled "Percentage of Voltage or Current" and the horizontal side, "Time in Time Constants." Curves A and B are plotted by taking the instantaneous voltage or current for the capacitor at an infinite number of time intervals, starting from time zero. As we have seen, the capacitor never fully charges to E_t. Thus, curves A and B never reach 100%, or 0%.

Before using the chart, we consider another time element—the time the capacitor is allowed to charge or discharge. The *charge time* may or may not be more than five time constants. The numbers along the horizontal side of the chart represent charge or discharge time, in time constants.

We can express the relationship between time constants and time of charge mathematically:

$$\#TC = \frac{t}{RC}$$

where

$\#TC$ = number of time constants
t = time allowed for charge or discharge, in seconds
R = resistance of circuit, in ohms
C = capacitance of circuit, in farads

Examine the universal time constant chart. Table 22-1 summarizes the functions of curves A and B as they apply to series RC circuits.

TABLE 22-1
Series RC Circuits

Curve A
E_C on capacitor charge
Curve B
E_R on capacitor charge
I_R on capacitor charge
E_C on capacitor discharge
E_R on capacitor discharge
I_R on capacitor discharge

Using the Universal Time Constant Chart

Transient problems fall into one of the following general categories:

1. Calculation of circuit current and voltages after a specified elapsed time.
2. Calculation of the time required for circuit current or voltage to reach predetermined levels.

3. Calculation of component values that provide specific voltages and currents at the end of a predetermined time.

The easiest way to learn to use this chart is to solve problems with it. We use the values given in Figure 22-5 to solve problem 1.

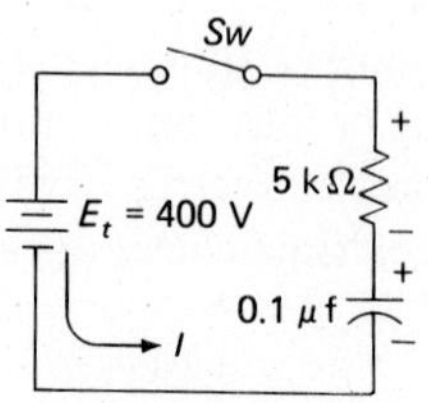

FIGURE 22-5

Problem 1

Find the value of E_C, 1200 μs, after the switch is closed.

Step 1: List givens and wanteds.

Given	Wanted
$E_t = 400$ V $R = 5$ kΩ $C = 0.1\ \mu$f $t = 1200\ \mu$s	Find E_C

Step 2: Determine the number of TC in the charge time, t.

Mathematical solution:

$$\#TC = \frac{t}{RC}$$

$$\#TC = \frac{1.2 \times 10^{-3}}{(5 \times 10^{3})(1 \times 10^{-7})}$$

$$\#TC = \frac{1.2 \times 10^{-3}}{5 \times 10^{-4}}$$

$$\#TC = 2.4$$

Calculator solution:

Action	Entry	Display
ENTER	1200,EE,+/−,6	1200 − 06
PRESS	÷	1.2 − 03
PRESS	(	1.2 − 03
ENTER	5,EE,3	5 03
PRESS	×	5 03
ENTER	.1,EE,+/−,6	0.1 − 06
PRESS	)	5 − 04
PRESS	=	2.4 00

Either solution results in 2.4 time constants.

Step 3: Locate point 2.4 on the horizontal (bottom) line of the time constant chart.

Step 4: On charge E_C is read on the A curve. From point 2.4 on the bottom line, move straight up until curve A is reached. Mark this point.

Step 5: From the point marked in step 4, move left until the vertical (percentage) line is intersected.

Step 6: Read the percentage at the point of intersection. For this circuit, the percentage is 91.

Step 7: Calculate voltage E_C, E_R, and I_R.

$$E_C = E_t \times 91\%$$

$$E_C = 400\text{ V} \times 0.91$$

$$E_C = 364\text{ V}$$

$$E_R = E_t - E_C$$

$$E_R = 400\text{ V} - 364\text{ V}$$

$$E_R = 36\text{ V}$$

$$I_R = \frac{E_R}{R_t}$$

$$I_R = \frac{36\text{ V}}{5 \times 10^3}$$

$$I_R = 7.2\text{ mA}$$

Current is equal at all points in a series circuit; therefore, I_C equals 7.2 mA.

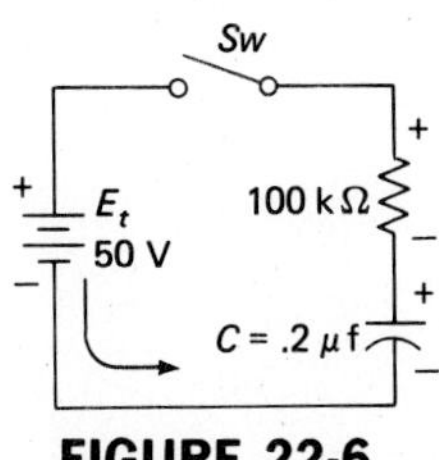

FIGURE 22-6

In this problem, we used the time constant chart to solve for circuit voltages and currents at a specified time. Now we use the chart to solve the number of time constants that the capacitor requires to charge to a predetermined percentage of E_t. Refer to Figure 22-6.

Problem 2

Step 1: List givens and wanteds.

Given	Wanted
$E_t = 50$ V $R = 100$ kΩ $C = 0.2$ μf	$\#TC$ for $E_C = 30$ V

Step 2: Determine the percentage of charge on the capacitor.

$$\% \text{ of charge} = \frac{E_C}{E_t} \times 100$$

$$\% \text{ of charge} = \frac{30 \text{ V}}{50 \text{ V}} \times 100$$

$$\% \text{ of charge} = 0.6 \times 100$$

$$\% \text{ of charge} = 60\%$$

Action	Entry	Display
ENTER	30	30
PRESS	÷	30
ENTER	50	50
PRESS	=	0.6
PRESS	×	0.6
ENTER	100	100
PRESS	=	60

Step 3: Locate the 60% mark on the left vertical side of the time constant chart. Since the capacitor is charging, move right until the 60% line intersects curve A. From this point, drop straight down to the bottom horizontal line and read the number of time constants that have elapsed. For this problem,

$$\#TC = 0.9$$

Problem 3

In the circuit illustrated by Figure 22-7, closing switch$_1$ allows the capacitor to charge to E_t. If switch$_1$ is then opened, the capacitor has a charge $E_C = E_t$ stored in its electrostatic field. Closing switch$_2$ provides a discharge path through R. Calculate E_C, E_R, and I_R after 130 μs of discharge.

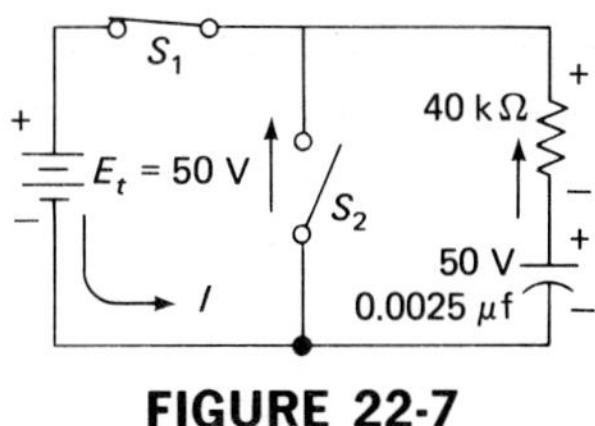

FIGURE 22-7

Step 1: List givens and wanteds.

Given	Wanted
$E_t = 50$ V	$E_C =$
$E_C = 50$ V	$E_R =$
$E_R = 0$ V	$I =$
$R = 40$ kΩ	
$t = 130\ \mu s$	
$C_t = 0.0025\ \mu f$	

Step 2: Determine $\#TC$ in 130 μs.

$$\#TC = \frac{t}{RC}$$

$$\#TC = \frac{1.3 \times 10^{-4}}{(40 \times 10^3) \times (2.5 \times 10^{-9})}$$

$$\#TC = 1.3$$

Step 3: Locate $TC = 1.3$ on the bottom line. Move up to the intersection with curve B (C is discharging). Then move left until the percent line is intercepted. Read 27%.

Step 4: Calculate 27% of original voltage.

$$E_C = 50 \text{ V} \times 0.27$$

$$E_C = 13.5 \text{ V}$$

Step 5: The capacitor is acting as the power source; thus E_R must equal E_C. Therefore,

$$E_R = 13.5 \text{ V}$$

Step 6: Current is equal at all points in a series circuit; therefore,

$$I = \frac{E_R}{R}$$

$$I = \frac{13.5 \text{ V}}{40 \text{ k}\Omega}$$

$$I = 0.338 \text{ mA}$$

Problem 4

The capacitor in the circuit in Figure 22-8 charges to 50 V in 1000 μs. What is the size of R?

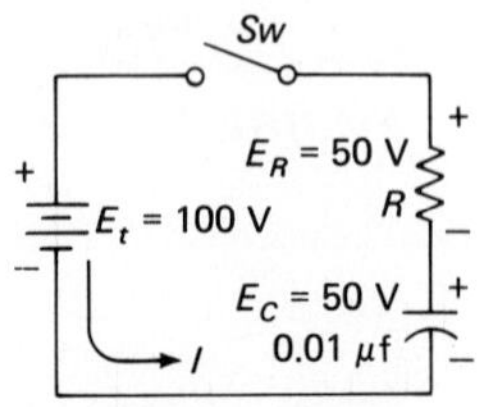

Sw
$E_R = 50$ V
R
$E_t = 100$ V
$E_C = 50$ V
0.01 μf
I

FIGURE 22-8

Stcp 1: List givens and wanteds.

Given	Wanted
$E_t = 100$ V $E_C = 50$ V $t = 1000$ μs $C = 0.01$ μf	$R =$

Step 2: Calculate percentage of E_t dropped as E_C:

$$\% \text{ of charge} = \frac{E_C}{E_t} \times 100$$

$$\% \text{ of charge} = \frac{50 \text{ V}}{100 \text{ V}} \times 100$$

$$\% \text{ of charge} = 50\%$$

Step 3: Use the chart. The capacitor is charging. Therefore, curve A is used. Locate the 50% point on the left side. Move right until curve A is intersected. Move down to the horizontal and check the number of TC. In this problem,

$$\#TC = 0.7$$

Step 4: Transpose the equation and calculate R:

Equation:

$$\#TC = \frac{t}{RC}$$

Transposed:

$$R = \frac{t}{\#TC(C)}$$

Solution:

$$R = \frac{1 \times 10^{-3}}{0.7(1 \times 10^{-8})}$$

$$R = 142.8 \text{ k}\Omega$$

Problem 5

Use Figure 22-9 to solve problem 5. The switch is placed in position A until the resistor voltage has decreased to 30 V. What size must the capacitor be if its voltage is 35 V 1500 μs after the switch is moved to position B?

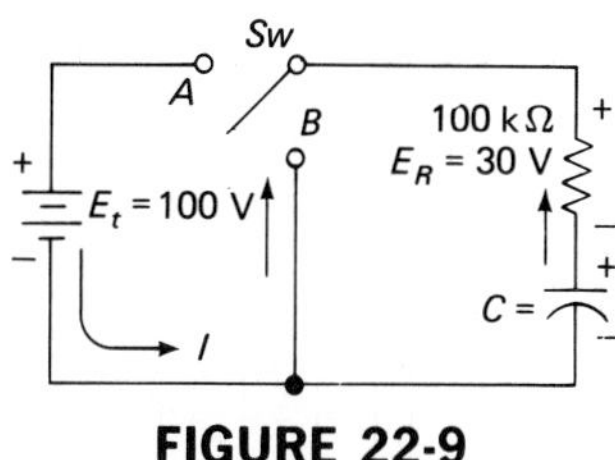

FIGURE 22-9

Step 1: List givens and wanteds:

Given	Wanted
$E_t = 100$ V $E_R = 30$ V $R_t = 100$ kΩ $t = 1500\ \mu$f	$C =$

Step 2: Use Kirchhoff's voltage law to calculate E_C when the switch is changed from position A to position B.

$$E_C = E_t - E_R$$

$$E_C = 100 \text{ V} - 30 \text{ V}$$

$$E_C = 70 \text{ V}$$

Step 3: Calculate the percentage of voltage across C.

$$\text{\% of charge} = \frac{35 \text{ V}}{70 \text{ V}} \times 100$$

$$\text{\% of charge} = 50\%$$

Step 4: Find the number of time constants as was done in finding 50% in step 3. With C discharging, we use curve B to find:

$$\#TC = 0.7$$

Step 5: Transpose the equation and calculate the value of C.

Equation:

$$\#TC = \frac{t}{RC}$$

Transposed:

$$C = \frac{t}{\#TC(R)}$$

Solution:

$$C = \frac{1.5 \times 10^{-3}}{0.7(1 \times 10^{5})}$$

$$C = 2.143 \times 10^{-8}$$

This converts to 0.0214 μf.

Problem 6

Refer to Figure 22-10. When the switch is closed, the capacitor in this circuit charges to 86 V after 2000 μs. What is E_t for the circuit?

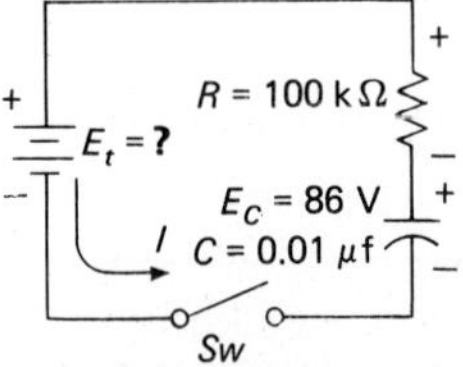

FIGURE 22-10

Step 1: List givens and wanteds.

Given	Wanted
$E_C = 86$ V $R_t = 100$ kΩ $C_t = 0.01$ μf $t = 2000$ μs	$E_t =$

Step 2: Calculate the number of TC for a capacitor charge of 86 V.

$$\#TC = \frac{t}{RC}$$

$$\#TC = \frac{2 \times 10^{-3}}{(1 \times 10^{5})(1 \times 10^{-8})}$$

$$\#TC = 2$$

Step 3: Locate two time constants on the horizontal. Move up to curve A, then left. Determine the percent of charge. In this circuit, the percent of charge is 86%.

Step 4: The capacitor is charged to 86% of E_t. To calculate total voltage, use:

$$E_t = \frac{E_C \times 100}{\%\ \text{charge}}$$

$$E_t = \frac{86\ \text{V} \times 100}{86\%}$$

$$E_t = 100\ \text{V}$$

Note: The percent of charge is the whole number 86, not the decimal number 0.86.

Self-Check

Answer each item by indicating whether the statement is true or false.

1. ____ One time constant is equal to the product of resistance and capacitance.

2. ____ The rate of charge for a capacitor in a series RC circuit is directly proportional to the size of the resistance.

3. ____ Increasing the voltage applied to a series RC circuit results in the capacitor taking longer to charge.

4. ____ The equation for calculating the number of time constants is $\#TC = R \times C$.

5. ____ With 50 V applied to a series RC circuit, E_C equals 50 V after 3 time constants.

6. ____ Curve B on the universal time constant chart is used to calculate capacitor charge in an RC circuit.

7. ____ A capacitor is charged to 60 V. It is allowed to discharge through a 10-kΩ resistor for 2.0 time constants. At that time, E_R equals 14 V.

8. ____ In an RC circuit with E_t = 100 V, the capacitor has charged to 86.3 V. That means two time constants have elapsed.

9. ____ A 1-μf capacitor is connected in series with a 10-kΩ resistor. The circuit's time constant is 10 μs.

10. ____ Series RC circuits have maximum current flow immediately after the switch is closed.

Series RL Circuits

Series RC and series RL circuits are quite similar. It is a good idea to review RC circuits before discussing RL. In the RC circuit:

1. A charged capacitor opposes voltage supplied by a power source.
2. A capacitor charges to 63% of the available voltage during a time constant.
3. In a series RC circuit, the sum of the voltage drops E_R and E_C equals applied voltage.
4. The rate of charge depends on the size of the resistor and the capacitor.
5. Changes in E_t do not affect the time it takes for a capacitor to charge to E_t, but they do affect the rate of charge.
6. The equation for one time constant is 1 $TC = R \times C$.

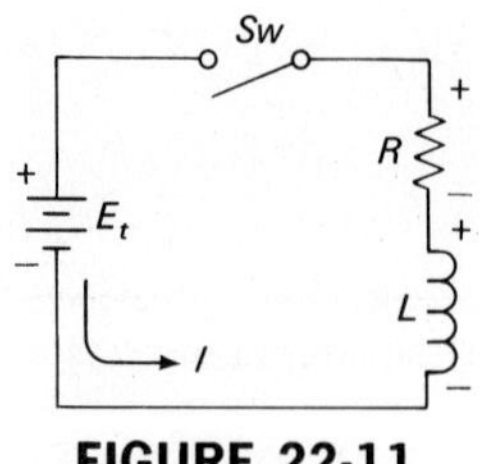

FIGURE 22-11

For the RL circuit illustrated in Figure 22-11, the following conditions are true:

1. A charged coil opposes any change in current.
2. A coil charges to 63% of the available current during each time constant.
3. The sum of the voltage drops on R and L equals applied voltage.
4. The rate of charge depends on the size of the resistor and the inductor.
5. Changes in E_t do not affect the time constant, but they do affect the rate of charge.
6. The formula for 1 TC is 1 $TC = L/R$.

The voltage drop on a capacitor is proportional to the charge stored in its electrostatic field. The current flowing through an inductor is proportional to the energy stored in its electromagnetic field. Neither, however, can change instantaneously when the applied voltage is changed.

For an example of the current action in a coil, observe Figure 22-12a. Here, we assume that the circuit has zero resistance. When the switch is closed, current is zero and $E_L = E_t$. Current immediately tries to flow. This causes a magnetic field to expand around the coil whose magnetic lines of force induce a counter-EMF in the coil. This CEMF will oppose the applied voltage. With zero resistance in the circuit, current increases at a rate which causes the induced voltage to equal the applied voltage. To maintain a constant induced voltage, the coil's magnetic field and current must increase at a constant rate. Theoretically, should the switch be left in the closed position indefinitely, the current will increase to an infinite value (Figure 22-12b). That, of course, is not realistic; all circuits must have some resistance. This resistance will limit the current flow; and no battery can supply infinite current.

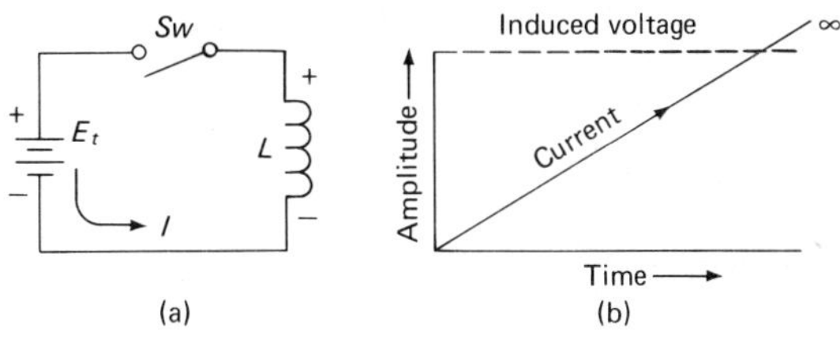

FIGURE 22-12

Because of the resistance in an RL circuit, current will not be allowed to increase at a linear rate, as it would in a purely inductive circuit. When the current increases, the voltage drop on R increases, causing the voltage across the coil to decrease at a gradual rate. This decrease in inductor voltage causes *current change* in the RL circuit to decrease at a continually *decreasing rate,* with the result that the rate of change becomes zero. At that point, current reaches its maximum level and $E_R = E_t$. On the universal time constant chart shown in Figure 22-13, curve A can be used to determine the buildup of current in the coil of an RL circuit. Current and voltage are always in phase across the resistor, which allows curve A to be used for E_R buildup. E_R constantly increases, and E_L constantly decreases. Curve B represents the change in E_L during the time the coil is charging.

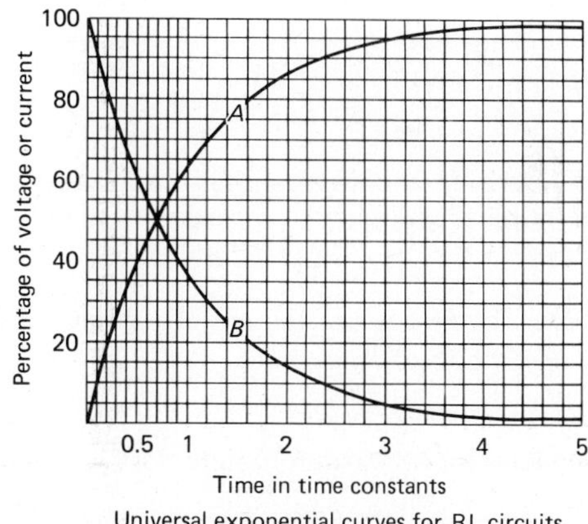

Universal exponential curves for RL circuits

FIGURE 22-13

At the first instant after voltage is applied to an RL circuit, E_L equals E_t, I_t is zero, and E_R equals zero.

After voltage has been applied to the RL circuit for five time constants, E_L equals zero, I_t is maximum, and E_R equals E_t.

A coil that is allowed to discharge acts as a power source during the time it takes the electromagnetic field to decay. When the coil acts as a power supply, the resistor must drop the entire voltage, causing E_L and E_t to be equal. Therefore, curve B can be used to calculate the circuit current, E_R and E_L, during discharge. The use of each curve is summarized in Table 22-2.

TABLE 22-2
Series RL Circuits

Curve A
E_R during field expansion
I_R during field expansion
Curve B
E_L during field expansion
E_L during field collapse
E_R during field collapse
I_R during field collapse

Now that we have established the relationship of curves A and B, let us examine the factors that control transient response: (1) inductance and (2) resistance.

When inductance is increased, the coil offers more opposition to a change in current. With the larger inductance, the time for the coil to become fully charged and the time for current to reach maximum is increased:

The time for current buildup in a series RL circuit is directly proportional to the amount of inductance in the circuit.

When resistance is increased, current reaches a smaller maximum value. If current is less, the time required to charge to that maximum is also less. Therefore,

The time for current buildup in a series RL circuit is inversely proportional to the amount of resistance in the circuit.

One time constant in a series RL circuit is equal to inductance divided by resistance:

$$1\ TC = \frac{L}{R}$$

where

TC = time for one time constant, in seconds
L = inductance, in henries
R = resistance, in ohms

During one time constant, current in the coil increases to 63% of the maximum current available. During each subsequent time constant, current increases by 63% of the available current.

During charge or discharge of the coil in a series RL circuit, current changes 63% of the available current during each time constant.

The charge and discharge sequence of series RL and RC circuits is similar, in that

1. Charge on the reactive component equals 63% of total available voltage or current after one time constant.
2. After two time constants, the charge is 86.3%.
3. After three time constants, the charge is 94.93%.
4. After four time constants, the charge is 98.1%.
5. After five time constants, the charge is 99.3%.

During discharge, the opposite is true. Voltage or current decreases by 63% of that available during each time constant. In a series RL circuit, current changes along the same lines as voltage changed in series RC circuits. Therefore,

1. During charge, current reaches maximum after five time constants of charging.
2. During discharge, current is zero after five time constants of discharge.

We have established the time element involved with the buildup or decay of current in a series RL circuit and pointed out that the universal time constant chart can be used with RL circuits. Now we consider the other time element—the actual time required for charge or discharge to occur. This is the number of TC and is plotted on the bottom line of Figure 22-13. The equation is

$$\#TC = \frac{t}{\frac{L}{R}}$$

In simpler form,

$$\#TC = \frac{R \times t}{L}$$

where

$\#TC$ = number of time constants
t = time allowed, in seconds
R = resistance, in ohms
L = inductance, in henries

Series RL Circuit Problem Solutions

Again, we use sample circuits and the universal time constant chart to solve problems for time, percentage of charge, and component values.

Problem 1

Refer to Figure 22-14.

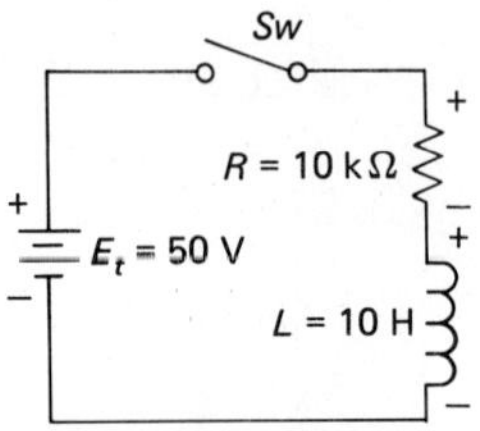

Sw
R = 10 kΩ
E_t = 50 V
L = 10 H

FIGURE 22-14

Step 1: List the givens and wanteds:

Given	Wanted
$E_t = 50$ V	$I =$
$L = 10$ H	$E_R =$
$R = 10$ kΩ	$E_L =$
$t = 2800\ \mu$s	

Step 2: Calculate the number of TC allowed.

$$\#TC = \frac{R \times t}{L}$$

$$\#TC = \frac{1 \times 10^4 \times 2.8 \times 10^{-3}}{1 \times 10^1}$$

$$\#TC = 2.8$$

Step 3: Use the time constant chart to

(a) Locate the 2.8 TC point on the horizontal axis.
(b) Move upward from this point to curve A.
(c) Move left to the percent-of-charge line.
(d) Read 94%.

Step 4: Calculate maximum available current:

$$I_{max} = \frac{E_t}{R_t} = 5 \text{ mA}$$

Step 5: Calculate 94% of 5 mA:

$$I = 94\% I_{max} = 0.94 \times 5 \text{ mA} = 4.7 \text{ mA}$$

Step 6: Calculate E_R:

$$E_R = I \times R = 4.7 \text{ mA} \times 10 \text{ k}\Omega = 47 \text{ V}$$

Step 7: Calculate E_L:

$$E_L = E_t - E_R = 50 \text{ V} - 47 \text{ V} = 3 \text{ V}$$

Problem 2

Refer to Figure 22-15. After the switch has been closed 1200 μs, E_L is 52.5 V. What is the inductance of L?

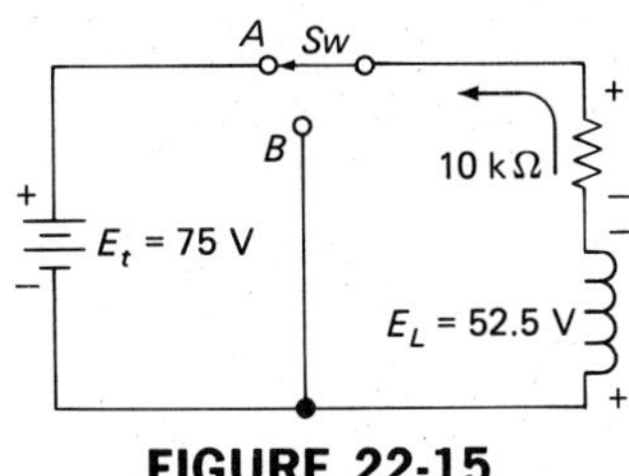

FIGURE 22-15

Step 1: List givens and wanteds:

Given	Wanted
$E_t = 75$ V $E_L = 52.5$ V $R = 10$ kΩ $t = 1200$ μs	$L =$

Step 2: Calculate the percent of available voltage:

$$\% \text{ of charge} = \frac{E_L}{E_t} \times 100$$

$$\% \text{ of charge} = \frac{52.5 \text{ V}}{75 \text{ V}} \times 100 = 70\%$$

Step 3: Find the number of TC used to change to 70% from the universal time constant chart.

(a) Locate the 70% point on the left side.
(b) Move right to curve B.
(c) Drop down to the bottom line.
(d) Read 0.35 TC.

Step 4: Calculate the amount of inductance:

(a) Transpose the equation $\#TC = \dfrac{R \times t}{L}$

(b) The new equation is $L = \dfrac{R \times t}{\#TC}$

(c) $L = \dfrac{1 \times 10^4 \times 1.2 \times 10^{-3}}{0.35}$

(d) $L = 34.3$ H

Problem 3

In Figure 22-16, E_L equals 40 V 1500 μs after the switch is closed. What is the ohmic value of R?

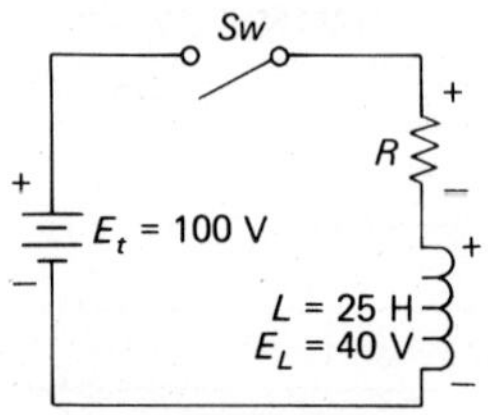

FIGURE 22-16

Step 1: List givens and wanteds:

Given	Wanted
$E_t = 100$ V $E_L = 40$ V $L = 25$ H $t = 1500\ \mu$s	$R =$

Step 2: Determine the percent of charge represented by E_L:

$$\% \text{ of charge} = \frac{E_L}{E_t} \times 100$$

$$\% \text{ of charge} = \frac{40 \text{ V}}{100 \text{ V}} \times 100 = 40\%$$

Step 3: Find the number of *TC* for E_L to change from 100 V to 40 V:

(a) Locate 40% point on the chart.
(b) Move right until curve B is intersected.
(c) Move down to bottom line.
(d) Number of *TC* = 0.9.

Step 4: Calculate the ohmic value of *R*:

Transpose equation:

$$\#TC = \frac{R \times t}{L}$$

$$R = \frac{\#TC \times L}{t}$$

$$R = \frac{0.9 \times 25 \text{ H}}{1.5 \times 10^{-3}} = 15 \text{ k}\Omega$$

Self-Check

Answer each item by indicating whether the statement is true or false.

11. ____ A coil blocks DC current flow in a circuit.

12. ____ The formula for one time constant for a series RL circuit is $1\ TC = L \times R$.

13. ____ When E_t is increased, the inductor charges at a faster rate than it does with the lower voltage.

14. ____ The coil drops 63% of E_t at the end of one time constant in a series RL circuit.

15. ____ On the universal time constant chart, curve A is used to calculate E_R in a series RL circuit during charge.

16. ____ The equation for the number of time constants in a series RL circuit is $\#TC = R \times L$.

17. ____ In a series RL circuit, four time constants have elapsed; E_R is a very small percent of E_t.

18. ____ Forty percent of E_t is dropped across the coil in a series RL circuit after 0.6 time constant.

19. ____ The coil in a series RL circuit is allowed to charge for 200 μs. If $R = 50$ kΩ and the number of *TC* is 1.5, the inductor is 2 mH.

20. ____ A series RL circuit is allowed to charge for 1.2 time constants. If E_t is 10 V, E_R equals 3 V.

Classification of Time Constants

Time constants are classified in three categories, long, medium, and short. Is a week a long time? Not compared to 100 years. Is 1 second a long time? Compared to 1 μs, yes. For measurements to be meaningful, they must be based on the same standard or condition. When you are waiting for someone who is late for an appointment, 15 minutes is a long time; when you are engrossed in something pleasant, 15 minutes is a short time. This does not mean 15 minutes is shorter, or that the other times are any longer. It does mean that when using a different standard, time can be expressed in different ways.

Time constants can be expressed in the same manner. Time constants in RC circuits depend on the value of *R* and *C*. In RL circuits, they depend on the value of *R* and *L*. Whether a time constant is short or long depends on the time it is compared to.

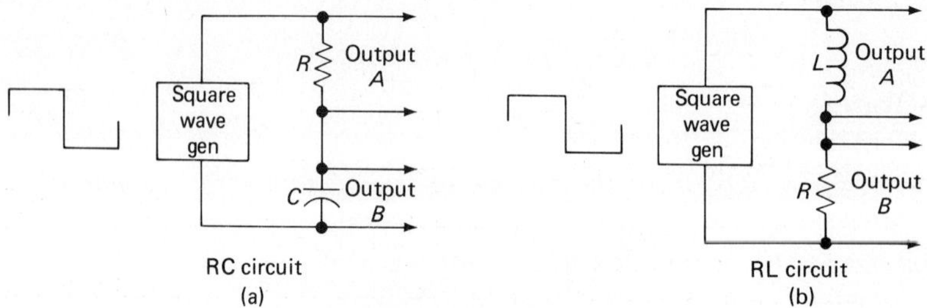

FIGURE 22-17

In examining the circuit in Figure 22-17, you can find a square-wave generator used as a voltage source. Comparison of time constants for this circuit are based on the time required for one alternation of the square wave. If this is a 1-kHz generator, the time for an alternation, *t*, is 500 μs. If the time constant is long compared to 500 μs, it is a *long time constant;* if the time constant is short compared to 500 μs, it is a *short time constant.* Standards in electronics for determining the length of time constants in a given circuit are:

1. A *short* time constant has a comparison ratio of 0.1 or less when *t* is compared to 1 *TC*.
2. Time constants between 0.1 and 10 are *medium* time constants when *t* is compared to 1 *TC*.
3. *Long* time constants are those whose *t* is 10 times (or more) the time for 1 *TC*.

Remember that time constants are stated as being long, medium, or short after comparing one time constant to the amount of time allowed for charge or discharge. Should a unit that affects either t or TC change, the time-constant length must be restated. Here, the frequency of the power source can be critical, because it provides the charge or discharge time for the circuit.

When a square wave is applied to the circuit in Figure 22-17a,b, the waveshapes shown in Table 22-3 result. In Figure 22-17a, an output can be taken across either component, R or C; in Figure 22-17b, the output can be taken across either R or L. The output at any point depends on whether the time constant is short, medium, or long.

Observe Table 22-3, where the output waveshapes that result from application of a square wave to the circuits are displayed.

TABLE 22-3

	Output A E_R in RC circuit E_L in RL circuit	Output B E_C in RC circuit E_R in RL circuit
Input wave		
Example 1 Long time constant t/RC or Rt/L = 0.1 or less		
Example 2 Medium time constant		
Example 3 Short time constant t/RC or Rt/L = 10 or more		

Examine example 1. The waves illustrate the outputs from a circuit with a long time constant. Output A is taken across R in the RC circuit and across L in the RL circuit. Notice that the output wave is almost identical to the input wave. Both amplitudes and shape are similar. In example 1, output B is taken across the capacitor in a series RC circuit. Notice that this waveshape bears little resemblance to the input wave; it is a triangular wave with low amplitude.

In example 2, medium time constant, output A is distorted, with a peak-to-peak amplitude that is larger than the amplitude of the input wave. Output B is also distorted, but less than that of output A.

Example 3 demonstrates a short time constant. Output A is distorted to the point where it is a "peaked" wave with a peak-to-peak amplitude double that of the input. Output B, slightly distorted, is similar to the input, with its amplitude equal to the input amplitude.

By careful examination of Table 22-3, we see that either type of circuit, RL or RC, can be used to develop waveshapes of desired shape and/or amplitude. All that is required is to apply a square wave input and use components of the correct size.

Self-Check

Answer each item by indicating whether the statement is true or false.

21. ___ A peaked-wave output is taken across the coil in a series RL circuit.

22. ___ To determine whether the output wave is peaked, we vary the size of the resistor.

23. ___ Output waves on capacitors and coils are identical for all time constants.

24. ___ A circuit has a long time constant. During its charge time, it becomes fully charged.

25. ___ A peaked wave has a peak-to-peak amplitude equal to the amplitude of the input wave.

Differentiator and Integrator Circuits

Many electronic circuits require precise timing. For this timing to be accurate enough, circuits must be designed that can accurately produce timed triggers for use in the regulation of timing. Two processes are commonly used to develop timing waves—*integration* and *differentiation.* These are mathematical terms taken from calculus. An understanding of the underlying math is not necessary to understand the circuits. Square wave, integrated wave, and differentiated wave are defined in the following statements and are shown in Figures 22-18a,b,c.

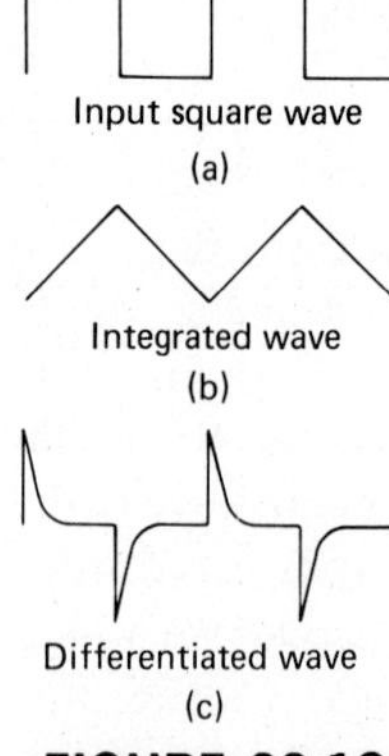

FIGURE 22-18

A square wave is a symmetrical wave whose positive and negative alternations have the same duration.

Figure 22-18a illustrates a square wave.

> ***An integrated wave is a wave whose amplitude starts at zero and increases to a maximum value.***

Figure 22-18b illustrates an integrated wave.

> ***A differentiated wave's amplitude starts at a maximum value and decreases to zero.***

Figure 22-18c illustrates a differentiated wave.

To define these two waveshapes further, we consider the four statements:

(a) An RC circuit is a differentiator if its time constant is short and the output wave is taken across the resistor.
(b) An RL circuit is a differentiator if the time constant is short and the output wave is taken across the coil.
(c) An RC circuit is an integrator if its time constant is long and the output wave is taken across the capacitor.
(d) An RL circuit is an integrator if its time constant is long and the output wave is taken across the resistor.

Integrating Circuits

An integrating circuit produces an output voltage that is proportional to the area under the input waveform. *Area* equals *voltage* multiplied by *time*. An integrated wave is similar to curve A of the universal time constant chart.

A practical way to produce an integrated waveform is to use a long-time-constant RC circuit similar to that in Figure 22-19. As shown in the diagram, the output wave is taken across the capacitor. The same waveshape can be developed using a long-time-constant RL circuit where the output is taken across the resistor. Integrators and differentiators are usually RC circuits because they are less expensive to build.

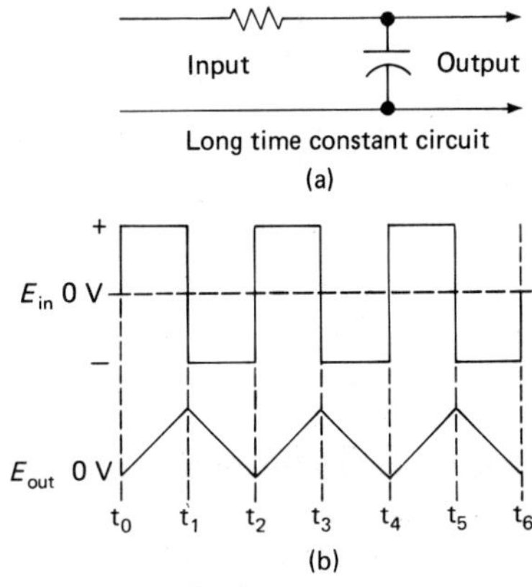

FIGURE 22-19

Examine Figure 22-19 fully. Note that output terminals are shown on either side of the capacitor. The E_{in} waveshape is a square wave, where E_{out} is a sharply integrated wave. The sharpness of the peaks can be varied by changing the value of the resistor. In practical applications, it is possible to place a rheostat in the resistor position, connect an oscilloscope across the output, and adjust the rheostat for the desired amount of integration while measuring the wave on the scope.

Differentiating Circuits

Differentiating circuits produce output waves that are proportional to the *rate of change* of the input. A differentiating circuit uses a short time constant and the output is taken across the resistor in an RC circuit, or across the coil in an RL circuit. Figure 22-20 illustrates an RC differentiating circuit. The differentiated output is similar to curve B of the universal time constant chart.

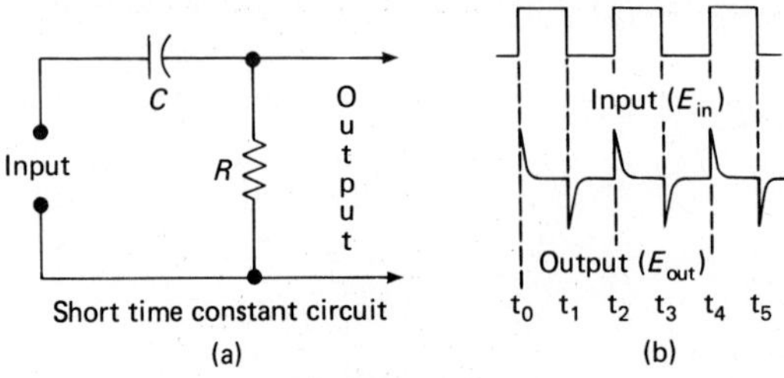

FIGURE 22-20

The circuit in Figure 22-20 illustrates the effect of a differentiating circuit on a square-wave input. Note that the output wave changes rapidly from maximum to zero. Again, the speed at which it drops to zero depends on the time constant and the size of *R*. Differentiating circuits usually are RC circuits. Again, the resistor can be a rheostat that can be varied to produce desired differentiation. During the negative alternation of the input signal, the differentiator provides a negative output pulse of equal amplitude and duration to the positive pulse available when the positive alternation is present.

Self-Check

Answer each item by indicating whether the statement is true or false.

26. ____ RL circuits are desirable for use as integrators when compared to RC circuits.

27. ____ Integrator circuits operate with long time constants.

28. ____ In a short-time-constant RC circuit, the output taken across the capacitor is a differentiated wave.

29. ____ Differentiated waves have sharp leading edges.

30. ____ Using a variable resistor in an RC differentiating circuit decreases the output wave amplitude when resistance is increased.

Summary

In this chapter we reviewed the characteristics of resistors, capacitors, and inductors. We learned that a transient voltage or current is one that changes from one steady state to another. The time required for this change is the *transient interval.* A graph of transient interval and time results in a waveshape.

In a series RC circuit, we saw that charge and discharge of the capacitor are directly proportional to the size of either resistance or capacitance. Increasing the size of either causes the charge time to increase. The product of resistance and capacitance were found to equal one time constant of the charge time, where five time constants are required to fully charge or discharge the capacitor. The capacitor charges or discharges 63% of the available voltage during each time constant.

The universal time constant chart is used with either RC or RL circuits to solve problems pertaining to charge and discharge of the coil or capacitor. Figure 22-21 is a universal time constant chart that contains a summary for use with both RC and RL circuits.

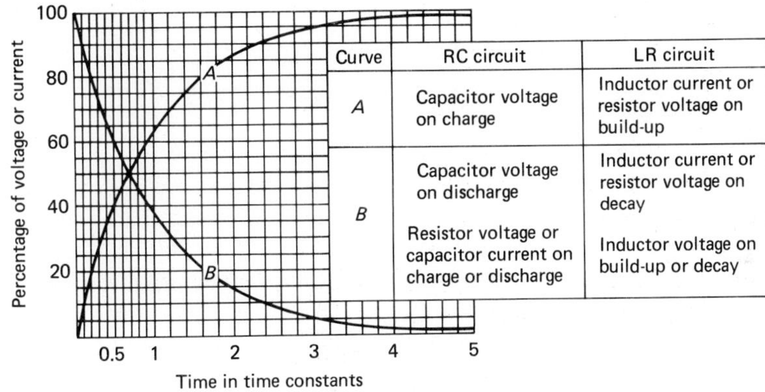

FIGURE 22-21

Formulas, procedures, and solutions were illustrated, which demonstrate solving these circuits for value and time.

The similarity between the action in series RC circuits and that in series RL circuits was discussed. The coil charges to 63% of available current in each time constant; it too is fully charged or discharged after five time constants. The time constant for a series RL circuit is found by dividing the inductance by resistance.

We noted the effect these circuits have on a symmetrical square wave when the square wave is used as the applied voltage. Time constants were classified as long, short, or medium. RC circuits operating with short and long time constants were explained, thus illustrating their use as integrator or differentiator circuits.

An understanding of these circuits is necessary in order to learn the timing sequences of complex systems.

Review Questions and/or Problems

1. Refer to Figure 22-22. Immediately upon moving Sw_1 to position A, the capacitor
   (a) acts like an open.
   (b) drops 63% of E_t.
   (c) acts like a short.
   (d) shares E_t with the resistor.

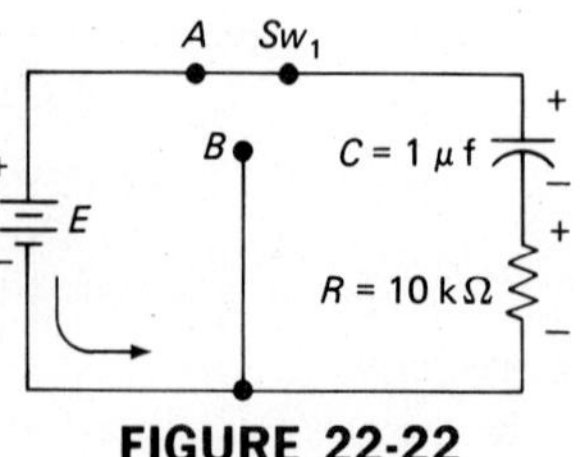

FIGURE 22-22

2. In Figure 22-22, the switch has been in position A for three time constants. E_C equals
   (a) 37% of E_t.
   (b) E_R.
   (c) 86% of E_t.
   (d) 95% of E_t.

3. Curve A on the universal time constant chart is used to calculate
   (a) I_C during charge.
   (b) E_C during charge.
   (c) E_C during discharge.
   (d) both a and b.

4. Refer to Figure 22-22. If $E_t = 50$ V, $R = 10$ kΩ, and $C = 0.02\ \mu$f, how long will it take for the capacitor to charge to 50 V?
   (a) 1 ms
   (b) 0.2 ms
   (c) 10 ms
   (d) 2 ms

5. Refer to Figure 22-23. Sw_1 is closed for 6 ms. What is the voltage on the capacitor?
   (a) 60 V
   (b) 45 V
   (c) 50 V
   (d) 65 V

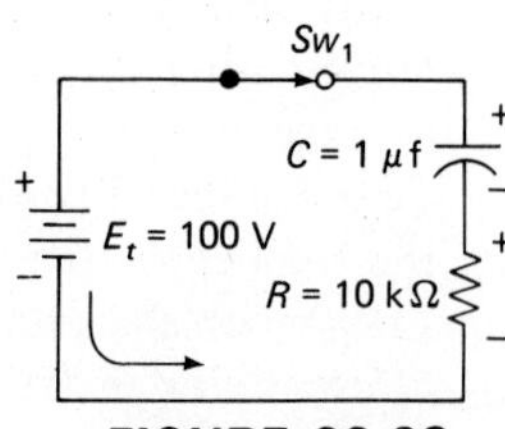

FIGURE 22-23

6. In a series RC circuit, curve B on the universal time constant chart is used to calculate which of the following?
   (a) E_C and I_t during charge
   (b) E_C *only* during charge
   (c) E_C *only* during discharge
   (d) E_C, E_R, and I_t during discharge

7. The switch in Figure 22-24 has been placed in position A for 10 ms, then moved to position B for 1.6 ms. What is E_C?
   (a) 1.6 V
   (b) 64 V
   (c) 16 V
   (d) 6.4 V

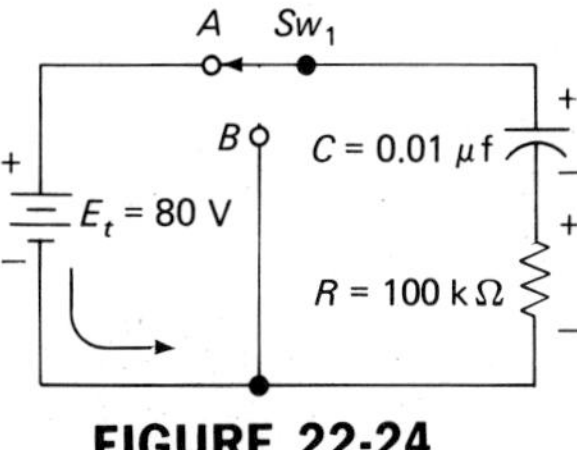

FIGURE 22-24

8. The switch in Figure 22-24 is placed in position A for 1000 μs. If $R = 80$ kΩ and $\#TC = 0.5$, what is the value of C?
   (a) 0.025 μf
   (b) 2.5 μf
   (c) 25 μf
   (d) 0.25 μf

9. Refer to Figure 22-25. The switch is placed in position A for 250 μs. At the end of this time, $E_R = 80$ V. What is the size of the resistor?
   (a) 125 Ω
   (b) 125 kΩ
   (c) 12.5 kΩ
   (d) 1.25 kΩ

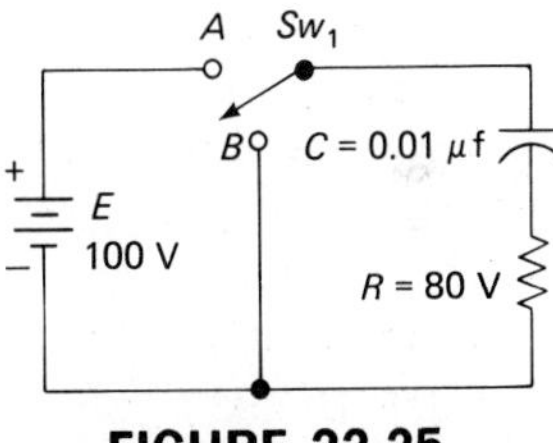

FIGURE 22-25

10. A capacitor charges to 30 V in 1.4 time constants. What is the total applied voltage?
    (a) 400 V
    (b) 150 V
    (c) 15 V
    (d) 40 V

11. The inductor in a series RL circuit
    (a) opposes DC current.
    (b) opposes any change in voltage.
    (c) opposes any change in current flow.
    (d) acts like an open after five time constants.

12. Refer to Figure 22-26. The instant the switch is closed,
    (a) E_t and E_L are equal.
    (b) E_t and E_R are equal.
    (c) E_R equals one-half E_t.
    (d) E_L equals zero volts.

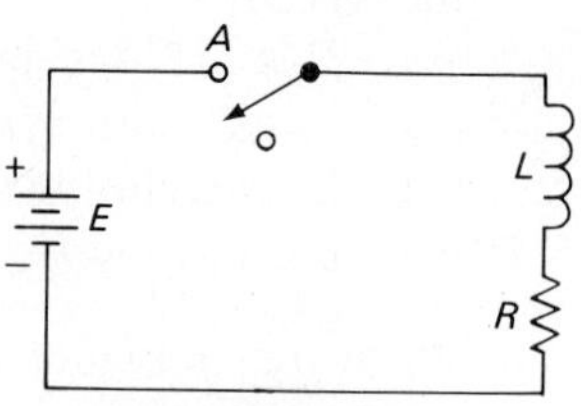

FIGURE 22-26

13. In a series RL circuit, curve A on the universal time constant chart can be used to calculate which of the following?
    (a) E_L during charge
    (b) E_R during charge
    (c) E_L during discharge
    (d) I_L during discharge

14. Examine Figure 22-27. After 1750 μs of charge time, the voltage on the resistor is
    (a) 25 V.
    (b) 50 V.
    (c) 75 V.
    (d) 125 V.

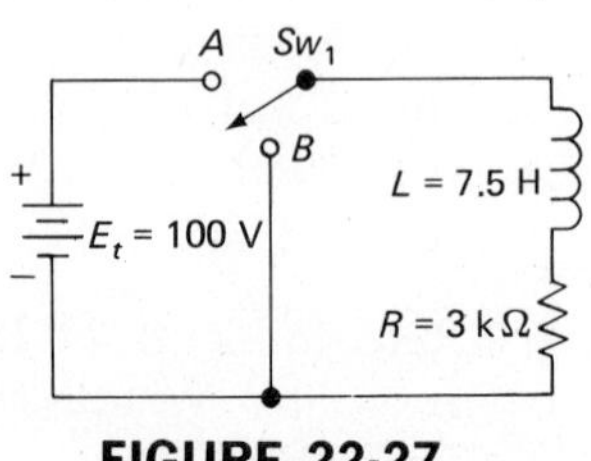

FIGURE 22-27

15. Refer to Figure 22-28. When the coil is fully charged, the switch is placed at B for 1000 μs. How long does it take for the coil to discharge completely?
    (a) 100 ms
    (b) 1 μs
    (c) 10 μs
    (d) 10 ms

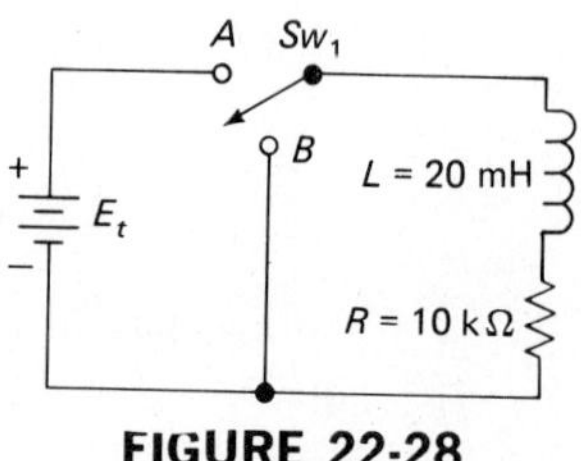

FIGURE 22-28

16. During a long time constant, the least distorted wave is available across the resistor of a series RC circuit.
    (a) true
    (b) false

17. In a series RL differentiating circuit, the differentiated output is taken across the resistor.
    (a) true
    (b) false

18. A short-time-constant, series RC circuit has the least distorted output across its capacitor.
    (a) true
    (b) false

19. Integration occurs across the capacitor in a series RC circuit.
    (a) true
    (b) false

20. For the output wave across the resistor in a series RC circuit to be differentiated, the circuit time constant must be short.
    (a) True
    (b) False

23

Resonant Circuits

Introduction

In this chapter we continue our study of RCL circuitry. At this point you will begin to study circuits suitable for use in radio, TV, and other communication systems. Without the resonant effect of RCL circuits, electronic communications would not exist.

We have learned that the effects of X_C and X_L are opposite and that,

> ***If frequency applied to the RCL circuit is increased, capacitive reactance decreases and inductive reactance increases.***

When the frequency applied to series and parallel RCL circuits causes X_C to equal X_L, the circuit is operating at *resonance*. In this chapter we perform analyses that explain the effect of changes in frequency on the operation of circuits.

As part of our study of resonance, we consider resonant frequency, half-power points, bandwidth, bandpass, and selectivity—terms used to describe a resonant circuit. Resonant circuits are used in radios and TV sets to select the desired station, in telephones to provide communications, and in television to control the frequency transmitted.

Objectives

Each student is required to:

1. Calculate the resonant frequency of series- and parallel-resonant RCL circuits.
2. Analyze the effect on current flow when a circuit is operating
   (a) At resonance.
   (b) Above resonance.
   (c) Below resonance.
3. Determine whether the circuit is operating inductive, capacitive, or resistive when the frequency is
   (a) At resonance.
   (b) Above resonance.
   (c) Below resonance.

4. Draw current, voltage, and impedance vector diagrams for series RCL circuits operating.
   (a) At resonance.
   (b) Above resonance.
   (c) Below resonance.
5. Define *quality factor* associated with resonant circuits and identify its symbol.
6. Explain the effect of resistance to reactance ratio on the quality factor of parallel and series resonant circuits.
7. Define.
   (a) Selectivity.
   (b) Resonant frequency.
   (c) Half-power point.
   (d) Bandwidth.
   (e) Bandpass.
8. Use the appropriate response curve of a resonant circuit to determine
   (a) Upper and lower half-power points.
   (b) Bandwidth.
   (c) Bandpass.
9. Explain the effect on current, phase angle, and frequency in series and parallel RCL circuits operating above, below, and at resonance when any of the following is varied:
   (a) Frequency.
   (b) Resistance.
   (c) Capacitance.
   (d) Inductance.
10. Explain how current curves for parallel circuits differ from those for series circuits.
11. Solve for Q in a parallel RCL circuit.
12. Draw and explain impedance curves for parallel RCL circuits.

Resonant Circuits

All resonant circuits contain resistance, capacitance, and inductance. Figure 23-1a represents a circuit of this type. Notice that X_C and X_L are equal. When the vector diagram for X_C and X_L is drawn (Figure 23-1b), we see that $Z = R$. At resonance, reactances are equal but opposite, and thus cancel. That leaves resistance as the only opposition to current flow. Minimum impedance is present. Knowing that $Z = R$, we can calculate total current:

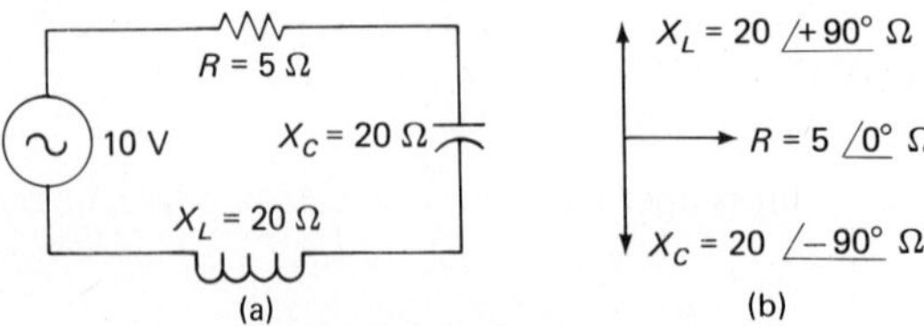

FIGURE 23-1

$$I_t = \frac{E_t}{R_t}$$

$$I_t = \frac{10 \text{ V}}{5 \ \Omega}$$

$$I_t = 2 \text{ A}$$

This circuit has several characteristics:

(a) $X_L = X_C$

(b) $Z_t = R_t = \text{minimum}$

(c) $I_t = \dfrac{E_t}{R_t} = \text{maximum}$

(d) $E_R = E_t$

(e) $E_C = E_L$

(f) $f_r = \dfrac{1}{2\pi\sqrt{LC}}$

When the circuit is operating at the frequency where $X_C = X_L$, the circuit is said to be operating at its *resonant frequency*. When the circuit is operating at resonant frequency, it is a *resonant circuit*. If a variable frequency power source is applied to the circuit, and its output frequency is varied, only one frequency causes the circuit to resonate.

There is only one resonant frequency for any single RCL circuit.

The characteristics of a series-resonant circuit are:

$$f_r = \frac{1}{2\pi\sqrt{LC}}$$

$$X_C = X_L$$

$$Z = \text{minimum}$$

$$Z = R$$

$$I_t = \text{maximum}$$

$$P_A = P_T$$

$$P_F = 1$$

The characteristics of a parallel-resonant circuit are:

$$f_r = \frac{1}{2\pi\sqrt{LC}}$$

$$I_L = I_C$$

$$X_L = X_C$$

$$P_A = P_T$$

$$P_F = 1$$

I_t is minimum; Z is maximum.

Self-Check

Answer each item by indicating whether the statement is true or false.

1. ____ At resonant frequency, Z_t equals $\sqrt{R}$.
2. ____ Current is maximum when the series circuit is operating at resonant frequency.
3. ____ At resonance, X_C equals R.
4. ____ Total voltage is dropped across the resistor in a series RCL resonant circuit.
5. ____ E_C and E_L may be extremely large voltages when a circuit is at resonance.

Series-Resonant Circuits

Resonant Frequency

In a series RCL circuit, there is only one frequency where X_C equals X_L—*resonant frequency*. The formula for calculating resonant frequency is:

$$f_r = \frac{1}{2\pi\sqrt{LC}}$$

Let us calculate resonant frequency for a circuit that contains a 20-mH coil and a 200-pf capacitor.

Mathematical solution:

$$f_r = \frac{1}{2\pi\sqrt{LC}}$$

$$f_r = \frac{1}{2 \times 3.14\sqrt{(2 \times 10^{-2})\,(2 \times 10^{-10})}}$$

$$f_r = \frac{1}{1.2566 \times 10^{-5}}$$

$$f_r = 7.9577 \times 10^{4} \text{ or } 79{,}577 \text{ Hz}$$

Action	*Entry*	*Display*
Note: Clear radical first		
ENTER	20,EE,+/−,3	20 − 03
PRESS	×	2 − 02
ENTER	200,EE,+/−,12	200 − 12
PRESS	=	4 − 12
PRESS	$\sqrt{X}$	2 − 06
PRESS	×	2 − 06
ENTER	2	2
PRESS	×	4 − 06
PRESS	π	3.1416 00
PRESS	=	1.2566 − 05
PRESS	1/X	7.9577472 04

Converted to a prefix, 7.9577 04 equals 79,577 Hz.

As has been mentioned, anytime frequency is changed, both X_C and X_L are affected. As frequency increases, X_L increases and X_C decreases. Figure 23-2 used to illustrate the affect of frequency on these reactances.

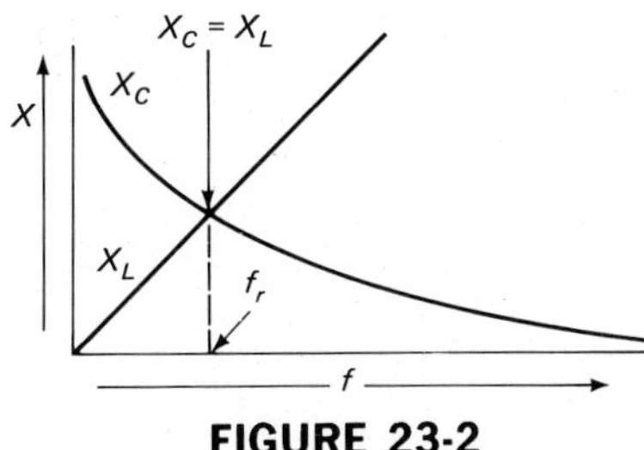

FIGURE 23-2

Let us examine Figure 23-2. At 0 Hz, X_C is maximum (an open) and X_L is zero (a short). As the frequency increases, X_C decreases, while X_L increases; but X_C is larger. At resonance, the two lines cross, signifying that X_C and X_L are equal. As the frequency continues to increase, X_C continues to decrease and X_L to increase. Above resonance, X_L is larger than X_C. At very high frequencies, X_C is very low and acts like a short. At the same frequency, X_L is very high and acts like an open.

The solid line in Figure 23-3 represents current in the series RCL circuit as frequency increases; the dashed line represents impedance. Note that at resonance, I_t is maximum and Z_t is minimum. Below resonance, X_C is larger, and the circuit acts capacitively. Above resonance, X_L is largest, and the circuit acts inductively.

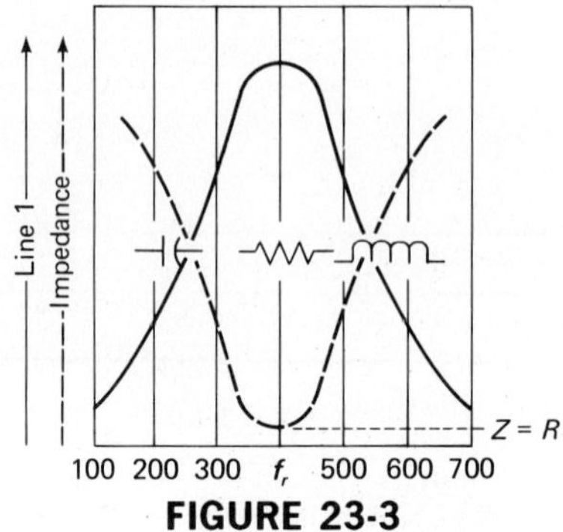

FIGURE 23-3

Self-Check

Answer each item by indicating whether the statement is true or false.

6. ____ Above resonance, X_C is largest, and the circuit acts capacitively.

7. ____ At resonant frequency of a series RCL circuit, Z_t is maximum.

8. ____ As frequency decreases, X_L decreases.

9. ____ At high frequencies, a capacitor behaves like a short.

10. ____ An inductively acting series RCL circuit has a positive phase angle.

Above Resonant Frequency

The circuit in Figure 23-4 has different values for X_C and X_L. If reactances are not equal, the circuit *cannot* operate at resonance. We already know that:

> ***Inductive reactance is directly proportional to frequency, and capacitance reactance is inversely proportional to frequency.***

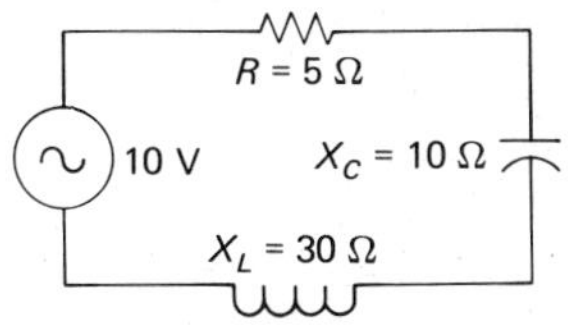

$R = 5\ \Omega$

10 V $X_C = 10\ \Omega$

$X_L = 30\ \Omega$

FIGURE 23-4

For this circuit to be brought to its resonant frequency, X_L must decrease or X_C must increase. If the operating frequency of the power supply is decreased, X_L decreases and X_C increases. At some frequency, X_L equals X_C. This is the circuit's resonant frequency. The equation for calculating resonant frequency has already been given.

The circuit can also be brought to resonance by decreasing either capacitance or inductance. By changing either component, the circuit's resonant frequency is changed. Each combination of capacitor and coil has it own resonant frequency. When the correct combination is selected, the circuit's resonant frequency becomes the same as the operating frequency of the power source. To select a pair of components suitable for resonating at a specific frequency, we can use either equation:

$$C = \frac{1}{2^2 \times \pi^2 \times L \times f_r^2}$$

$$L = \frac{1}{2^2 \times \pi^2 \times C \times f_r^2}$$

Below Resonant Frequency

In the circuit in Figure 23-5, X_C is larger than X_L. This circuit is operating below resonance. To bring it to resonance, the applied frequency must be increased.

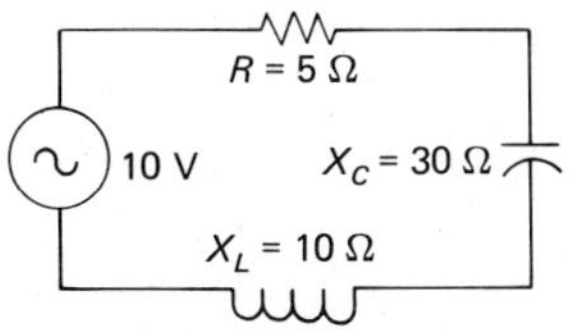

$R = 5\ \Omega$

10 V $X_C = 30\ \Omega$

$X_L = 10\ \Omega$

FIGURE 23-5

Figure 23-6 illustrates the vector diagrams for each circuit already discussed. For the resonant circuit, X_C and X_L cancel, leaving the impedance, $Z = R$ at 0°. The impedance angle is the phase angle and is the angle θ. This is the same angle θ that was used in series-reactive circuits. Then, we learned that power factor equals cos θ. Finding cosine θ, using a calculator reveals:

Power factor, P_F, $= \cos\theta = 1$

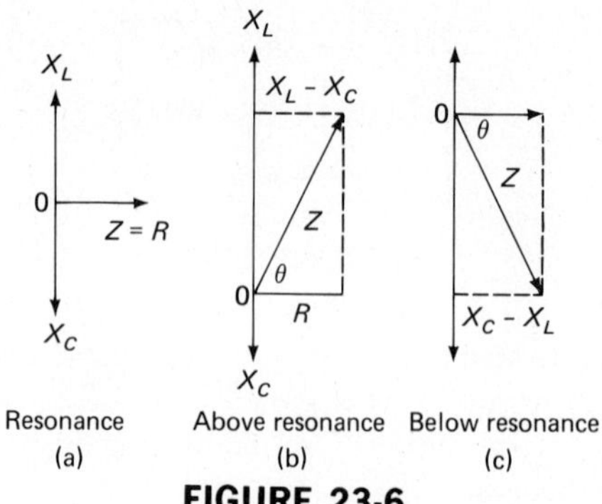

FIGURE 23-6

Any series RCL circuit operating above resonance has a positive impedance phase angle; the circuit that is below resonance has a negative phase angle. In either case, $\cos \theta$ is less than 1, indicating that P_F is less and that power apparent is no longer equal to power true, as it was at resonance. This gives us two more items, which we can add to the list of characteristics for a series resonant circuit:

$$\theta = 0°$$

$$P_F = \cos \theta = 1$$

Self-Check

Answer each item by indicating whether the statement is true or false.

11. ___ Series RCL circuits operating above resonance have X_C larger than X_L.

12. ___ Series RCL circuits operating below resonance have negative phase angles.

13. ___ For X_C to increase, the applied frequency must be decreased.

14. ___ At resonance, X_C and X_L are equal.

15. ___ A circuit is at resonance when $P_F = 1$.

Voltage and Current Relationships

In examining Figure 23-7a, we find an AC ammeter inserted in the circuit. Regardless of where this meter is inserted, its indication is the same. We already know that current is the same at all points in a series circuit and that voltage *leads* current in a coil and *lags* current in a capacitor, whereas the voltage is in phase with the current in the resistor. When the circuit is at resonance, X_C and X_L are equal and cancel. This means E_t, I_t, and E_R are all in phase (see the vector diagram in Figure 23-7b).

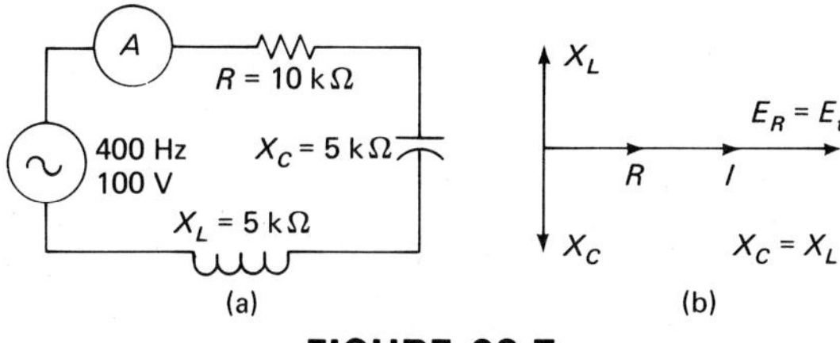

FIGURE 23-7

Figure 23-8 shows a circuit and its vector diagram. This circuit is capacitive because current leads E_t. The phase angle is 45° because $X_C - X_L = R$.

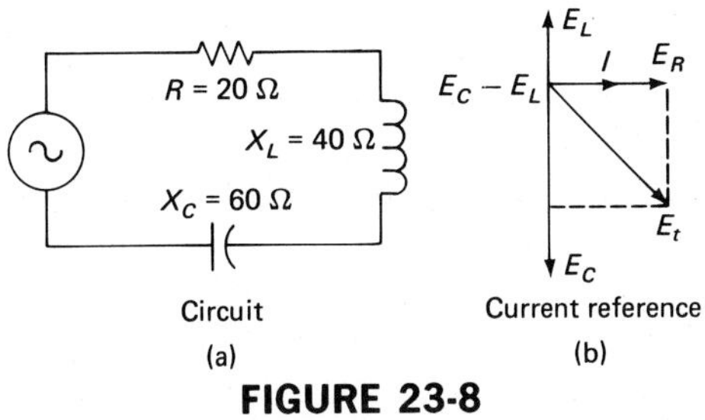

FIGURE 23-8

Figure 23-9 contains a circuit operating above resonance, as well as its vector diagram. Note that the current vector lags the E_t vector. If the current vector lags the voltage vector, the circuit is acting inductively. The phase angle equals 45° because the reactance and resistance vectors are equal in length. Total current, I_t, is used as the reference vector, as was done in series RC, RL, and RCL circuits.

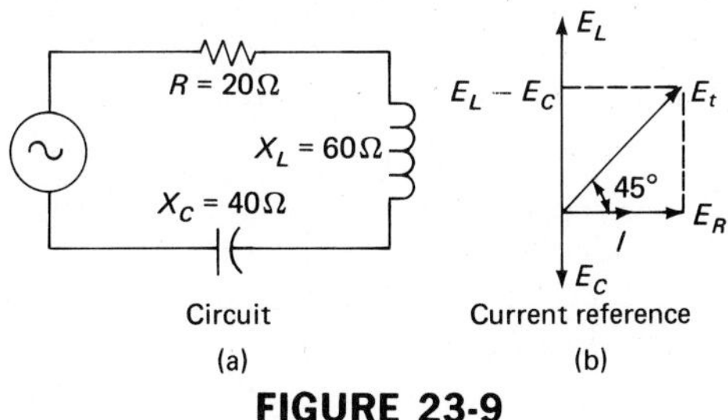

FIGURE 23-9

Figure 23-10 represents a typical current plot for a series RCL circuit. From its peak at f_r, it decreases as frequency changes above or below resonance. There are two points on this curve where current is 70.7% of maximum and power equals 50% of peak power. These points are called *half-power points*. The point above resonance is the *upper half-power point* (UHPP), and the one below is the *lower half-power point* (LHPP). Half-power points are calculated:

$$HPP = 0.707 \times I_{pk}$$

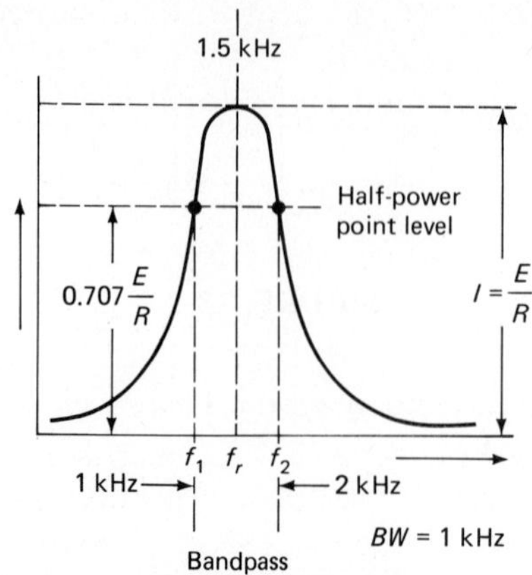

FIGURE 23-10

The number of frequencies between LHPP and UHPP make up the *bandwidth* (BW). Individual frequencies included in the BW are the *bandpass* (BP). The bandwidth is calculated by subtracting the frequency at the LHPP from the frequency at the UHPP. The term *bandpass* refers to the frequencies that "pass" between these points. Consider:

$$f_r = 1.5 \text{ kHz}$$

$$f_{UHPP} = 2 \text{ kHz}$$

$$f_{LHPP} = 1 \text{ kHz}$$

$$BW = f_{UHPP} - f_{LHPP}$$

$$BW = 2 \text{ kHz} - 1 \text{ kHz}$$

$$BW = 1 \text{ kHz}$$

BP equals all frequencies from 1 kHz to 2 kHz.

In realizing that the variation of certain values in a resonant circuit can affect its BW and BP, we come to another item of interest. The fewer frequencies a circuit's BW and BP have, the better it can select a specific band of frequencies. This characteristic is the *selectivity* of a circuit.

Selectivity is the relative ability of a circuit to select a frequency from all those available.

Self-Check

Answer each item by indicating whether the statement is true or false.

16. ____ The bandpass of the circuit in Figure 23-10 is 1 kHz.

17. ____ The bandwidth has resonant frequency as its center frequency.

18. ___ Half-power points represent specific frequency and current values.

19. ___ Bandwidth is found by subtracting the UHPP from the LHPP.

20. ___ Resonant frequency can be calculated by taking the reciprocal of $2\pi \sqrt{LC}$.

The Effect of Changes in Component Size

Increasing the resistance of a series RCL circuit has no effect on resonant frequency. Changing R, however, affects the amount of current that can flow at resonance. By changing the peak current that can flow at resonance, we can also change the half-power points, BW and BP.

When we change peak current, we change the *quality factor* of the circuit Q. Q is defined as the ratio of energy stored (by the reactive components) to the energy dissipated by resistance. This ratio can be stated in one of two ways:

$$Q = \frac{X_L}{R}$$

$$Q = \frac{X_C}{R}$$

When the value of the series resistor is increased, Q is decreased. The affect of changing R and Q is illustrated in Figure 23-11. Current flowing in this circuit must pass through both the coil and the capacitor; therefore, either reactance can be used to calculate circuit Q.

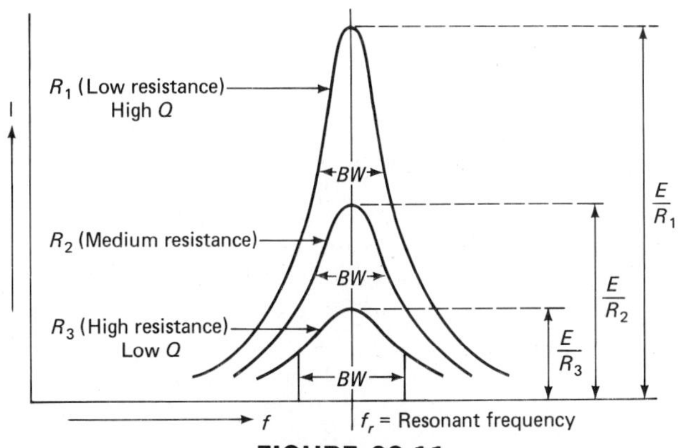

FIGURE 23-11

Because resistance and Q are inversely proportional, increasing the size of R decreases Q; decreasing R causes Q to increase. This is important. The higher the value of Q, the narrower a circuit's BW and the fewer frequencies its BP has. Without affecting resonant frequency, we can change the size of R to make the circuit operate as we would like. Circuits with high Q have better selectivity than those with low Q. This means the number of frequencies included in BW and BP are controlled more precisely. When high selectivity is required, the circuit usually does not contain a resistor. The total resistance of a circuit is the resistance of the wire used both to wind the coil and as conductors. The relationship of Q to BW is stated by the equation,

$$BW = \frac{f_r}{Q}$$

Figure 23-12 is a high-Q circuit. Its impedance is low at resonance, increasing rapidly as frequency varies away from resonance. The Q of this circuit is found by:

$$Q = \frac{X_L}{R} \text{ or } \frac{X_C}{R}$$

$$Q = \frac{1000}{10}$$

$$Q = 100$$

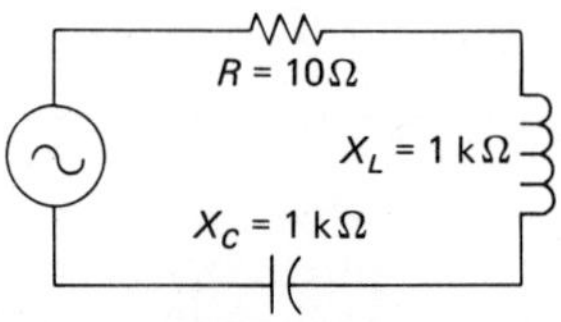

$R = 10\Omega$

$X_L = 1\ \text{k}\Omega$

$X_C = 1\ \text{k}\Omega$

FIGURE 23-12

Changes in Capacitance or Inductance

Resonant frequency is determined by the size of L and C. When either is changed, the resonant frequency also changes. When a radio is tuned, a variable capacitor is being changed, which, in turn, causes the resonant frequency of a circuit to change; thus the station is "selected." Variable coils are also used to tune circuits. The fine-tuning control on most TV sets is a tuned coil that changes resonant frequency. Figure 23-13 contains a series RCL circuit with variable capacitance and inductance.

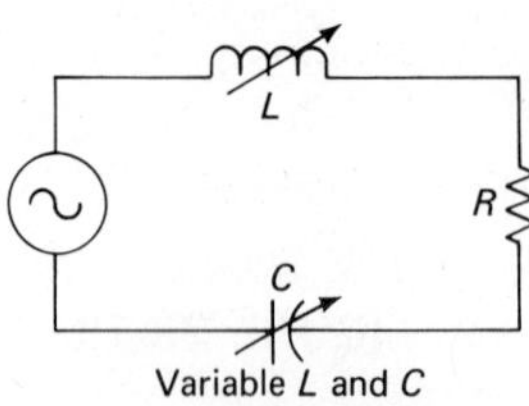

Variable L and C

FIGURE 23-13

Let us examine the series RCL circuit to determine the effect of changes in capacitance, inductance, frequency, and resistance. Table 23-1 displays the effect of these changes for a circuit that is operating below resonance.

TABLE 23-1
Circuit Below Resonance

Increases in	*Current*	*Impedance*	*Phase Angle*
Frequency	INC	DEC	DEC
Resistance	DEC	INC	DEC
Capacitance	INC	DEC	DEC
Inductance	INC	DEC	DEC

Table 23-2 illustrates the effect of changes on circuit operation for a series RCL circuit operating above its resonant frequency.

TABLE 23-2
Circuit Above Resonance

Increases in	*Current*	*Impedance*	*Phase Angle*
Frequency	DEC	INC	INC
Resistance	DEC	INC	DEC
Capacitance	DEC	INC	INC
Inductance	DEC	INC	INC

Self-Check

Answer each item by indicating whether the statement is true or false.

21. ____ Frequency and phase angle are directly proportional.

22. ____ As capacitance increases, resonant frequency also increases.

23. ____ To calculate BW, we can use the equation f_r/Q.

24. ____ Increases in resistance affect resonant frequency.

25. ____ Increases in inductance affect the Q of a coil.

Parallel-Resonant Circuits

Resonance in parallel RCL circuits occurs when $I_C = I_L$. When the frequency applied causes X_L to equal X_C, the current through these branches must be equal.

Resonant Frequency

Figure 23-14 contains a parallel RCL circuit. This circuit will resonate at a frequency that can be calculated.

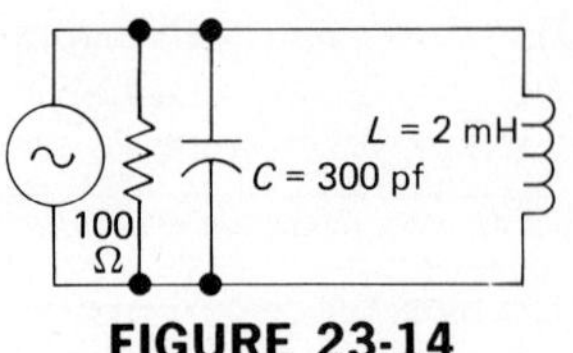

FIGURE 23-14

Mathematical solution:

$$f_r = 1/2\pi\sqrt{LC}$$

$$f_r = \frac{1}{2 \times 3.14 \times \sqrt{3 \times 10^{-10} \times 2 \times 10^{-3}}}$$

$$f_r = \frac{1}{2 \times 3.14 \times 7.75 \times 10^{-7}}$$

$$f_r = \frac{1}{4.87 \times 10^{-6}}$$

$$f_r = 2.05 \times 10^5 \text{ or } 205 \text{ kHz}$$

Calculator solution:

Action	Entry	Display
Note: Clear Radical First		
ENTER	2,EE,+/−,3	2 − 03
PRESS	×	2 − 03
ENTER	300,EE,+/−,12	300 − 12
PRESS	=	6 − 13
PRESS	$\sqrt{X}$	7.746 − 07
PRESS	×	7.746 − 07
ENTER	2	2
PRESS	×	1.5492 − 06
PRESS	π	3.1416 00
PRESS	=	4.867 − 06
PRESS	1/X	2.055 05

2.055 05 converts to 205.5 kHz.

The difference in results occurs because some values are rounded off in the mathematical solution. Both results are within tolerance at 205 kHz; therefore, the circuit is resonant at 205 kHz.

The parallel RCL circuit in Figure 23-15 has three branches, each of which drops voltage equal to E_t or 100 V. The amount of current flowing in each branch depends on the opposition in that branch and is determined, using Ohm's law. Remember that in parallel RCL circuits, we cannot add these currents arithmetically. They must be added vectorially, using the Pythagorean theorem, or they must be graphed. Branch currents are *not* in phase. Let us define the relationships that apply:

I_R is in phase with E_R in the resistive branch.
Current leads voltage by 90° in the capacitor.
Current lags voltage by 90° in the coil.

Because E_t is in phase across all branches, E_t can be used as the reference to state: I_R is in phase with E_t; I_C leads E_t by 90°; I_L lags E_t by 90°.

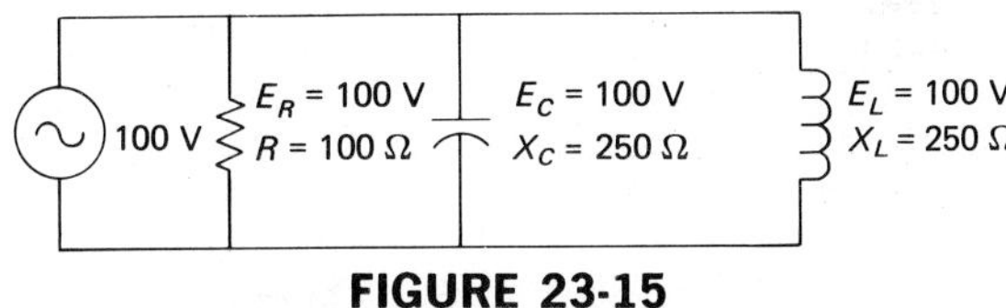

FIGURE 23-15

To solve for branch currents, we use Ohm's law:

$$I_R = \frac{E_t}{R} = \frac{100\text{ V}}{100\ \Omega} = 1\text{ A}$$

$$I_C = \frac{E_t}{X_C} = \frac{100\text{ V}}{250\ \Omega} = 0.4\text{ A}$$

$$I_L = \frac{E_t}{X_L} = \frac{100\text{ V}}{250\ \Omega} = 0.4\text{ A}$$

When these vectors are plotted as in Figure 23-16, we see that reactances and reactive currents are equal but opposite. Thus, the reactive currents cancel, leaving the current drawn from the power source at 1 A. This current is called *line current*. Now that we know total current, we can determine total impedance:

$$Z_t = \frac{E_t}{I_t}$$

$$Z_t = \frac{100\text{ V}}{1\text{ A}}$$

$$Z_t = 100\ \Omega$$

This illustrates that at resonance, $Z = R$, Z is maximum, and I is minimum.

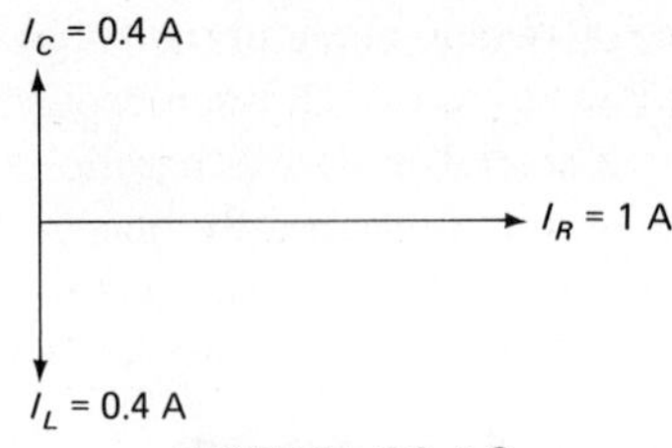

FIGURE 23-16

Parallel resonant circuits present very high impedance to the power source, resulting in a very small current drain on the power source. In a parallel-resonant circuit, this current is known as *line* current, I_l. Let us examine the action in the parallel tank circuit operating at its resonant frequency.

Examine Figure 23-17. With the switch in the position shown, the capacitor quickly charges to the applied voltage. Once the capacitor is fully charged, current flow stops. The capacitor has stored DC energy in its electrostatic field.

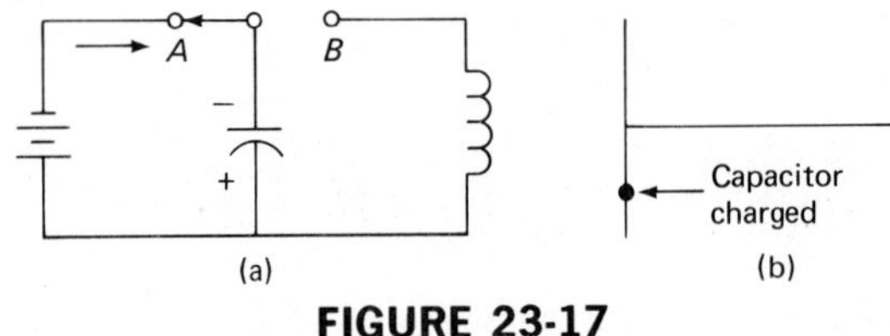

FIGURE 23-17

When the switch is moved to point B, the coil is connected in series with the charged capacitor. The electrons find a path for discharge through the coil, and current starts to flow. The coil opposes the change in current flow and begins to store the energy in its expanding magnetic field.

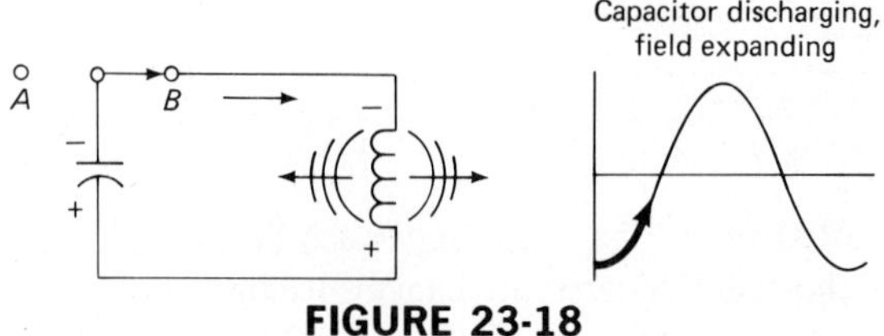

FIGURE 23-18

Starting at Figure 23-18 and ending at Figure 23-25, we see a heavy dark line; this line represents the charge and discharge of the capacitor. During the time covered by Figures 23-18 and 23-19, the energy stored in the capacitor is transferred to the coil and stored in the magnetic field. When the capacitor can no longer sustain current flow (Figure 23-19), the coil senses a decrease in current and immediately acts to oppose the change. During the time covered by Figures 23-20 and 23-21, the coil's field collapses and transfers its energy to the capacitor. At this time (Figure 23-21), the coil can no longer sustain current flow in the direction current has been flowing, and all current stops momentarily. During the time covered by Figures 23-22 and 23-23, the capacitor again acts as a power source, charging the coil. If we look at Figure 23-23, we see that the coil's fields have expanded completely. When the capacitor can no longer sustain current flow, the coil senses current trying to stop and acts to oppose the change. During the time covered by Figures 23-24 and 23-25, the coil supplies energy to the circuit; it is in the process of recharging the capacitor. In Figure 23-25, the capacitor has been recharged and current is stopped momentarily.

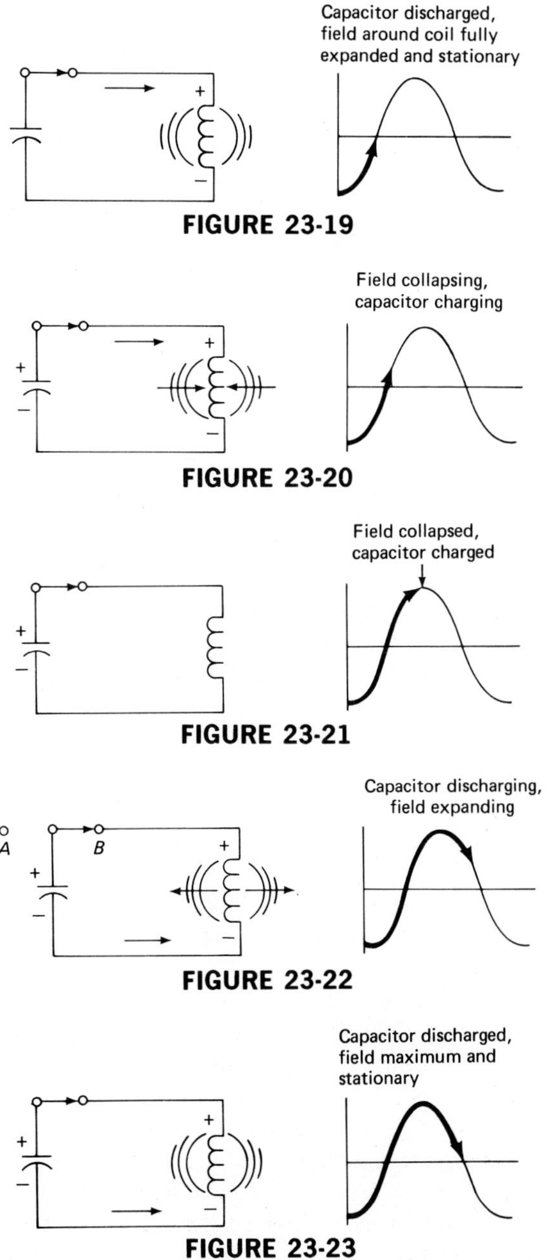

FIGURE 23-19

FIGURE 23-20

FIGURE 23-21

FIGURE 23-22

FIGURE 23-23

If this circuit were working in a vacuum, with zero resistance, it should oscillate indefinitely. But nothing is perfect, and the resistance of the circuit causes the power to dissipate over a period of time. In fact, there is a gradual decrease in the amplitude of each sine wave, with the result that over a period of time all of the energy is dissipated as heat. The number of times per second that the circuit goes through the complete sequence (Figures 23-17 to 23-25), is the *natural resonant frequency* of that particular tank circuit.

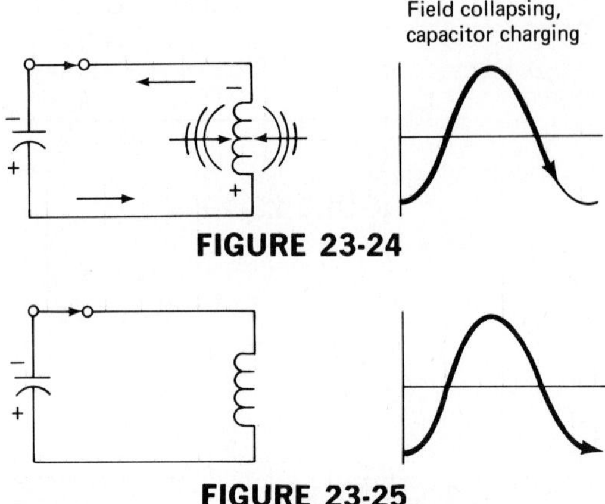

FIGURE 23-24

FIGURE 23-25

A parallel tank circuit resonates at its natural frequency after the tank has been charged. To maintain resonance, it is only necessary to recharge the capacitor periodically. When connected to an alternating current source, however, it resonates when the applied frequency equals the tank's natural resonant frequency.

Each wave decreases in size, because of the heat dissipated by circuit resistance. The reduction in amplitude leads to the conclusion that the sine wave has been *damped.* The damped wave shown in Figure 23-26 is a result of the dissipation of stored power during the period. These waveshapes can be viewed with an oscilloscope.

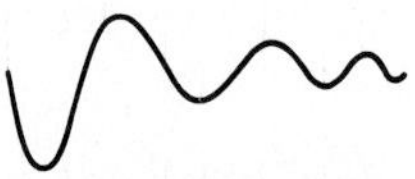

FIGURE 23-26

We know that if two batteries of equal voltage are connected so that their potential opposes, current will be zero (Figure 23-27). Notice that a lamp is connected between the two negative poles; the potential at each pole is -12 V. When a voltmeter is connected across the lamp, it indicates 0 V. A difference in potential is a requirement for current flow. Since 0 V is present, 0 A flows.

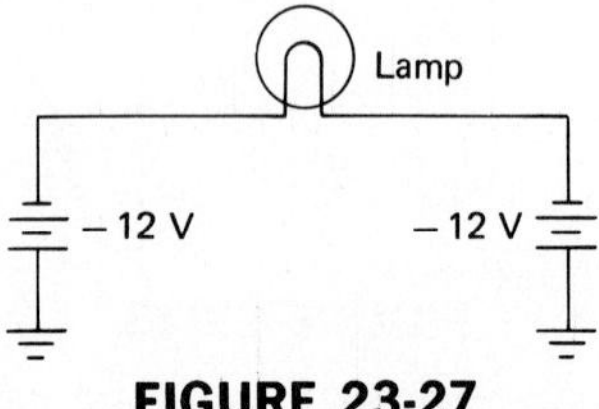

FIGURE 23-27

The result would be identical if the two power sources were AC generators with the same frequency and amplitude. Refer to the circuit in Figure 23-28. A voltmeter connected across this lamp would also read 0 V, and no current would flow. In the resonant circuit, however, the tank's sine wave is damped, or attenuated. Figure 23-29 depicts the two sine waves present in a parallel-resonant, AC-operated circuit. If these sine waves were used to replace those shown in Figure 23-28, the result would be different. Because the tank sine wave is attenuated, a small potential would be applied to the lamp, and a very small current would flow.

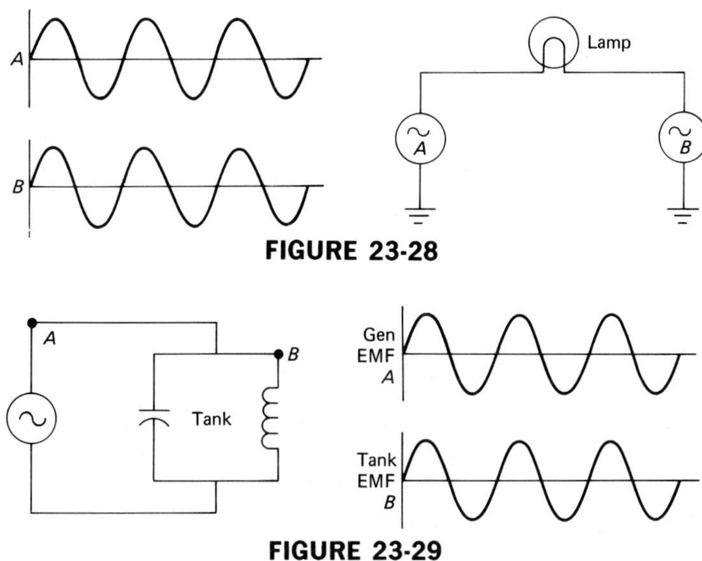

FIGURE 23-28

FIGURE 23-29

For current to be held to a small amount, impedance of the circuit must be very high.

Impedance in a parallel tank circuit is maximum.

Self-Check

Answer each item by indicating whether the statement is true or false.

26. ____ The equation used to calculate resonant frequency is the same for series circuits and parallel circuits.

27. ____ Line current in a parallel-resonant circuit is minimum.

28. ____ In the parallel-resonant circuit, $I_C = I_L$.

29. ____ The parallel tank circuit can resonate from power supplied by a DC source.

30. ___ The energy stored by the capacitor is transferred to the coil, and vice versa, in a parallel tank circuit.

Above Resonant Frequency

Recall that when the frequency applied to reactive components is increased, the following occurs:

X_C decreases, and X_L increases.

As the frequency applied to a resonant RCL circuit is increased, X_C decreases and X_L increases. Since the reference for a circuit of this type is current, the capacitor can conduct more current than the coil. Thus, when the circuit is operated *above resonance,* it acts *capacitively*.

Refer to Figure 23-30a. With the frequency above resonance, the reactance values are as shown. X_C offers 50 Ω of opposition, with 2 A of current; X_L offers 10 kΩ and has 10 mA of current flow. Because the capacitor current is strong enough to more than cancel the inductor current, the circuit acts capacitively.

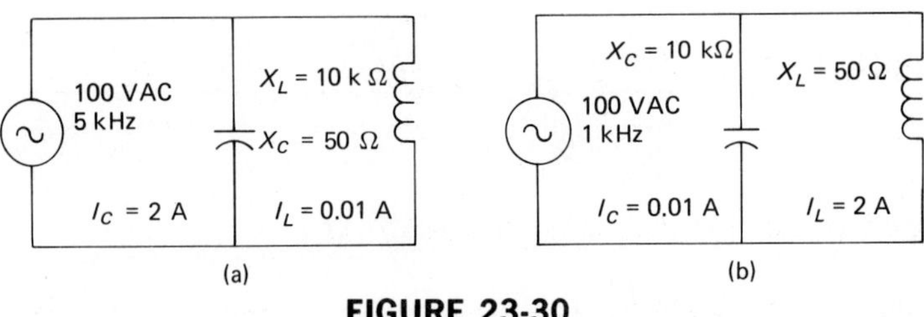

FIGURE 23-30

Below Resonant Frequency

The opposite is true when frequency is *below resonance*. In the circuit in Figure 23-30b, $X_C = 10$ kΩ and $X_L = 50$ Ω. Current flow in the coil is 2 A; in the capacitor, current is 10 mA. Because the coil has the highest current, the effect of capacitance is cancelled, and the circuit operates *inductively*.

Voltage and Current Relationships

With the resonant frequency applied by the power source, the tank circuit offers high impedance to the line current. Therefore, line current is minimum.

Keep in mind that in a parallel RCL circuit, current leads voltage by 90° in a capacitor; in a coil, current lags voltage by 90°. At resonance, these two currents are equal and opposite. When the capacitor is discharging, the discharge current flows through the coil. The coil stores energy in its electromagnetic field. When the capacitor is fully discharged, the field around the coil collapses, causing current to flow in the capacitor. This current causes energy to be stored in the capacitor's electrostatic field.

The tank circuit has a small amount of resistance due to the wire used in its construction. This resistance dissipates power (energy), which means a small amount of line current is allowed to flow to replenish power dissipated in the form of heat ($I^2 \times R$). If the line current is minimum at resonance, impedance must be maximum.

Self-Check

Answer each item by indicating whether the statement is true or false.

31. ___ When a parallel RCL circuit operates below resonance, it is operating inductively.

32. ___ The wire resistance of a parallel tank circuit causes the output wave to be damped.

33. ___ Current leads voltage by 90° in the capacitor in a parallel-resonant circuit.

34. ___ As frequency increases in a parallel circuit, X_L increases.

35. ___ When a parallel RCL circuit operates above resonance, it is operating inductively.

The Effect of Changing Component Size

To illustrate the effect of changing the component size, we change only one component at a time. Recall that:

$$I_C = \frac{E_t}{X_C}$$

$$I_L = \frac{E_t}{X_L}$$

$$f_r = \frac{1}{2\pi\sqrt{LC}}$$

$$X_L = 2\pi fL$$

$$X_C = \frac{1}{2\pi fC}$$

$$Z_t = \frac{E_t}{I_t}$$

In the case of circuits operating at resonance, changes in circuit characteristics have specific effects on the operation of the circuit (Table 23-3).

TABLE 23-3
Circuit at Resonance

Increases in	*Current*	*Impedance*	*Phase Angle*
Frequency	INC	DEC	INC
Resistance	DEC	INC	No change
Capacitance	INC	DEC	INC
Inductance	INC	DEC	INC

A decrease in either frequency or component value has the opposite effect to that shown.

When the circuit is operating below its resonant frequency, these changes differ noticeably. Examine Table 23-4.

TABLE 23-4
Below Resonance

Increases in	*Current*	*Impedance*	*Phase Angle*
Frequency	DEC	INC	DEC
Resistance	DEC	INC	INC
Capacitance	DEC	INC	DEC
Inductance	DEC	INC	DEC

A decrease in either frequency or component value has the opposite effect to that shown.

In circuits operating above resonance, the effect is quite different. Table 23-5 presents those effects.

TABLE 23-5
Above Resonance

Increases in	*Current*	*Impedance*	*Phase Angle*
Frequency	INC	DEC	INC
Resistance	DEC	INC	INC
Capacitance	INC	DEC	INC
Inductance	INC	DEC	INC

Q Factor, BW, and BP for Parallel RCL Circuits

Quality, bandwidth, and bandpass are the characteristics of a parallel circuit of interest to us. The circuit values suitable for use in calculating Q are given in Figure 23-31. (*Note:* The resistor shown in series with the coil represents wire resistance contained in the tank circuit.) In this circuit, we see the same relationship as that in series RCL circuits.

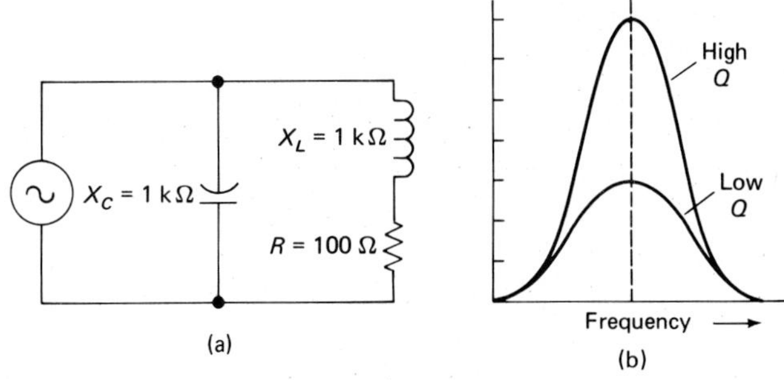

FIGURE 23-31

The Q of a parallel resonant tank is the ratio of inductive reactance to resistance.

In equation form, this is stated:

$$Q = \frac{X_L}{R}$$

When the resistance of a *parallel tank* is low, the Q is high. Remember, however, that Q is affected by changes in frequency. What is the Q of the tank shown in Figure 23-31a?

Mathematical solution:

$$Q = \frac{X_L}{R}$$

$$Q = \frac{1000}{100}$$

$$Q = 10$$

Any number 10 or higher is a high Q.

When a coil has a high Q, it also has low resistance. Examination of the Q equation reveals that coils with low resistance have a higher Q. If coil resistance is high, Q is low (Figure 23-31b).

Up to now, we have considered only the resistance in the tank circuit during the capacitor-coil action. Figure 23-32 shows a resistor in series with the generator. This represents resistance in the path of "line current." In considering this resistance, we notice that the circuit has become a series-parallel circuit and that the voltage across each branch of the tank is no longer equal to E_t. The voltage drop on the tank becomes

$$E_{\text{tank}} = E_t - E_{\text{Rseries}}$$

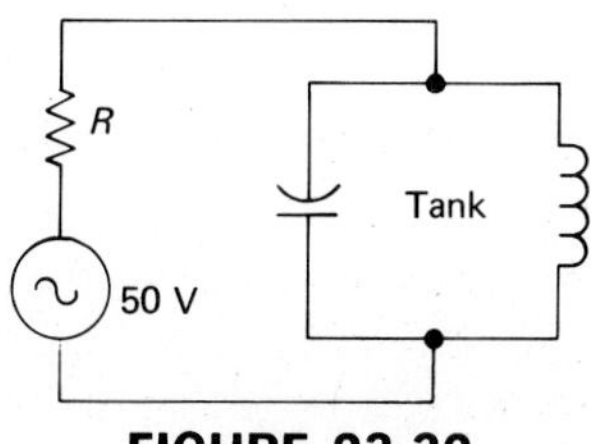

FIGURE 23-32

When the input frequency is varied from below resonance to above resonance, the voltage curve for the parallel tank resembles the curve shown in Figure 23-33.

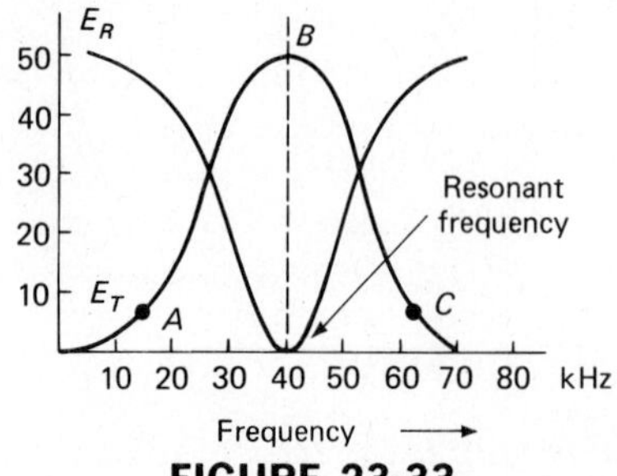

FIGURE 23-33

We assume that the variable power supply has an output of 50 V and that its frequency can be varied from 0 to 100 kHz. Below resonance, X_C is larger than X_L and I_C is smaller than I_L. Below resonance, the circuit acts like a resistor and a coil in series. As we approach resonant frequency, I_L continuously grows smaller and I_C increases. Above resonance, I_C is larger than I_L. This circuit behaves like a series RC circuit. As frequency increases, I_C continually increases, dropping more voltage on the series resistor.

When plotting the voltage curves shown in Figure 23-33, keep in mind that voltages add vectorially and that their sum is 50 V. At resonant frequency (point B), I_C and I_L cancel, and the voltages can be added directly. At resonance (40 kHz), the voltage drop (E_R) across the resistor is minimum.

The voltage response curve plotted in Figure 23-33 can be used to calculate bandwidth and bandpass for the circuit. The half-power points are found by:

$$HPP = 0.707 \times E_{pk}$$

$$HPP = 0.707 \times 50$$

$$HPP = 35.35 \text{ V}$$

Reminder: Power is 50% of peak power at each half-power point.

We locate the two half-power points on the graph by marking the 35.35-V positions on the E_t curve of the voltage plot. We can determine the frequency by drawing a line from each point to the frequency scale directly below.

Note, in Figure 23-34, that the half-power points have been located and the perpendiculars drawn to intersect the frequency line of the graph. To determine the bandwidth, we use:

$$BW = f_{UHPP} - f_{LHPP}$$

$$BW = 52 \text{ kHz} - 28 \text{ kHz}$$

$$BW = 24 \text{ kHz}$$

$$BP = \text{all frequencies between 28 and 52 kHz}$$

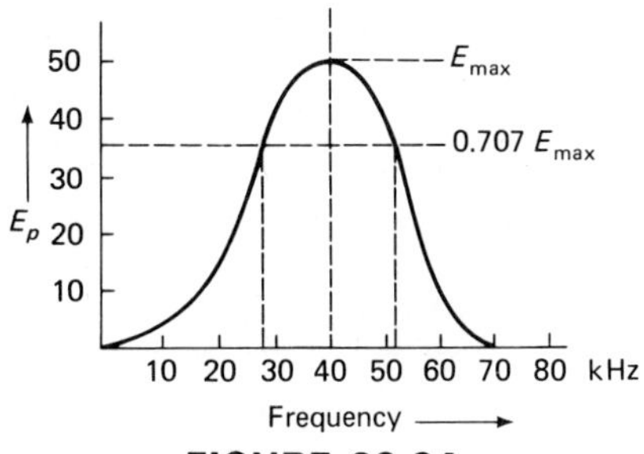

FIGURE 23-34

We can also use the impedance curve to compute BW and BP. In this method, we use 0.707 × Z_{max} to determine the half-power points. These points are located on the curve, and the perpendiculars are dropped to intersect the frequency (reference) line. In Figure 23-35, the frequency intersects are 1200 and 2200 Hz. Subtracting the numbers reveals a BW of 1 kHz. The BP contains all frequencies between 1200 and 2200 Hz.

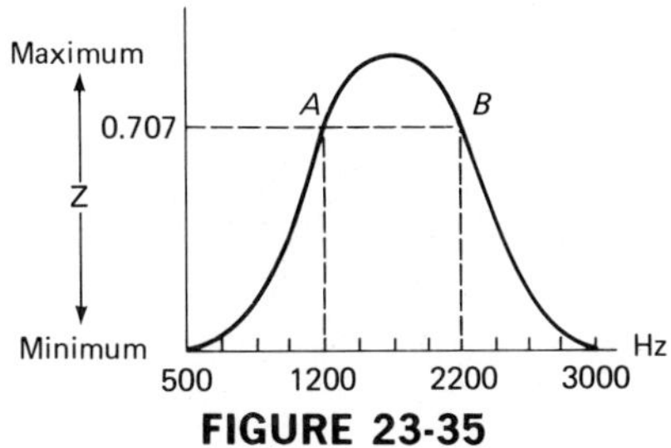

FIGURE 23-35

Three branch parallel-resonant circuits differ from the two branch circuits we have been discussing, because they contain a fixed parallel-resistive branch. This circuit has three possible current paths (Figure 23-36a). The LC network in the circuit resonates at the same frequency as it would if it were a two-branch network with no parallel resistance. The difference is in the amount of current flowing through the resistive branch. Notice that no resistance is shown in series with the coil. The vector diagram for this circuit is shown in Figure 23-36b. When I_C equals I_L, the circuit appears to be purely resistive; its impedance value equals resistance. To illustrate the characteristics of this type of circuit, let us analyze the circuit in Figure 23-37a.

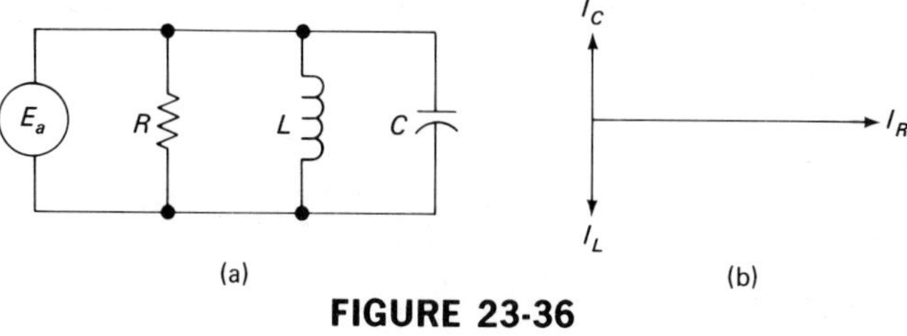

FIGURE 23-36

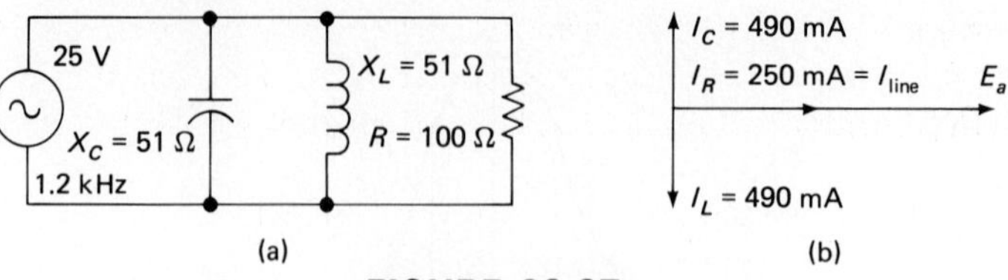

FIGURE 23-37

Given Values	Solve for
(a) $E_t = 25$ V	$I_C =$
(b) $X_C = 51\ \Omega$	$I_L =$
(c) $f_r = 1.2$ kHz	$Q =$
(d) $R = 100\ \Omega$	$I_{line} =$
(e) $X_L = 51\ \Omega$	$I_{tank} =$

Solutions are:

$$I_C = \frac{E_t}{X_C}$$

$$I_C = \frac{25\text{ V}}{51\ \Omega}$$

$$I_C = 490\text{ mA}$$

$$I_L = \frac{E_t}{X_L}$$

$$I_L = \frac{25\text{ V}}{51\ \Omega}$$

$$I_L = 490\text{ mA}$$

$$I_R = \frac{E_t}{R}$$

$$I_R = \frac{25\text{ V}}{100\ \Omega}$$

$$I_R = 250\text{ mA}$$

To determine the circuit current, plot the vector diagram shown in Figure 23-37b. Reactive currents are equal and opposite; thus they cancel. At resonance, circuit current equals the resistor current of 250 mA:

$$I_{line} = I_R = 250\text{ mA}$$

The term *tank current* describes the current that revolves inside the tank when energy is being transferred back and forth between the coil and the capacitor. To calculate this current (I_{tank}), we calculate either reactive current, since only one component supplies current at any given moment. Therefore,

$$I_{tank} = \frac{E_t}{X_L}$$

$$I_{tank} = \frac{25\text{ V}}{51\ \Omega}$$

$$I_{tank} = 490\text{ mA}$$

The Q of this circuit can be solved by one of the equations that follow. Note that in this circuit, reactance and resistance are in a position opposite that of the circuits already studied.

$$Q = \frac{R}{X_L}$$

$$Q = \frac{100\ \Omega}{51\ \Omega}$$

$$Q = 1.96$$

$$Q = \frac{I_{tank}}{I_{line}}$$

$$Q = \frac{490\text{ mA}}{250\text{ mA}}$$

$$Q = 1.96$$

Bandwidth and bandpass are applied to the three branch RCL circuits. The equation for bandwidth is the same as that for series resonant circuits:

$$BW = \frac{f_r}{Q}$$

$$BW = \frac{1200}{1.96}$$

$$BW = 612\text{ Hz}$$

Increasing the parallel resistance causes the line current to decrease. As the parallel resistance approaches infinity, the line current approaches zero. As resistance increases, the bandwidth decreases and circuit selectivity increases. The use of three branch resonant circuits allows us to adjust the bandwidth, selectivity, and bandpass by changing the parallel resistance. The inverse relationship between resistance and bandwidth is illustrated in Figure 23-38.

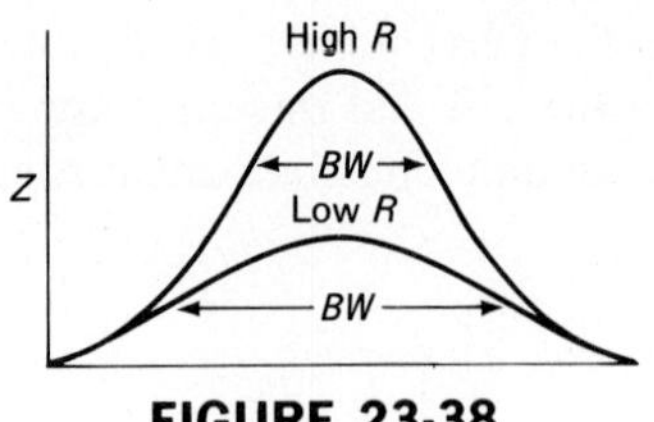

FIGURE 23-38

Self-Check

Answer each item by indicating whether the statement is true or false.

36. ___ Increasing series resistance in a parallel-resonant circuit causes the phase angle to increase.

37. ___ Increasing frequency in a parallel-resonant circuit causes the line current to increase.

38. ___ Increasing capacitance in a parallel-resonant circuit causes the line current to decrease.

39. ___ A parallel RCL circuit is operating *below resonance*. Decreasing inductance causes line current to increase.

40. ___ A parallel RCL circuit is operating *above resonance*. One way to tune it to resonant frequency is to decrease its inductance.

41. ___ The RCL circuit is below resonance. Increasing capacitance causes its operating frequency to approach resonant frequency.

42. ___ The Q of a two-branch circuit is calculated using the same equation as is used for a three-branch circuit.

43. ___ Bandwidth is calculated by adding the frequencies of the two half-power points.

44. ___ Parallel resistance and Q are directly proportional in three-branch RCL circuits.

45. ___ In a parallel-resonant circuit, impedance is maximum.

Summary: Comparison of Series and Parallel Resonant Circuits

It is important to understand the relationship of series-resonant circuits and parallel-resonant circuits. To better understand this relationship, consider Table 23-6.

TABLE 23-6
Resonant Circuit Comparisons

Series Resonance	*Parallel Resonance*
$f_r = \dfrac{1}{2\pi\sqrt{LC}}$	$f_r = \dfrac{1}{2\pi\sqrt{LC}}$
$X_L = X_C$	$I_L = I_C$
Z is minimum	Z is maximum
I_{line} is maximum	I_{line} is minimum
Bandwidth depends on Q	Bandwidth depends on Q

Let us review the current curves for each circuit. Current curves are illustrated in Figure 23-39. In Table 23-6, we see that in a series circuit, I_{line} is maximum and that in a parallel circuit, I_{line} is minimum.

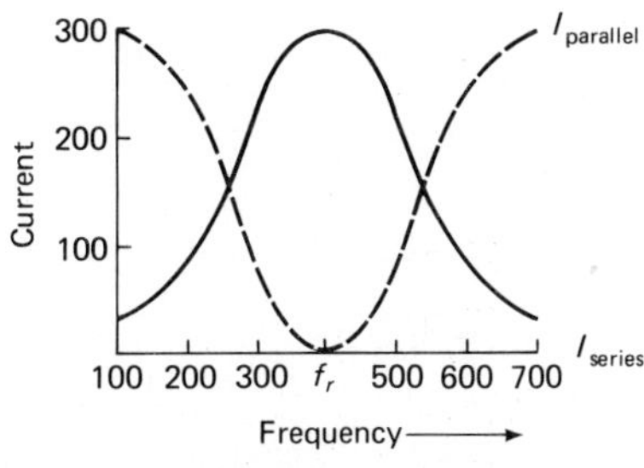

FIGURE 23-39

The diagram in Figure 23-40 is a comparison of impedance curves for the two types of circuit. Note that in all cases, the curves are opposite.

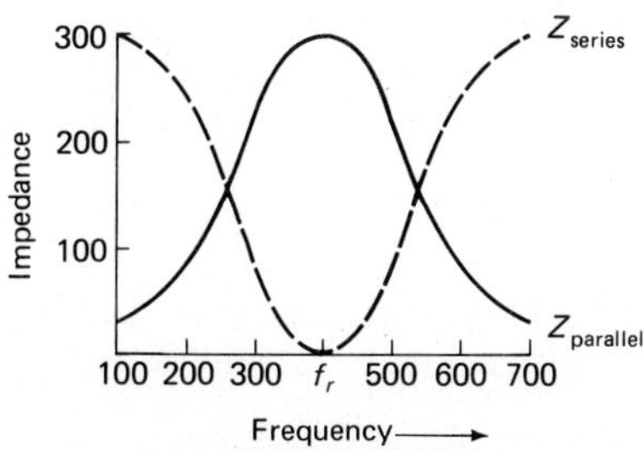

FIGURE 23-40

Review Questions and/or Problems

1. In a series circuit, increasing either L or C causes resonant frequency to
   (a) increase.
   (b) decrease.
   (c) remain the same.

2. A series-resonant circuit contains $X_L = 1\ \text{k}\Omega$ and $R = 5\ \Omega$. What is the circuit's Q?
   (a) 0.0005
   (b) 5
   (c) 200
   (d) 1000

3. As a circuit is tuned nearer resonance, the phase angle
   (a) increases.
   (b) decreases.
   (c) remains the same.
   (d) approaches 90°.

4. The bandwidth of a circuit is calculated by adding the frequencies at upper and lower half-power points.
   (a) true
   (b) false

5. To increase the Q of a series RCL circuit, we
   (a) increase the resistance.
   (b) decrease the resistance.
   (c) increase the frequency.
   (d) decrease the frequency.

6. Bandpass is the sum of all frequencies between the half-power points.
   (a) true
   (b) false

7. A circuit with good selectivity has a wide bandwidth.
   (a) true
   (b) false

8. Increasing frequency in a resonant circuit causes
   (a) X_C to increase.
   (b) X_L to increase.
   (c) f_r to increase.
   (d) None of these

9. At the resonant frequency, the Z of a parallel circuit is
   (a) minimum; line current is maximum.
   (b) maximum; line current is maximum.
   (c) minimum; line current is minimum.
   (d) maximum; line current is minimum.

10. The impedance of a series RCL circuit is
   (a) maximum at resonance.
   (b) minimum at resonance.
   (c) equal to X_L.
   (d) equal to X_C.

11. The current in a parallel RCL circuit is
   (a) maximum at resonance.
   (b) minimum at resonance.
   (c) equal to I_C.
   (d) equal to I_L.

12. A parallel circuit is operating at resonance; line current can be increased by
   (a) increasing frequency.
   (b) decreasing resistance.
   (c) decreasing frequency.
   (d) All are correct.

13. In a series circuit at resonance,
   (a) $Z = R$.
   (b) I_t is maximum.
   (c) $X_C = X_L$.
   (d) All are correct

14. Given the circuit in Figure 23-41, solve for frequency at resonance.
   (a) 4 kHz
   (b) 40 kHz
   (c) 400 Hz
   (d) 40 Hz

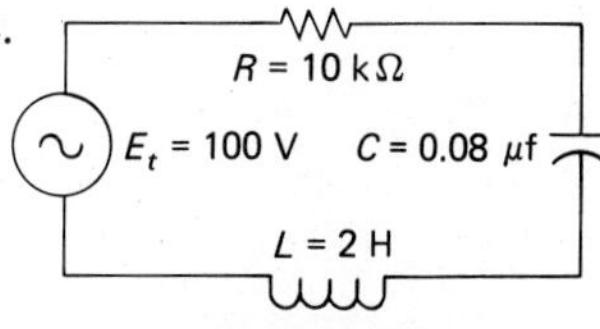

FIGURE 23-41

15. Given the circuit shown in Figure 23-42, solve for frequency at resonance.
   (a) 125 Hz
   (b) 12.5 Hz
   (c) 1250 Hz
   (d) 12.5 kHz

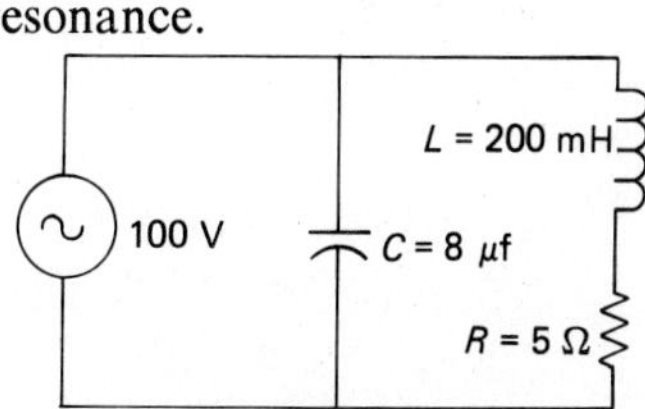

FIGURE 23-42

24

Filters and Coupling

Introduction

The principles involved in RC, RL, and RCL circuits can be used to ensure passage or rejection of a band of frequencies. First, we must know what these circuits are and why they are used. The circuits used to select the band of frequencies to be passed or blocked are called *filters;* circuits used to make sure unwanted frequencies are not coupled from one circuit to another are *coupling circuits.* By selective use of capacitors and coils of specific size, we can couple desired frequencies from one circuit to another or ensure that unwanted frequencies are not coupled. A filter may include several capacitors and coils, whereas a coupling circuit may use a conductor, resistor, capacitor, coil, or transformer.

Usually, both types of circuits depend for their operation on the effect of X_C and X_L at selected frequencies. Filters make use of the variations of inductive and capacitive reactance which changes with frequency. The variation of impedance in a series RCL circuit can be used to pass or reject certain frequency bands. *Bandpass* is the range over which frequencies are passed freely. The *attenuation band* is the range across which frequency passage is poor. The frequency at which attenuation begins is called *cutoff frequency.* An audio transformer is a good example; it is designed to pass all frequencies up to 20 kHz. Its cutoff frequency must be at least 20 kHz. Filter circuits designed to block, or attenuate, a specific band of frequencies are called *band-reject filters.*

Coupling may be direct, resistive, capacitive, or inductive. Later in this chapter we discuss direct coupling, RC coupling, impedance coupling, and transformer coupling.

Objectives

Each student is required to:

1. Define:
   (a) Low-pass filter.
   (b) High-pass filter.

(c) Bandpass filter.
(d) Band-reject filter.
(e) Direct coupling.
(f) RC coupling.
(g) LC coupling.
(h) Transformer coupling.

2. Calculate the cutoff frequency for a low-pass filter.
3. Describe the operation of each circuit listed in number 1.

Filter Circuits

Before we go any further, we need to review the basic principles of frequency response. Capacitors and inductors react differently to changes in frequency. The equations for capacitive and inductive reactance are:

$$X_C = \frac{1}{2\pi fC}$$

$$X_L = 2\pi fL$$

When frequency is increased, X_C decreases, but X_L increases. As frequency increases, the capacitor approaches zero opposition (a short) and the inductor approaches infinite opposition (an open). The opposite occurs when frequency is decreased: X_C increases (approaches an open) and X_L decreases (approaches a short). See Figure 24-1.

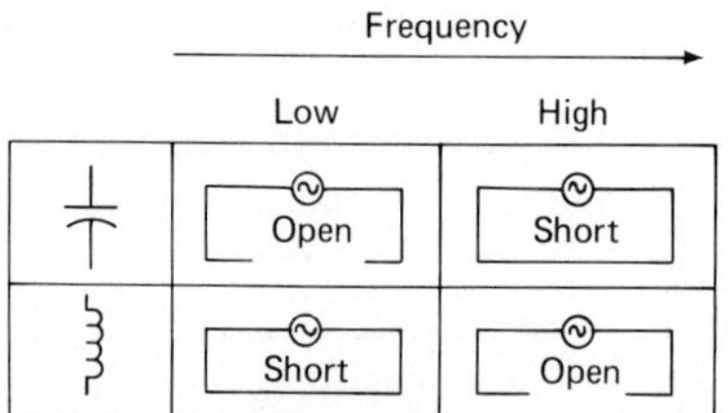

FIGURE 24-1

Refer to Figure 24-2. Circuit a is a series RC circuit where the capacitor is in series with the output. At low frequencies, the capacitor blocks the output; but at high frequencies all of the input is allowed to pass to the output. Circuit b is a series RL circuit where the inductor is in series with the output. At all low frequencies, the coil has low opposition and all of the input is passed to the output; at high frequencies, however, X_L is high and blocks passage of the input. Circuit c is a series LC circuit where X_C is in parallel with the output and X_L is in series with the output. At low frequencies, X_C is high and the input frequency is coupled to the output. At these same frequencies, the coil has low opposition; since it is in series with the input, it blocks a minimum of input. At high frequencies, X_L is very large and blocks much of the input signal. Because X_L is in series with the load, most of the input is blocked before it reaches the output. For the small signal that may pass through the coil, X_C is very low and shorts the remaining signal to ground. The result is that circuit c makes an excellent *low-pass* filter for passing frequencies up to a given cutoff frequency. Circuit d is a series LC circuit with X_L in parallel with the load and X_C in series. At low frequencies, X_C is high and blocks the input from the output. At low frequencies, X_L is very low, shorting any signal that arrives at its top to ground. To high frequencies, X_C is low and shorts the signal to the output, whereas X_L is large and develops all the signal for coupling to the output. Circuit d is a *high-pass* filter.

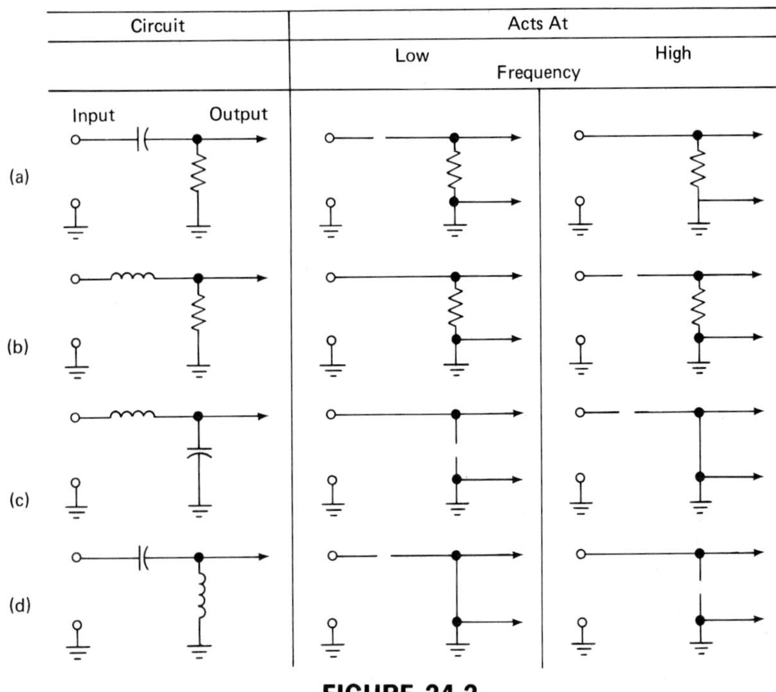

FIGURE 24-2

How, then, can we use a series-resonant circuit as a filter? Examine Figure 24-3. Each circuit in the figure represents a series-resonant circuit. At resonance, E_C and E_L are equal but opposite (180° out of phase). Impedance (Z) is minimum because X_C and X_L are equal, and cancel. This means that current is maximum. If voltage is measured across the resistor, we find that $E_R = E_t$. Tuning the frequency to either side of resonance causes X_C and X_L to no longer be equal. With one reactance larger than the other, the result of subtraction shows that a reactance is in series with the resistor. Now the reactance drops part of the applied voltage; causing the voltage drop on the resistor to decrease (represented by the response curve in Figure 24-4).

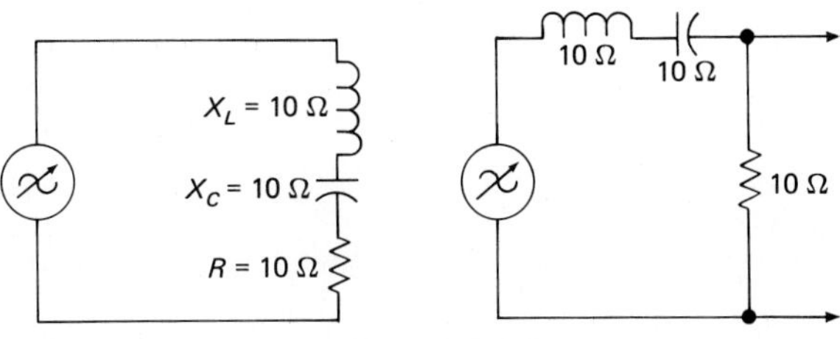

FIGURE 24-3

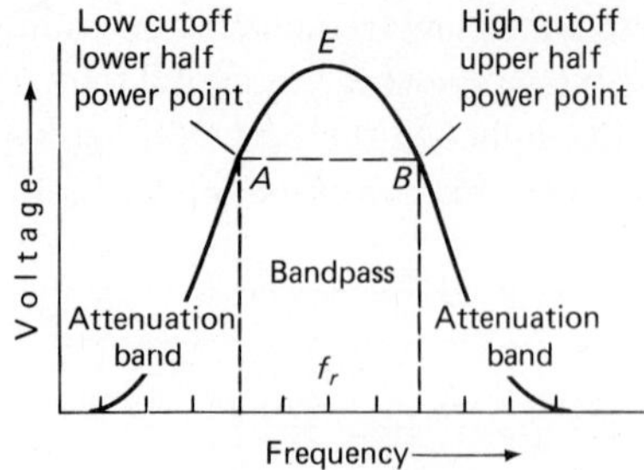

FIGURE 24-4

In Figure 24-4, the changes in voltage drop across the resistor are plotted. The curve represents a *bandpass filter;* frequencies between the half-power points are passed to the next circuit. All other frequencies are attenuated, or "filtered out," and are not passed. Each half-power point is a *cutoff frequency.*

If we design a circuit like the one shown in Figure 24-5a, and take the output across the coil and capacitor, the response curve will resemble that in Figure 24-5b. In this configuration, the circuit is used as a *band-reject filter.* Band-reject filters are designed to reject a specific band of frequencies and pass all others. Frequencies between half-power points occur when the combined opposition of X_C and X_L is very low, causing the signals to short around the output, to ground. To frequencies both above and below the half-power points, impedance is higher; an increasing proportion of their voltage is dropped across the part of the circuit that is in parallel with the output. This type of circuit is used extensively in biomedical equipment to remove voltage variations caused by the 60 cycle commercial power the instruments use.

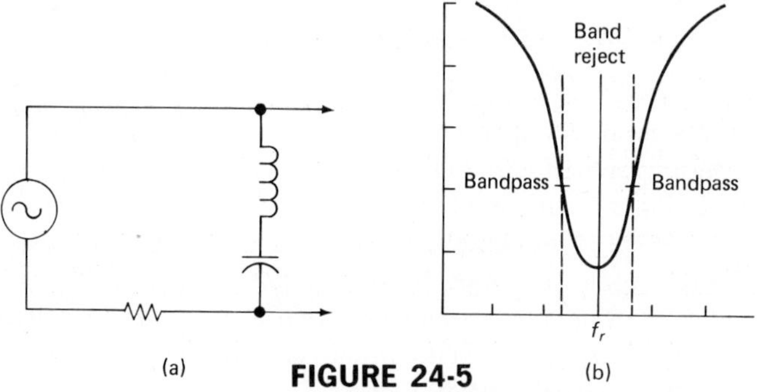

FIGURE 24-5

Another important filter is the *low-pass filter.* This circuit does just what its name implies: passes low frequencies and rejects high ones. Sample circuits are presented in Figure 24-6.

In Figure 24-6, circuit a is a series LC circuit with the capacitor in parallel with the output. At low frequencies, X_C is high and X_L low. This results in most of the input being coupled to the output. At high frequencies, X_C is low and X_L high. The coil tends to block the input, and the capacitor tends to short the input to ground. This causes the filter to pass low frequencies and reject high ones. Circuit b is a series RC circuit with the capacitor in parallel with the output. It operates much like the circuit above, except that R is constant and does not change with frequency. At high frequencies, X_C is low, and the signal is shorted to ground. Circuit c is a series RL circuit where the coil is in series with the output. At low frequencies, X_L is low, and much of the input is dropped across the resistor. Because R is in parallel with the load, E_R and E_L must be equal. At high frequencies, X_L is high and drops most of the input signal. That leaves little if any voltage to be dropped across the resistor and output. All three circuits are low-pass filters.

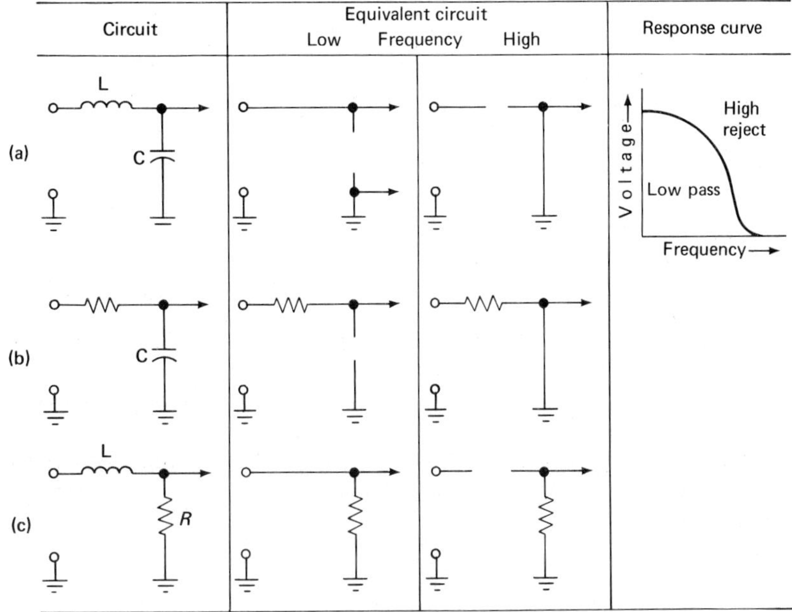

FIGURE 24-6

The other type is a *high-pass filter* (Figure 24-7). It is designed to pass high frequencies and reject low ones. Its design and operation are opposite that of low-pass filters.

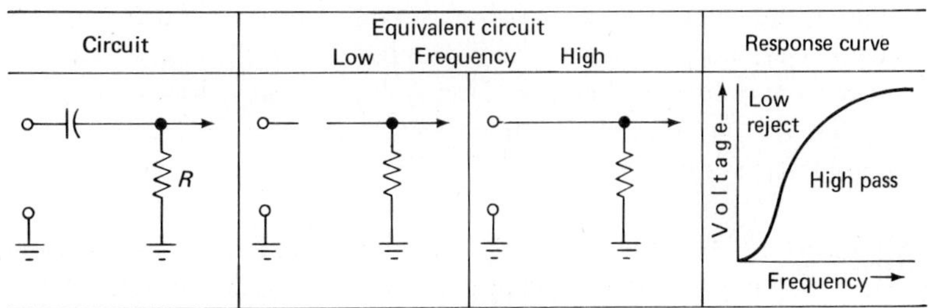

FIGURE 24-7

In this series RC circuit, the capacitor is in series with the output. At low frequencies X_C is large and drops most of the input voltage, which leaves little to be dropped across the resistor and load. At high frequencies, X_C is a short, and all the input is dropped across the resistor and the load.

Remember that the circuits we have discussed are merely representative of what happens; the actual circuits are often much more complex. Filters are constructed in the following forms, L-section, T-section, and pi-section. They attentuate undesirable frequencies and improve system operation.

Low-Pass Filters

Figure 24-8 contains schematic drawings for pi-, T-, and L-section, low-pass filters. The series arm of this type of filter contains inductance; the parallel arm contains capacitance. As frequency increases, X_L increases and X_C decreases. Note the response curve in Figure 24-8d, where the circuit is characterized by a gradual cutoff. To determine the cutoff frequency for this circuit, we use:

$$f_{co} = \frac{1}{\pi \sqrt{LC}}$$

Regardless of the type of filter, the amounts of capacitance and inductance for a specific cutoff frequency are the same. Therefore, when these components are included in two sections, their totals are taken into account when selecting component sizes.

Remember:

1. Capacitors in series act like resistors in parallel; capacitors in parallel act like resistors in parallel.
2. Coils in series act like resistors in series; coils in parallel act like resistors in parallel.

The LC low-pass filter easily passes all frequencies below the cutoff frequency. It also attenuates all frequencies above the cutoff point.

Figure 24-8a shows the L-section, LC, and low-pass filter. In this circuit, L and C form a frequency-sensitive voltage divider. At low frequencies, the series reactance X_L is low and the parallel reactance X_C is high. The capacitor drops most of the input voltage causing it to be passed to the output. At high frequencies, most of the voltage is dropped across the coil, which allows only a small voltage to be passed to the output. Starting at a low frequency, and gradually increasing the frequency, we find that the output remains fairly constant until the cutoff frequency is reached. Above cutoff, output decreases rapidly.

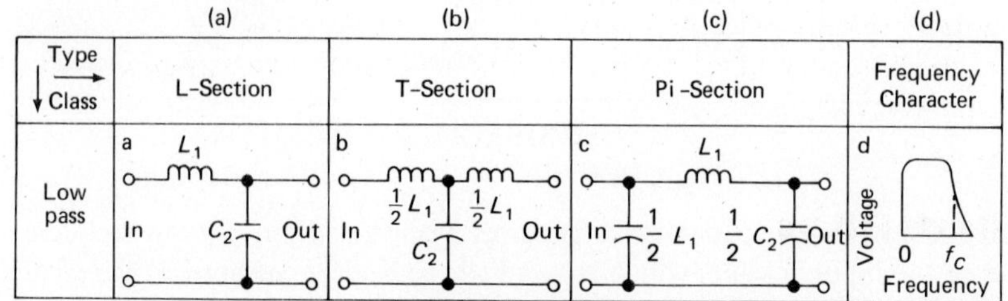

FIGURE 24-8

Figure 24-8b contains a T-section, LC, low-pass filter. The inductance is divided into two coils, and the capacitor is in parallel with half of the inductance and load. At low frequencies, X_L is small and X_C large. As frequency increases, X_L increases and X_C decreases, resulting in a low-pass filter. The response curve for this circuit is shown in Figure 24-8d.

Figure 24-8c is an illustration of a pi-section filter. Here, the coil divides capacitance in two parts. High frequencies meet a low X_C at the first half of the capacitance and a high X_L in the series coil. Both act to reject frequencies above cutoff. The second half of the capacitance effectively shunts to ground high frequencies that make it through the coil. The advantages of using pi- and T-section filters is that they can be connected to the circuit from either end and have the same effect.

An L-section filter must be connected so the capacitor is between the output and the ground. When we view this type of filter from the coil end, we see that it has a high impedance to high frequencies; when we look into the filter from the capacitor end, we see a low impedance to high frequencies. In the case of pi- and T-section filters, the impedance to high frequencies is the same from either end.

High-Pass Filters

Figure 24-9 illustrates L-, T-, and pi-section, high-pass filters. This family of filters is used to pass all frequencies higher than cutoff frequency.

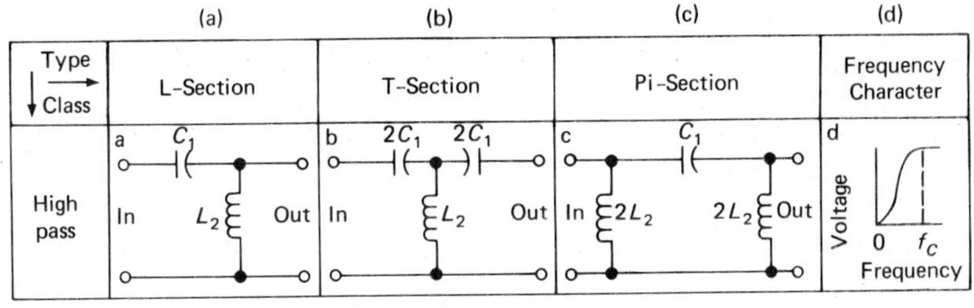

FIGURE 24-9

The filters have the same configuration as those used as low-pass filters, the only difference being that the positions of L and C have been reversed. The capacitance, in series with the output, acts as a short to high frequencies. The coil, which shunts the output, has a high impedance to high frequencies, and that, in turn, results in the voltage for high frequencies being dropped across the coil in parallel with the load. The response curve is shown in Figure 24-9d.

The T-section, high-pass filter is shown in Figure 24-9b. Notice that, in this circuit, capacitance is twice the capacitance used in either the L-section or the pi-section filters. Capacitors in series act like resistors in parallel; therefore, to have the same amount of capacitance as we do in the other two, we must use two capacitors with twice the value.

In the case of the pi-section filter, to have an L_t equal to the value of L used in Figure 24-9a, the inductance of each coil is doubled.

Bandpass Filters

L-, T-, and pi-section bandpass filters are shown in Figure 24-10, their response curve in Figure 24-10d.

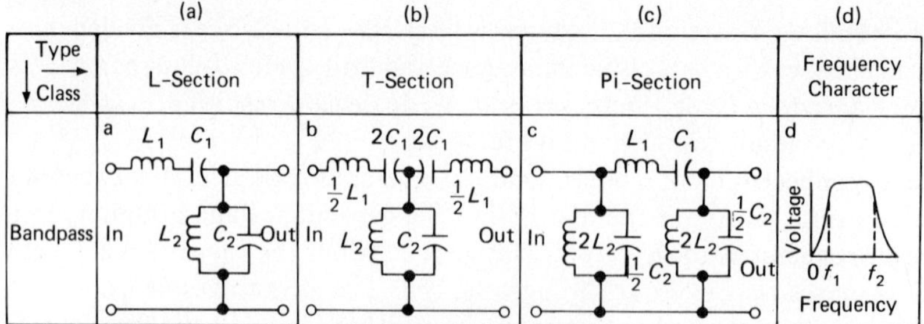

FIGURE 24-10

In Figure 24-10a, notice that the schematic represents an L-section bandpass filter. L_1 and C_1 form a series-resonant circuit, whereas L_2 and C_2 form a parallel-resonant circuit. Component sizes have been selected so each circuit resonates at the same frequency.

At resonance, the series circuit offers minimum impedance to the frequency; but the parallel circuit offers maximum impedance: the maximum signal is passed to the load. At all other times, the impedance of the two circuits is such that the output is less than it is at resonance. Upper and lower cutoff frequencies are listed as f_1 and f_2. These points coincide with the half-power points and are determined by the size of the circuits Q.

The pi- and T-section filters illustrated in Figures 24-10b and 24-10c are bandpass filters that operate the same way that the L-section filter already discussed operates. The difference in their operation lies in the way values for L and C are selected. Also, these filters react to frequency identically, regardless of the input end.

Band-Reject Filters

Band-reject filters are illustrated by Figure 24-11. Figure 24-11d shows the frequency-response curve for circuits of this type.

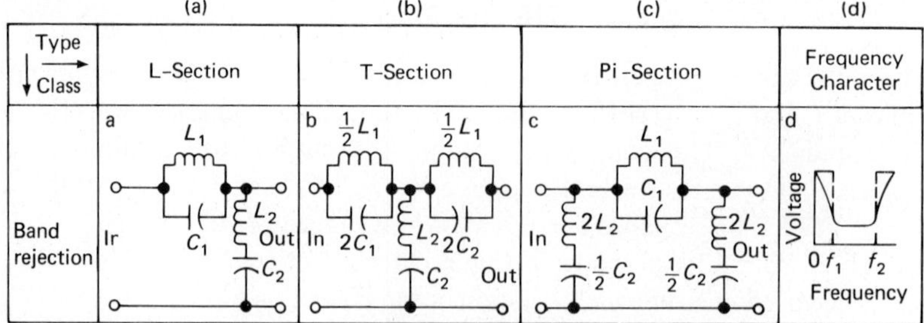

FIGURE 24-11

Figure 24-11a represents an L-section band-reject filter. The filters are designed to reject a specific frequency band. Again, two resonant circuits with the same resonant frequency are included. They differ in that the parallel circuit in the series path offers maximum opposition at resonance. Should any current make it through the parallel tank, the series RCL circuit is at minimum impedance and will short the remaining signal to ground.

Referring to the frequency response curve, we see that the band-reject filter also has two cutoff frequencies. Frequencies below the lower cutoff frequency, as well as above the upper cutoff frequency, are easily passed to the output.

Self-Check

Answer each item by indicating whether the statement is true or false.

1. ____ A series-connected capacitor blocks high frequencies.
2. ____ A series-resonant circuit has minimum impedance at resonance.
3. ____ A coil connected in parallel with the output shunts high frequencies to ground.
4. ____ A band-reject filter passes all frequencies between its upper and lower cutoff frequencies.
5. ____ A coil connected in series with the output can be used to block low frequencies.
6. ____ A high-pass filter passes frequencies above its cutoff frequency.
7. ____ A capacitor connected in shunt with the output shorts low frequencies to ground.
8. ____ A low-pass filter attenuates all frequencies below its cutoff frequency.
9. ____ A bandpass filter passes all frequencies that lie between its cutoff frequencies.
10. ____ A parallel-resonant circuit blocks the output when connected in parallel with the output.

Coupling Circuits

Coupling circuits have much in common with the filters already discussed. Some also depend on the effect of changing X_C and X_L to accomplish their purpose, while others do not.

Direct Coupling

Direct coupling (Figure 24-12a) can be as simple as using a conductor to connect two circuits. This results in a *direct* path for current flow between the two circuits. Direct coupling can provide an exact reproduction of the input signal at the output of the coupling circuit. This exact reproduction, called *high fidelity,* is desirable in such circuits as stereo amplifiers. Direct coupling, however, also connects the DC voltage of both circuits. In many cases, this can be a disadvantage.

Figure 24-12b shows a type of direct coupling in which a resistor is used in the coupling path. This resistor, in series with the signal path, also limits the current flow between the two circuits. The coupling resistor reduces the amplitude of an input signal. If the resistor alone were used, it would act as a current limiter, and we could say the signal is *resistance-coupled.*

Direct coupling can be used over a wide range of frequencies, starting at 0 Hz. Because the coupling circuit contains no capacitance or inductance, it does not cause a phase shift of the incoming signal. This circuit is considered "resistive" only; except for the current and voltage effects of the resistor, its operation is identical to that of the straight conductor.

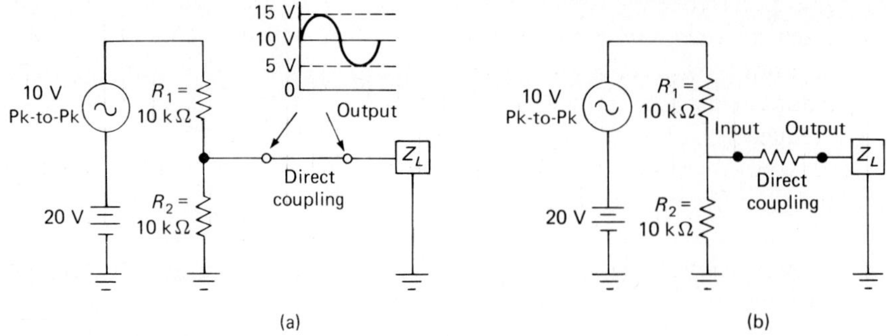

FIGURE 24-12

RC Coupling

Refer to Figure 24-13a. This circuit contains resistance-capacitance (RC) coupling (within the dashed lines). The capacitor is connected so it blocks DC current from one circuit to another. Input to the circuit is a 10-V, peak-to-peak signal with a +10-V reference. Notice that the output of the coupling circuit is 90° out of phase with the input when it reaches Z_L. Actually, the phase difference is less than this; but we use 90° for purposes of explanation. Remember that this is the effect of an RC series circuit.

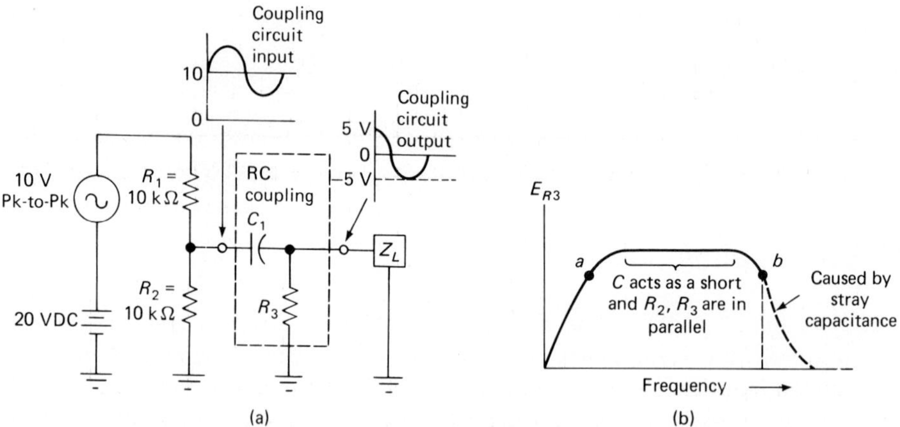

FIGURE 24-13

How does frequency affect this circuit? On one hand, the higher the frequency, the lower X_C becomes. The circuit operates more resistively, which places R_2 and R_3 in parallel. R_1, R_2, and R_3 serve as a series-parallel voltage divider between input signal and DC voltage. If R_1 and R_2 are equal, each voltage is cut approximately in half, as shown by the input. The point at which the coupling capacitor is considered a short is point *a* on the response curve of Figure 24-13b.

On the other hand, low frequencies are blocked by high X_C, and they fall below the cutoff frequency. At some high frequency (point b on the drawing), the response curve again drops back to zero. This is caused by *stray capacitance.* Remember that capacitance is present any time two conductors are separated by a dielectric. Since rubber, plastic, air, and many other materials are dielectrics, capacitance exists throughout a circuit. One conductor of this capacitance is the ground circuit. At high frequencies, the stray capacitance has an X_C of 0 Ω, at which point the entire signal is shorted to ground by stray capacitance. The range of frequencies passed by RC coupling is controlled by the amount of fixed and stray capacitance present.

At points *a* and *b* X_C equals *R*. This causes *R* to drop exactly half of the input voltage. These points are also the half-power points, and the circuit acts like a bandpass circuit with a very wide frequency band. With proper design, RC coupling can be used with very low frequencies or with radio frequencies.

Impedance (LC) Coupling

The impedance (LC) coupling circuit is suitable for operation at all RF frequencies. It is possible to extend the upper-frequency cutoff point considerably by using a coil in place of the resistor in the RC coupling circuit (Figure 24-14).

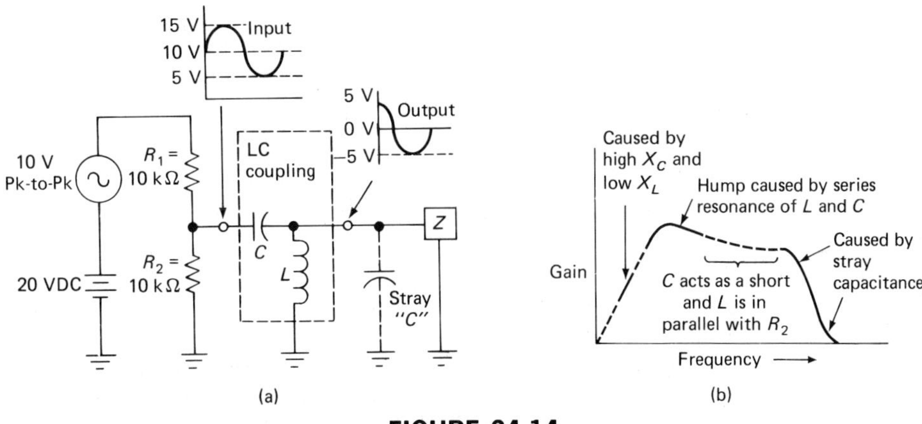

FIGURE 24-14

At extremely low frequencies, the capacitor acts as an open, but the coil acts as a short. Therefore, X_C prevents the input signal from being applied to *Z*. The half-power point occurs when $X_C = X_L$. At the point where *L* and *C* are resonant, it is possible that output is greater than input. This depends on the *Q* of the coil. As frequency continues to increase, however, the output drops back and remains fairly stable.

At some high frequency, X_C becomes zero and the capacitor acts as a short. At this point, X_L is high and is in parallel with R_2. As the frequency continues to increase, so does X_L. Since X_L is in parallel with R_2, the equivalent impedance of R_2 and X_L increases, causing a larger voltage to be applied to *Z*. Maintenance of a higher voltage causes the frequency response to be extended higher than with the RC coupling circuit.

Figure 24-14b is a typical response curve obtained from impedance-coupled circuits. This type of coupling is used in television and other high-frequency equipment.

Because stray capacitance is distributed throughout a circuit, we must continually be aware that upper-frequency signals short to ground. In the case of very high frequencies, such factors as the placement of conductors and the type of chassis determine the upper cutoff frequency of a circuit.

Transformer (Inductive) Coupling

Earlier, we considered three types of transformers: power, audio, and RF. Audio and RF transformers, are used in coupling circuits, to couple energy by mutual induction. Depending on the frequency of operation, a transformer may or may not have an iron core. The primary winding connects to the input signal and the secondary winding to the load. Why would we use transformers for coupling instead of RC or LC circuits?

Figure 24-15 shows two circuits connected by *transformer coupling*. Transformer coupling has features that are not available in other types. We can get a step-up or step-down of the voltage or use the mutual induction present as a block against the DC that may be present in either circuit. We can wind the transformer so the impedance of its windings best suits the circuit within which it operates, and a transformer can be tuned to pass a specific band of frequencies. By winding the transformer to specific impedances, it can be used to *match impedance* between two circuits. Selectivity is much greater with transformer coupling. This type of coupling is used in the intermediate frequency (IF) circuits of radios, television, and radar.

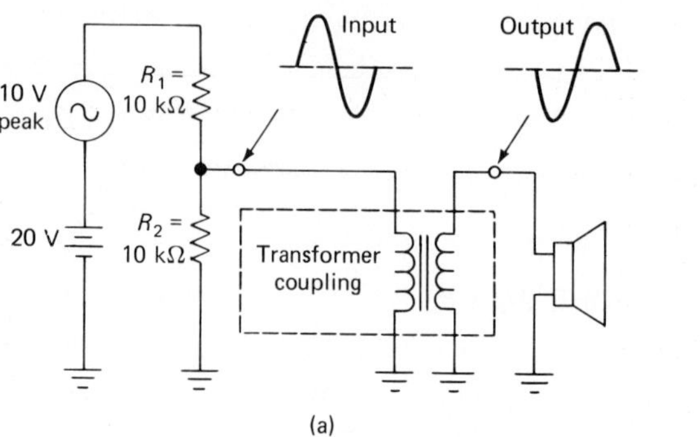

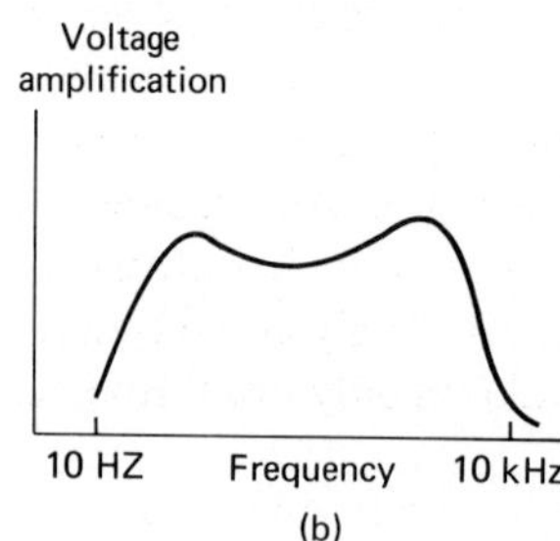

FIGURE 24-15

It has disadvantages, however. For example, the transformer adds cost and weight to the finished product, and frequency response may be poorer when a transformer is used. Additional magnetic shielding is often necessary to protect adjacent circuits from the field built up around the transformer.

Self-Check

Answer each item by indicating whether the statement is true or false.

11. ___ RC coupling provides good selectivity.

12. ___ Transformer coupling can be used to match circuit impedances.

13. ___ LC coupling circuits contain a coil that is in series with the output.

14. ___ Direct coupling provides good high fidelity.

15. ___ RC-coupled circuits have a high-frequency decrease in output, which is caused by stray inductance.

Summary

Filter circuits were discussed in this chapter. A filter circuit is constructed from an arrangement of coils and capacitors. These circuits are designed to pass or reject specific bands of frequency. The types covered were low-pass, high-pass, bandpass, and band-reject.

Coupling circuits were also discussed. They are used to connect two circuits. Through this connection the signal from one circuit is coupled to the input of another circuit. Direct, resistive, RC, impedance, and transformer coupling circuits were discussed, and their advantages and disadvantages were noted.

Review Questions and/or Problems

1. Connecting a capacitor in parallel with a load
   (a) shorts all AC frequencies to ground.
   (b) allows only high frequencies to pass.
   (c) allows only audio frequencies to pass.
   (d) allows only low frequencies to pass.

2. To pass high frequencies to the output, we
   (a) connect a capacitor in parallel with the load.
   (b) connect a coil in series with the load.
   (c) connect a coil in parallel with the load.
   (d) use direct coupling.

3. Pi-section filters can be designed to pass either high or low frequencies.
   (a) true
   (b) false

4. A low-pass filter is designed to pass up to 20 kHz. Which of the following frequencies is attenuated the most?
   (a) 15 kHz
   (b) 0.3 kHz
   (c) 0.4 kHz
   (d) 1 kHz

5. To design a high-pass filter, we must connect the capacitor in series with the load.
   (a) true
   (b) false

6. The cutoff frequency of a filter indicates
   (a) 70.7% of output voltage and 100% of input power.
   (b) 70.7% of input voltage and 50% of input power.
   (c) 50% of input voltage and 70.7% of input power.
   (d) 50% of input voltage and 100% of input power.

7. A bandpass filter has two cutoff frequencies and attenuates all frequencies between the cutoffs.
   (a) true
   (b) false

8. A series-connected series-resonant LC filter passes all frequencies above and below its two cutoff frequencies.
   (a) true
   (b) false

9. A series-connected parallel-resonant LC filter offers high impedance to frequencies above and below the two cutoff frequencies.
   (a) true
   (b) false

10. A high-pass filter receives frequencies from 0 to 100 kHz. Which frequency is attenuated the most?
    (a) 90 kHz
    (b) 0.60 kHz
    (c) 50 kHz
    (d) 40 kHz

11. A band-reject filter is designed by placing
    (a) a parallel-resonant tank circuit in parallel with the load.
    (b) a series-resonant circuit in series with the load.
    (c) a series-resonant circuit in series with a parallel-resonant tank circuit.
    (d) a parallel-resonant tank circuit in series with the load.

12. A T-section filter with two coils connected in series makes an excellent low-pass filter.
    (a) true
    (b) false

13. A band-reject filter rejects all frequencies between its two cutoff frequencies.
    (a) true
    (b) false

14. A series-resonant LC circuit connected in series with the load serves as a
    (a) bandpass filter.
    (b) high-pass filter.
    (c) low-pass filter.
    (d) band-reject filter.

15. A parallel-resonant LC circuit connected in parallel with the load serves as a
    (a) bandpass filter.
    (b) high-pass filter.
    (c) low-pass filter.
    (d) band-reject filter.

25

Generators and Motors

Introduction

Our study of magnetic principles has been quite varied up to this point, but we must still explore two important aspects of magnetism and electricity: motors and generators.

A great majority of the electrical power used worldwide is developed by mechanical generators. In industry and in homes, most mechanical work is done by electric motors. For instance, a refrigerator uses an AC motor; the blower in an automobile's air-conditioning system has a DC motor. Both AC and DC motors are suitable for use in many applications.

In this chapter, we review some important principles of magnetism that we have already covered. Then we explore the areas of AC and DC generators and AC and DC motors.

Objectives

The student is required to:

1. List the three requirements for induction.
2. State Lenz's law and explain its application to generator and motor circuits.
3. When given the position of a magnetic field and the direction of movement a conductor takes through the magnetic field, determine the direction of current flow in the conductor.
4. Explain the function of the following in a generator:
   (a) Pole pieces.
   (b) Field winding.
   (c) Rotating conductor (armature).
   (d) Commutators.
   (e) Brushes.

   - (f) Slip rings.
   - (g) External circuit.
   - (h) Stator.
   - (i) Rotor.
5. List three factors that determine the magnitude of an induced voltage.
6. Label the components used in constructing AC and DC generators.
7. Define the following terms and identify the waveshape for each:
   - (a) Steady DC voltage.
   - (b) Pulsating DC voltage.
   - (c) Ripple voltage.
8. When given the schematic diagram of each type of generator, identify:
   - (a) Series-wound generators.
   - (b) Shunt-wound generators.
   - (c) Compound-wound generators.

Requirements for Induction

In the chapters on inductance and transformers, we saw that a current can be induced in a coil if there is relative motion between a magnet and the coil. The three requirements for induction are:

1. A conductor.
2. A magnetic field.
3. Relative motion between the two.

In the examples cited, motion is caused by inserting or withdrawing a bar magnet in the core area of a coil. This motion can just as easily be caused by making the magnet stationary and moving the coil (a method called *generator action*).

Generating an Induced Voltage and Current

Earlier, we discussed Lenz's law. In defining this law, Lenz stated:

> ***In all cases, the induced current is in such a direction as to oppose the motion that generates it.***

In Figure 25-1 we find two magnets, both of which have the same direction of movement (down). Each magnet is associated with a separate coil of wire. In both examples the magnetic field developed around the coil has its north pole at the top and opposes the field of the bar magnet moving in the coil. The current flow represented by the arrow, however, results in current flow through the resistor in opposite directions—a result of the way the coils are wound. Note that the top winding of coil a passes behind the core, whereas the top winding of coil b passes in front of the core.

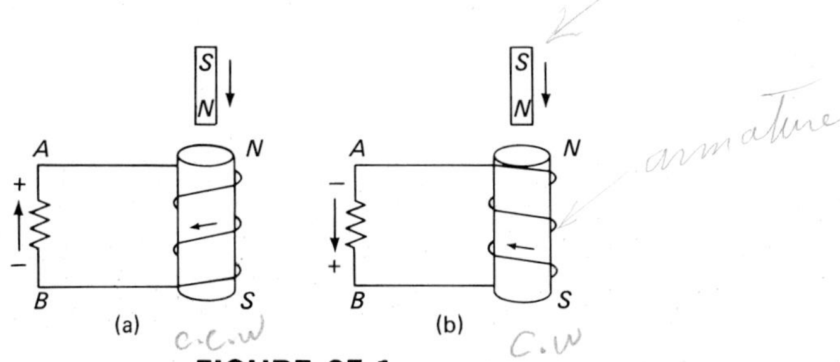

FIGURE 25-1

To determine the direction of current flow in a coil of this type, the following steps are required:

1. Note the magnetic polarity of the bar magnet.
2. Note the direction of movement.
3. Lenz's law states that the induced field opposes the field around the bar magnet. For it to do this, the following must be true:
   (a) In a magnet moving away from the coil, the end of the coil nearest the magnet must attract the coil. Thus its pole must be opposite the magnet's pole. Unlike poles attract, which causes them to oppose the motion of a magnet moving away from the coil.
   (b) In a magnet moving toward the coil, the end of the coil nearest the magnet must have the same polarity as the near end of the magnet. Like poles repel (or oppose) each other, which causes them to oppose the motion of the magnet toward the coil.
4. Once we know the coil's polarity, we can use the left-hand rule to determine the direction of current flow through its windings. The steps for this determination are:
   (a) Use your left hand.
   (b) Point your *left* thumb in the direction of the north pole.
   (c) Grasp the coil in your fingers.
   (d) Your fingers point in the direction current takes in flowing across the face of the coil.
5. Trace the current from the coil to the resistor. Current flows out of the coil, through the resistor, and reenters the coil at the opposite end.

If we reverse the magnet and insert the south pole in the top of the coil, the opposite happens. When we apply the rules just stated, we see that the magnetic polarity of the coil is the same as that shown in Figure 25-2. Note that here, current arrows are opposite their positions in Figure 25-1. In both figures, if we cause the magnets to move in the opposite direction to that shown, the effect is exactly opposite. This results in a reversal of current flow in the load resistor.

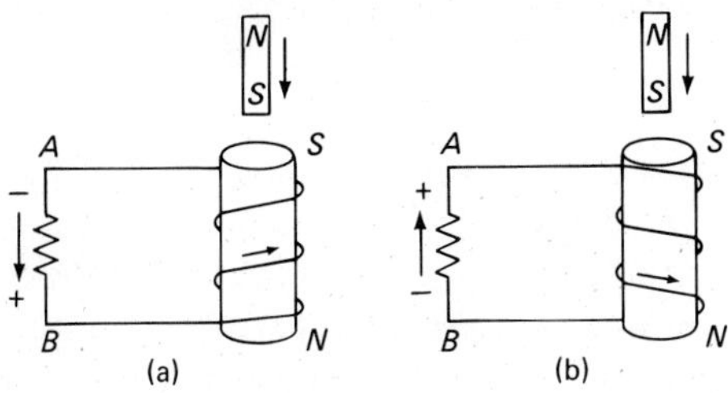

FIGURE 25-2

Direction of Induced Current

Figure 25-3 is an illustration of two fixed magnetic fields with conductors moving into them from opposite directions. Which direction does induced current flow in each conductor?

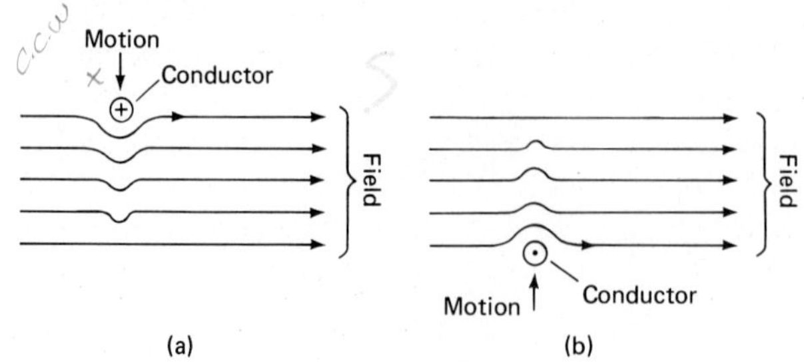

FIGURE 25-3

In Figure 25-3a, notice that the lines of magnetism run from left to right. In the chapter on magnetism, we saw that magnetic lines leave the north pole and return to the south pole. Therefore, the arrows represent the south pole. With conductor motion *downward,* the conductor distorts the magnetic lines downward. By following this depression around the conductor, we can determine the direction of rotation of the induced magnetic field around the conductor. The field rotates counterclockwise. We use the left-hand rule to determine that current is flowing *into the page.*

Figure 25-3b is identical to Figure 25-3a, except that in (b), conductor motion is in the opposite direction. The magnetic lines are distorted in the opposite direction, resulting in a magnetic field that rotates clockwise. By application of the left-hand rule, we determine that current flow in the conductor is *out of the page.*

Note that a change in direction causes induced current to reverse its direction of flow. The number of reversals of direction of movement determine the number of times that current flow reverses.

The conductor in Figure 25-4 moves parallel to the magnetic lines. Since no lines are cut, no current is induced. If we move a conductor at a given speed of rotation through a magnetic field, the current induced depends on the angle of motion between the conductor and the magnetic line (Figure 25-5). A conductor is positioned through four different positions and angles. Because no lines are cut, the conductor's movement from the location shown to point A causes no voltage to be induced. When the conductor is moved from the position shown to point B, two lines of force are cut, and a small voltage is induced in the conductor. If the conductor is moved from the position shown to point C, four lines are cut. A higher voltage is induced by this movement than when it moved to point B. Moving the conductor from its position straight up to point D causes five lines to be cut. This movement causes the largest voltage to be induced in the conductor.

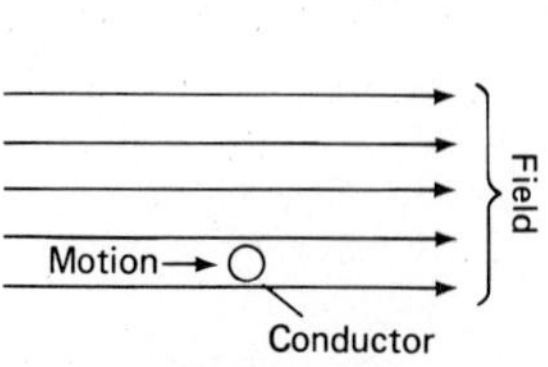

FIGURE 25-4

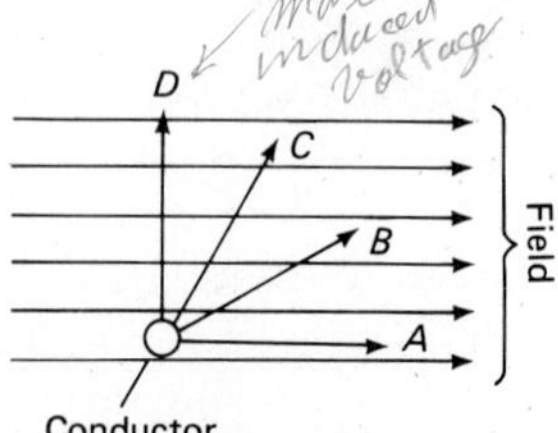

FIGURE 25-5

From this discussion, we see that the greater the number of lines cut, the larger the voltage induced in the conductor. Induced voltage is zero when movement is parallel to the magnetic lines; it is maximum when the conductor is moving at 90° to the field.

Now we analyze what happens when a loop of wire is rotated through the magnetic field. Figure 25-6a–c is used. Conductors A and B represent two ends of the loop. When the loop is positioned at these points, in Figure 25-6a, no lines are cut, and zero voltage is induced. As the loop rotates to positions A2 and B2, it begins to cut lines, until at one point it is moving at 90° to the lines, thus inducing maximum voltage in the conductor. At position A2, current flows into the page, and at B2 it flows out of the page. With a continuous loop behind these points, current coming into the loop flows in at A2 and out at B2. When the loop rotates another 90°, it is positioned at points A3 and B3 (Figure 25-6b). The conductors do not cut lines, and zero voltage is induced. Conductors A and B, however, have changed positions from the start of rotation.

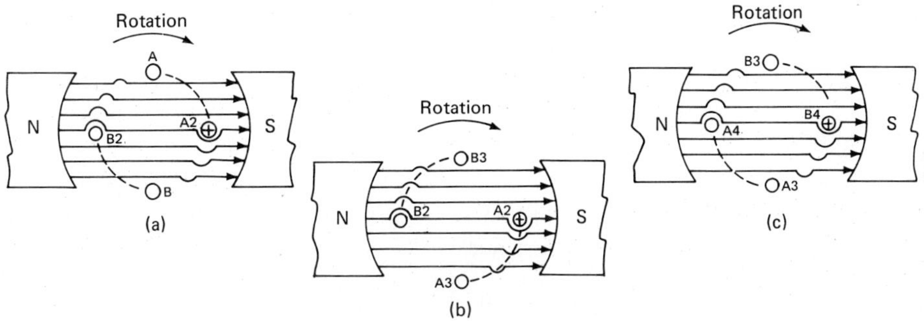

FIGURE 25-6

As the loop continues to rotate (Figure 25-6c), B moves down and A moves up. During this 180° of movement, current flows into the loop through conductor B and out through conductor A.

Here is a review of what happened:

1. When the loop started to rotate, current flowed into A and out of B.
2. As the movement approached 90°, the number of lines cut increased and the current increased to its maximum.
3. The next 90° of rotation found the number of lines being cut reducing to zero, along with a decrease in current, until current stopped flowing.
4. At this point, conductor B is at the top and A is at the bottom.
5. The next 90° of rotation causes conductor B to move down and A to move up. This causes the number of lines cut to increase and current flow in the loop to increase. Current flow in the loop now flows in the opposite direction from its previous direction. Current flows in B and out A.
6. Completion of the 360° rotation again causes the number of lines cut to decrease to zero. Current also drops to zero.

If the loop continues to rotate, this sequence of events is repeated for each 360° of rotation. The output waveshape is illustrated in Figure 25-7a; note that it is a sine wave.

The sine wave has continually changing amplitude and periodic reversals in direction.

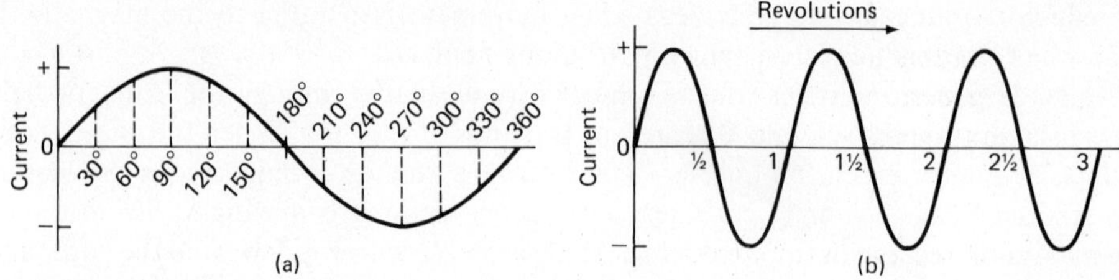

FIGURE 25-7

This output wave matches the description. The output of the device is an alternating current. The frequency of output depends on the loop's rate of rotation. Rotating the loop through three complete revolutions results in the waveshape shown as Figure 25-7b.

Self-Check

Complete each item by inserting the words and/or numbers needed to make a true and complete statement.

1. List three requirements for induction.
   (a) conducta
   (b) magnetic field
   (c) relative motion

2. Lenz's law states that induce current (action → reactions).

3. A sine wave has continually changing amplitude and ________ in current flow.

4. When using the left-hand rule to locate the poles of a coil, your thumb points to the north pole.

5. The magnetic polarity of an energized coil can be determined by using the left hand rule *

Alternating Current Generators.

In introducing the sine wave in Chapter 14, the reason for calling the output of an AC generator a sine wave was explained. Therefore, we do not go into a long explanation here. We do, however, review it briefly.

In a generator, a mechanical device rotates a conductor or group of conductors through a magnetic field. The voltage induced in the conductors is used to supply current to a circuit which does work.

A generator is a device used to convert mechanical energy into electrical energy.

The sine-wave output of an AC generator results from calculation of the *instantaneous* voltage available at the generator output for an infinite number of points during 360° of rotation. The equation for instantaneous voltage is:

$$E_{\text{inst}} = E_{\text{peak}} \times \sin \theta$$

Figure 25-8a–b represents the development of a sine wave. Points are selected along the 360° rotation in Figure 25-8a. Each point is then related to its corresponding point along the output sine wave in Figure 25-8b. The result of the rotating radius vector is a *sine wave.*

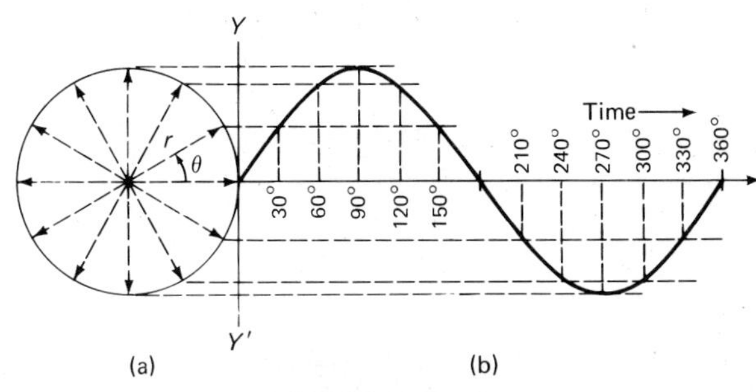

FIGURE 25-8

Factors Determining the Magnitude of Induced Voltage

We have already discussed the effect of a single loop rotating through a fixed magnetic field. The relationship of the rotating radius vector and a rotating conductive loop should now be obvious. When the conductor moves at 90° to the field, the maximum number of lines is cut and peak voltage is induced.

The three things necessary for induction of a voltage are: (1) a conductor, (2) a magnetic field, and (3) relative motion between the two. What happens if we change any one of them? The magnitude of the output voltage can be increased by increasing

1. The number of loops in the conductor.
2. The speed of rotation.
3. The strength of the magnetic field.

Components of a Basic AC Generator

Up to now, we have discussed the generation of AC voltage in simple terms. We should therefore have a basic understanding of how AC voltage is produced and can move on to investigating what an alternating current generator, or *alternator,* consists of.

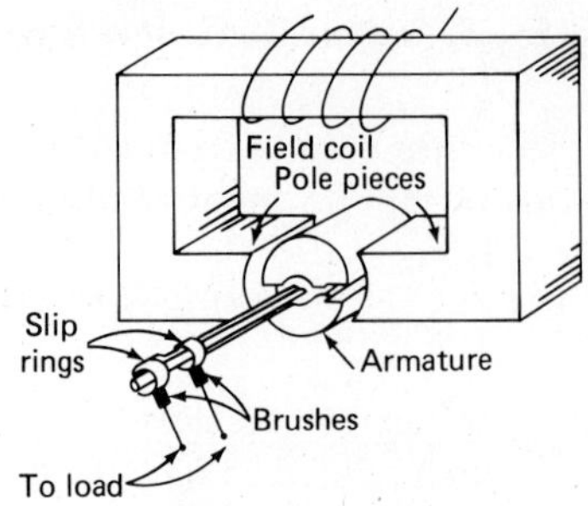

FIGURE 25-9

The basic alternator shown in Figure 25-9 has five parts:

pole pieces
North and south poles of the magnet, used to establish a magnetic field. Although this magnet can be a permanent magnet, it is usually an electromagnet.

field winding
A coil wound around a core of soft iron. The core is shaped so that its ends serve as the poles for the magnetic field. Current flowing in the field winding causes a magnetic field to be created between the pole pieces.

armature
The part containing the conductor loops that are rotated through the magnetic field. The movement of these conductors results in current being induced in the armature winding.

slip rings
The ends of coils wrapped around the armature are connected to slip rings, which are continuous metal bands that rotate with the armature. They serve to couple induced current to an external circuit which uses the current to do useful work.

brushes
Rectangular blocks of conductive material (graphite) used to connect the slip rings to an external circuit. The brushes remain stationary while the slip rings rotate. A curved surface on the brushes is held against the slip rings by spring tension to ensure sufficient contact for current to flow.

The external circuit, while not part of the generator itself, is a critical part of the system. The external circuit receives the induced current from the brushes and conducts it through a resistive device, which performs work. In large generators such as those used to generate household current, this external circuit can be highly complex.

Self-Check

Answer each item by indicating whether the statement is true or false.

6. ____ A sine wave represents one revolution of 360° in an AC generator.

7. ____ The output voltage of an AC generator equals $0.5 \times E_{peak}$ when rotated 30° from zero.

8. ___ Increasing the speed of rotation of an AC generator causes peak voltage output to increase.

9. ___ Commutators are used to connect an AC generator to an external circuit.

10. ___ A generator is a device capable of converting electrical energy into mechanical energy.

Direct Current Generators

We have just completed a brief introduction to the AC generator. Now we do the same thing with the basic DC generator.

In Figure 25-10, we see that a DC generator is similar to an AC generator. In fact, the only difference is that a DC generator uses commutator segments instead of slip rings. Commutator segments serve the same purpose as slip rings: coupling the generator's output to the brushes. Another function of commutator segments is:

Commutator segments switch the connection to the external circuit at the end of each 180° of rotation. This is necessary to cause current in the external circuit to flow in the same direction at all times, something that is necessary for DC.

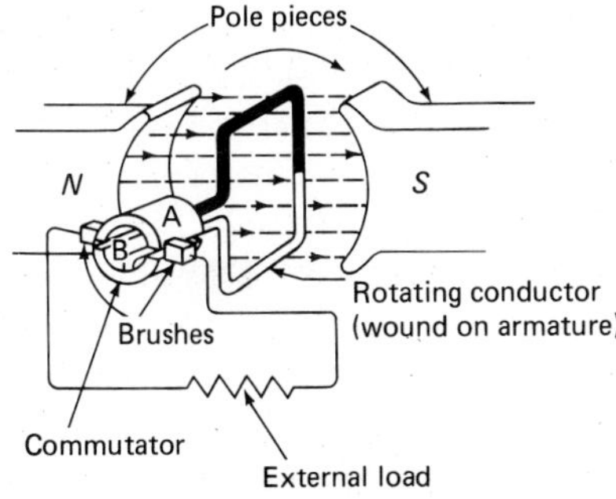

FIGURE 25-10

Pole pieces, field winding, brushes, and armature all serve the same purpose as they do in AC generators. Notice that the brushes are at opposite sides of the commutator. This allows the commutator to reverse the armature connection and maintain the flow of direct current. The external circuit is provided direct current, and the current flows in the load in the same direction at all times.

Types of Direct Current

The three types of DC are:

1. Steady
2. Pulsating
3. Rippling

In Figure 25-10, an armature loop is shown in the position where zero volts are induced. At this time, the brushes rest on the insulating material separating the commutator segments, and no current flows to the external circuit.

As the loop rotates away from this point, the number of magnetic lines cut increases until we have the situation illustrated in Figure 25-11. At that point, the conductors are cutting the maximum number of lines and the induced voltage is maximum; also, the brushes are set to couple the output. Current flows out of the left brush and returns to the right brush while current through the load resistor flows from left to right.

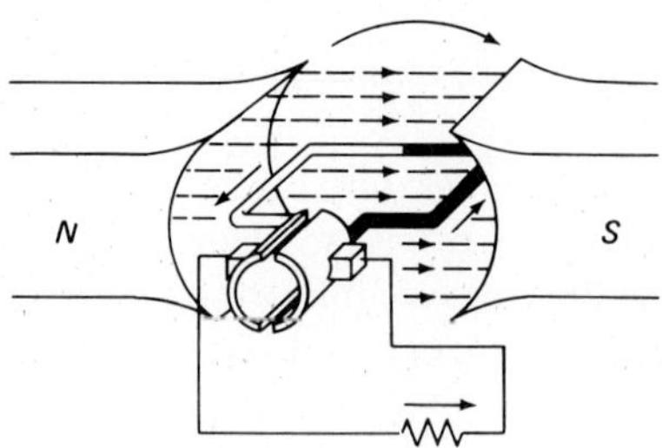

FIGURE 25-11

After a further rotation of 90°, the conductors are situated such that no lines are cut, zero voltage is induced, and the brushes again rest on the insulating material separating the commutator segments. The position of each segment, however, is reversed from the start. When the loop rotates to 270°, the current induced flows out of the left brush, through the load resistor from left to right, and returns through the right brush. When a rotation of 360° has been completed, the commutator is back in its original position: Zero voltage is induced, zero current flows, and the brushes rest between segments.

Because of this action, the voltage applied to the resistor does not reverse; it merely varies between zero and maximum (Figure 25-12).

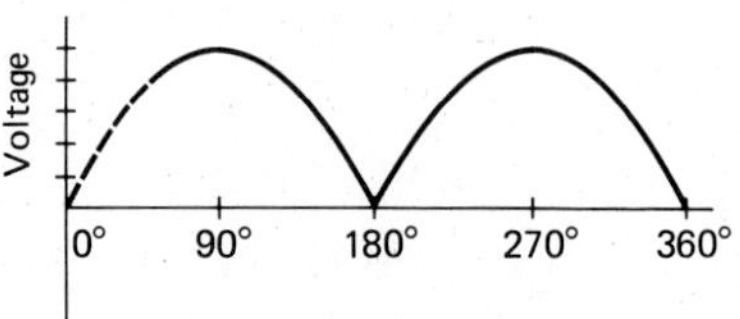

FIGURE 25-12

Voltage with this characteristic is called *pulsating direct current.*

Pulsating direct current voltage is a DC voltage that periodically drops to zero.

Pulsating DC is of little use, since the drop to zero is not desirable; a voltage of the same amplitude at all times is much more useful. How can we improve the output of a DC generator to reduce the pulsating effect?

Figure 25-13 is a DC generator whose armature has two loops. Each loop is exactly 90° from the other. The commutator has also been divided into four segments, with one pair connected to a loop and the other pair to the second loop. Because this armature is rotated through the magnetic field, each loop supplies only half the output voltage.

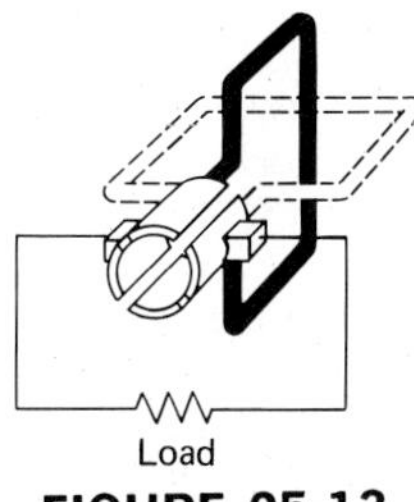

FIGURE 25-13

Figure 25-14a illustrates the output of such an armature. Notice that the voltage never drops to zero. The voltage developed by this generator is called *ripple voltage.* A DC ripple voltage varies but never drops to zero.

Ripple voltage is voltage that has amplitude variations that never drop to zero.

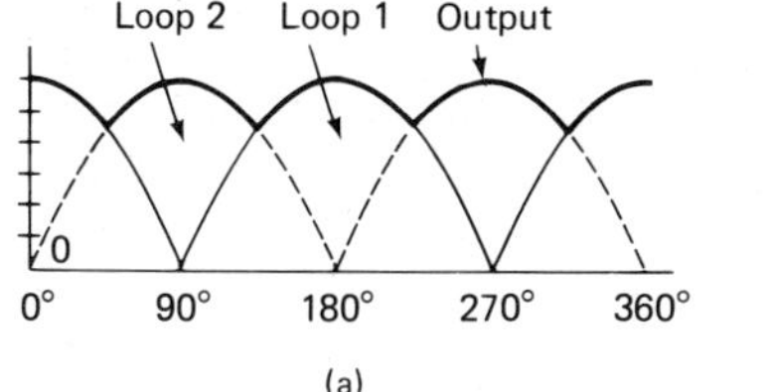

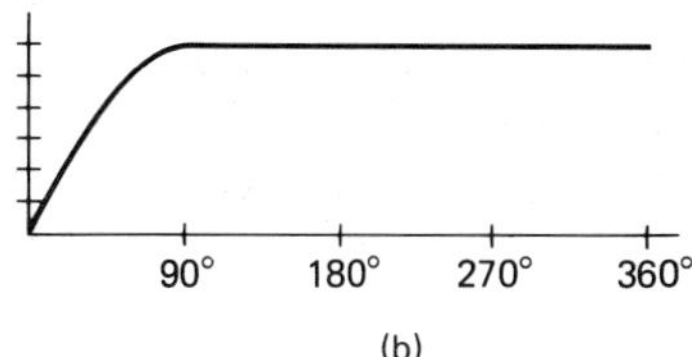

FIGURE 25-14

To produce a DC voltage with smaller variations in amplitude, we simply add more loops to the armature. Each loop must be connected to another pair of commutator segments. For example, a generator with 18 loops has 36 commutator segments. Each loop supplies output voltage for 10° at a time, which means output voltage never drops below 99.6% of peak output voltage. For practical purposes, this output is equivalent to a steady DC such as that shown in Figure 25-14b.

Types of DC Generators

DC generators are classified in one of three categories—series, shunt, and compound. Each class is discussed briefly. In each case, a class depends on the relationship of the field and the armature windings.

Figure 25-15a illustrates a *series-wound generator.* Notice that the field winding is in series with the armature. Another feature of the series-wound generator is the rheostat connected across the field winding. This rheostat is used to regulate field current, which, in turn, controls the output of the generator.

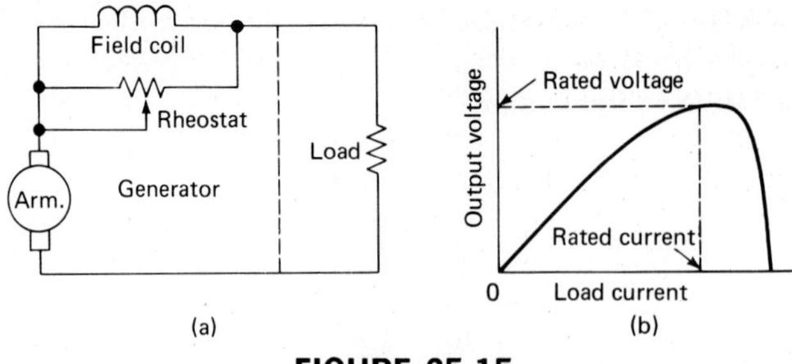

FIGURE 25-15

Field winding and armature are both in series with the load; no current can flow unless the generator is connected to a load. Figure 25-15b illustrates output voltage in contrast to output current. As the load current increases, so does the strength of the magnetic field, and the armature cuts more lines of force. This causes an increase in the output voltage. The rheostat connected across the field shunts part of the load current around the field coil, thus regulating the strength of the magnetic field, and that can be used to control the output voltage.

Figure 25-16a shows a *shunt-wound generator*. This generator has a field coil that is in parallel with the armature and load. Field current can flow even when no load is connected. Figure 25-16b shows that output voltage is maximum when the load current is zero. A rheostat is connected in series with the field coil; this allows limiting the field current to the level desired. By controlling field current, we can limit magnetic field strength and control output voltage. As the load current increases, output voltage decreases.

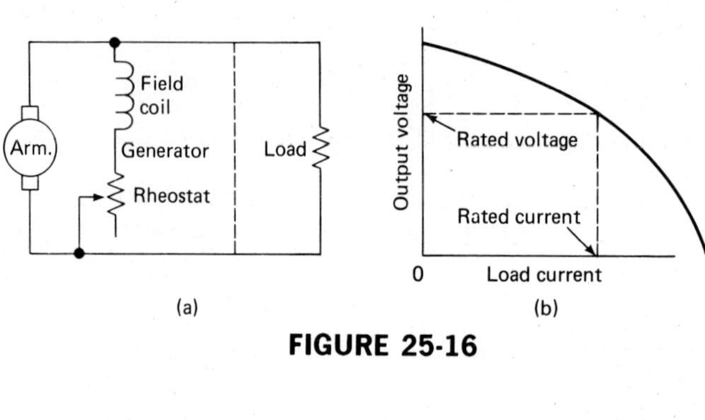

FIGURE 25-16

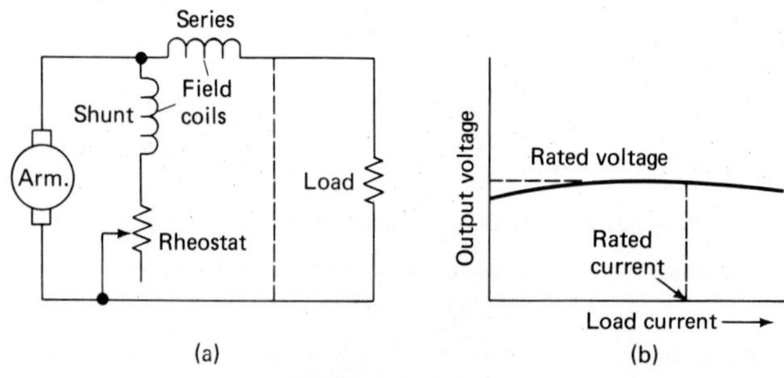

FIGURE 25-17

A *compound-wound generator* is illustrated in Figure 25-17a. Notice that this circuit has two field windings, one series and one parallel. A rheostat is connected in series with the shunt winding, an arrangement that combines the best qualities of both series and shunt fields.

Figure 25-17b illustrates the effect of compound winding on generator-output voltage. As can be seen, load current and output voltage are much better regulated.

Self-Check

Answer each item by indicating whether the statement is true or false.

11. ___ An AC generator uses slip rings to couple an output voltage, but a DC generator uses commutator segments.

12. ___ The field winding of a generator is used to supply power to an external circuit.

13. ___ The output of a DC generator is a pulsating direct current.

14. ___ The armature and field windings of a series-wound generator are connected in series.

15. ___ Adding commutator sections to the armature of a generator increases the peak output voltage of a generator.

Electric Motors

In analyzing generators, we studied a conductor that physically moves through a magnetic field, thus voltage was induced in the conductor. A magnetic field, a conductor, and motion are used to convert mechanical energy into electrical energy.

As an electronics technician, you will not normally be expected to repair motors. You should, however, have a working knowledge of them and their operation. Many characteristics of generators apply to motors as well.

Motors are used to convert electrical energy into mechanical energy.

The action of a motor results from the interaction of two magnetic fields. When these fields interact, a physical motion is produced. If the force exerted by the fields is strong enough, one of the magnets is moved by that force. Remember:

Like magnetic poles repel; unlike magnetic poles attract.

Magnetic lines that move in the same direction, repel;

Magnetic lines that move in the opposite direction, attract.

The latter two predict what happens in *motor action.* A series of diagrams is presented in Figure 25-18 that illustrate motor action. Diagrams a and b represent magnetic fields; diagram b has a conductor resting inside its field, while c shows the effect of current flowing into the page. A counterclockwise magnetic field builds around the conductor. At the top of the conductor, the two fields have lines running in opposite directions; this creates an attracting force. At this same time, lines at the bottom of the conductor are in the same direction, which creates a repelling force. The conductor is forced to move upward. In diagram d, the opposite current flow results in the conductor moving down.

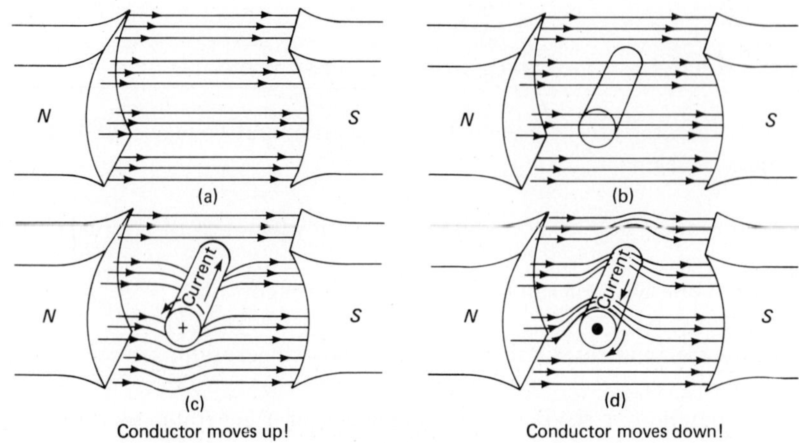

FIGURE 25-18

The interaction of fields is the same as in generators, with the exception that force is generated in a motor, whereas voltage is generated in a generator. The interaction of fields results in:

1. The maximum effect occurs when the conductor is at 90° to the magnetic field.
2. Zero effect occurs when the conductor is parallel to the external magnetic field.
3. Between these two points, the force varies from zero to maximum.

Direct-Current Motors

Figure 25-19 illustrates the components of a basic DC motor: pole pieces, armature, brushes, and commutator. Note that the components are the same as they are in a DC generator, with each component performing the same function it performs in a generator.

Figures 25-20 is an analysis of the same basic motor and the development of *torque* (motor action) that develops in the magnetic field. Diagram A shows the motor in its maximum torque position, with clockwise rotation. Diagrams B and C complete the rotation through 180°, illustrating the reversal of the commutator position, which provides reversal of the DC current applied to the armature. This reversal of polarity causes torque to be applied continually to the armature.

Figure 25-21 illustrates a multiple-coil armature and multiple-commutator segments. The torque developed by the motor is more constant than is the torque of the motor just described. Adding coils and commutator segments improves the motor's operational characteristics. Note Figure 25-22, which shows the main parts of a basic DC motor. This motor has several commutator pairs.

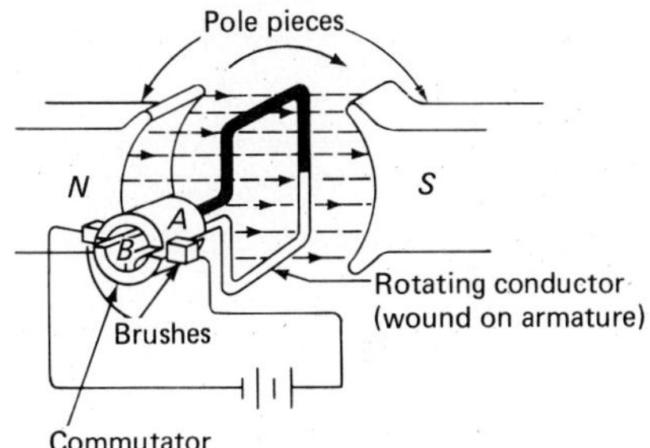

FIGURE 25-19

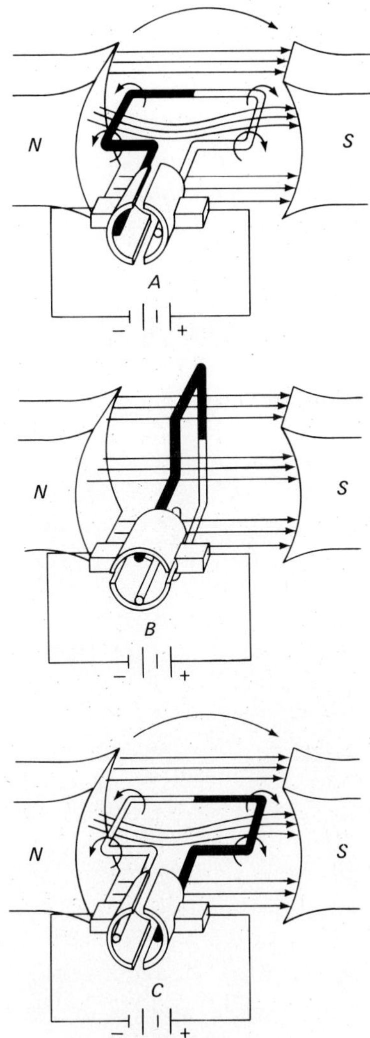

FIGURE 25-20

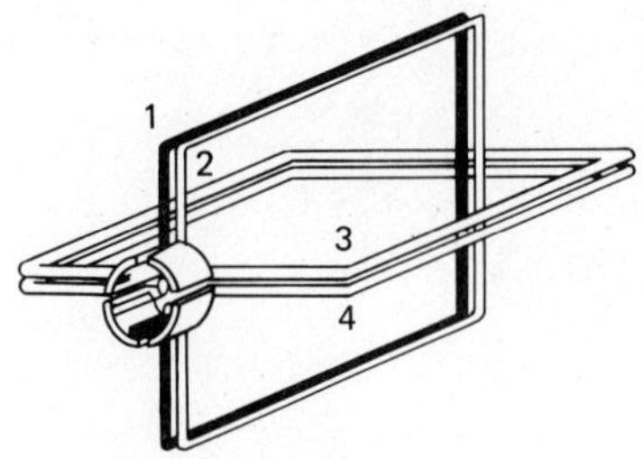

Armature windings simplified

FIGURE 25-21

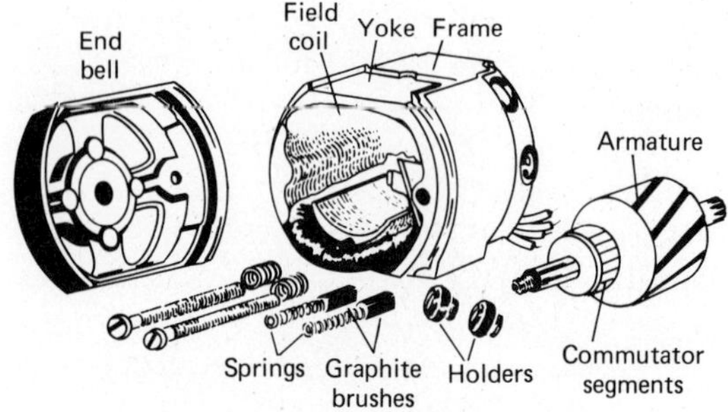

FIGURE 25-22

Alternating-Current Motors

FIGURE 25-23

The following terms are often used to describe parts of a motor:

1. *Stator* A stationary coil or group of coils.
2. *Rotor* A rotating coil or group of coils.

Up to now, we have treated the armature as a rotor; but either part, the field or the armature, can be made to rotate. The stationary component becomes the stator.

AC-powered electric motors are somewhat more complex than DC-powered motors. The motor illustrated in Figure 25-24 is called a *universal motor*. Note that it has brushes and a commutator. This motor can be used with either AC or DC power. It has an armature and field coils in series. Although the torque developed by a universal motor is not constant, all the torque developed causes the motor to rotate in one direction. The main disadvantage of an AC universal motor is its tendency to arc between brushes and commutator.

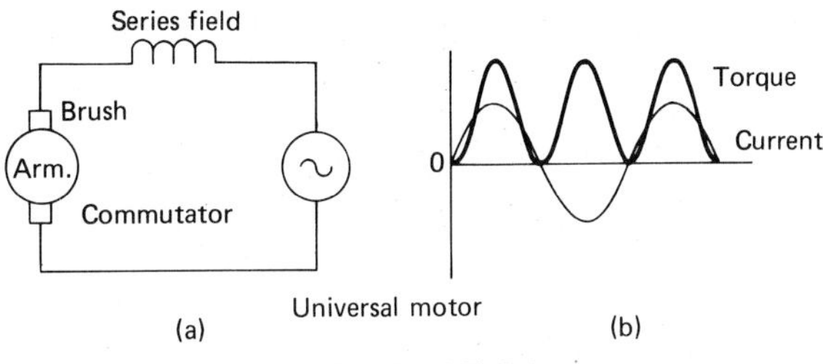

FIGURE 25-24

AC motors that do not use brushes and commutators are called *induction motors*. They depend on the interaction of the two magnetic fields in the field and armature windings. An induction motor does not have the arcing problem of brushes and commutator. In it, the field winding is connected directly to the AC source; there is no connection between stator and armature. The field winding, however, induces current to flow into the armature winding. The armature's magnetic field, the result of current induced, opposes the existing field. Thus, torque is developed.

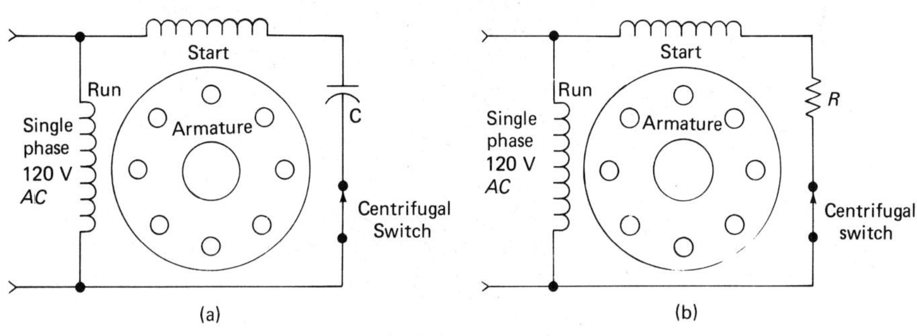

FIGURE 25-25

Many motors operate on single-phase AC current. For them to develop starting torque, a phase shift across the armature must exist. This shift can be created by connecting a resistor and a capacitor in series and connecting them so that energy passing through this branch arrives across the armature 90° later than the original wave. A series-connected capacitor and resistor are shown in Figure 25-25. Note that the start winding is separated from the main winding. Current in the start winding is shifted so its phase differs from that of the main winding. The effect on the armature is that torque is developed which is strong enough to make the armature rotate. A centrifugal switch removes the start winding as soon as the armature is turning fast enough. These motors are called *split-phase motors*.

Self-Check

Answer each item by indicating whether the statement is true or false.

16. ___ Motors are devices that convert mechanical energy into electrical energy.

17. ___ A DC motor uses commutators to couple current to the field winding.

18. ___ An AC motor is called a split-phase motor.

19. ___ Greatest torque is developed in an armature coil when its conductors move parallel to the magnetic field.

20. ___ The more commutator segments a DC motor has, the more evenly is the torque that is developed.

21. ___ Magnetic lines of force flowing in the same direction will repel.

22. ___ In a DC motor, increasing the number of coils wound on the armature results in the development of torque that is more even.

23. ___ Torque is developed when two magnetic fields interact.

24. ___ To reverse the direction of rotation for a DC motor, the polarity of the voltage connected to the brushes is reversed.

25. ___ A universal motor can be used with either AC or DC.

Summary

In this chapter we studied generators and motors. Both devices are important in electrical operations. Generators produce most of the electrical power used throughout the world.

Motors have so many applications, it is impossible to name them all.

Generators convert mechanical energy into electrical energy.

Motors convert electrical energy into mechanical energy.

Generators are used to produce both AC and DC voltage. Whether it is AC or DC, their construction is similar. The difference is that a DC generator uses commutator segments to relay energy to the external circuit, and an AC generator uses slip rings to couple the energy. A universal motor can be used with either AC or DC.

Review Questions and/or Problems

1. Which illustration in Figure 25-26 is a correct representation of Lenz's law?
   (a) a
   (b) b
   (c) c
   (d) d

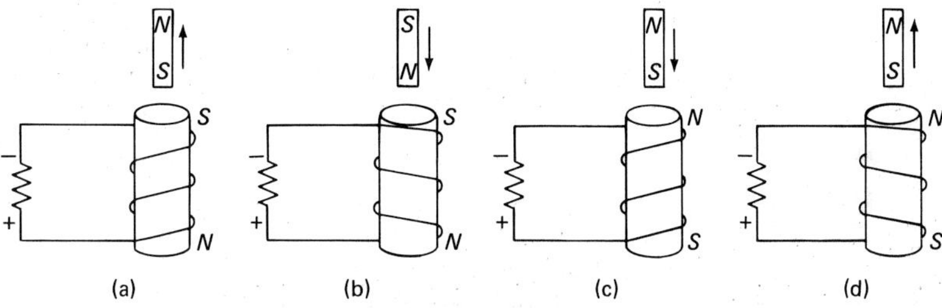

FIGURE 25-26

2. For induction to occur, which must be present?
   (a) Coil, core, and relative motion
   (b) Conductor, magnetic field, and relative motion
   (c) Electromagnet, conductor, and slip rings
   (d) Magnetic field, slip rings, and relative motion

3. In Figure 25-27, which illustration displays correct polarities?
   (a) a
   (b) b
   (c) c
   (d) d

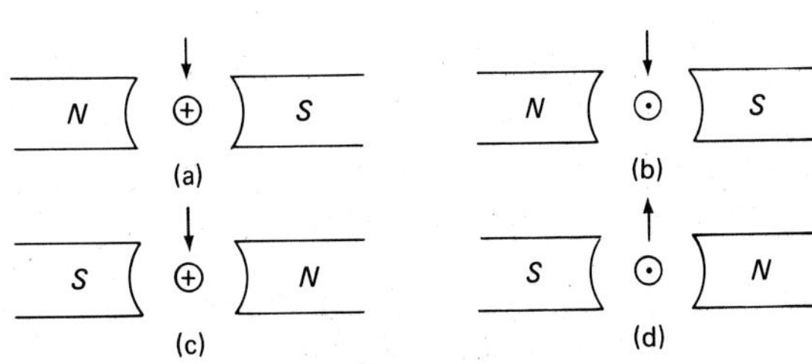

FIGURE 25-27

4. An external circuit and load make it possible for an AC generator to
   (a) produce higher output.
   (b) accomplish useful work.
   (c) reverse its rotation.
   (d) produce a higher frequency.

5. Which illustration in Figure 25-28 shows the correct direction of current flow?
   (a) a
   (b) b
   (c) c
   (d) d

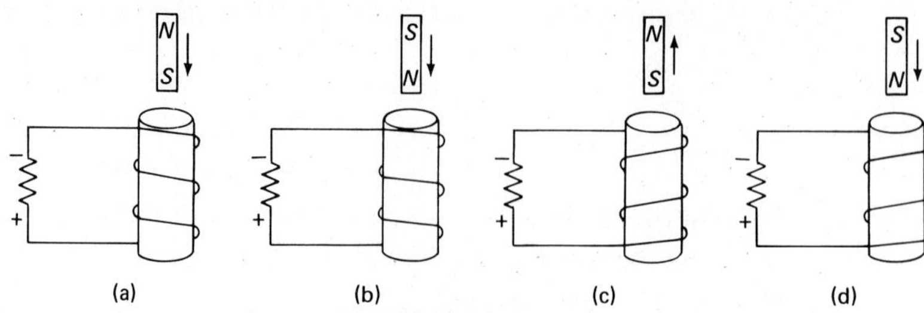

FIGURE 25-28

6. The output of an AC generator can be increased by
   (a) using a weaker magnetic field.
   (b) adding more conductors to the armature.
   (c) slowing the speed of armature rotation.
   (d) increasing the number of slip rings.

7. In Figure 25-29, slip rings are represented by the letter
   (a) A.
   (b) B.
   (c) C.
   (d) D.

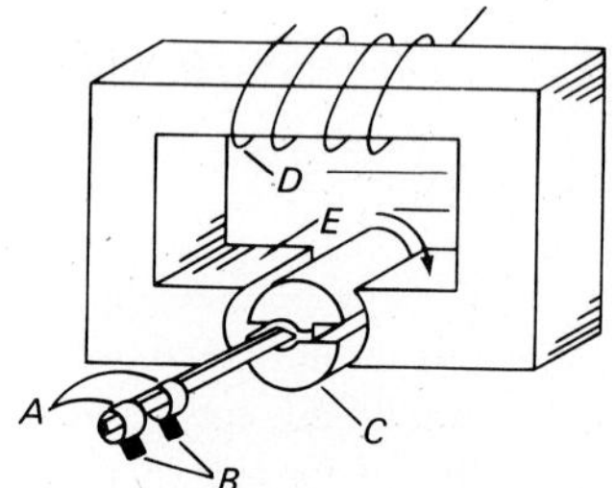

FIGURE 25-29

8. In Figure 25-29, B represents
   (a) pole pieces.
   (b) brushes.
   (c) armature.
   (d) field winding.

9. Refer to Figure 25-29, C represents
   (a) the armature.
   (b) the rotating conductor.
   (c) the field winding.
   (d) the slip rings.

10. The component of a generator that rotates is called the
   (a) field winding.
   (b) brushes.
   (c) armature.
   (d) pole pieces.

11. The component in a DC generator that is different from the component of an AC generator is the
    (a) brushes.
    (b) commutator.
    (c) pole pieces.
    (d) armature.

12. The amount of ripple in the output of a DC generator can be reduced by
    (a) increasing the speed of the generator.
    (b) creating a stronger magnetic field.
    (c) increasing the number of coils in the armature.
    (d) increasing the number of conductors per coil.

13. How many commutator segments are there for each coil on the armature?
    (a) one
    (b) two
    (c) three
    (d) four

14. In Figure 25-30, which schematic represents a a shunt generator?
    (a) a
    (b) b
    (c) c
    (d) d

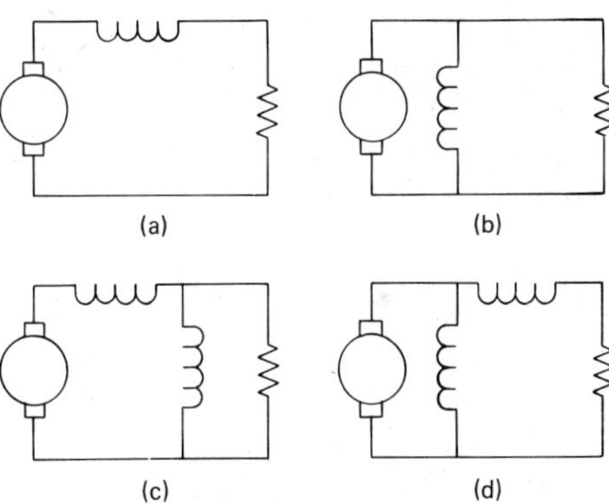

FIGURE 25-30

15. A device that converts electrical energy into mechanical energy is a
    (a) conductor.
    (b) motor.
    (c) generator.
    (d) power supply.

16. The force required to turn a motor is called
    (a) an armature.
    (b) reversed polarity.
    (c) counter-EMF.
    (d) torque.

17. When a coil is moving perpendicular to a fixed magnetic field,
    (a) torque is zero.
    (b) coil movement is stopped.
    (c) current in the coil is maximum.
    (d) the coil is saturated.

18. The universal-wound motor can be used with either AC or DC.
    (a) true
    (b) false

19. Induction motors use a split-phase starting system.
    (a) true
    (b) false

20. A compound-wound generator has both series and shunt field coils.
    (a) true
    (b) false

21. The pole pieces of a generator have a magnetic field between them.
    (a) true
    (b) false

22. Generators must be driven by an external force.
    (a) true
    (b) false

26

Electricity in the Home

Introduction

We all have some knowledge of household electrical systems, but some of us may be familiar only with common appliances. In this chapter we consider the voltage and circuitry involved in operating these appliances.

In the United States, 60 Hz is the standard frequency for transmission of commercial electricity. A voltage of 120 V is commonly used. Higher voltages are used in special applications requiring large amounts of power. Stoves, dryers, air conditioners, and furnaces usually are operated on 240 V.

Regardless of the voltage and frequency used, distribution systems have many similarities. In this chapter we will discuss the U.S. 120-V system and the wiring circuits commonly used in the home.

Objectives

Each student is required to.

1. List the procedures to be followed in connecting lights, convenience outlets, and switches.
2. Draw simple schematics for
   (a) A SPST switch-operated light.
   (b) A SPST switch that controls the operation of two parallel lamps.
   (c) Two convenience outlets connected in parallel, located: (1) in the same wall; and (2) on opposite sides of a door paralleled from a single source.
   (d) A single lamp controlled by two three-way switches.
   (e) A single lamp controlled by two three-way and one four-way switches.
3. Explain the purpose of an *equipment ground* in a system.

Three-Wire Single-Phase Service

The average house in the United States requires considerable power to operate its appliances. Most require 120-V service. Other appliances, however, such as an electric range, require two 120-V services, or a total of 240 V. Both voltages are available from a three-wire system. The system shown in Figure 26-1 is a three-wire system. The transformer shown is the one supplied by the power company; the three wires connect the transformer to the house power panel. From this panel, voltage and current are delivered by a wiring network. Many circuits are duplicated in the system, since all circuits must be parallel off this three-wire entrance.

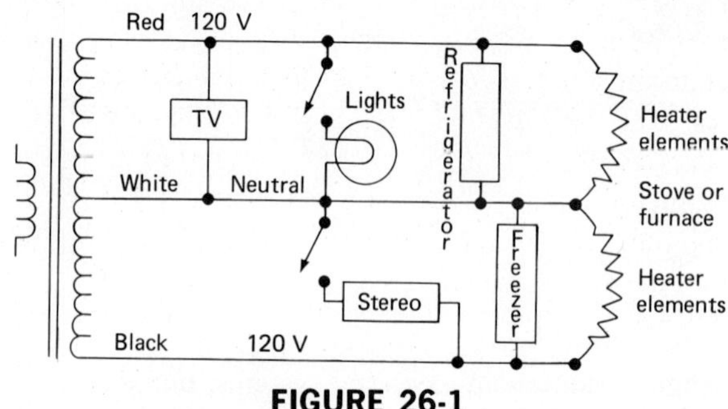

FIGURE 26-1

Note the red, white, and black wires in the transformer secondary. The white wire is neutral, but the other two are hot; 120 V is applied through them. Lights and appliances that require 120 V can be connected between either red and white or black and white conductors.

Actually, devices requiring 240 V usually are connected as two 120-V with a common neutral connection. For example, the cooking elements of an electric stove often have two resistive devices so connected. Many appliances requiring 240 V are connected between red and black conductors, with the center tap of their current-drawing elements tapped into the white conductor (Figure 26-1).

Electrical service to a typical house is established at 240 V, with a maximum of 200 A. Applying the power equation reveals that this makes available continuous 48 kW.

$$P = I \times E$$

$$P = 100 \text{ A} \times 240 \text{ V}$$

$$P = 24{,}000 \text{ W or } 24 \text{ kW}$$

To handle currents of this type, the entrance cable must have a large diameter; but 120-V circuits that run from the power panel throughout the house are usually wired with 10-, 12-, or 14-gage conductors. By checking the wattage and voltage requirements of a particular appliance, and with this equation, we can predict the demand for current:

$$I = \frac{P}{E}$$

Once we know the current requirements, and by consulting the copper wire table in Appendix G, we can determine the size wire needed. Always use a wire with greater current capability than that determined by using the power equation. The extra capability allows appliances to be added without the risk of overload.

Distribution in a House

It is common to use the black conductor as a circuit ground where DC is involved. When wiring AC circuits, the *green* conductor is used as an equipment (circuit) ground. Both grounds serve the same purpose. With the higher voltages and currents available in AC, however, there is greater danger of electrical shock. In that case, it is necessary to take precautions. This is done by connecting all systems to an equipment ground, using the green conductor and green screws on all convenience outlets. In the 120-V system the black wire is the *hot* wire; the white wire is neutral.

Each circuit of a power panel begins either with a fuse or with a circuit breaker, which serves as protection against current overload. The black wire is connected to the output end of this device, and both white and green conductors are connected to a common *buss* in the power panel. The cable is routed through the ceiling and/or walls of the house to the desired location. At that point, the three wires are connected to a lamp, switch, convenience outlet, or other device.

Lamp Connections

Lamp circuits can be controlled by one of several switches. Many lamps have built-in switches for controlling them at the point of use; others are controlled by one, two, three, or more wall switches. Earlier in these studies you learned about switches; electricians tend to call them by other names:

SPST.
Single-control-point switch

DPST.
Three-way switch; the lamp can be turned on or off at two different locations

DPDT.
Four-way (traveler) switches; used with three-way switches to extend the number of control points to three, four, five, or more points, as desired

Single-Control-Point Lamp Circuits

Each circuit used in a house is identical to a series, parallel, or series-parallel circuit already discussed. The first circuit discussed here is the single-lamp circuit shown in Figure 26-2a. Note that the circuit contains one switch and one lamp, with interconnecting conductors. Figure 26-2b represents the same circuit in basic schematic form.

Figure 26-3b shows a circuit containing one switch and two parallel-connected lamps. The schematic drawing for this circuit is shown in Figure 26-3a.

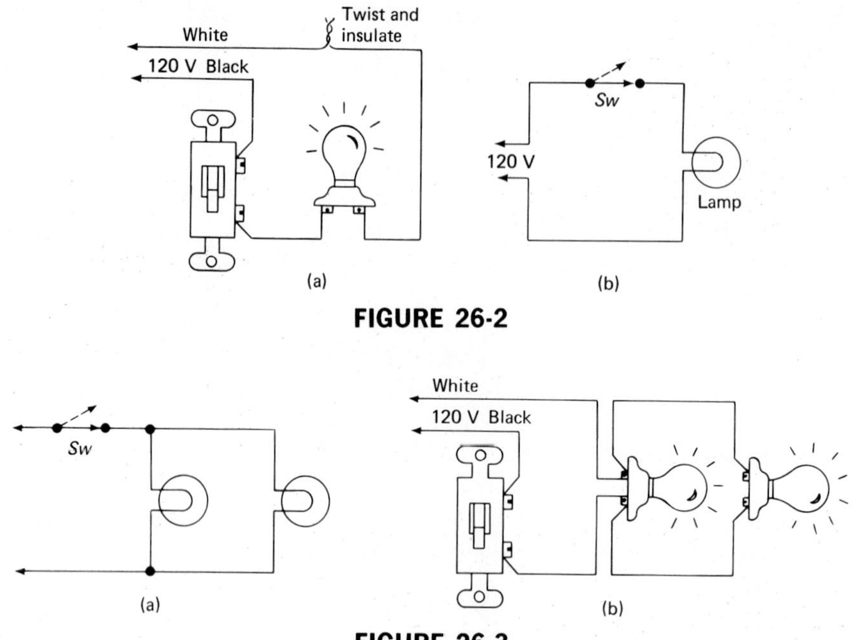

FIGURE 26-2

FIGURE 26-3

Multiple-Control-Point Lamp Circuits

Figure 26-4a presents the schematic of a two-point-control lamp circuit. Figure 26-4b contains two three-way switches which are used to control one lamp. This is the type of circuit used to control a lamp from either end of a hallway.

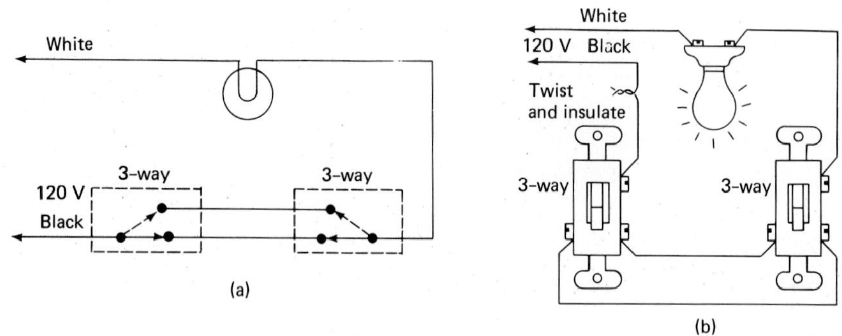

FIGURE 26-4

Figure 26-5a shows a pictorial diagram for the circuit used to modify the circuit in Figure 26-4b to provide a third control point. This point is provided by a four-way switch. To add control points, it is necessary to add other four-way switches to the circuit. This circuit is often used in living rooms where control is desired at various points. Figure 26-5b shows the schematic diagram.

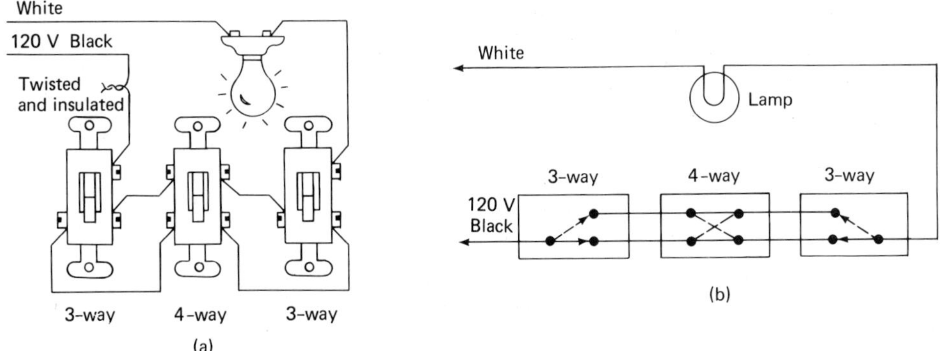

FIGURE 26-5

Convenience Outlet Connections

Convenience outlets can be installed in walls to provide electrical service to appliances as needed. Portable appliances tend to wear and become dangerous. Lamp sockets such as those used in table lamps are themselves dangerous; therefore, the equipment ground (green wire) is extremely important. A point of contact is incorporated in the design of convenience outlets. All screws used to connect the equipment ground are green for easy identification. Table lamps that use two-prong plugs are a source of danger when not connected to the system with the correct alignment. To minimize the danger, convenience outlets are color coded; the black (hot) wire is connected to a gold screw, and the white (neutral) wire is connected to a white (silver) screw. Color coding is especially important with convenience outlets, because an electrician cannot predict the future use of an outlet.

Figure 26-6 shows two convenience outlets wired in parallel. Note that the wiring diagram shows the wiring required to run from an overhead ceiling box to a wall box, then to another wall box. This circuit can be paralleled from box to box by running the cable along or through an existing wall.

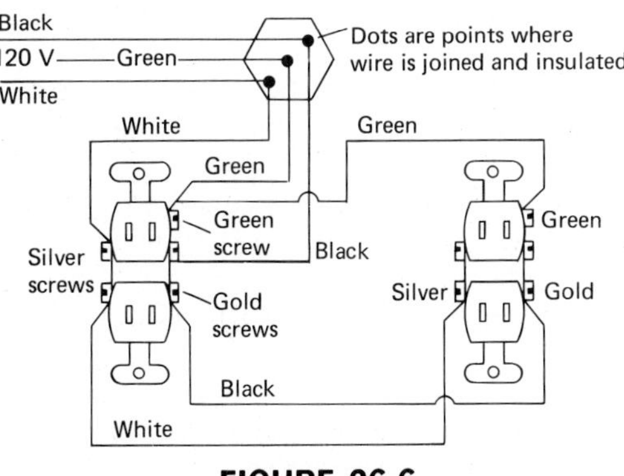

FIGURE 26-6

Figure 26-7 illustrates a circuit where two convenience outlets are paralleled off the ceiling box to different areas. This is often necessary when convenience outlets are desired in walls where a door prevents running wires through the wall or where there is some other obstruction.

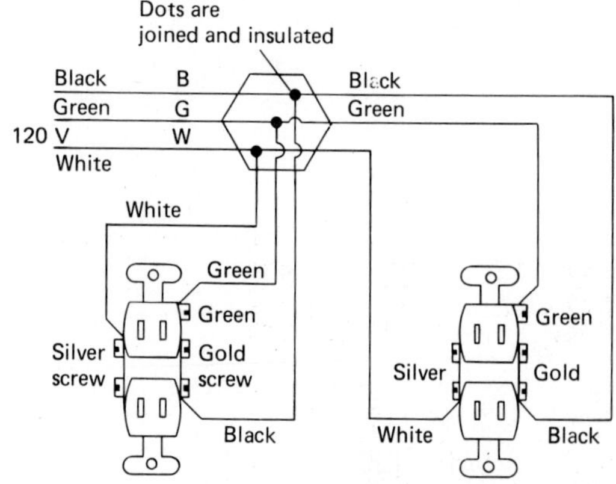

FIGURE 26-7

All electrical connections must be made in some type of box.

Never connect more than four cables in the same box.

When running new electrical circuits, use an NM (nonmetallic) sheathed cable like that shown in Figure 26-8. At each end of the run, extend the cable 6 to 8 in. through the box; that provides enough extra cable to connect each wire, insulate the connection, and fold the cable into the box for mounting.

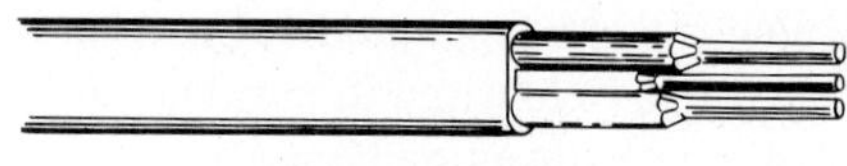

FIGURE 26-8

Summary

This chapter is not intended to make the reader an electrician; it is intended to provide a quick look at basic housewiring circuits and to present a representative sampling of a few basic circuits.

Review Questions

Answer each item by indicating whether the statement is true or false.

1. ___ An SPST switch is used in a circuit designed to provide two-point control of a lamp.
2. ___ An equipment ground provides protection in a system.
3. ___ It is legal to parallel five or six circuits from one ceiling box.
4. ___ Four-way switches provide a third control point for a light circuit.
5. ___ On a convenience outlet, the black wire should be connected to a gold screw.

27

Safety and First Aid

Introduction

In any learning experience, major emphasis must be placed on the safe performance of equipment. Electronics is no different! Electrical shock is potentially one of the most dangerous hazards.

The ability to work safely and provide at least minimal first aid are necessary for everyone involved in electronics. Future technicians must become familiar with the causes of accidents and take precautions to identify the hazards encountered and use the proper first-aid procedures.

Objectives

Each student is required to:

1. Name the major causes of accidents.
2. Identify electrical hazards in an electronics shop.
3. List the common safety practices for eliminating hazards.
4. List the steps in treating a victim of electrical shock.
5. List three types of fire and explain the procedures for extinguishing them.
6. Cite first-aid procedures for the control of bleeding.

Causes of Accidents

Cause and effect are the two areas of concern in accident prevention. Effects are why accident prevention is so important to the technician. Some effects readily apparent are injury, loss of pay, or even loss of life.

‹ best way to prevent accidents is to identify and remove the causes. Research shows that 98% of ‹ents can be avoided. For this reason, everyone is expected to contribute to the safe operation of ‹pment.

Two areas have been identified as causes of accidents:

1) *Unsafe Acts* Unsafe acts by individuals account for about 88% of accidents.
(2) *Material failure* The failure of materials adds about 10%.

In an actual situation, determination of the cause of an accident is so difficult that it is almost impossible to separate unsafe practices from the failure of materials.

What *can* be controlled are unsafe acts. Inattention, impatience, carelessness, ignorance, and anxiety all contribute to causing accidents. Watch people who exhibit these traits, and you will see that they seem to be involved in more than their share of accidents.

Accidents caused by material failure are much harder to predict. Often a machine will be defective, but the defect goes undetected for some time before an accident occurs. It is difficult if not impossible to anticipate this type of accident. Careful inspections and good preventive maintenance can reduce them, however.

The dangers associated with electricity cannot be overemphasized. Common sense and safe work habits can help avoid electrical accidents. Electrical shock is caused by *high voltage*, severe burns by *high current*. Since one does not exist without the other, shock victims usually suffer both types of injury.

All technicians *must* know and use safety procedures. The first rule of electrical safety is:

Never work on electrical equipment when the power is on, unless it is absolutely necessary!

Equipment manufacturers often include safety devices designed to stop current flow in case of an overload. These devices are not intended to protect people, only equipment. For safety purposes, the slogan is:

Be sure, be safe!

Fuses are installed in electrical circuits to protect equipment; they are not designed to protect people. The fuse is current-sensitive; it stops current when its rating is exceeded. The fuse size, or rating, is selected to fit the current requirement of the equipment. A fuse that has "blown" (stopped current) is not a fault to be corrected. It indicates that a more serious problem exists which must be corrected before the fuse is replaced.

Excess current flowing in a circuit can damage costly equipment beyond repair. The rule is:

Never replace a fuse with one that has a higher current rating or a lower voltage rating than the one being replaced!

A very small amount of current passing through the chest cavity of a person is enough to cause severe injury or even death. Examples of current amounts and effects are given in Figure 27-1.

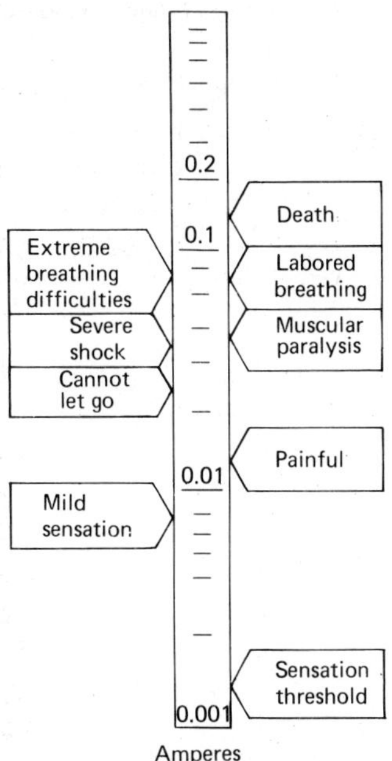

FIGURE 27-1

Remember, these effects are caused by currents too small to blow the circuit's fuse. In fact, the current needed to light a 50-W bulb is many times that shown for causing death.

A fuse is not designed to protect—and does not protect—human life!

Factors Involved in Electrical Shock

No doctor will state a definite level of current as enough to kill a person. The statements in Figure 27-1 are averages. Actual amounts vary from person to person; further, they vary for each person from one period to another. Physicians know that a person's ability to withstand a shock varies widely over short periods of time. Factors that also affect the severity of a shock are the part of the body involved, the amount of current, and the duration of contact.

Electrical shock results in two types of injuries, burns and paralysis. Either ranges from minor, temporary harm to death. People in excellent health can die from small currents. Those less healthy are at even greater risk. Remember: these are very small amounts of current.

A shock involving the lower arm is usually least damaging. The arm muscles contract, causing a person to break contact with the circuit. Small currents may cause only a small tingle; higher currents are much more dangerous. High currents can burn entry and exit holes through the skin; if the current is high enough, it destroys all cellular structure through which it passes. Higher current can also cause muscular contraction, which prevents the victim from releasing the source which results in more severe shock (Figure 27-2).

FIGURE 27-2

Muscular contraction can cause injuries that have nothing to do with those caused by the shock. The legs may collapse, resulting in a fall that breaks bones. Or the victim might bump into someone or something, causing further injury.

As already stated, electrical shock can result in minor injury or death. The severest shocks involve the brain and the chest cavity. It is possible to survive shocks to other parts of the body that would be fatal in these areas. The brain is shielded by the skull to some extent, but the chest is not. The danger of shock to the chest area cannot be overstated. To limit the possibility of a chest-area shock, be a *one-armed* technician.

Never use both hands when working on electrical equipment, unless both hands are absolutely necessary.

Some key points to remember about electrical shock are:

current.
Higher current results in injury that is more severe.

area.
The danger increases when the affected area of the body contains vital organs.

time.
The longer current flows, the more severe the shock.

power.
Turn power *off*. Unless it is absolutely necessary, never work on a "hot" circuit.

jewelry.
Avoid wearing metal jewelry. Metal increases the likelihood of shock.

moisture.
Never work on electrical equipment when you are wet or sweaty. Body resistance is much lower then.

never experiment.
Do not touch live wires to see if they are "hot." Use a meter.

Self-Check

Complete each item by inserting the words and/or numbers needed to make a true and complete statement.

1. Look at Figure 27-1; 0.2 A of electrical current would probably result in ______________.

2. Define:
   (a) unsafe act.
   (b) cause and effect.
   (c) material failure.
   (d) electrical shock.

3. Areas of the body that are most vulnerable to electric shock are
   (a)
   (b)

4. When working around high-voltage equipment, be a ______________ technician.

5. The danger of ______________ is increased by wearing metal jewelry.

First-Aid Procedures

When someone suffers a shock, immediate first aid is necessary. First-aid measures should be applied as soon as possible. The first step is:

Turn the power off as quickly as possible!

If you cannot turn the power off, you must remove the victim from contact with the live circuit. When doing this, be sure not to get shocked by the hot circuit. To remove the victim, use a nonconductive material, such as a dry rope, belt, or wood pole. Put the object over an arm or leg and use it to drag the person free.

Severe electrical shock usually causes a person to stop breathing. It is crucial that breathing is restarted at once. Oxygen is delivered to the brain by the heart. When the body fails to receive a continuous supply of oxygen, damage to the brain occurs even if the victim survives. The second step is:

Make sure the heart is beating!

If the heart is not beating, get it started again. Quite often, the beat can be restarted with a sharp blow to the chest area. If you cannot restart the heartbeat, cardiopulmonary resuscitation (CPR) is required.

Once the heartbeat is confirmed, begin artificial respiration at once. This leads to step 3:

Use mouth-to-mouth resuscitation.

Success often depends on how quickly resuscitation is started. In case of an electrical shock to a fellow worker:

Do not waste time, but be careful. Send someone for medical assistance. Remove the victim from the power source.

1. ***Turn off or unplug the power source.***
2. ***Use insulated material when removing the victim.***
3. ***Make sure the heart is beating.***

Start artificial respiration. The four types are:

1. Prone pressure.
2. Back pressure, arm lift.
3. Back pressure, hip lift.
4. Mouth-to-mouth resuscitation has been shown to be the best, so use it! The correct procedures are:

   (a) Place the victim on his back.
   (b) Use your fingers to clear the air passage of dentures and foreign matter.
   (c) Make sure the victim has not swallowed his tongue. If he has, use your finger to return it to the mouth.
   (d) Tilt the head back as far as possible (Figure 27-3). Called the "sword-swallowing position," this position ensures maintenance of clear passage of air.
   (e) Insert the left thumb in the victim's mouth. Grasp the lower teeth with your thumb and lift the lower jaw (Figure 27-3).
   (f) Use your right hand to clamp the nose tightly shut.
   (g) Take a deep breath.
   (h) Cover the victim's mouth with your mouth, forming an airtight seal.
   (i) Blow forcefully into the victim's mouth. Observe whether the chest rises. If the victim is an infant, blow gently (Figure 27-4).
   (j) Remove your mouth and allow the chest to exhale.
   (k) Repeat this action 12 to 20 times per minute.
   (l) Continue the process until the victim has started breathing or until you are relieved by trained medical personnel.

Since your body uses only 25% of the oxygen you inhale, there is enough left in the air that you exhale to maintain the victim's oxygen level.

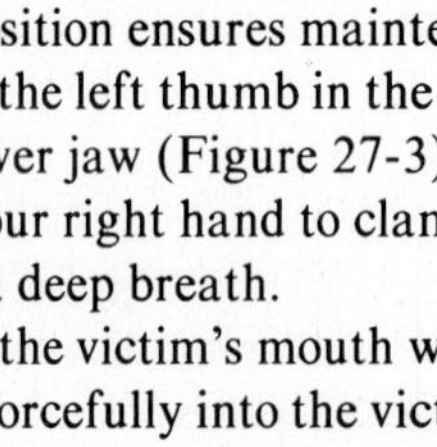

FIGURE 27-3

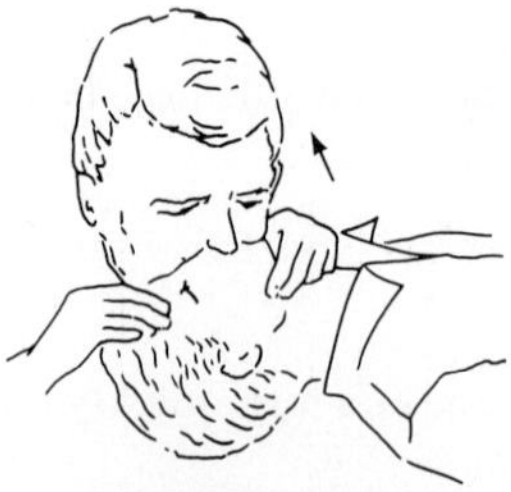

FIGURE 27-4

Self-Check

Complete each item by inserting the words and/or numbers needed to make a true and complete statement.

6. Name at least two actions you could take to disable a live circuit.
   (a)
   (b)

7. To remove someone from a hot wire, ______________.

8. The first action to take after a shock victim is removed from an electrical source is determining whether the victim's ______________ is ______________.

9. For performance of mouth-to-mouth resuscitation, the victim is placed in the ______________ position.

10. When administering resuscitation, the lungs should be inflated ______________ times per minute.

First Aid for Bleeding Wounds

One other first-aid procedure should be mentioned: control of bleeding. The best first-aid method is to apply pressure. Often, the application of pressure directly to the wound is enough to stem the flow. There are several pressure points on the body that are useful in controlling excess bleeding. They are illustrated in Figure 27-5. Select a point between the injury and the heart, and apply firm pressure. The bleeding should slow. In either case, continue pressure until the bleeding stops or a doctor relieves you.

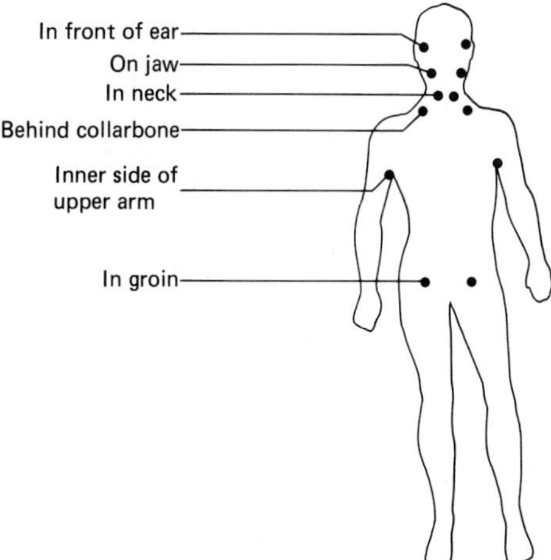

FIGURE 27-5

Caution: A tourniquet should not be used to treat bleeding injuries unless it is clearly a matter of life or death. Once a tourniquet is in place, do not disturb it. Tourniquets should be removed only by a doctor.

Self-Check

Complete each item by inserting the words and/or numbers needed to make a true and complete statement.

11. For most bleeding wounds, ______________ is the best first aid.

12. For a wound in the lower leg, the best pressure point is probably in the ______________.

13. A tourniquet is used only in ______________ situations.

Fire Safety

The best fire protection is prevention. Careful attention to the work area can reduce the chance of fire.

Good housekeeping is the first line of defense. Trash, rags, cleaning fluids, and paint should not be allowed to accumulate, for all are ideal sources of fire.

Flammable materials must be stored away from the work area!

Fires are divided into three classes. Fire extinguishers are identified as to their use on specific classes of fires:

Class A.
Wood, paper, rags, etc. Water is an effective extinguisher.

Class B.
Gasoline, diesel fuel, solvents, and other petroleum distillates. Water splashes the liquid and spreads the flames. To extinguish fires of this class, they must be denied oxygen. This is done using carbon dioxide, foam, or pressurized dry-powder extinguishers.

Class C.
Electrical fires. The added hazard of electrical shock makes this class particularly dangerous. Avoid using water, soda acid, or other current-carrying agents as extinguishers. Carbon dioxide and pressurized powder extinguishers are ideal for coping with class C fires.

In recent years, the pressurized dry-powder extinguisher has been perfected. It is suitable for fighting fires of all three classes. New extinguishers should be of this type. Regardless of the type of extinguisher, it is imperative that the class of fire be identified, as well as the correct extinguisher for fighting the fire and the proper technique for using the extinguisher.

Recently the Halon extinguisher has been introduced for use on computer and other electrical fires.

Dry powder causes problems for months after use because of its penetrating into critical areas. With Halon the circuit can be operated immediately if no other damage is present.

Self-Check

Complete each item by inserting the words and/or numbers needed to make a true and complete statement.

14. The best form of fire protection is ______________.

15. Highly flammable materials should be stored in a ______________.

16. Electrical fires are in class ______________.

17. The ______________ extinguisher is suitable for use with all types of fires.

Summary

Electrical safety, fire safety, and first-aid procedures for electrical shock and bleeding were discussed. Although electronics technicians are not expected to be medically qualified for treatment, they should know how to perform first aid until professional medical help arrives.

Review Questions and/or Problems

Match each lettered statement with a numbered statement.

1.	____ Stops electrical current	(a)	Best method of artificial respiration
2.	____ 12 to 20 times per minute	(b)	Careful inspection schedule
3.	____ Unsafe acts	(c)	Wood, paper, and rag fires
4.	____ Class A fires	(d)	88% of all accidents
5.	____ Best fire protection	(e)	Adequate prevention
6.	____ Sword-swallowing position	(f)	Electrical circuitry
7.	____ Mouth-to-mouth	(g)	Best type of fire extinguisher
8.	____ Exhaling forcefully	(h)	Blown fuse
9.	____ Reduces equipment failures	(i)	Part of mouth-to-mouth resuscitation
10.	____ Dry-powder extinguisher	(j)	Rate of breathing for mouth-to-mouth resuscitation

Select the correct option from the four listed and answer each question.

1. Most accidents are caused by
   (a) accident-prevention programs.
   (b) unsafe acts of individuals.
   (c) faulty equipment.
   (d) acts of God.

2. A fatal electrical shock involves
   (a) the length of time shock is experienced.
   (b) a person's health.
   (c) the area of the body involved.
   (d) all of the above.

3. Fuses are used in electrical equipment to
   (a) regulate current to a desired level.
   (b) protect operating personnel.
   (c) reduce circuit current to safe levels.
   (d) protect electrical equipment.

4. When a shock victim is "frozen" to a hot wire, the first thing you must do is
   (a) grab the victim and pull him loose.
   (b) apply mouth-to-mouth resuscitation.
   (c) turn off the power.
   (d) notify a doctor.

5. Which is *not* part of mouth-to-mouth resuscitation?
   (a) helping the victim exhale by pressing down on the chest
   (b) clamping the nose shut to form an airtight seal
   (c) clearing the mouth and throat by using a finger
   (d) placing the victim in the sword-swallowing position

Glossary

AC generator A rotating electrical machine that uses slip rings to convert mechanical energy to alternating-current electrical energy.

Air capacitor A capacitor that uses air as the dielectric material separating its plates.

Air-core coil A number of turns of spiral wire formed into a coil that has air as its nonmagnetic core.

Air-core transformer A transformer (usually RF) having two or more coils wound around a nonmetallic core.

Alternating current (AC) Current that is continually changing in amplitude and periodically reversing in polarity.

Alternating current generator *See* AC generator.

Alternation Variation, either positive or negative, of a waveform from zero to maximum and back to zero; equals one-half cycle.

Alternator *See* AC generator.

Ammeter An instrument used to measure the amount of current in amperes.

Ampere (A) Practical unit for current; 1 A flows through 1 Ω of resistance when 1 V is applied; 1 A flows when 6.28×10^{18} electrons (one coulomb) pass a point within a circuit during 1 second.

Ampere-hour The quantity of electricity that passes through a circuit in 1 hour when the rate of flow is 1 ampere.

Ampere-turns A magnetomotive force that equals the product of the number of turns in a coil and the current in amperes that flows in the coil.

Analog Operates by angular rotation. A device whose output is a continuous function of the input.

Armature The moving element in an electromechanical device such as a motor, generator, or relay.

Atom Smallest portion of an element which exhibits all the properties of the element.

Attenuate To decrease the amplitude of.

Attraction The force that tends to make two objects approach each other.

Audio frequency (AF) Any frequency corresponding to a normally audible sound wave. The range of frequency extending from approximately 20 Hz to 20 kHz.

Audio-frequency transformer An iron-core transformer designed to couple audio frequencies from one circuit to another.

Autotransformer A transformer with a single winding (electrically); in which part of the winding acts as a primary and all of the winding as a secondary (step-up transformer operation), or in which all the winding acts as a primary and part as the secondary (step-down transformer operation).

Bandpass (BP) The number of hertz separating the limiting frequencies between the half-power points in a filter or a resonant circuit.

Bandwidth (BW) The range of frequencies between the half-power points of a resonant circuit or a filter circuit.

Battery A device containing 2 or more cells that converts chemical energy into electrical energy.

Bidirectional Responsive in opposite directions.

Bleeder resistor A resistor used to draw a fixed current; a resistor connected across a power supply to discharge filter capacitors.

Break The open in a circuit caused by separating contacts of a relay, switch, or other device.

Breakdown voltage Voltage at which an insulator or dielectric ruptures and allows current to flow.

Bridge circuit A circuit with series-parallel branches that are connected by a common bridge.

Capacitance (C) The ability of a device to store electrical energy in an electrostatic field; measured in farads.

Capacitive reactance (X_C) The opposition of a capacitor to a change in current. Expressed by

$$X_C = \frac{1}{2\pi fC}$$

Capacitor A device made by separating two conductors with an insulating material (dielectric); which when charged will oppose a change in voltage. When charged, a capacitor blocks DC and passes AC.

Charge Electrical energy stored in a capacitor's electrostatic field, within a battery, or on a charged body.

Charged bodies Objects that have an excess or a deficiency of electrons.

Charging current Current that charges a capacitor or a coil.

Ceiling box A box made of metal or some other material that is mounted in the ceiling of a room, used to make circuit connections for a wiring system.

Circuit A completed path through which current can flow between two points.

Circuit breaker A device that opens a circuit automatically, usually by electromagnetic means, when current exceeds the safe limits.

Circuit diagram A schematic drawing that shows the electrical connections of a circuit.

Circular mil (Cmil) A universal term used for defining cross-sectional areas; equal to 1 mil (0.001 in.) in diameter.

Coefficient of coupling The degree of coupling that exists between two circuits.

Coil A device made by wrapping turns of a conductor around an air core or a soft iron core; a device that opposes any change in current.

Commutator A device used to connect a DC generator to an external circuit.

Component An essential functional part of a system or a circuit.

Conductance (G) The ability of a material to support current flow; the reciprocal of resistance, having the symbol ℧ (ohm); the equation is:

$$G = \frac{1}{R}$$

Conductor Any material (usually a metal) that conducts electricity through the transfer of orbital electrons.

Contact One of the current carrying parts of a switch, relay, or connector that are engaged or disengaged to open or close the associated electrical circuit.

Continuity A continuous path for the flow of current in an electrical circuit.

Convenience outlet A wall receptacle used in a house wiring system that provides electrical power at convenient points in the system.

Conventional current An explanation of current based on the belief that current in an external circuit flows from the positive pole of a battery to its negative pole.

Copper losses Power losses in motors, generators, and transformers which result from current acting on resistance; often called the I^2R loss.

Core The central area of a coil or transformer around which the windings are wound; usually either air or soft iron.

Cosine (cos) In trigonometry, the ratio of the adjacent side of a right triangle to the hypotenuse.

Coulomb A measure of charge equal to 6.28×10^{18} electrons.

Coulomb's law The law which states that an electrostatic attraction or repulsion exists between charged bodies. Formula is

$$\text{Force} = \frac{Q_1 \times Q_2}{d^2}$$

Like charges repel and unlike charges attract.

Counterelectromotive force (CEMF) An electromotive force that is created by a chemical or magnetic effect upon a circuit and that acts in opposition to the voltage applied to the circuit.

Crystal A thin piece of a natural substance, such as quartz or tourmaline, capable of producing AC voltage at a specific frequency when excited by DC voltage; operates on the piezoelectric principle.

Cutoff frequency The lowest and/or highest frequency that a filter circuit will pass. The frequency at which output voltage equals 0.707 times the peak output voltage.

Cycle The change of an alternating wave from zero to a positive peak to zero to a negative peak and back to zero. One 360° rotation of a generator or sine wave.

Cycles per second (cps) The number of cycles that occur in 1 second; also called frequency. Now replaced by hertz.

d'Arsonval meter movement An analog meter movement used in instruments designed for precision measurement of DC.

DC *See* Direct current.

DC generator A rotating electric device for converting mechanical energy into DC electrical energy.

DC resistance The opposition a device offers to direct current.

Decay A gradual reduction of a quantity.

Deenergize To disconnect a device from its power source.

Dielectric The insulating (non-conducting) medium between two plates of a capacitor.

Difference in potential The algebraic difference between the voltage at two points.

Differentiating circuit A circuit the output of which is substantially in proportion to the rate of change of the input voltage or current.

Digital Use of numbers expressed in digits and in certain scale notation to represent a set of variables.

Direct coupling The connection of two circuits by a direct connection or a low-value resistor.

Direct current (DC) A current that flows in only one direction.

Distributed capacitance (Stray) The capacitance that exists between adjacent conductors where capacitance was not intended.

Distributed inductance Inductance along the entire length of a current carrying conductor as distinguished from the inductance contained in a coil.

Double-pole-double-throw switch (DPDT) A switch that has six terminals and is used to connect one pair of movable terminals to either of the other two pairs of fixed terminals.

Double-pole-single-throw switch (DPST) A switch that has four terminals and is used to connect or disconnect two pairs of terminals simultaneously.

Dry cell A voltage-generating cell having an immobilized electrolyte.

Duration *See* Period.

Dynamo An alternate name for a DC generator.

Dynamometer A measuring instrument that operates on the electromagnetic principle of rotating torque between two coils; used as a wattmeter.

Eddy current losses The power loss that occurs in the core of a magnetically operated device because of eddy currents.

Eddy currents Those currents induced in a body or a conducting mass by a variation in magnetic flux.

Effective voltage An amount of AC voltage or current, in sine wave form, that has the same heating effect as an equal amount of DC voltage or current. For a sine wave effective voltage is equal to 0.707 times the peak voltage.

Electrical charge (Q) The number of excess electrons or the deficiency of electrons on a charged body. Excess electrons are called negative charge, a deficiency of electrons is called a positive charge.

Electrolytic capacitor A capacitor consisting of two conducting electrodes, with the anode having a metal oxide film formed on it. The film acts as the dielectric or insulating material.

Electromagnet A temporary magnet consisting of a coil and an iron core. A magnetic field exists only when current flows through the coil.

Electromagnetic coupling Two circuits coupled together by one magnetic field.

Electromagnetic field *See* Magnetic field.

Electromotive force (EMF) The force which causes electricity to flow when there is a difference in potential between two points. The unit of measurement is the volt.

Electron A negatively charged particle that orbits the nucleus of an atom.

Electron current (I) The movement of electrons from the negative side of a power source to the positive side of the source through an external conductor.

Electron drift The orderly movement of electrons in a conductor.

Electrostatic charge An electric charge stored in the dielectric of a capacitor or on a charged body.

Electrostatic field The space around a body in which the body's influence is felt.

Element In chemistry, a substance that cannot be divided into simpler substances by chemical means and whose atoms all have the same atomic number.

Energize To supply the power necessary for normal operation, for example, energizing a relay.

Energy In physics, the capacity for performing work.

Equipment ground A conductor used as a common voltage reference point for all equipment and convenience outlets in a wiring system.

Equivalent circuit An electrical circuit, usually drawn, which is electrically equivalent to another circuit; most often used in the analysis of complex circuits.

External circuit All wires and other conductive devices which are connected outside the power source.

Farad (f) The unit of measure for capacitance; 1 farad is present when 1 Coulomb raises the potential by 1 volt.

Formula: $C = \frac{Q}{E}$

Filter A circuit made of capacitors and coils that is used to attenuate and/or pass certain bands of frequencies.
Filter capacitor A capacitor used in a filter circuit. This term is often used when referring to the capacitors in a low-voltage power supply.
Fixed capacitor A capacitor having a fixed capacitance.
Fixed resistor A resistor having a fixed resistance.
Flux density (B) The number of magnetic lines per cubic centimeter.
Four-way switch A double-pole-double-throw switch used in house wiring systems to provide three or more control points for a circuit's operation.
Free electrons Electrons that have broken their bond with an atom and are free to be used as electron current.
Frequency (f) The number of cycles per second of rotation for a generator, motor, or other circuit. Frequency is stated in Hertz (Hz).

Formula: $f = \frac{1}{t}$

FSD *See* Full-scale deflection.
Full-scale deflection (FSD) The movement of the pointer of an analog meter movement to the maximum current position; normally the right side of the scale.
Galvanometer An analog measuring device capable of measuring small currents of either polarity, with zero at its center scale.
Generator A device that converts mechanical energy into electrical energy.
Giga (G) A prefix meaning 1 billion, or 10^9.
Ground The common voltage reference point for a circuit.
Henry (H) The unit of measure for inductance. One henry is present when one ampere per second induces one volt in a coil.
Hertz (Hz) The unit of measure for frequency; 1 hertz equals 1 cycle per second.
High pass filter A filter circuit designed to attenuate all frequencies below a cutoff frequency and to pass all those above that frequency.
High Q A high ratio of inductive reactance to resistance.
Hot A circuit in which voltage has been applied. A terminal that is not grounded.
Hysteresis (H) The property of a magnetic substance that causes magnetization to lag behind the force that produced it.
Hysteresis loss Loss in a circuit due to magnetic hysteresis.
Impedance (Z) The total opposition to the flow of alternating current that results from combined resistance and reactance.
Impedance coupling A method of coupling using an impedance as the coupling device to connect two circuits.
Impedance matching The coupling of circuits by means where the output impedance of one circuit matches the input impedance of the next; for example, transformer coupling.

Impedance matching transformer A transformer used to match impedance between two circuits.
Induced Produced by exposure to a varying electric or magnetic field.
Induced current Current created by induction.
Induced voltage Voltage created by induction.
Inductance (L) The property of an electric circuit which opposes any change in current flow.
Inductive reactance (X_L) Opposition to a changing current that results from inductance. Its unit of measure is ohms.

Formula: $X_L = 2\pi fL$.

In phase The condition where two waves have the same frequency and pass through all points simultaneously.
Input impedance The impedance that a circuit presents to a power source or another circuit.
Instantaneous value Any value between zero and maximum depending on the instant of selection.
Insulation A material in which current flow is negligible; used to isolate a conductor from another conductor; or for other safety purposes.
Insulator A material, or a combination of materials, having an atomic structure which provides very few free electrons.
Integrating circuit A circuit whose output waveform is substantially the time integral of its input waveform.
Internal resistance Resistance within a battery or other power source.
Intensity The strength or amplitude of a quantity.
Inverse scale A scale having zero at its right end; for example, the scale used with an analog ohmmeter.
Ion An atom with an imbalance of electrons and protons.
Ionization process The process of adding or removing electrons in an atom.
IR drop *See* Voltage drop.
Iron-core coil A coil that has an iron core as part of its magnetic circuit.
Iron-core transformer A transformer that has an iron core as part of its magnetic circuit.
I^2R loss Power loss in transformers and other conductors that results from current flow (I) through resistance (R) of the conductors.
Joule A unit of energy equal to 1 watt/second.
Kilo (k) A prefix that indicates 1000, or 10^3.
Kilohertz (kHz) 1000 cycles per second, or 1 kHz.
Kirchhoff's current law A law stating that the algebraic sum of all the currents at any point in a circuit equals zero.
Kirchhoff's voltage law A law stating that the algebraic sum of all the voltages in a closed loop equals zero.
Lagging angle An angle by which current lags voltage in a coil or by which voltage lags current in a capacitor.
Lambda (λ) The symbol for wavelength.
Laminations Thin sheets of metal used in the cores of coils, transformers, motors, and generators.
Laws of electrostatic charges (1) Like charges repel. (2) Unlike charges attract.
Laws of magnetism (1) Like poles repel. (2) Unlike poles attract.
LC circuit A circuit containing coils and capacitors only.
Leading angle The angle that voltage leads current in a coil or current leads voltage in a capacitor.

Left-hand rule A method by which the left hand is used, to determine the polarity of an electromagnetic field or the direction of current flow.

Lenz's law The EMF that is induced in any circuit is always in such a direction as to oppose the effect that produces it.

Lines of force Graphic representation of electrostatic and magnetic fields showing direction and intensity.

Lodestone (also called loadstone) A natural magnet consisting chiefly of a magnetic oxide of iron called magnetite.

L-section filter A filter consisting of a coil and a capacitor that is connected in an inverted-L configuration.

Magnet A substance which has the property of magnetism.

Magnetic circuit A complete path through which magnetic lines of force travel between the *north* and *south* poles of a magnet.

Magnetic field The area around a magnet where its field is present.

Magnetic lines of force Magnetic lines along which a compass will align itself.

Magnetic material Any substance that can be attracted by a magnet.

Magnetism The property that allows one material to attract other magnetic materials.

Magnetomotive force (mmf) The force by which a magnetic field is produced.

Magnetic poles The points of maximum attraction on the faces of a magnet; designated as *north* and *south* poles.

Make The closing of contacts in a relay, a switch, or some other device.

Matter Anything that has weight and occupies space.

Mega (M) Prefix used to denote one million or 1×10^6.

Megahertz (MHz) 1 million hertz or 1 million cycles per second.

Meter (1) Unit of metric length equal to 39.37 in. (2) An instrument that can be used to measure current, voltage, ohms, or watts.

Meter shunt A small resistor that is connected in parallel with a meter movement that carries all the current that exceeds the movement's full-scale current rating.

Mho (℧) Unit of measurement of conductance.

Mica capacitor A capacitor that has mica as its dielectric.

Micro (μ) Prefix meaning one millionth or 10^{-6}.

Microfarad (μf) One millionth of a farad.

Microhenry (μH) One millionth of a henry.

Microsecond (μs) One millionth of a second.

Mil One thousandth of an inch (0.001 in.).

Milli (m) Prefix for 1 thousandth or 10^{-3}.

Milliammeter An ammeter used to measure current in milliamperes.

Milliampere (mA) One thousandth of an ampere (0.001 A).

Millihenry (mH) One thousandth of a henry (0.001 H).

Millisecond (ms) One thousandth of a second (0.001 s).

Molecule Smallest division of a compound. Further division results in elements.

Motor A device that converts electrical energy into mechanical energy.

Multimeter A combination volt-ohm-ammeter circuitry.

Mutual inductance The linking of two coils by the magnetic field that surrounds one of the coils.

Nano (n) Prefix meaning 1 billionth or 10^{-9}.

Natural magnet *See* Lodestone.
Negative (neg) (1) A terminal having an accumulation of free electrons. (2) A designation used to represent the opposite of positive.
Negative ion An atom that has gained one or more electrons and is negatively charged.
Neutron A particle found in the nucleus of an atom which has a neutral electrical charge.
Nucleus The core of an atom; containing protons and neutrons.
Ohm (Ω) The practical unit of resistance; 1 Ω is present when 1 V causes 1 A to flow.
Ohmic value Resistance expressed in ohms.
Ohmmeter Any meter that is used to measure ohms of resistance.
Ohm's law In an electrical circuit; current is directly proportional to voltage and inversely proportional to resistance.

Formula: $I = \dfrac{E}{R}$

Ohms per volt (Ω/V) An indication of voltmeter sensitivity.
Open An incomplete path for current flow.
Oscilloscope An instrument that makes possible the visual inspection of electrical waveforms.
Out of phase Waveshapes that are the same frequency but that have different timing.
Output impedance The impedance present at the output of a power supply or other curcuit.
Padder capacitor An adjustable capacitor connected in series with other capacitance that is used to fine tune a circuit.
Parallel circuit A circuit that contains two or more paths for current flow which is supplied by a single power supply.
Parallel resonance A reactive circuit that has inductance and capacitance connected in parallel which is operating at the frequency where X_L equals X_C.
Parallel-series circuits A circuit in which two or more components are connected in series and then connected in parallel with other branches to form a parallel circuit.
Peak negative The maximum negative point of a sine wave.
Peak positive The maximum positive point of a sine wave.
Peak-to-peak value (pk-pk) The measured value of a sine wave from peak-positive voltage to peak-negative voltage.

Formulas: $E_{pk\text{-}pk} = 2.828 \times E_{eff}$ or $E_{pk\text{-}pk} = 2 \times E_{pk}$

Peak value The maximum instantaneous value of a variable; for example, an alternating current or voltage.
Peak voltage (current) The absolute maximum amplitude of a changing voltage or current.

Formula: $E_{pk} = 1.414 \times E_{eff}$

Period The time for the completion of one cycle.
Phase The angular relationship between current and voltage in AC circuits.
Phase angle θ The angle obtained by taking the inverse (arc) tangent of the ratio obtained when the reactance component of a circuit is divided by the resistance component.
Pico (p) Prefix indicating 1 millionth of 1 millionth or 10^{-12}.
Picofarad (pf) One millionth of one millionth of a farad.

Piezoelectric effect The ability of certain crystalline substances, when excited by a DC voltage, to change shape and generate AC voltage.
Pi-section filter A filter connected in a pi configuration; usually made from capacitors and coils.
Positive charge The condition that exists when the number of protons is greater than the number of electrons.
Positive ion An atom that has lost one or more electrons and is positively charged.
Potential The difference in voltage between two points.
Potential difference The algebraic difference between the voltages at two points.
Potentiometer A three-terminal variable resistor used to select a potential for application to a subsequent circuit.
Power (P) The time rate of doing work.

Formula: $P = I \times E$

Power apparent (P_A) Total power that appears to be dissipated in an RC, RL, or RCL circuit.

Formula: $P_A = I_t \times E_t$

Power factor (P_F) The relationship of power true to power apparent in circuit.

Formula: $P_F = \dfrac{P_T}{P_A}$

Power frequencies The frequencies (50 Hz, 60 Hz, and 400 Hz) at which electric power is generated and distributed. In the United States the standard frequency for power distribution is 60 Hz.
Power supply An electronic circuit that supplies electrical power to another circuit.
Power transformer A transformer used at power frequencies.
Power true (P_T) Power dissipated by the resistor in an RC, RL, or RCL circuit.

Formula: $P_T = I_R \times E_R$

Primary cell (1) A cell that produces electrical current through chemical action. (2) A cell that cannot be recharged.
Primary winding The transformer winding that receives energy from the power supply and transfers it to the secondary windings by mutual induction.
Printed circuit The conductor pattern that is formed from copper foil that has been deposited on either phenolic or epoxy glass board.
Proton A positively charged particle that is bound into the nucleus of an atom.
Pulsating DC A DC voltage that varies between maximum value and zero.
Q (1) A unit of charge equal to one Coulomb. (2) The figure of merit (quality) of a circuit containing resistance and inductance.

Formula: $Q = \dfrac{X_L}{R}$

Radio frequency (RF) Any frequency at which coherent electromagnetic radiation of energy is possible.
Radio-frequency band A band of frequencies extending from approximately 20 kHz to 300 GHz.
RC coupling The use of resistance and capacitance to couple a signal from one circuit to another.
RC network A circuit containing capacitors and resistors.

RCL network A circuit containing capacitors, coils, and resistors.

Reactance (X) Opposition to alternating current that results from capacitance reactance or inductance reactance.

Reactive power (VAR) The power that appears to be used by reactive components.

Reciprocal The number 1 (unity) divided by a quantity, for example, the reciprocal of four is ¼.

Reflected impedance The impedance that appears to exist in a transformer primary due to current flow in the secondary.

Relay (K) A remotely controlled electromagnetically operated switch.

Reluctance Opposition to the flow of magnetic lines of force.

Residual magnetism Magnetism that remains in a material after the magnetizing force is removed.

Resistance (R) Opposition to current flow.

Resistive coupling The use of a resistor in coupling the signal from one circuit to another circuit.

Resistor (R) A device that opposes current flow.

Resonant frequency (f_r) (1) The frequency at which a tuned circuit oscillates; see Tuned circuit. (2) The frequency at which X_L will equal X_C in an RCL circuit.

Retentivity The ability of a material to retain magnetism after the magnetizing force has been removed.

RF *See* Radio frequency.

Rheostat A two-terminal variable resistor used to vary current within a circuit.

Ripple voltage A DC voltage whose amplitude varies between peak and a value other than zero.

RL network A circuit containing coils and resistors.

RMS The abbreviation for Root-Mean-Square.

RMS voltage *See* Effective voltage.

Rosin-core solder A tin-lead alloy wire that melts easily and whose center is filled with rosin. The rosin acts as a flux for cleaning the surfaces to be soldered.

Rotary switch A switch whose contacts rotate as they are moved to make and break. This switch can have several different fixed contacts for each movable contact and can be mounted to the same shaft to provide several switches on one control. Used in stereo and test equipment.

Rotor The rotating part of a motor or generator.

Schematic The diagram of an electronic circuit showing electrical connections and identification of components.

Secondary cell A cell used to produce voltage that can be recharged by forcing current to flow through it in a direction opposite to the direction of discharge current.

Secondary winding The coil that receives energy from the primary winding of a transformer by mutual induction and then delivers energy to a load.

Selectivity The relative ability of a circuit to select a desired frequency and reject all others.

Self-inductance EMF that is induced into the windings of a coil by the magnetic field of the coil.

Sensitivity The relative ability of a circuit to respond to small voltages.

Series circuit A circuit that contains only one possible path for current flow.

Series-parallel circuit Components connected in parallel networks that are connected in series with other networks or components.

Series resonance A series circuit containing capacitance and inductance that is operating at a frequency such that $X_L = X_C$.

Shield A metal circuit installed to conduct magnetic lines.

Short circuit A zero-resistance path for current flow.

Shorted component A component whose resistance is 0 Ω.

Shunt Connect in parallel with.

Sine (sin) In trigonometry; the ratio of the length of the side opposite an acute angle to the length of a right triangle's hypotenuse.

Sine wave The graphical representation of an AC signal whose strength is proportional to the sine of an angle that is a linear function of time and distance.

Single-pole-double-throw switch (SPDT) A switch that has one movable contact and two fixed contacts that can be used to alternately control two separate circuits.

Single-pole-single-throw switch (SPST) A switch having one fixed contact and one movable contact that is used for control of a single circuit.

Slip rings Metal rings connected to the rotating armature of an alternator that couple the energy from the alternator to the external circuit.

Solder *See* Rosin core solder.

Soldering iron A tool used to apply heat to the point of connection while soldering.

Solenoid A current carrying coil of wire that has magnetic characteristics.

Split phase motor A single phase induction motor, which develops starting torque by phase displacement between field windings.

Static charge A charge that is at rest, not moving.

Static electricity Electricity at rest, as opposed to current.

Stator The stationary part of a motor or a generator.

Step-down transformer A transformer in which the primary voltage is larger than the secondary voltage.

Step-up transformer A transformer in which the secondary voltage is higher than the primary voltage.

Storage battery The common name for a lead-acid battery, for example, an automobile battery.

Switch (Sw) A device used to control current flow in a circuit.

Tangent (tan) In trigonometry, the ratio of the opposite side (altitude) of a right triangle to the adjacent (base) side.

Tank circuit A parallel resonant circuit.

Tapped resistor A fixed wirewound resistor or a fixed-film resistor that has terminals connected along its length allowing it to serve as two or more resistors.

Temporary magnet A magnet that loses its magnetism rapidly when the magnetizing force is removed.

Three-way switch A single-pole-double-throw switch used in house wiring systems to provide two separate points of control for a circuit.

Throw In switches and relays, the movement of the movable contact with respect to a fixed contact.

Time constant The time required for coils and capacitors to charge to 63.2% of maximum value or to discharge to 36.8% of maximum value.

Tolerance The maximum permissible variation from a standard value that is allowed.

Torque A force that produces or tends to produce rotation.

Transformer A device that transfers electrical energy from one coil to another by mutual induction.

Transient Instantaneous voltage or current change from one steady state to another steady state.

Trimmer capacitor A small variable capacitor that is connected in parallel with a fixed capacitor that allows a circuit to be tuned by changing total capacitance.

T-section filter A network of resistor and reactive components connected in a T configuration and used as a filter circuit.
Tuned circuit A circuit containing capacitance, inductance, and resistance designed to operate at a specific resonant frequency.
Tuned tank circuit *See* Parallel resonance.
Turn One wrap of a conductor around a core.
Turns ratio The ratio of the number of turns in the primary winding of a transformer to the number of turns in the secondary winding.
Unidirectional Flowing in one direction, for example, DC.
Valence The measure of the extent to which an atom can combine with other atoms.
Valence electrons Electrons located in the outer band of an atom.
Variable resistor A resistor whose value can be changed.
Vector A straight line drawn to scale which shows both direction and magnitude.
Vector diagram A diagram showing the direction and magnitude of two or more vectors, such as X_L, X_C, R, and Z.
Vibrator A magnetically operated interrupter relay. Used to convert DC voltage to a pulsating DC or as a buzzer.
Volt (V* or *E) The unit of measure for an EMF or potential. The EMF present when 1 A flows through 1 Ω.
Voltage divider A circuit suitable for supplying two or more voltages from one voltage source.
Voltage drop The voltage measured across a resistive device.
Voltage rating The maximum voltage that a component can sustain without breaking down.
Voltmeter A meter used to measure voltage.
Voltmeter sensitivity The ratio of a meter's total resistance to a full-scale reading in volts; expressed as ohms per volt.
Volt-ohm-meter *See* VOM.
VOM A portable multimeter capable of measuring voltage, ohms, and milliamperes.
WATT (W) The unit of measure for electrical power; 1 watt of power is consumed when 1 V causes 1 A to flow.
Wattage rating The rating used to express the maximum power that a device can be expected to handle safely.
Watt/hour A unit of energy measurement equal to 1 watt per hour.
Wattmeter An instrument used to measure electrical power in watts.
Wavelength (λ) The distance electrical energy will travel during one cycle of the generator.
Waveshape A graph of an electrical wave as a function of amplitude, time, and distance.
Wet cell A cell that has a fluid electrolyte. *See* Secondary cell.
Wheatstone bridge A resistive bridge circuit that is used as a very accurate ohmmeter.
Winding A continuous coil formed by one or more turns of a conductor around a core.
Wiper The moving part of a switch, potentiometer or rheostat.
Wire A metal conductor used to carry current.
Wirewound resistor A resistor that is made by wrapping resistive wire around an insulating frame.
Work A force moving over a distance performs work.
X axis The horizontal axis of a graph.
Y axis The vertical axis of a graph.

Bibliography

Alerich, Walter N. 1971. *Electrical construction wiring.* Chicago: American Technical Society.

Angerbauer, George J. 1978. *Principles of DC and AC circuits.* North Scituate, Mass.: Duxbury Press.

Boylestad, R. and Nashelsky, L. 1977. *Electricity, electronics and electromagnetics.* Englewood Cliffs, N. J.: Prentice-Hall.

Buban, P. and Schmitt, M. L. 1972. *Technical electricity and electronics.* New York: McGraw-Hill.

Cooke, N. M. and Adams, H. F. R. 1970. *Basic mathematics for electronics.* 3d ed. New York: McGraw-Hill.

Department of Defense. *Air Force Manual 52-8.* Washington, D.C.: U.S. Government Printing Office.

Department of Defense. *Miscellaneous electronics training materials.* Washington, D.C.: U.S. Government Printing Office.

Department of Defense. *Technical orders 31-1-141-1 -18.* Washington, D.C.: U.S. Government Printing Office.

Gerrish, H. H. 1968. *Electricity and electronics.* South Holland, Ill.: Goodheart-Wilcox.

Gerrish, H. and Dugger, W. E., Jr. 1979. *Transistor electronics.* South Holland, Ill.: Goodheart-Wilcox.

Graf, R. F., ed. *Dictionary of electronic terms.* 1972. 4th ed. Fort Worth, Tex.: Tandy Corporation.

Grob, B. 1977. *Basic electronics.* New York: McGraw-Hill.

Hand, W. P. and Williams, G. 1980. *Basic electronics.* Encino, Ca.: Glencoe.

Lackey, J. E. "Why not teach the VOM?" *SCHOOL SHOP Serving Industrial Vocational-Technical Education 40 (March 1981):23–25.*

Lackey, J. E. "Designing a voltage divider." *SCHOOL SHOP Serving Industrial Vocational-Technical Education 41 (November 1981):30–31.*

Philco Corporation. 1960. *Basic concepts and DC circuits.* Philadelphia: Philco Corporation.

Philco Corporation. 1960. *Fundamentals of AC and AC circuit analysis.* Philadelphia: Philco Corporation.

Turner, Rufus. 1957. *Basic electricity.* New York: Rinehart.

Appendix A

Review of Basic Mathematics

Introduction

For you to be successful as an electronic technician, you must possess some skill in mathematics. All electrical circuitry can be stated mathematically. To understand the explanations of these circuits, you must know the math that supports the circuits used in your work. Scientific notation is used to simplify calculations involving large whole numbers and small decimal numbers. It is used extensively to express the operation of equipment and designs.

A calculator will help you develop the ability to use scientific notation. These subjects are discussed elsewhere in this book.

Objectives

Each student is required to:

1. Solve simple linear equations, using transposition and substitution.
2. Rearrange equations and isolate single unknowns.

Linear Equations

An equation is a mathematical expression used to state that two quantities are equal. To indicate this equality, the equal sign (=) is inserted between the quantities. It tells us that the quantities it separates have the same value. Examples are:

$$7 + 2 = 3 + 6$$

$$9 = 9$$

When the terms on each side of the equal are combined, the results are equal.

All operations performed in solving equations must be such that both sides of the equation are equal at all times. Mathematicians have long recognized seven rules for solving equations:

1. If equals are added to equals, the sums are equal.
2. If equals are subtracted from equals, the remainders are equal.
3. If equals are multiplied by equals, the products are equal.
4. If equals are divided by equals, the quotients are equal.
5. Quantities that are equal to the same quantity or to equal quantities are equal to each other.
6. Like powers of equal quantities are equal.
7. Like roots of equal quantities are equal.

These rules are used extensively in solving equations. Each time a rule is applied, however, it must be applied to both sides of an equation. With these rules, we can add, subtract, multiply, divide, raise to a power, or extract the root of one side of any equation, as long as we do exactly the same to the other side.

The seven rules listed are used when grouping knowns and unknowns within an equation.

Let us examine each rule to see how it applies in the solution of a problem.

Rule 1: If equals are added to equals, the sums are equal.

Given: $X - 5 = 13$; solve for X.

Equation:	$X - 5 = 13$
Add $+5$ to each side:	$+5 = +5$
Combine quantities:	$X = 18$

It appears the -5 has been removed from the left side of the equation; actually, it has been moved to the right, where it appears as $+5$. The process of moving quantities from one side of an equation to the other is called *transposition.*

Rule 2: If equals are subtracted from equals, the remainders are equal.

Given: $X + 8 = 17$; solve for X.

Equation:	$X + 8 = 17$
Subtract 8 from each side:	$-8 = -8$
Combine quantities:	$X = 9$

The $+8$ is transposed to the right side of the equation by changing its sign to minus. When the results are combined, $X = 9$.

Rule 3: If equals are multiplied by equals, the products are equal.

Given: $\frac{X}{5} = 12$; solve for X.

Equation: $\frac{X}{5} = 12$

Multiply both sides by 5: $\frac{X}{5}(5) = (12)(5)$

Combine quantities: $X = 60$

Rule 4: If equals are divided by equals, the quotients are equal.

Given: $5X = 60$; solve for X.

Equation: $5X = 60$

Divide both sides by 5: $\frac{5X}{5} = \frac{60}{5}$

Combine quantities: $X = 12$

Rule 5: Quantities that are equal to the same quantity or to equal quantities are equal to each other.

Given: Two statements, $X = A$ and $B = A$; illustrate rule 5.

We have two equations: $X = A$ and $B = A$

Since both X and B are equal to A, then $X = B$

Rule 6: Like powers of equal quantities are equal.

Given: $8 = 8$

Square each side: $8^2 = 8^2$

Combine quantities: $64 = 64$

Rule 7: Like roots of equal quantities are equal.

Given: $64 = 64$

Extract square roots: $\sqrt{64} = \sqrt{64}$

Combine quantities: $8 = 8$

Remember: All of these rules are used in transposing quantities. It is possible to transpose quantities without going through each step. To perform transposition use this shortcut:

1. Identify the quantity to be transposed.
2. Determine the sign of operation of the quantity to be transposed.
3. Change the sign of operation from that identified.
4. Perform this operation on both sides of the equation.

In other words, change the sign, move the quantity to the other side of the equation, and combine terms. Some examples are:

Sign of Operation Indicated	*Operation Performed on Opposite Side of Equation*
+	−
−	+
×	÷
÷	×
X^2	$\sqrt{X}$
$\sqrt{X}$	X^2

Self-Check

Answer each item by indicating whether the statement is true or false.

1. ___ The = symbol tells you that two terms are equal.
2. ___ Seven rules apply to the transposition of terms.
3. ___ To transpose a quantity you must change its sign of operation and move it to the opposite side of the equal sign.
4. ___ To square a number we multiply it times itself.
5. ___ Correct application of each rule stated maintains the equality of an equation.

Solving Equations

In complex equations it is often necessary to perform more than one transposition. To make a transposition of this type, you can perform any or all of the seven steps listed. This involves:

1. Identifying each quantity (known or unknown) to be transposed.
2. Identifying the sign of operation for each quantity.
3. Changing the sign of operation for each quantity and moving it to the opposite side of the equation.
4. Combining like terms.

Treating each quantity in this manner results in a solved equation. Some examples follow.

Example 1: Given the equation, solve for X.

Equation: $$5X + 7 + 9 = 3X - 4$$

Transpose: $$+ 7 + 9 = 3X$$

Change signs: $$5X - 3X + 7 + 9 - 7 - 9 = 3X - 3X - 4 - 7 - 9$$

Combine like terms: $$2X = -20$$

Divide both sides by two: $$\frac{2X}{2} = \frac{-20}{2}$$

Combine like terms: $$X = -10$$

Example 2: Given the equation, solve for X.

Equation:	$-3X - 8 = 8 - 7X$
Transpose -8 and $-7X$:	$-3X + 7X = 8 + 8$
Combine like terms:	$4X = 16$
Divide by 4:	$X = 4$

Example 3: Given the equation, solve for X.

Equation:	$\dfrac{3X}{4} = 27$
Multiply by 4:	$3X = 108$
Divide by 3:	$X = 36$

Example 4: Given the equation, solve for A.

Equation:	$3(4A - 5) - 4(A - 6) = 3(A + 1) + 1$
Simplify:	$12A - 15 - 4A + 24 = 3A + 3 + 1$
Transpose:	$12A - 4A - 3A = 3 + 1 + 15 - 24$
Combine terms:	$5A = -5$
Divide by 5:	$A = -1$

Many problems in electronics involve fractions. Such problems are easiest to solve when they have been converted to equations that do not contain fractions. To make this conversion, we must find the lowest common denominator (LCD) and multiply the LCD by each numerator in the equation. That cancels the denominators, leaving a straight-line equation with no fractions. From this point, the solution is identical to the solutions already explained. Some examples are:

Example 1: Given the equation, solve for X:

$$\frac{3X}{4} + \frac{5}{8} = \frac{X}{3} + 6$$

Find LCD = 24:	4, 8, and 3 all divide evenly into 24
Multiply all terms:	$(24)\dfrac{3X}{4} + (24)\dfrac{5}{8} = (24)\dfrac{X}{3} + (24)(6)$
Result:	$\dfrac{72X}{4} + \dfrac{120}{8} = \dfrac{24X}{3} + 144$
Reduce fractions to eliminate denominators:	$18X + 15 = 8X + 144$
Transpose:	$18X - 8X = 144 - 15$
Combine terms:	$10X = 129$
Divide by 10:	$X = 12.9$

Example 2: Given the equation $\frac{5}{2Y} + \frac{1}{2} = \frac{7Y - 1}{5Y}$, solve for Y:

LCD = $10Y$: $$(10Y)\frac{5}{2Y} + (10Y)\frac{1}{2} = (10Y)\frac{7Y - 1}{5Y}$$

Multiply by LCD: $$\frac{50Y}{2Y} + \frac{10Y}{2} = \frac{70Y^2 - 10Y}{5Y}$$

Reduce fractions: $$25 + 5Y = 14Y - 2$$

Transpose: $$5Y - 14Y = -2 - 25$$

Combine like terms: $$-9Y = -27$$

Divide by -9: $$Y = 3$$

Another method used to convert fractions to straight-line form is *cross-multiplication.* This method uses fewer steps and is more convenient for solving simple equations.

Example 3: Given the ratio, solve for X: $\frac{1}{3X} \times \frac{5}{6}$

Cross-multiply: $$5\,(3X) = (6)(1)$$

Combine terms: $$15X = 6$$

Divide by 15: $$X = \frac{6}{15}$$

Reduce fraction $$X = \frac{2}{5}$$

Notice the dashed lines in the example. They help us identify the quantities to be multiplied and thus help us avoid mistakes. The arrows tell us to multiply $5 \times 3X$ and 6×1.

Formulas

Normally, formulas are equations which contain more than one unknown. Electronics problems often contain two or more unknowns. In most cases, we know the values of all components except one. Because most items are identified as unknowns in a formula, it is often necessary to transpose the formula in order to isolate the unknown quantity from those that are known.

Example 1: Given the formula, solve for R.

Formula: $$E = I \times R$$

Transpose by dividing by I: $$\frac{E}{I} = \frac{I \times R}{I}$$

Result: $$R = \frac{E}{I}$$

Example 2: Given the equation, solve for L.

Formula: $$X_L = 2\pi f L$$

Dividing by $2\pi f$ results in: $$L = \frac{X_L}{2\pi f}$$

Example 3: Given the equation, solve for C.

Equation: $$\frac{A}{B} = \frac{C}{D}$$

Cross-multiply: $$AD = BC$$

Divide by B: $$C = \frac{AD}{B}$$

Example 4: Given the equation, solve for F.

Equation: $$2\pi F C X_C = 1$$

Divide by $2\pi C X_C$: $$F = \frac{1}{2\pi C X_C}$$

Self-Check

Answer each item by indicating whether the statement is true or false.

6. ____ Cross multiplication simplifies the solution of ratio equations.

7. ____ Transposition is limited to the use of any one of the seven rules listed.

8. ____ Grouping of knowns and unknowns in an equation is part of its eventual solution.

Review Problems

1. Solve for the unknown in each equation:
   (a) $X + 10 = 18$
   (b) $2X + 12 = 24$
   (c) $3X - 9 = 6$
   (d) $7X + X = 16$
   (e) $7X - X = 49 - X$
   (f) $10X - 15 + 5 = 110$
   (g) $3X - 10 = 2$
   (h) $2X - 19 = 1$
   (i) $4X - 12 = 3X - 10$
   (j) $7X - (+4) = 3X + 9 + 15$
   (k) $10X - 6 = 100 - 16$
   (l) $2X + 9 + 1 = 3X - 9$

2. Solve for A in the equation $\frac{A}{B} = \frac{C}{D}$.

   (a) $A = \frac{BC}{D}$ (c) $A = \frac{DC}{B}$

   (b) $A = \frac{BD}{C}$ (d) $A = \frac{CD}{B}$

3. Solve for frequency (f) in the formula $X_C = \frac{1}{2\pi f C}$.

   (a) $f = 2\pi C\,(X_C)$ (c) $f = \frac{1}{2\pi C}$

   (b) $f = \frac{1}{2\pi C X_C}$ (d) $f =$ none of the above

Appendix B

Soldering Techniques

Learning to solder properly is an important part of becoming a good technician. Each job you do requires a different technique. With experience, you can identify what is needed in each case. In this section, the emphasis is on developing initial soldering skills.

The most important tool for any soldering job is a soldering iron suitable for the job. For work in modern circuitry, an iron of 25 to 35 W is suitable. Figure B-1 is an illustration of a 30-W iron that is available commercially. Irons of this type are available with a copper tip or with an iron-plated tip. An iron-plated tip requires less maintenance but is more expensive. A copper tip holds solder well and has excellent heat-transfer capability. Care must always be taken to keep the copper tip clean and tinned with solder. Failure to do so causes the copper to become pitted and damages the tip. An iron-plated tip does not transfer heat quite as well as a copper tip but is less likely to become pitted.

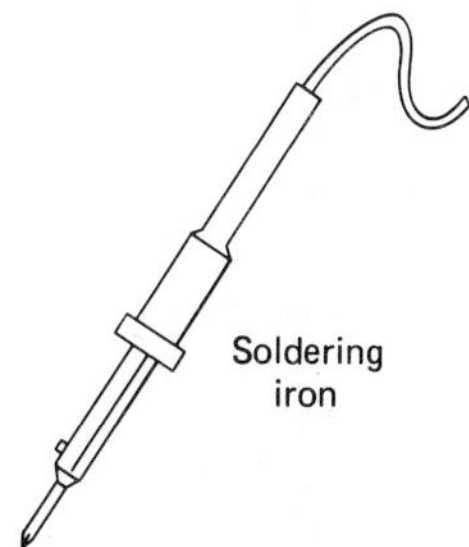

FIGURE B-1

In electronics circuitry, rosin-core solder should always be used. Solder of small diameter with a content of 60% tin and 40% lead is best for most jobs. This solder begins to melt at 361° F and is molten at 370° F. The molten metal is used to coat all metal surfaces of the point to be joined. During cooling, the solder returns to its solid state and forms a zero-resistance continuous conductor over the area joined. Never use acid-core solder on electronics circuitry; the acid can corrode and damage the circuitry. All warranties are voided once this type of solder is used.

Caution: Never use acid-core solder on electronic circuitry.

A soldering iron must be checked and prepared for use each time you use it. In doing this, make sure the tip is not pitted. If it is pitted, use a file to cut the metal down until it is smooth. In either case, make sure a good coat of solder clings to the heated iron. The term used to describe this characteristic is *tinned*. If the tip has a coat of solder, it has been tinned. When tinning is needed, connect the iron to the voltage source and allow it to heat. When it is heated, apply solder until all bare areas are tinned. A well-tinned tip transfers heat better and is less likely to become pitted.

Never file an iron-plated tip; doing so can damage the tip.

When using a soldering iron, keep the tip clean and tinned at all times. A damp sponge or cloth or a dampened piece of paper towel can be used to wipe the tip. This should be done every two to three minutes, and the tip should be checked for a good coat of solder. When a heated tip has been sitting idle for a period of time, check it carefully before use.

The drawing in Figure B-2 shows a soldering tip that is in contact with the metal parts to be joined. The resistor is to be connected to the terminal lug at the point shown. Both surfaces should be cleaned and the resistor lead should be connected snugly to the lug. The iron is then placed in the position shown, to preheat the connection. Preheating should continue until the resistor lead is hot enough to melt solder.

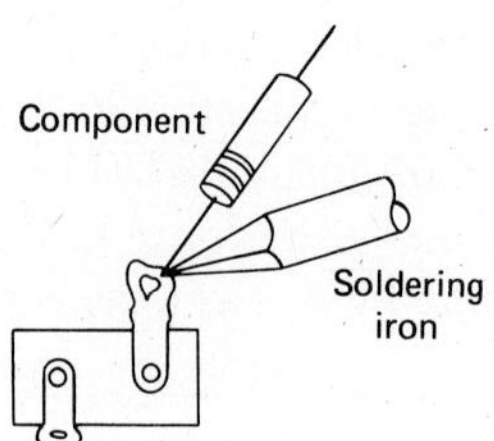

FIGURE B-2

When the metal has been heated to the solder's melting point, touch the solder to the resistor lead (Figure B-3). Placing the solder against the resistor lead ensures that the solder will not melt until all metal has reached the correct temperature.

Do not use too much solder.

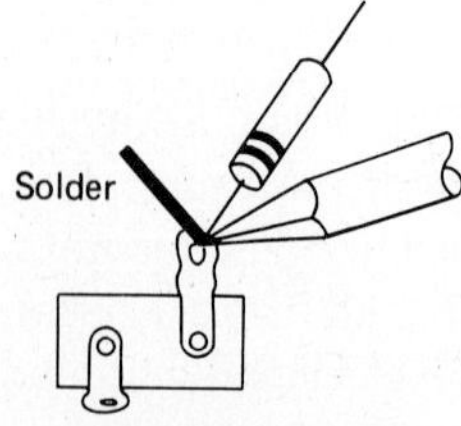

FIGURE B-3

With the right amount of heat and solder, removing the tip should result in a connection like that shown in Figure B-4. Do not move any part of the connection while the solder is still molten. Doing so could result in a bad solder joint and cause problems later. Good solder joints have a bright silvery shine. They are smooth and are not grainy.

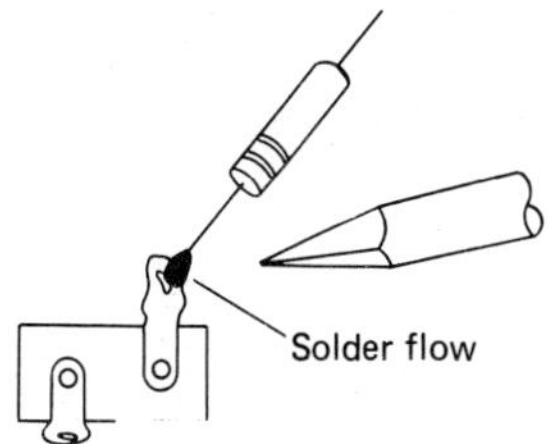

FIGURE B-4

Much of the soldering done by the modern technician is on printed-circuit boards. Care must be taken not to break the foil circuits etched on the PC board. Excessive heat can cause strips of foil to lift and be broken. Another problem is overuse of solder. In some cases, this can bridge from one foil conductor to another and cause a short between circuits.

Figure B-5 shows a board correctly soldered. All connections have a smooth surface and a bright silvery appearance.

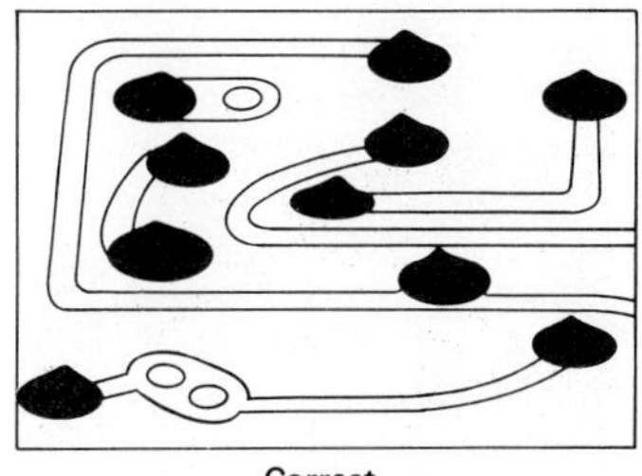

Correct

FIGURE B-5

Figure B-6 shows the result of too much heat, a tip that is too large, or too much solder. To correct this mistake, the joint must be reheated and the excess solder removed. Then the joint must cool.

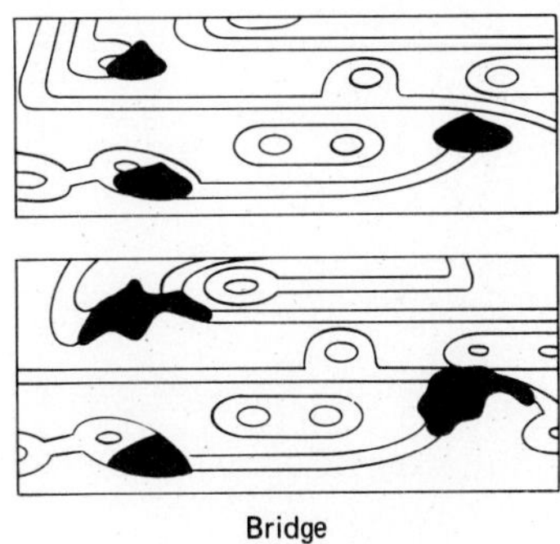

Bridge

FIGURE B-6

In Figure B-7, we see connections where the solder did not flow properly. This can occur because the tip and/or foil were not cleaned properly or the tip was not allowed to remain in contact long enough. Regardless of the reason, the cause is insufficient heat. The joints must be reheated; then the solder will flow correctly. This mistake is common among people learning to solder; they are afraid they will damage the circuit by applying too much heat, and instead do not apply enough.

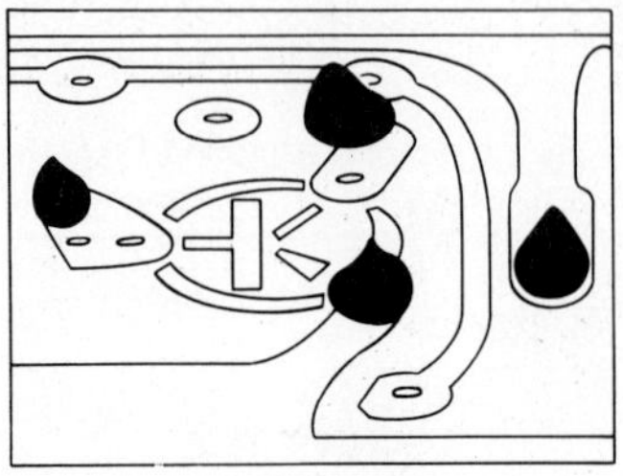

FIGURE B-7

Figure B-8 illustrates another problem, one that results from too much heat and/or solder. It is the opposite problem of that in Figure B-6. Too much heat causes the copper foil to lift off the board, where it can break away from the rest of the circuit. Excess heat can also build up in large mounds on the copper pads, resulting in the bridge already described.

Incorrect

FIGURE B-8

Practice preparing soldering tips, preheating, and applying solder. The ability to do that can make you a better technician. This is a case in which practice does make for perfection.

Appendix C

Decimal Multipliers (Prefixes)

A conversion chart and symbols for use in converting prefixes into numerical data and powers of 10.

Symbol	Prefix	Quantity: Amount (name)	Quantity: Power of 10
p	pico	one-trillionth	10^{-12}
n	nano	one-billionth	10^{-9}
μ	micro	one-millionth	10^{-6}
m	milli	one-thousandth	10^{-3}
c	centi	one-hundredth	10^{-2}
d	deci	one-tenth	10^{-1}
units	unity	unity	10^{0}
dk	deka	ten	10^{1}
h	hekto	hundred	10^{2}
k	kilo	thousand	10^{3}
M	Mega	million	10^{6}
G	Giga	billion	10^{9}
T	Tera	trillion	10^{12}

Appendix D

Inch-Decimal-Metric Conversions

A chart for use in converting fractions, inches, and metric measurements.

Inch-Decimal-Metric Equivalents

Fractions of an Inch	*Decimal Equivalents*	*Metric Equivalents (Millimeters)*
1/64	0.0156	0.397
1/32	0.0313	0.794
3/64	0.0469	1.191
1/16	0.0625	1.588
5/64	0.07813	1.984
3/32	0.0938	2.381
7/64	0.1094	2.778
1/8	0.1250	3.175
9/64	0.1406	3.572
5/32	0.1563	3.969
11/64	0.1719	4.366
3/16	0.1875	4.763
13/64	0.2031	5.159
7/32	0.2188	5.556
15/64	0.2344	5.953
1/4	0.2500	6.350
17/64	0.2656	6.747
9/32	0.2813	7.144
19/64	0.2969	7.541
5/16	0.3125	7.938
21/64	0.3281	8.334
11/32	0.3438	8.731
23/64	0.3594	9.128
3/8	0.3750	9.525

Chart (continued)

Fractions of an Inch	*Decimal Equivalents*	*Metric Equivalents (Millimeters)*
25/64	0.3906	9.922
13/32	0.4063	10.319
27/64	0.4219	10.716
7/16	0.4375	11.113
29/64	0.4531	11.509
15/32	0.4688	11.906
31/64	0.4844	12.303
1/2	0.500	12.700
33/64	0.5156	13.097
17/32	0.5313	13.494
35/64	0.5469	13.891
9/16	0.5625	14.288
37/64	0.5781	14.684
19/32	0.5938	15.081
39/64	0.6094	15.478
5/8	0.6250	15.875
41/64	0.6406	16.272
21/32	0.6563	16.669
43/64	0.6719	17.066
11/16	0.6875	17.463
45/64	0.7031	17.859
23/32	0.7188	18.256
47/64	0.7344	18.653
3/4	0.7500	19.050
49/64	0.7656	19.447
25/32	0.7813	19.844
51/64	0.7969	20.241
13/16	0.8125	20.638
53/64	0.8281	21.034
27/32	0.8438	21.431
55/64	0.8594	21.828
7/8	0.8750	22.225
57/64	0.8906	22.622
29/32	0.9063	23.019
59/64	0.9219	23.416
15/16	0.9375	23.813
61/64	0.9531	24.209
31/32	0.9688	24.606
63/64	0.9844	25.003
1 inch	1.0000	25.400

Appendix E

Greek Alphabet and Designators

This table is a list of common Greek words and their use as symbols in the mathematical analysis of electronic circuitry.

Name	*Upper Case*	*Lower Case*	*Used to Designate*
alpha	A	α	Angles, area, coefficients
beta	B	β	Angles, flux density, coefficients
gamma	Γ	γ	Conductivity, specific gravity
delta	Δ	δ	Variation, density
epsilon	E	ϵ	Base for natural logarithms
zeta	Z	ζ	Impedance, coefficients, coordinates
eta	H	η	Hysteresis, coefficient, efficiency
theta	Θ	θ	Temperature, phase angle
iota	I	ι	
kappa	K	κ	Dielectric constant, susceptibility
lambda	Λ	λ	Wavelength
mu	M	μ	Micro, amplification factor, and permeability
nu	N	ν	Reluctivity
xi	Ξ	ξ	
Omicron	O	o	
Pi	Π	π	Ratio of a circle's circumference to its diameter: 3.1415927
rho	P	ρ	Resistivity
sigma	Σ	σ	Summation
tau	T	τ	Time constant
upsilon	Υ	υ	
phi	Φ	ϕ	Magnetic flux
chi	X	χ	
psi	Ψ	ψ	Phase difference
omega	Ω	ω	Uppercase: ohms Lowercase: angular velocity

Appendix F

Electromagnetic Wave Spectrum

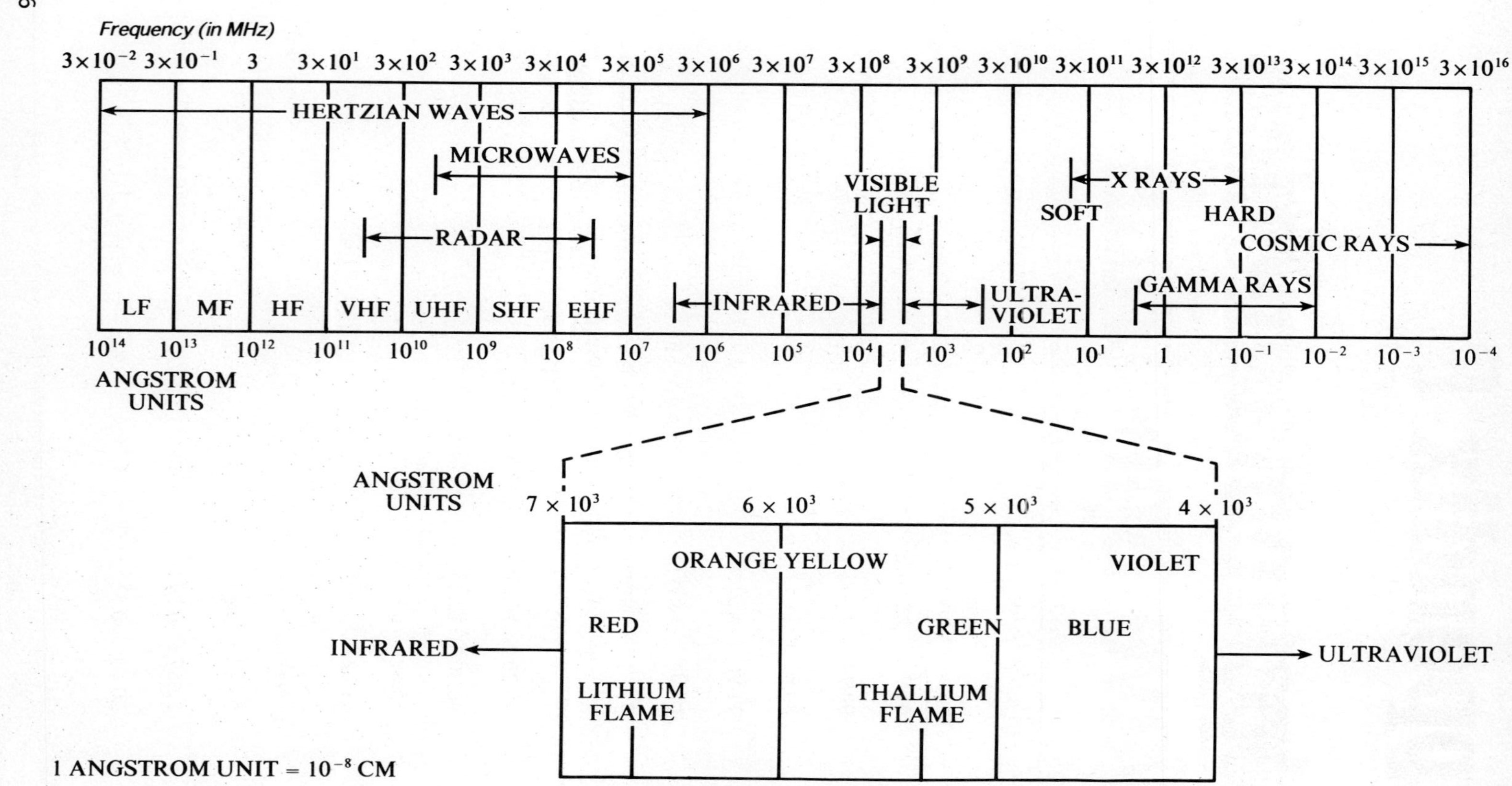

Frequency (in MHz)
3×10⁻² 3×10⁻¹ 3 3×10¹ 3×10² 3×10³ 3×10⁴ 3×10⁵ 3×10⁶ 3×10⁷ 3×10⁸ 3×10⁹ 3×10¹⁰ 3×10¹¹ 3×10¹² 3×10¹³ 3×10¹⁴ 3×10¹⁵ 3×10¹⁶
HERTZIAN WAVES
MICROWAVES
RADAR
VISIBLE LIGHT
X RAYS
SOFT
HARD
COSMIC RAYS
GAMMA RAYS
INFRARED
ULTRA-VIOLET
LF
MF
HF
VHF
UHF
SHF
EHF
10¹⁴ 10¹³ 10¹² 10¹¹ 10¹⁰ 10⁹ 10⁸ 10⁷ 10⁶ 10⁵ 10⁴ 10³ 10² 10¹ 1 10⁻¹ 10⁻² 10⁻³ 10⁻⁴
ANGSTROM UNITS
ANGSTROM UNITS
7 × 10³
6 × 10³
5 × 10³
4 × 10³
ORANGE YELLOW
VIOLET
RED
GREEN
BLUE
INFRARED
ULTRAVIOLET
LITHIUM FLAME
THALLIUM FLAME
1 ANGSTROM UNIT = 10⁻⁸ CM

Appendix G

Copper-Wire Table

Gage AWG or B&S	*Diameter*		*Area Circular Mils*	*Ohms per 1000 ft (70°/167° F)*	*Feet per Ohm (70° F)*	*Current-Carrying Capacity (700 cm/AMP)*
	Mils	Milli-meters				
10	101.9	2.588	10380	1.018 1.216	987.0	14.8
11	90.74	2.305	8234	1.284 1.532	778.8	11.8
12	80.81	2.053	6530	1.619 1.931	617.7	9.33
13	71.96	1.828	5178	2.042 2.436	489.7	7.40
14	64.08	1.628	4107	2.575 3.071	388.4	5.87
15	57.07	1.450	3257	3.247 3.873	307.9	4.65
16	50.82	1.291	2583	4.094 4.884	244.3	3.69
17	45.26	1.150	2048	5.163 6.158	193.7	2.93
18	40.30	1.024	1624	6.510 7.765	153.6	2.32
19	35.89	0.9116	1288	8.210 9.792	121.8	1.84
20	31.96	0.8118	1022	10.35 12.35	96.60	1.46
21	28.46	0.7230	810.1	13.05 15.57	76.60	1.16
22	25.35	0.6438	642.4	16.46 19.63	60.70	0.918
23	22.57	0.5733	509.5	20.76 24.76	51.90	0.728
24	20.10	0.5106	404.0	26.17 31.22	38.20	0.577

Gage AWG or B&S	Diameter Mils	Diameter Millimeters	Area Circular Mils	Ohms per 1000 ft (70°/167° F)	Feet per Ohm (70° F)	Current-Carrying Capacity (700 cm/AMP)
25	17.90	0.4547	320.4	33.00	30.30	0.458
				39.36		
26	15.94	0.4049	254.1	41.62	24.10	0.363
				49.64		
27	14.20	0.3606	201.5	52.48	19.10	0.288
				62.59		
28	12.64	0.3211	159.8	66.17	15.10	0.228
				78.93		
29	11.26	0.2859	126.7	83.44	11.90	0.181
				99.52		
30	10.03	0.2546	100.5	105.2	9.53	0.144
				125.5		
31	8.928	0.2268	79.70	132.7	7.57	0.114
				158.2		
32	7.950	0.2019	63.21	167.3	5.98	0.090
				199.5		
33	7.080	0.1798	50.13	211.0	4.74	0.072
				251.6		
34	6.305	0.1601	39.75	266.0	3.76	0.057
				317.3		
35	5.615	0.1426	31.52	335.0	2.92	0.045
				400.0		
36	5.000	0.1270	25.00	423.0	2.37	0.036
				504.5		
37	4.453	0.1131	19.83	533.4	1.88	0.028
38	3.965	0.1007	15.72	672.6	1.49	0.022
39	3.531	0.0897	12.47	848.1	1.18	0.018
40	3.145	0.0799	9.88	1069	0.94	0.014

Appendix H

Transformer Color Codes

Power Transformer Color Code

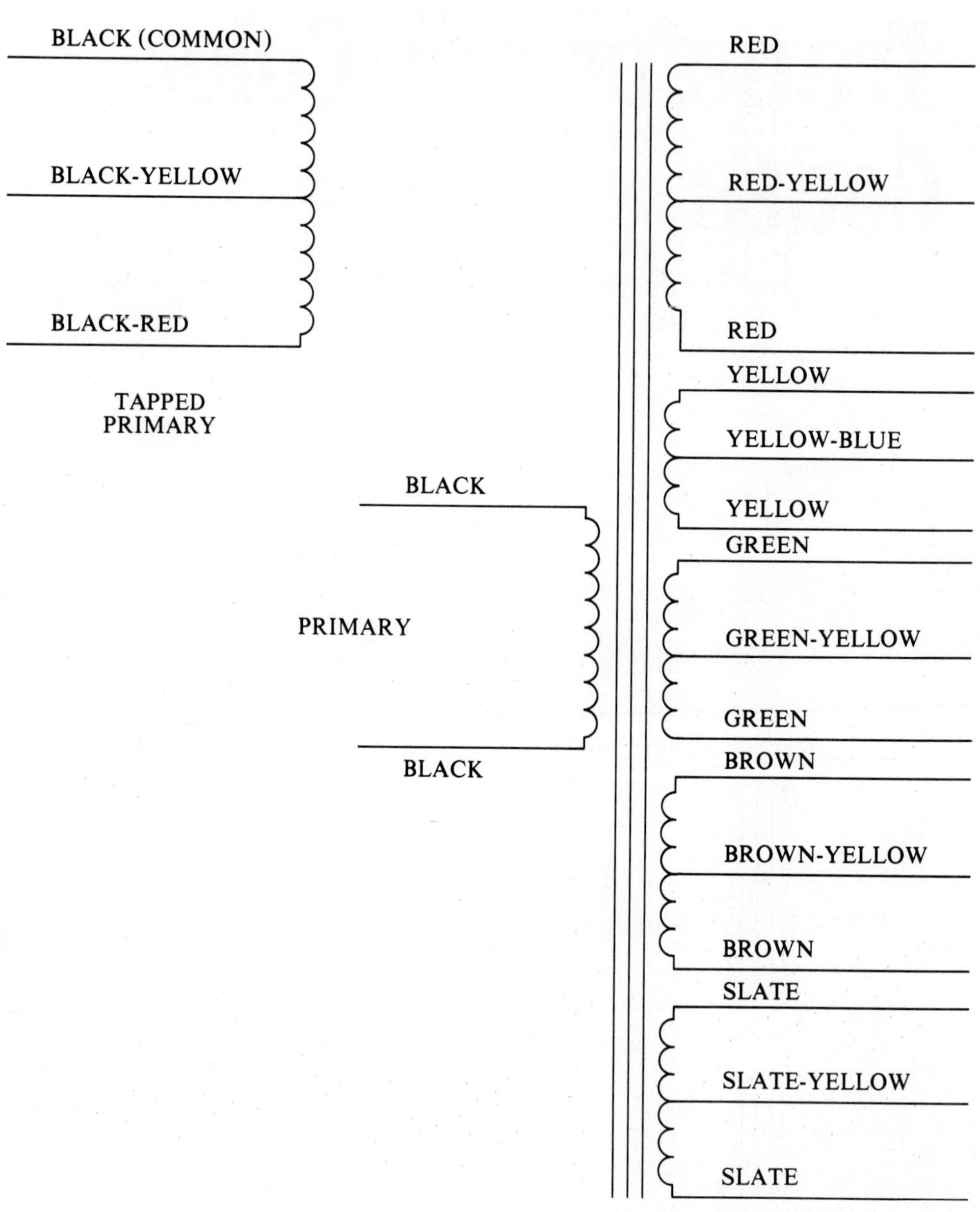

Transformer Color Codes continued:

Audio-Frequency Transformer Color Code

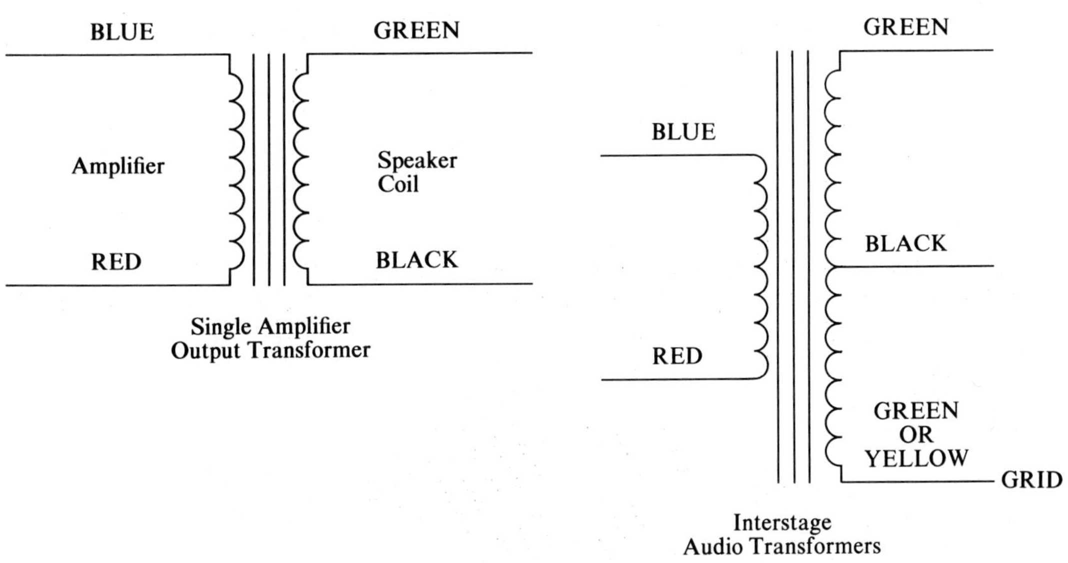

Single Amplifier
Output Transformer

Interstage
Audio Transformers

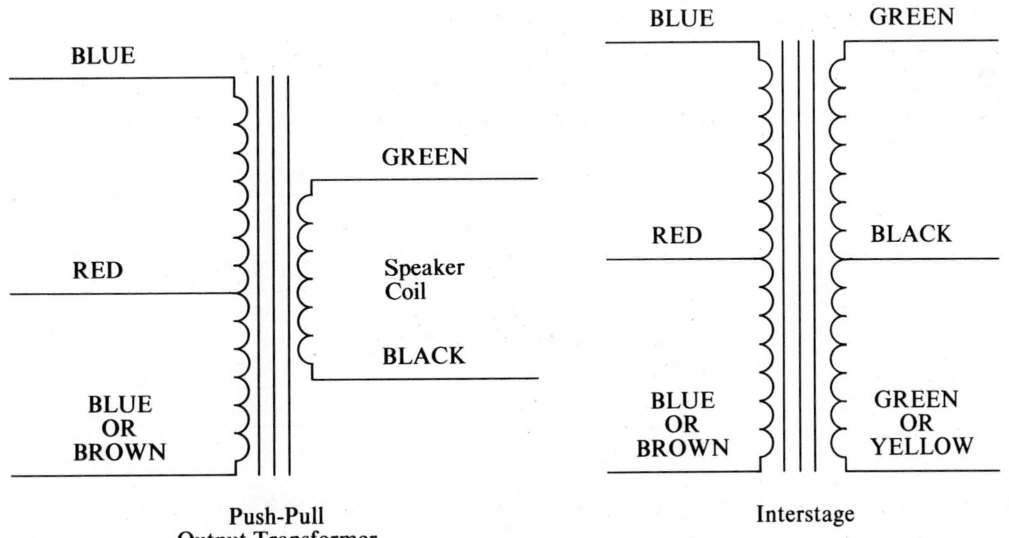

Push-Pull
Output Transformer

Interstage

Intermediate-Frequency Transformer Color Code

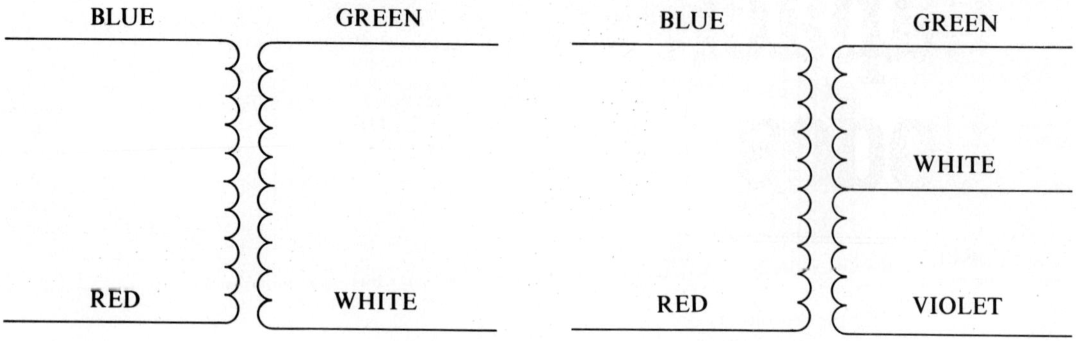

Appendix I

Capacitor Color Codes

In modern circuitry, it has become possible to shrink the size of most capacitors to the point where they require little space. With this reduction, many older capacitors are no longer suitable for use because of the small space available.

Many electrolytic capacitors presently in use are of the tantalum type. They can be mounted on a printed-circuit board in a space 1/2 in. in diameter and are less than 1 in. tall. Their capacitance ranges from 0.1 μf to 1200 μf and their voltage ratings range from 4 V to 125 V. They provide high stability of operation, have low DC leakage, and have a low dissipation factor. The capacitance of capacitors is stamped on the body of the unit.

Electrolytic capacitors of greater capacitance are available in sizes ranging from 2 μf to 400,000 μf, with operating voltages of 1 V DC to 600 V DC. Size and polarity are marked on the body of the capacitor.

Film capacitors are available in ranges of 0.001 μf to 1.75 μf, with voltage ratings of 50 V to 2000 V. Length is 0.5 in. to 2 in. in length. This size is sufficient for their operating characteristics to be stamped on their bodies.

Ceramic capacitors are available in capacitances ranging from 1 pf to 0.47 μf, with voltage ratings from 100 V DC to 7500 V DC. Many ceramic capacitors are large enough for their capacitance to be stamped on them; others use any of five markings. These markings are noted on the drawing that follows this explanation. The color code used is listed in the accompanying table.

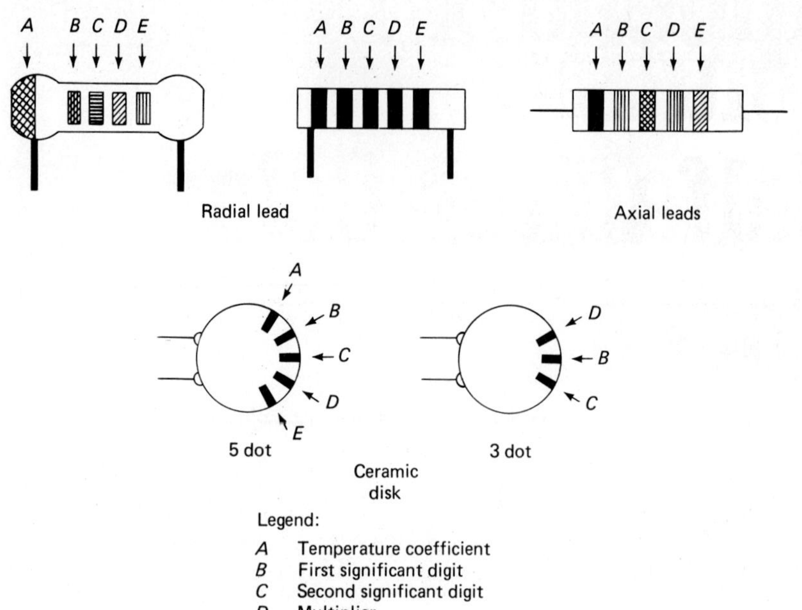

Ceramic Capacitor Color Code

				Tolerance		
Color	*First Digit*	*Second Digit*	*Multiplier*	More than 10 pf in %	Less than 10 pf in %	*Temperature Coefficient*
Black	0	0	1	±20	±2.0	0
Brown	1	1	10	±1		−30
Red	2	2	100	±2		−80
Orange	3	3	1,000			−150
Yellow	4	4	10,000			−220
Green	5	5		±5	±0.5	−330
Blue	6	6				−470
Violet	7	7				−750
Gray	8	8	0.01		±0.25	+30
White	9	9	0.1	±10	±1.0	+120 to −750*
						+300 to −330**
Silver						+100**
Gold						Bypass or coupling*

**Electronic Industries Association (EIA) code.*

***Joint Army Navy (JAN) code. Used to mark components for military applications.*

Solutions to Self-Checks

Chapter 1

1. Protons, neutrons, electrons **2.** Positive **3.** Solids, liquids, gases **4.** Ionization **5.** False **6.** True **7.** True **8.** True **9.** False **10.** True **11.** False **12.** False **13.** True **14.** True

Chapter 2

1. 10, ten **2.** Decimal **3.** Exponents **4.** One (1), ten (10) **5.** Negative **6.** Exponent

7. Prefix	Numerical Value
a. mega	1,000,000
b. kilo	1000
c. milli	.001
d. micro	.000001
e. pico	.000000000001

8. 123 kilo **9.** Micro symbol is μ. **10.** 15 mega converts to 15,000,000. **11.** Completed table is:

Item Number	*Decimal Quantity*	*Scientific Notation*	*Assigned Prefix*
a	1500	1.5×10^3	1.5 k (kilo)
b	.000005	5×10^{-6}	5 μ (micro)
c	4,000,000	4×10^6	4 M (mega)
d	.0036	3.6×10^{-3}	3.6 m (milli)
e	.0095	9.5×10^{-3}	9.5 m (milli)
f	.000000000001	1×10^{-12}	1 p (pico)

12. False **13.** True **14.** True **15.** True **16.** False

Chapter 3

1. Conventional **2.** Negative–positive **3.** 2 (two) **4.** I (intensity) **5.** Ampere (A) **6.** Electromotive force, difference in potential **7.** Ampere (A) **8.** Volt (V) **9.** Mechanical, chemical, thermoelectric (heat), photoelectric (solar), and piezoelectric (crystal) **10.** Chemical **11.** False **12.** True **13.** True **14.** False **15.** True **16.** True **17.** False **18.** False **19.** False **20.** False **21.** True **22.** False **23.** False **24.** False **25.** True **26.** True **27.** True **28.** True **29.** True **30.** False **31.** False

Chapter 4

1. True **2.** True **3.** False **4.** True **5.** False **6.** False **7.** True **8.** False **9.** True **10.** False **11.** True **12.** True **13.** True **14.** False **15.** False **16.** False

Chapter 5

1. True **2.** False **3.** False **4.** False **5.** True **6.** Directly, inversely **7.** Current, resistance **8.** Increase **9.** 5 (five) ohms (Ω) **10.** 25 V (volts) **11.** Decrease to one-half **12.** 2 A (two amperes) **13.** Resistance **14.** True **15.** True **16.** False **17.** True **18.** False **19.** True **20.** False **21.** True **22.** False **23.** True **24.** True **25.** Zero (0) **26.** 50 V **27.** Individual voltages, total (applied) **28.** 55 V **29.** Decrease **30.** Increase

Chapter 6

1. Negative, positive **2.** Increase **3.** Voltage **4.** Sum **5.** Current, resistance **6.** False **7.** True **8.** False (This is kilowatt hour or 1000/hour.) **9.** True **10.** True **11.** Troubleshooting **12.** Infinite **13.** Parallel **14.** Zero (0) **15.** Ammeter **16.** Ohmmeter **17.** Zero (0)

Chapter 7

1. True **2.** False **3.** False **4.** True **5.** False **6.** 6 kΩ. *Solution:*
Solve R_3 & R_4:

$$R_{eq1} = \frac{R}{N} = \frac{40}{2} = 20 \text{ k}\Omega$$

Solve R_2 & R_{eq1}:

$$R_{eq2} = \frac{R_2 \times R_{eq2}}{R_2 + R_{eq2}} = \frac{600}{5} = 12 \text{ k}\Omega$$

Solve R_1 & R_{eq2}:

$$R_{eq3} = \frac{R}{N} = \frac{12}{2} = 6 \text{ k}\Omega$$

$R_t = R_{eq3} = 6 \text{ k}\Omega$

7. 18 kΩ. *Solution:*

$$R_t = \frac{R}{N} = \frac{54}{3} = 18 \text{ k}\Omega$$

8. 36 kΩ. *Solution:*

$$R_t = \frac{R_1 \times R_2}{R_1 + R_2} = \frac{5400}{150} = 36 \text{ k}\Omega$$

9. Decrease **10.** More than one **11.** False **12.** True **13.** True **14.** False **15.** False **16.** False **17.** False **18.** True **19.** True **20.** False **21.** True **22.** False **23.** True **24.** True **25.** True **26.** Infinite **27.** Zero **28.** Open **29.** Shorted **30.** Series **31.** Isolate (disconnect)

Chapter 8

1. False **2.** True **3.** False **4.** True **5.** True **6.** False **7.** R_7 **8.** Total resistance (R_t), total current (I_t), and total voltage (E_t) **9.** Resistance, current **10.** Series **11.** Total current (I_t) **12.** Increase **13.** Causes it to decrease **14.** Causes it to increase **15.** Causes it to increase **16.** Causes it to decrease **17.** Infinite ohms (∞ Ω) **18.** Zero ohms (0 Ω) **19.** Total voltage (E) **20.** zero volts (0 V) **21.** Voltage drop of the remaining branches. Note this is a series-parallel circuit. **22.** Zero volts (0 V)

Chapter 9

1. Rheostat, potentiometer **2.** Three (3) **3.** Current **4.** Voltage, circuit **5.** Variable resistor **6.** Two (2) **7.** Negative **8.** Series resistive **9.** a. Voltage reference point b. Common connection **10.** Positive **11.** Supply more than one voltage from a single source **12.** Using the common reference point B, voltages will be: a. +50 V, point A to ground b. −40 V, point C to ground c. −120 V, point D to ground d. −150 V, point E to ground **13.** Common connection point or common reference point **14.** True **15.** True **16.** False **17.** False **18.** True

Chapter 10

1. Wheatstone **2.** Current **3.** Balanced **4.** Calibrate (balance) **5.** Speed

Chapter 11

1. True **2.** False **3.** False **4.** False **5.** False **6.** True **7.** False **8.** False **9.** False **10.** True **11.** False **12.** True **13.** False **14.** True **15.** True **16.** True **17.** False **18.** True **19.** True **20.** True **21.** True **22.** True **23.** True **24.** True **25.** True **26.** True **27.** True **28.** True **29.** False **30.** True

Chapter 12

1. True **2.** True **3.** True **4.** False **5.** True **6.** True **7.** False **8.** False **9.** False **10.** True **11.** True **12.** True **13.** False **14.** True **15.** False **16.** True **17.** False **18.** True **19.** False **20.** True **21.** True **22.** True

23. False **24.** True **25.** True **26.** False **27.** True **28.** False **29.** True **30.** False **31.** False **32.** True

Chapter 13

1. The five effects of electricity are: a. Heating effect b. Electromagnetic c. Electrostatic d. Chemical e. Physiological **2.** Analog **3.** Two types of heat-effect meters. a. Thermocouple b. Hot wire **4.** Digital **5.** Two advantages of hot-wire types. a. Measures either AC or DC. b. Measures high AC currents. **6.** True **7.** True **8.** True **9.** True **10.** True **11.** True **12.** False **13.** True **14.** False **15.** False **16.** Series **17.** Better sensitivity (higher sensitivity) **18.** Decrease **19.** Increase **20.** On **21.** Ohms Zero Adjust **22.** Inverse **23.** 24 kΩ. *Solution:*
Step 1: $R_t = E_t / I_{meter} = 25\text{ V}/1\text{ mA} = 25\text{ k}\Omega$
Step 2: limiter $R = R_t - R_{meter} - R_{zero} = 25\text{ k}\Omega$
Step 3: limiter $R = 25\text{ k}\Omega - 100\ \Omega - 900\ \Omega = 24\text{ k}\Omega$
24. Short **25.** Internal resistance **26.** Multimeter **27.** Voltage, current, resistance **28.** Range, function **29.** Three (3) **30.** Seven (7) **31.** Electromagnetism **32.** Alternating current **33.** Moving coil

Chapter 14

1. True **2.** False **3.** False **4.** True **5.** False **6.** False **7.** True **8.** False **9.** True **10.** True **11.** False **12.** True **13.** False **14.** True **15.** False **16.** True **17.** True **18.** True **19.** True **20.** True **21.** True **22.** True **23.** False **24.** True **25.** True

Chapter 15

1. True **2.** False **3.** True **4.** True **5.** False **6.** False **7.** False **8.** False **9.** True **10.** True **11.** False **12.** False **13.** False **14.** True **15.** False

Chapter 16

1. Coil, choke **2.** Soft iron **3.** Magnetic field **4.** Mutal induction **5.** Counter EMF **6.** Four factors are: a. Number of turns. b. Length-to-diameter ratio. c. Material used as the core. d. Method used for winding. **7.** Directly **8.** Layered **9.** Sum of the individual inductances **10.** Inversely **11.** True **12.** True **13.** False **14.** False **15.** True **16.** False **17.** True **18.** False **19.** False **20.** True **21.** True **22.** True **23.** True **24.** False **25.** False **26.** False **27.** False **28.** True **29.** True **30.** True

Chapter 17

1. Transformer action **2.** Up **3.** Coefficient of coupling **4.** 8 (secondary current = 8 A). *Solution:*

Ratio:	$N_p \times I_p = N_s \times I_s$
Substitute values:	$4 \times 2 = 1 \times I_s$
Combine values:	$I_s = 8\text{ A}$

5. 16 kΩ (secondary impedance). *Solution:*

Ratio:	$N_p^2 \times Z_s = N_s^2 \times Z_p$
Substitute values:	$(1)^2 \times Z_s = (4)^2 \times 1000\Omega$
Solve:	$1 \times Zs = 16 \times 1000\Omega$
Result:	$Z_s = 16\text{ k}\Omega$

6. Four types of transformers: a. Audio frequency b. Radio frequency c. Power frequency d. Autotransformer **7.** Iron, air **8.** Current **9.** Soft-iron **10.** 60 **11.** Three power frequencies: a. 50 Hz b. 60 Hz c. 400 Hz **12.** True **13.** False **14.** True **15.** False **16.** False

Chapter 18

1. True **2.** False **3.** True **4.** True **5.** True **6.** False **7.** False **8.** True **9.** True **10.** False **11.** True **12.** True **13.** False **14.** True **15.** True **16.** True **17.** True **18.** False **19.** False **20.** True **21.** True **22.** False **23.** True **24.** True *Solution:*

Effect of C on $X_C \uparrow X_C = \dfrac{1}{2\pi f C \downarrow}$ Smallest capacitor has largest X_C. Largest X_C drops most voltage; therefore, smallest capacitor drops largest V.

25. True **26.** True **27.** False **28.** False **29.** False **30.** True

Chapter 19

1. False **2.** True **3.** True **4.** True **5.** False **6.** False **7.** False **8.** True **9.** True **10.** False **11.** True **12.** True **13.** True **14.** False **15.** True **16.** True **17.** True **18.** True **19.** True **20.** True **21.** True **22.** True **23.** True **24.** True **25.** False (theorem magnitude—tangent for direction) **26.** False **27.** True **28.** False **29.** True **30.** False **31.** False **32.** True **33.** False **34.** True

35. False **36.** False $\sin\theta = \dfrac{X\text{ (reactance)}}{Z\text{ (impedance)}}$

37. False **38.** False $\cos\theta = \dfrac{R\text{ (resistance)}}{Z\text{ (impedance)}}$

39. True **40.** True. *Solution:*
Step 1: Cos θ = resistance divided by impedance
Step 2: Resistance = cos θ X impedance

Chapter 20

1. b $\left(Z = \dfrac{E_t}{I_t}.\right)$

2. b (false) **3.** b (false) **4.** a (true) **5.** a (true) **6.** False **7.** False **8.** False (parallel circuits only with all branches known) **9.** True **10.** False **11.** False (applies to right triangles only) **12.** True **13.** True **14.** False **15.** True **16.** False **17.** False **18.** False **19.** False **20.** True

Chapter 21

1. True **2.** False **3.** False **4.** False **5.** True **6.** True **7.** True **8.** True **9.** False **10.** False **11.** False **12.** True **13.** False **14.** True **15.** False (results from plotting these)

Chapter 22

1. True **2.** False **3.** False **4.** False **5.** False **6.** False **7.** False **8.** True **9.** False **10.** True **11.** False **12.** False **13.** True **14.** False **15.** True **16.** False **17.** False **18.** False **19.** False **20.** False **21.** True **22.** True **23.** False **24.** False **25.** False **26.** False **27.** True **28.** False **29.** True **30.** False (affects time of charge, no E peak)

Chapter 23

1. False **2.** True **3.** False **4.** True **5.** True **6.** False **7.** False **8.** True **9.** True **10.** True **11.** False **12.** True **13.** True **14.** True **15.** True **16.** False **17.** True **18.** True **19.** False **20.** True **21.** False **22.** False **23.** False **24.** False **25.** True **26.** True **27.** True **28.** True **29.** True **30.** True **31.** True **32.** True **33.** True **34.** True **35.** False **36.** False (phase angle theta) **37.** True **38.** False **39.** False **40.** True **41.** True **42.** True **43.** False **44.** True **45.** True

Chapter 24

1. False **2.** True **3.** False **4.** False **5.** False **6.** True **7.** False **8.** False **9.** True **10.** False **11.** False **12.** True **13.** False **14.** True **15.** False

Chapter 25

1. Three requirements for induction: a. magnetic field b. Conductor c. Relative motion between the two **2.** Any induced field will oppose the field that caused it to be induced. **3.** Constantly changing amplitude and periodic current reversals **4.** Left thumb **5.** Left hand rule **6.** True **7.** True **8.** False **9.** False **10.** False **11.** True **12.** False **13.** True **14.** True **15.** False **16.** False **17.** False **18.** True **19.** False **20.** True **21.** True **22.** True **23.** True **24.** True **25.** True

Chapter 27

1. Death **2.** Definitions: a. *Unsafe act*—Any act of carelessness that could result in an accident.
b. *Cause and effect*—Cause is what results in an accident; Effect is the result of an accident.

c. *Material failure*—Machine or other equipment defect that results in failure and an accident.
d. *Electrical shock*—The result of electrical current flowing through living tissue. **3.** Vulnerable areas of the body: a. Chest cavity b. Skull and brain area **4.** One-handed **5.** Electrical shock **6.** Assure that the power is turned off. **7.** Drag the person free. **8.** Heart is beating. **9.** Sword swallowing **10.** 20 **11.** Direct pressure **12.** Knee or groin **13.** Life threatening **14.** Prevention **15.** Separate area **16.** Class C **17.** Pressurized dry powder

Index